Materials Chemistry at High Temperatures

Volume 2

Processing and Performance

Materials Chemistry at High Temperatures

Volume 2

Processing and Performance

Editor

John W. Hastie

National Institute of Standards and Technology
Gaithersburg, MD

Humana Press • Clifton, New Jersey

Based on papers presented at the Sixth International Conference on High Temperatures—Chemistry of Inorganic Materials, held April 3–7, 1989 at The National Institute of Standards and Technology, Gaithersburg, Maryland
Sponsored by: The National Institute of Standards and Technology and The International Institute of Pure and Applied Chemistry
With the Participation of: The Air Force Office of Scientific Research

Volume 27
ISBN: 0-89603-187-x

This publication is printed on acid-free paper.

Many of these papers were selected from the Proceedings of the Sixth International Conference on High Temperatures—Chemistry of Inorganic Materials. Originally published in *High Temperature Science, An International Journal,* vols. 26–28. The plenary papers were originally published by IUPAC in *Pure and Applied Chemistry,* **62,** No. 1, January 1990.

Introduction

Conference Overview and the Role of Chemistry in High-Temperature Materials Science and Technology

LEO BREWER

Department of Chemistry, University of California, and Materials and Chemical Sciences Division, Lawrence Berkeley Laboratory, 1 Cyclotron Road, Berkeley, CA 94720

I don't want to compete with the fascinating historic account that John Drowart gave us, but I would like to go through the history of high-temperature symposia. I hope I don't get the reaction that I get from some of my classes when I say, "Remember when such-and-such happened during the War?" And I get this blank look, and one of the students will say, "I wasn't born until after the Korean War." Nevertheless, during World War II, many people in the high-temperature field had their first initiation. But there was one handicap. Owing to security measures, they were not able to interact with one another. Following the War, it was recognized that the high-temperature field was going to expand to meet the demands for materials with unique properties. To meet the demands for new fabrication techniques, it was important to establish better communications among various people. High-temperature symposia were established at that time and have continued very frequently, and I'd like to point out why they are especially important for this field.

One problem is that it is not easy to work at high temperatures. There are many unusual aspects of high-temperature science that often are not recognized by those who have not had extensive experience in the field. It's very important to be acquainted with the instrumentation and techniques that might be useful. One of the purposes then of these symposia was to acquaint people with the new techniques that had developed. This conference has been very successful in that regard. We've had both poster and oral presentations that were clearly presentations of new techniques. They emphasize the value of these techniques that weren't even in use at the time following the War, for example, the lasers. One of the areas in which high-temperature chemistry is particularly handicapped is the problem of containers. How do you contain these materials without contamination? The use of fast-pulse heating and levitation are certainly techniques that are going to be greatly expanded to overcome this. But there are some other problems that require adequate communication between people in the field. Another important aspect of these high-temperature symposia are not only the formal presentations; it is very easy to get together during coffee breaks, at meals, and during social hours to get acquainted with other people in the field and with what facilities they have. As we saw in many of the presentations here, there is collaboration between people at various institutions. Because it is often very difficult to have these very specialized pieces of equipment in all the laboratories, collaboration in use of facilities is very important.

There are some other aspects of high temperature chemistry that make these symposia particularly important. Fritz Franzen mentioned my definition of high-temperature chemistry. At what temperature should research be labeled as high temperature? Everyone is accustomed to the behavior of materials at room temperature and usually assume that they can extrapolate to higher temperatures. However, in many systems, quite unexpected behavior is observed upon raising the temperature. Such behavior marks the boundary of high-temperature science. So the definition of high-temperature science is "when things start behaving the way you don't expect them to behave," and I usually cite the case of boron chemistry where its high-temperature behavior starts below room temperature. So we do have to recognize that we have things happening that people don't normally expect or believe. An example of this came after the War when people heard that there were people interested in studying gases at high temperatures. The usual reaction was, "Oh, that's a waste of time. If you go up to high temperatures, everything will decompose to atoms and there won't be anything to study." But what they didn't recognize is that these gases are usually in contact with condensed phases. One can show through simple thermodynamics that the higher you go in temperature, the more complex the vapor becomes. It becomes more complex with respect to the number of different of species that are significant, as well as the size and complexity of the molecules that are there. Even today, it frustrates people that this is so. The lack of recognition of the complexity of high-temperature vapors has handicapped many industrial processes where unexpected reactions were taking place at high temperatures. On setting up new pilot plants, they've really gone astray because things have not behaved at high temperatures as they would at room temperature. The clarification of the role of complex gaseous species in high-temperature systems has not been an easy chore and much more work is needed to characterize the structures and properties of these species. The presentations at this conference illustrate the recognition of the importance of this problem. Both equilibrium species, as well as clusters formed by expansion of vapors, have been covered. Over 30 papers and posters deal with the problem of characterization of complex species in high-temperature gases.

This complexity of high-temperature systems requires a very broad outlook on the part of investigators of these systems. One must consider a wide variety of species to gain an adequate understanding of high-temperature behavior. High-temperature science is going to be important, but it is going to be important in unexpected directions. I think the high-temperature superconductor is an illustration of why we must maintain a broad outlook on how we are going to use high-temperature science, because we can't anticipate the leads that will come up and the new directions that will arrive. Students often come in to talk to me and ask, "In what direction should I concentrate? What's going to be the important direction? I have a mobile in my office that has seven hands pointing in different directions. I tell them, "Those are the directions you should be interested in." You want a broad base, because you can never anticipate what things will be important, and you should be prepared to aim for the new things that are coming up in the future. High-temperature research is difficult. One not only has to consider the possibility of many unusual species, but the experimental conditions are difficult to control. A common difficulty is the matter of containment without excessive contamination. One of the areas that was covered very extensively in this symposium is the problem of interactions between vapors and solids. Clearly, not only in erosion and corrosion, but also in preparation of materials, this is a very important topic, and we had 50 presentations at this symposium. This was very successful in presenting the topics that are important for this field.

Understanding of the kinetics of reactions at interfaces is particularly important and is an area that needs much more work to develop a broader understanding of high-temperature reactions, whose rates vary rapidly with temperature. The emphasis on thermodynamics is that we have a foundation that ties it together, whereas in kinetics, it is very diverse and they are really handicapped there. Obviously, from many of the presentations, the lack of understanding in kinetics is a critical barrier that is going to be a very difficult obstacle to overcome. There will certainly have to be more work in those directions. Even the problem of adequately characterizing the temperature can be a difficult job. The complexity of the systems and the difficulty of making reliable measurements set severe limits on the amount of information that can be obtained by direct experiment. But present day technology requires a great deal of information about high-temperature behavior to prepare materials by high-temperature processes and improve the efficiency of energy utilization and other high-temperature operations.

It's obviously very important to get data. Because of the fact that the behavior is unexpected, we need measurements of what really goes on at high temperatures. But if one considers the various combinations of the elements just in binary and ternary systems, we are up to astronomical numbers. And as we go to higher components, it is clear that nobody is going to be able to measure all these properties that we would like to know, particularly considering the difficulties in many cases of doing these measurements at high temperatures. So it becomes very important to pick decisive measurements, not just to pick anything at random and measure it. You must be systematic in that when you work with a phase diagram, you not only want to pick one phase, but pick related phases and how they relate to one another. It is particularly important to develop predictive models since it is not possible to hope to get all the data one needs. In other words, the measurements can be increased in value many fold if they will help improve our ability to predict reactions that haven't been measured, or, in some cases, can't be conveniently measured. So it is very important then to do measurements, perhaps for a sequence of elements in the periodic table, to give one an idea of what the trends are so that you can get some insight into the chemical bonding and then develop some better ways of predicting. We had, at this symposium, 40 presentations of various types of predictive models that will help improve our ability.

One of the areas that is most important for high-temperature science is thermodynamics. Thermodynamics is a most powerful tool in allowing you to take measurements that are made at low temperatures and use those to predict what will happen at high temperatures. One of my regrets is that there has been such a decline in room temperature calorimetry, because that has been a very important source of information for high temperatures. If you can determine the enthalpy of formation, even if you do it at room temperature, determine the entropies from spectroscopic data or by using the third law, and estimate high-temperature heat capacities, you can often make much more accurate predictions of high-temperature behavior than you could hope to measure. However, the importance of thermodynamics is well-recognized by the high-temperature community. This is illustrated by the more than 50 papers and posters that deal with characterization of the thermodynamics of systems of importance for high-temperature chemistry or deal with the important job of providing data banks.

The availability of data banks is most important. The lack of needed information is a serious problem for all disciplines, but it is a particularly critical problem for high-temperature scientists. Even information that may be available is not always readily accessible. What we need, and there have been attempts, is a large-scale effort to establish data bases. At the end of World War I, a substantial effort was made and produced a large amount of data, but it was not maintained. Then, the Bureau of Standards started an effort in this direction, but they were never adequately funded. One time when their funds were being reduced, when I was in Washington on other business, I had an appointment with the person in charge of their budget. He agreed to talk to me and I found out that he knew nothing whatever about what was going on in the Bureau. He called in some of the people under him that did know something about the Bureau and they discussed it. Their reaction was, "We don't think that the work is worth supporting because we never hear from industry that it is of any use to them.'' Well, these people in industry, of course, get these tables and they think that there is nothing to putting these numbers together, and they never point out to anyone how important it is to have these data.

What are we going to do? We want a centrally coordinated organization to provide a consistent data base. But it is impractical. We're never going to get the funding for it. One other problem is that we are never going to get enough qualified people together in one place to do all the things that would have to be done. So we have to compromise. I wish to discuss the responsibility of scientists in the matter of making information more readily available. Each one of us goes through the literature for data that is important for your particular measurement. What one has to recognize is that, after that effort has been made, it is important to make a little more effort to put those data in a form where they can be published so that other people can use them. What we have to do is really recruit everybody in the field to do their part in the preparation of these data bases.

There are some disadvantages. We are not going to have the consistency that we would like to have. But I don't think that we are going to be able to ever get one group that will put these all together and get them consistent. It will be useful to take these data, though they may be not quite consistent, and make some little shifts and get them on a consistent basis. But I think that to meet this demand for these critical data that people will need, we all have to get in here and try to contribute. The Bureau of Standards has had a program where they've tried to stimulate work of this type. I don't know whether future budgets will supply much money for it, but it is important to have some coordinating agency that will at least tie together these evaluations so that it will at least have a similar format for presenting the information.

It is impractical for most scientists and engineers to retrieve, critically evaluate, and put into usable form any significant fraction of the information they need. The retrieval of information from the literature is often a serious bottleneck. It is not only the tedious and time-consuming work of finding papers dispersed in many journals and published in many languages that stands in the way of utilizing the exponentially growing mass of information. A more serious obstacle is the need for a broad experience in the field to be able to evaluate critically the reliability of the measurements. A large fraction of the data in the primary literature is unreliable. This is particularly true of high-temperature literature because of the difficulty of obtaining measurements under extreme conditions. It requires a great deal of experience to extract reliable information from the literature. Most people are not familiar enough with the experimental or theoretical techniques to be able to critically evaluate the various, often conflicting, values. Reliable data are even more important when no measurements exist for the compounds of interest, and one must use various predictive correlation models to obtain the needed information. If the input data for the model are in error, the predictions can be greatly in error.

Bob Lamoreaux and I found many examples where a large fraction of the published data were seriously in error in the review of the thermodynamic data and phase diagrams for the binary systems of molybdenum that was published in *Atomic Energy Review Special Issue No.7* by the International Atomic Energy Agency. As an example, we found 13 papers all using the same technique, high-temperature solid-electrolyte cells, to determine the Gibbs energy of formation. We critically evaluated those papers: what sort of containers they used; how pure the materials were; did they use techniques like the simple technique of varying the gas flow through their system to see if degassing was affecting the potential and various other techniques of that type; many of the papers gave no indication that they had done that. So that is an example where it is easy to take measurements, but the measurements can often be useless. Of the 13 papers, we found only three that we felt were accurate. It is very important not only to review all the data, but to apply appropriate weighting factors that can vary by many orders of magnitude. One of the other examples was molybdenum diiodide, where we had to reject every publication of measured values, because we did have some predictive models as to what would be reasonable bonding energies for MoI_2. All these measurements used transpiration experiments that indicated they were getting MoI_2. But from our previous experience, and it was possible to confirm this later, we recognized that they were working in quartz tubes that would yield enough oxygen so that they were measuring $MoOI_2$, not MoI_2. In some cases, you have to throw away all of the data. And finally, one other example, the chromium–molybdenum system, where there have been seven determinations of the liquidus and solidus. We went through and reviewed again the purities of materials, the containers they used, and previous work by the same investigators to see whether there was similar consistency as far as doing careful experiments. We concluded that we would reject six of them, and only take one of them.

Thus, it is clear that it is most important to have data banks available that have been prepared by persons with adequate background to critically evaluate the data. A critical evaluation is not an easy chore. One must often reject most or all of the reported results. One must review carefully the experimental procedures used and examine the publication or previous publications by the same authors to determine whether they have been taking the effort to pin down systematic errors. It is necessary to utilize any reliable predictive models to determine if the experimental data are reasonable even within an order of magnitude. It is most important to give the reader some feeling for the possible uncertainty of the final accepted values. The ± values assigned should not be just the statistical uncertainty (which is a measure of the scatter around the average). The uncertainty cited should include an estimate of possible systematic errors. It is particularly important to not arrive at the recommended value by just averaging all the reported results; that would just mix the reasonably correct results with those in serious error.

There is a tremendous economic saving if each user of data need not take the time to search the literature independently. There is an even greater saving if the user is directed to the most reliable data and is not misled by the erroneous values. The wrong choice between divergent values could make the difference between success and failure of a proposed process. Each of us has to critically evaluate data of interest to our particular subfield. It is important to make the results of our evaluations generally available to others who would not be as expert in our subfield to arrive at a reliable evaluation.

Because of the difficulty of obtaining reliable data for high-temperature systems, there are two approaches to compilations of thermodynamic data. One is the compilation in which the data are critically evaluated in terms of the experimental methods used and possible systematic errors and evaluation of other results from the same laboratory compared to laboratories with established reliability. Such compilations can be carried out independently of any models to establish what appear to be the best experimental thermodynamic results. Another type of compilation is carried out in the context of some model where the data are examined with respect to the expectations of the model. An example where this is now routinely done is in the treatment of aqueous electrolyte data in terms of the Debye-Hückel theory. In this instance, the model is of very great importance because its greatest validity is in the extremely dilute range where it is particularly difficult to obtain reliable experimental data.

Another example is the use of pair-interaction theory as a basis for use of a second virial-coefficient equation of state for extrapolation of gaseous thermodynamic data to the low pressure gaseous standard state. A similar pair-interaction theory is often used as the basis for a Gibbs energy of dilution equation for solutions corresponding to a linear deviation from the Henry's Law slope. In using models of these types in compilations, one must be cautious about the implications of some of the assumptions of the model. In evaluating uncertainty limits for the compiled results, one must include uncertainties introduced by the model. As an example, one would not use an equation of state based on just a second virial coefficient for an ionized gas just as a linear deviation from Henry's Law is not applicable to an aqueous electrolyte. In the extrapolation of the thermodynamic data to the infinitely dilute or solute standard state, one must consider which systems might have long-range forces that would cause substantial deviations from Henry's Law even at the most dilute concentrations, which can be studied experimentally. There is also the problem that the component being assumed may be dissociated at the low concentrations, such as to make Henry's Law inoperable even in the limit of infinite dilution. It is important to use bonding models as a guide that warns when difficulties can be expected and provides for reformulation of the thermodynamic data in terms of a component that is actually the major species at the lowest concentrations. These are standard thermodynamic problems, but it is very important that the role of these essential models be clearly defined by the compiler.

Models are of value to the compiler aside from the problem of extrapolating to the infinitely dilute standard state or for carrying out the Gibbs-Duhem integration to infinite dilution. For example, it is sometimes convenient to express solution data in terms of the solubility parameter of the regular solution theory as a valuable means of tabulating the data more compactly. However, one must consider the implications of entropy assumptions that are implicit in the model. All of these models are useful in compilation efforts, but the possible influence of the model upon the resulting values must be clearly spelled out.

Much valuable data have been reported at this conference. These data are not only of value for the specific systems studied, but when used to develop or improve predictive models, the value of the data are increased many fold. At least 40 of the papers and posters at this conference were tied to models that were used to interpret and evaluate the measurements or the measurements were used to check existing models or develop new models that can yield useful data.

There are several other important areas covered in this conference. A most common problem in high-temperature processing is the understanding of interactions between condensed and gaseous phases. Over 50 of the presentations at this conference have dealt with various aspects of the problems of contact between condensed and gaseous phases. It is clear that the high-temperature community is addressing that problem. Because of the difficulty of obtaining reliable measurements at high temperatures, it is most important to have a variety of techniques available. Almost 30 presentations dealt with new techniques or improvements of previous techniques. The laser is becoming a widely-used tool for characterizing high-temperature systems.

For those of us who promoted high-temperature conferences in the past, the success of the present conference is a pleasant experience. It has been a wonderful opportunity for high-temperature workers around the world to get together and share new developments that will be useful to all. One can always ask for expansion of research in the high-temperature field to meet the many industrial demands for new materials, but we must face the budgetary restrictions that have been with us and might get worse. There is one area where we must press strongly for increased support. It is most important to have an expansion of the programs to develop data bases. Any money invested in these programs will be paid off many fold by the savings of time for the users who would not have to dig through the literature and who would not be misled by erroneous data that would be screened by the experienced people compiling the data bases. The other area that should be strongly supported is the development of chemical bonding models that would allow predictions of thermodynamic data for the multitude of systems where data do not exist. There are many fruitful opportunities in that direction. Of course, neither the data base compilations nor the development of models can be carried out adequately without the continued influx of new data measurements.

ACKNOWLEDGMENT

This work was supported by the Director, Office of Energy Research, Office of Basic Energy Sciences, Materials Sciences Division of the United States Department of Energy under Contract DE-AC03-76SF00098.

*Editors' Note: This account was given orally on behalf of IUPAC and is not available in document form.

Preface

The role of chemistry in materials processing and performance, at high temperatures, is of the utmost importance in today's technology. This is particularly the case for advanced inorganic materials, including the elements, their compounds, alloys, ceramics, composites—and specialized forms, such as films, coatings, powders, slags, and fluxes. The key combination of inorganic chemistry with high temperatures, known as the field of High Temperature Chemistry, has grown dramatically during the past 50 years. A number of books, symposia proceedings, and journals representing this field have appeared during this period. However, much of the early research emphasis and reports have been on the formation, characterization, and thermodynamic properties of high temperature species because of the intrinsic molecular nature of materials at high temperatures. In response to the need for improved processing and performance of materials, increased attention has recently been given to the physicochemical interplay between high temperature molecular species and solid or liquid interfaces and phases. New measurement techniques, models, and databases have been developed and applied to these heterogeneous systems. Even the previously hypothetical intermediate state of critical nuclei, or molecular clusters, is now amenable to detailed experimental (and theoretical) investigation. These, and related aspects of high temperature and materials science, are discussed here by prominent international experts. The purpose of these two volumes, then, is to represent the current status of fundamental research in the various disciplines that deal with the scientifically and technologically important interplay of high temperature and materials chemistry. In the first volume, two major chapters are offered that deal with (1) Advances in Measurement Techniques and (2) Thermochemistry and Models. The second volume contains major chapters on (3) Processing and Synthesis and (4) Performance Under Extreme Environments. The special Introduction, by Leo Brewer, discusses the importance of these particular areas to the overall field.

Many of the papers that follow were presented at an international conference, cosponsored by NIST and the IUPAC Commission for High Temperature and Solid State Materials, with additional support from the Air Force Office of Scientific Research. Other papers presented at the conference, but not included here, appear in volume 28 of *High Temperature Science, An International Journal.*

John W. Hastie
National Institute of Standards and Technology
(formerly the National Bureau of Standards)
Gaithersburg, MD

Table of Contents

Process Models and Materials by Design

Chapter Two: Performance Under Extreme Environments

Hot Corrosion

High Power Laser–Materials Interactions

Processing and Synthesis

Recent Advances in the Chemistry of Gas Phase Transition Metal Clusters

A. KALDOR* AND D. M. COX

Corporate Research Laboratory, Exxon Research and Engineering Co., Clinton Twp., Rt. 22E, Annandale, NJ 08801

ABSTRACT

Clusters containing a well-defined number of transition metal atoms have unique chemical, electronic, and magnetic properties that vary dramatically with the number of constituent atoms, the metal, and the charge on the cluster. We have been developing experimental techniques to probe these properties for gas phase clusters. This report summarizes and briefly reviews some of these results.

Index Entries: Transition metal atoms; constituent atoms; chemical properties; electronic properties; magnetic properties.

INTRODUCTION

Clusters containing a well-defined number of transition metal atoms have unique chemical, electronic, and magnetic properties that vary dramatically with the number of constituent atoms, the metal, and the charge on the cluster (*1*). Compared to atomic and bulk materials, our knowledge is surprisingly limited about these clusters in the 3–100 atoms size range. This is a challenging regime from both an experimental and theoretical perspective, because this is the transition from molecular description of matter to the solid- or liquid-like. The size regime is also important from a technological perspective, because this is where unique properties are exhibited by materials in applications as diverse as catalysis and "writing" metal contacts on microscopic semiconductor devices. Interest in clusters is fueled both by the potential offered by their intrinsic unique properties and what we can learn on the atomic scale about hard to probe, complex heterogeneous systems.

*Author to whom all correspondence and reprint orders should be addressed.

Honing our intuition is important since atomically rough heterogeneous systems are particularly difficult to study with atomic resolution owing to limitations of the available tools. Yet, in this same size regime, some catalysts are believed to exhibit strong size-dependent behavior (*2*), materials are often metastable and dynamic (*3*), and nanoscale materials, inhomogeneous on the scale of a few atoms, exhibit unique physical and chemical properties (*4*). Some fundamental questions in catalysis, e.g., active sites, adatom effects, promotion, poisoning, and so on, focus on effects localized on the atomic scale and these require models with similar specificity. It has been postulated that organometallic cluster compounds serve as atomic scale models for active sites in studies of surface catalysis (*5*). Indeed, this approach does yield insight about the mechanism of surface reactions. Organometallic complexes provide structural, spectroscopic, and thermodynamic data base for reaction intermediates and radicals bound to surfaces.

Much effort has been invested to synthesize cluster compounds with the right catalytic properties, either as homogeneous catalysts or, as supported, heterogeneous ones. A rich menu of multimetal cluster compounds has been made, and much has been learned about the metal–metal bonding, cluster structure, and coordination but only a few multimetal centered cluster compounds have emerged as interesting, catalytically-active materials (*6*). This is probably owing to the coordinative saturation and the high degree of symmetry inherent in many of these complexes. Often, these complexes have not survived thermal activation, i.e., loss of ligands even when supported.

We have been interested in approaching this problem from the perspective of gas phase, "naked" metal clusters. These are, at the outset, free of support interactions, matrix effects, and coordinatively unsaturated. The objective is to discover their physical and chemical properties as a function of cluster size and probe how these properties change as the degree of coordination is changed, i.e., at various stages of "dressing" of the clusters with ligands of differing properties. Finally, the ultimate challenge is to deposit the unique clusters on supports and explore the properties of these new materials. Recent advances have made it possible to perform such experiments in the gas phase using most metals and with a broad range of reagents. Now one can deposit monosized clusters of many metals on a variety of supports. In this report, we plan to briefly review the emerging trends in the field.

Rules that define the chemical bonding of transition metal clusters when coordinately saturated (*7*) need further development to deal with "naked" metal clusters though they have been useful as the clusters adsorb ligands (*8*). At this point, we know little of the structure of these materials, and our scientific intuition is not well developed. With so few atoms involved, one would expect that they are close-packed and globular. The cohesive energy of clusters as a function of the number of constituent atoms competes with other energetics. Clusters in the gas phase release energy by collision with rare gas carriers, vaporizing constituent atoms, reconstructing to different configurations and condensing onto a support. Some clusters are fragile upon ionization, some fragment/fission upon reaction, and some do not survive deposition onto a support. Although we want to understand why these losses occur, we are looking for the survivors.

It has been exciting to find so many diverse examples of cluster systems that exhibit dramatic, nonmonotonic, size selective reaction rates, as well as unusual levels of saturated coverage as a function of the metal, the charge on the cluster, and the reagent used (*1*). But it has been very surprising and difficult to understand why only certain cluster sizes have definable isomeric forms and why some are, essentially, unreactive (*9*).

The fact that some clusters have isomers suggests that the simple globular, close-packed structure is too simple a model. The small energy difference among possible structures encourages the concept of fluxional structural isomers and, in the limit, conjures the image of a "molten droplet." This is not a novel concept. Surfaces of solids are mobile well below the melting point. In fact, significant reconstruction of catalyst surfaces is known to occur at about ⅓ of the melting temperature of transition metals, known as the Huttig temperature (*10*). The melting point of clusters is known to drop as a function of cluster size, thus reconstruction may become more facile the smaller the cluster. This would be similar to an equilibrated metal surface with no structure-dependent, reactive properties though there could be size dependence owing to electronic effects. Structure dependence would be superimposed on size dependence for cluster with stable, or metastable structures (*11*).

Some clusters may be unreactive, because with highly symmetric, unusually stable configurations, they cannot accommodate the reaction products. This is only one example of accommodation induced size selectivity; the notion includes not only the ability of the cluster to accommodate the reagent spatially, but also energetically, including the ability to handle excess energy resulting from the chemisorption process. If the product cannot be stabilized, the reaction will not be completed. These accommodation factors may be the reason why for several metals, clusters of fewer than 30 atoms are usually less reactive than larger ones.

The electronic structure of clusters differs from the bulk surface. The density of states is more sparse, and there are more surface orbital energy and symmetry restrictions. Valence electron binding energies vary nonsystematically; cluster orbitals are oriented and have varying extensions depending on the metal and orbital involved. In order for a reaction to occur, there has to be a match between the energetics and symmetry of a cluster state and a reactant; there also has to be orbital overlap. This is a source of selectivity with cluster size, structure, and reagents. A further source of selectivity is the limited number of reacting states. Specific size selective reactivity will be observed if cluster valence electrons of the required energy, symmetry, and spatial extension are available and if the cluster can accommodate the reaction products, in the broadest definition, discussed above. The complex interaction of all these forces explains why it is so difficult, at this point, to formulate a comprehensive description of the diverse, often bewildering observations of size selective reactivity reported to date.

This discussion will focus on work done in our laboratory and, except by some reference, will not do justice to some excellent work done by others. There are several recent reviews that have covered the field in more depth than we will do here.

EXPERIMENTAL APPROACH

The experimental apparatus and the data analysis have been described elsewhere, and only a brief discussion is given here (*1*). Figure 1 shows the layout of the functional components. Laser vaporization in the throat of a supersonic pulsed valve has made studies of transition metal clusters much easier than conventional high temperature vaporization techniques. The pulsed valve introduces a high pressure of helium over the metal target and, at the right moment, a doubled Nd–Yag laser focused on the metal target is fired and generates a "cold" plasma. This is a soup of species at several thousand degrees, which consists of atoms, positively- and negatively-charged ions, as well as electrons. The carrier gas sweeps this "soup" through a narrow channel, causing nucleation and cluster growth by condensation and cooling. The length of this condensation zone can be adjusted to maximize cluster growth and bring the cluster temperature to equilibrium with the ambient. It is assumed in all experiments that the system had a chance to come to equilibrium and metastable structures, or states have all relaxed. Depending on the ion stabilities involved, not all charged clusters survive. For example, one is able to generate neutral and positively-charged iron clusters, but not negatively-charged ones; positively-charged palladium clusters have not been produced. This is a complex and poorly-characterized system, and while this source is an attractive, simple device for screening studies, it needs to be further developed for other applications. A lot of work is going into developing continuous, or pseudocontinuous sources to make possible experiments, such as EXAFS, that require high stability and high duty cycle.

Once the clusters or cluster ions are formed, they can be subject to a variety of studies. These are all performed while the clusters are in flight to take advantage of the inherent sensitivity of mass spectrometric detection since the absolute number of clusters at a specific size can be very few indeed. The development of ultrasensitive measurement techniques has made it possible to detect as few as a hundred clusters of a specific size. Thus, the experiments (see ref. *1* and references therein) probe the reactivity, dynamics, energetics, structure, and electronic properties of the clusters. Ionization thresholds have been measured using tunable lasers. Stern-Gerlach type magnetic deflection experiments have been performed on some clusters, as have polarizability experiments.

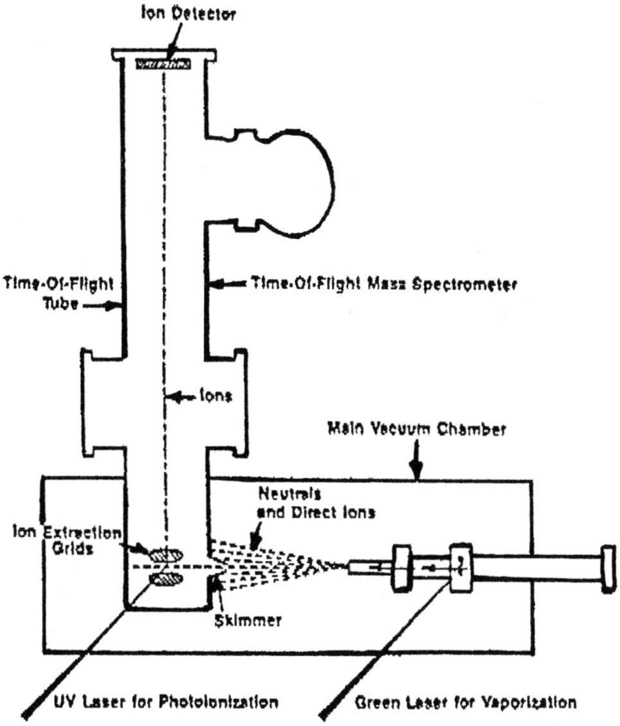

Fig. 1. Pulsed cluster beam apparatus.

Negatively-charged clusters have been used for electron affinity measurements as a function of cluster size, as well as to obtain photoelectron spectra on mass-selected clusters and cluster adducts. Other physical characterizations have involved heating the clusters with lasers to measure the binding energy of metal atoms and adducts. Tunable infrared lasers have been used to thermally desorb molecules bound to the clusters and, in the process, using mass spectrometric detection to obtain the vibrational spectra in one measurement of chemisorbed species, for the entire cluster distribution, as a function of cluster size.

Experiments to study the chemisorption of various molecules by clusters as a function of cluster size follow two approaches. One approach aims to measure the rate of reaction, by monitoring metal loss as a function of reaction conditions, and the other the products of the reactions, for example the degree of "surface coverage," or saturation achieved. Both approaches use the reactor configuration, shown in Fig. 2. The rector is attached to the face plate of the cluster source. The "pulse" of helium, with the cluster distribution imbedded, flows into the tube reactor. This consists of a tube with two symmetric pulsed valves, similarly configured. One valve injects pure helium and the other helium with the reactant mixed into it. The idea is that the pulse from the first valve is used to measure the physical, e.g., mixing effects of the helium pulse on the cluster beam and the overall measurement system. This

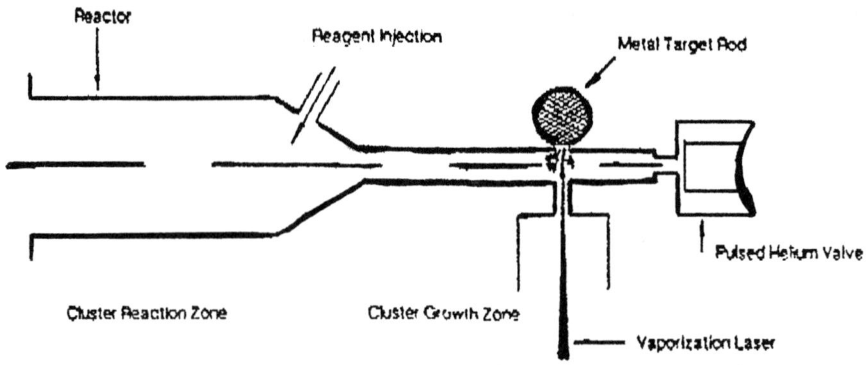

Fig. 2. Details of cluster source and reactor.

provides the base cluster concentration, and data for the entire distribution are stored in the computer. The second pulse carries the reactant, and this measurement is the residual, unreacted cluster concentration. Cluster depletion can now be determined as a function of cluster size. This, as a function of reagent partial pressure, yields the rate of the depletion with a specific reagent as a function of cluster size. Figure 3 shows such data for clusters of neutral and positively-charged iron clusters for reactions (1) and (2), respectively

$$Fe_x + H_2 \longrightarrow Fe_xH_2 \tag{1}$$

$$Fe_x^+ + H_2 \longrightarrow Fe_xH_2^+ \tag{2}$$

These results show dramatic variation of reaction rate with cluster size for both iron species. Similar measurements of the rates for deuterium, shown in Fig. 4 as a ratio of k(H)/k(D), indicate a huge, size-dependent kinetic isotope effect. Saturation levels are measured at progressively higher reagent partial pressure, and the product mass peaks are monitored (*11*). As shown in Fig. 5, multiple levels of surface coverage are observed, suggesting that intermediate stages of stable cluster structures may be present. One can also observe dehydrogenation reactions as a function of cluster size, as shown in Fig. 6. This is an example where the primary chemisorption reaction is not size selective, but the subsequent loss of hydrogen is (*12*).

Once the clusters are made and have been reacted, they are detected by time-of-flight mass spectrometry. The pulsed nature of the source is ideally suited for the detection of directly generated cluster ions and products of cluster ion reactions. Laser photoionization is used to detect neutral species utilizing low fluence (<1 mJ) excimer lasers operating at 7.87 or 6.42eV. Tunable lasers are used to measure IP-thresholds below 6.4 eV. Work is underway to push this to 12 eV using nonlinear mixing technology. Usually microJoule level radiation is adequate for single photon ionization, and in fact, one wants to avoid multiphoton ioniza-

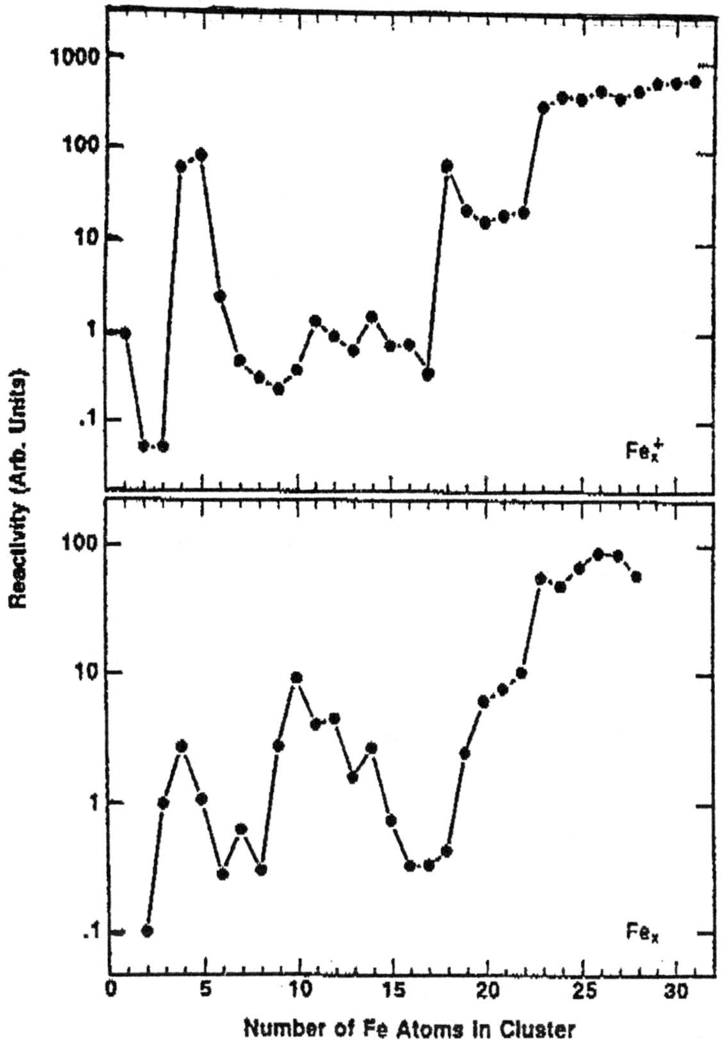

Fig. 3. Top, reactivity of Fe_x^+ with H_2, and Bottom, reactivity of Fe_x with H_2.

tion. MPI can cause other complications, such as photofragmentation, or photochemistry in the ionization zone. There are obvious limits to mass spectrometry, and use of isotopically-pure metals is a great help. Other creative uses of mass spectrometry permeate the field. These include sophisticated use of FT-ICR-MS and other mass selective ion reaction techniques.

Chemical Reaction Studies

Size selective dissociative chemisorption has been observed for several important reactions. These include activations of H–H, N–N, O–O, CO_2, RO–H, R–H, and RS–H bonds. We have been able to compare the

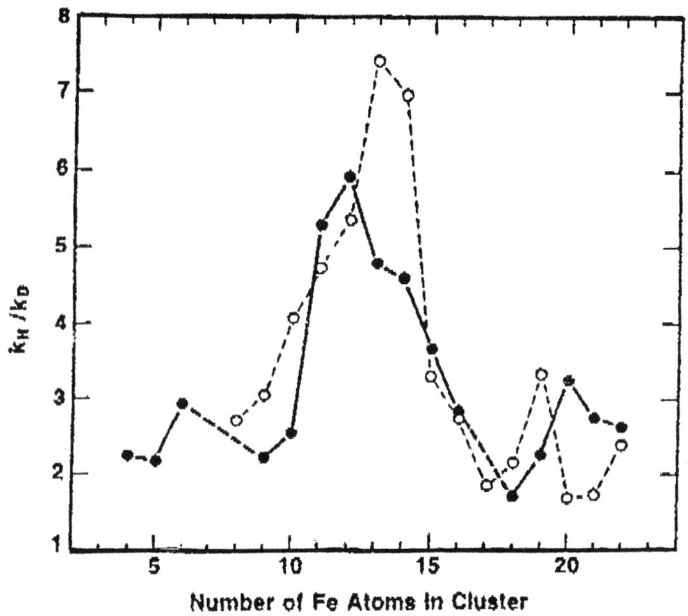

Fig. 4. H/D kinetic isotope effect for Fe_x (open circles), and Fe_x^+ (closed circles) as a function of cluster size.

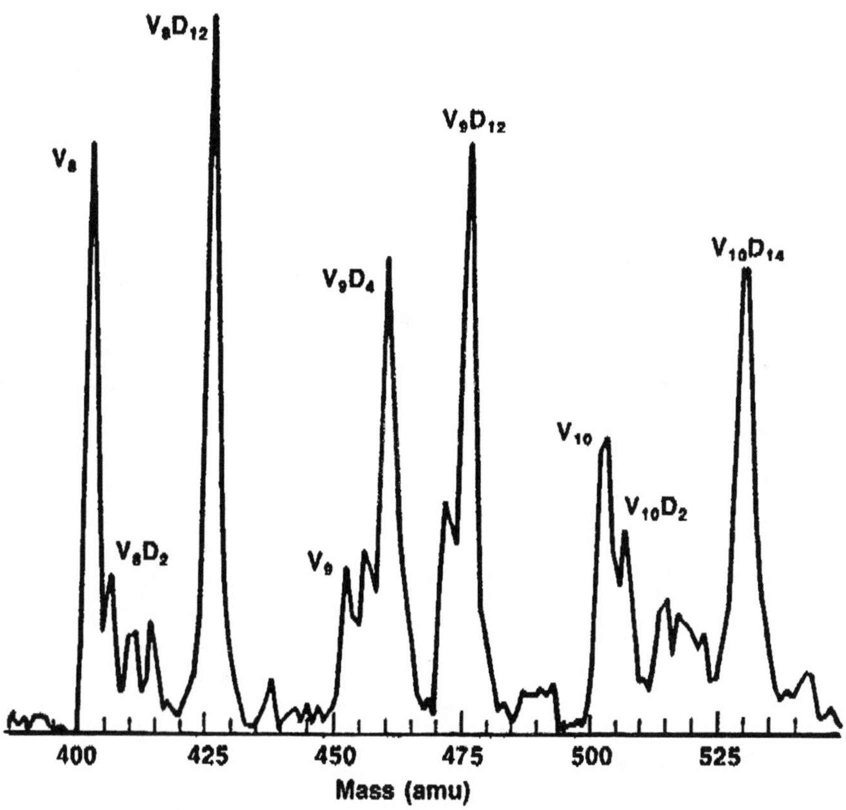

Fig. 5. Species observed in the mass spectrum upon saturated chemisorption of D_2 on V_x clusters.

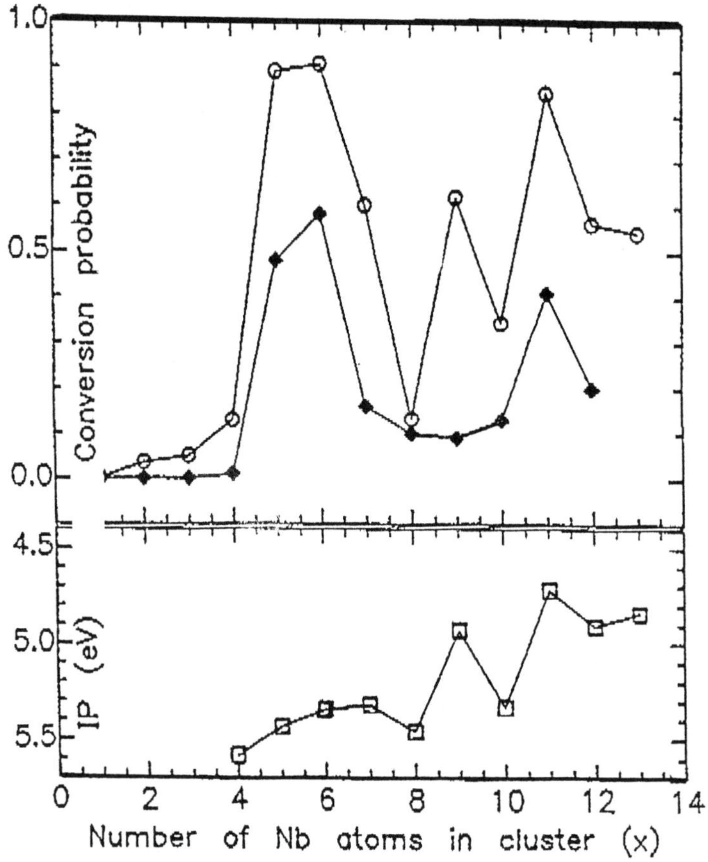

Fig. 6. Benzene chemisorbed on Nb clusters loses hydrogen to produce Nb_xC_6 species. Top panel, conversion probability to Nb_xC_7. Bottom panel, IP variations with cluster size.

effect of charge, adatoms such as oxygen, carbon, and so on, and the role of the constituent metal atoms. Theory is evolving to provide a framework to organize the observed results, but it is not at the stage where it is able to predict the multitude of observed dramatic size selective effects. It is providing general guidelines from a limited set of calculations (*13*). Here, we plan to discuss the chemisorption of hydrogen, methane, and benzene as examples.

Dihydrogen

One of the prototypical reactions is dissociative chemisorption of dihydrogen. This reaction has now been studied for V, Nb, Ta, Fe, Co, Rh, Ni, Pd, and Pt, as well as Al. In all cases, the reactivity varies nonmonotonically with cluster size, and this size dependent reactivity is unique for each metal. As shown in Fig. 7, several orders of magnitude variation is observed for the dissociative chemisorption of the first dihydrogen on iron clusters. It is interesting to note that for iron clusters with

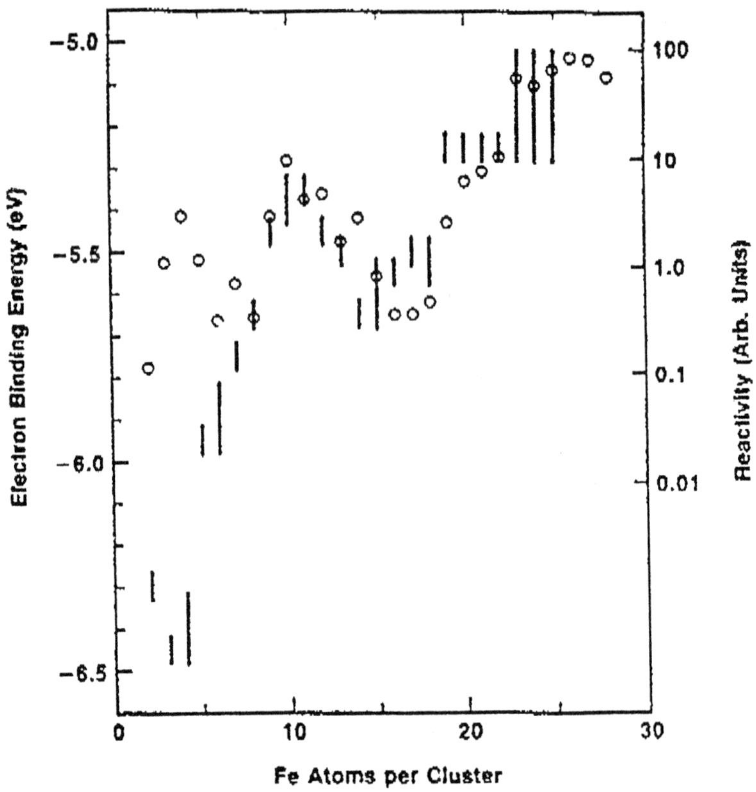

Fig. 7. Ionization potential of iron clusters (vertical lines), and relative reaction rate measurement of iron clusters with dihydrogen (open circles).

more than eight atoms, the reaction rate is greater for clusters where the ionization potential, i.e., the binding energy of the least strongly-bound electron is the smallest. For clusters smaller than eight atoms, the correlation is direct; the larger the IP, the faster the reaction. This correlation holds for reactions of niobium, vanadium, tantalum, and palladium, as well as aluminum, the systems for which some IP data now exists. Figure 3 compares the reaction rate of neutral with charged iron clusters as a function of cluster size. The effect noted for iron is much more dramatic than observed for niobium. The case of iron is a rare one; it is one of the few examples where, for several cluster sizes, the ion reaction rates are slower than for a neutral species. It is also clear that, by 25 atoms, the effect of charge is dissipated and the rates are comparable.

Most clusters will adsorb several molecules of hydrogen, but the rate of chemisorption of the first molecule is quite interesting. As already noted, this reaction exhibits a remarkable range of size selective behavior with most transition metal clusters and cluster ions. Iron, cobalt, niobium, vanadium, tantalum, and aluminum show the strongest size selectivity, whereas nickel and palladium are an intermediate case. Rhodium and platinum are relatively size independent. Equally interesting is the kinetic isotope effect (shown in Fig. 4) for iron, which shows a nearly identical size-dependent kinetic isotope effect, k(H)/K(D) for the neutral and positively-sized clusters, ranging from 2 to 7. This suggests that the reaction for both ions and neutrals has a similar potential surface as a function of cluster size. This is an important element in formulating a model for this chemistry. Palladium clusters show an inverse isotope effect k(H)/K(D)<1 as a function of cluster size, suggesting that other forces can interfere with the simple picture (*14*).

A somewhat different measure of size selective chemistry is the maximum number of hydrogen molecules a cluster can chemisorb as a function of size. Here, the difference among the metals is quite significant. Iron will accept only one hydrogen atom/surface metal atom even though there are other open surface sites that accept molecules, such as ammonia. Niobium, vanadium, and tantalum will accommodate a somewhat higher hydrogen content, up to an D/M = 1.4. Small clusters of Pd, as shown in Fig. 8, will accommodate up to H/M = 3, with some variation with size, falling to D/M = 1 for clusters larger than Pd_{15}. For cluster ions of Pt, Ni, and Rh, the small clusters show D/M = 5, 4, and 8,

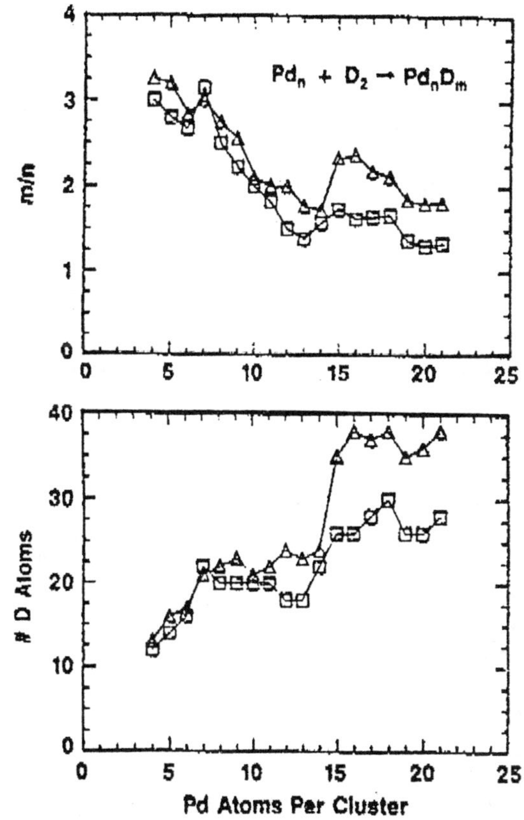

Fig. 8. Deuterium uptake as a function of cluster size. Upper panel, the ratio of D/M plotted against cluster size. Lower panel, the number of D atoms adsorbed on cluster vs cluster size. Squares refer to D_2 injected into reactor and triangles when injected upstream of the vaporization laser.

respectively. The D/M is size dependent, falling toward D/M = 1 for Ni_x^+ clusters with more than 13 atoms, but D/M = 1 is not reached until >20 atoms for Pt_x^+ (*15*). Reaction rate studies also provide insight about isomers. Fig. 9 shows the reaction rate for dihydrogen with niobium clusters and positively-charged cluster ions. The data indicate that, for the 12-atom cluster and cluster ion, at least two isomers coexist, whereas for the 6-atom cluster, only one species is present. It is interesting to note that the negatively-charged, 12-atom niobium cluster shows a linear reactivity pattern. Isomers of other niobium clusters have also been observed, e.g. Nb_9 and Nb_{19}^+.

The effects of adatoms have also been explored. Figure 10 shows the variation of the dinitrogen activation rate as a function of cluster size with and without an adatom of oxygen. By 8 atoms, the contribution of the oxygen atom is dampened, but for smaller clusters, the effect is dramatic, nearly an order of magnitude "mirror image" of the reaction rates. The adatom appears to act as if it was a tantalum atom, shifting reaction rates by one cluster size. For dihydrogen, the activation rates of TaO and Ta_x are similar.

Methane

Clusters of platinum, positive platinum ions, and palladium dissociatively chemisorb methane, whereas iron, niobium, rhodium, and aluminum do not. The strength of the C–H bond is comparable to dihydrogen, but other factors also may be important. Platinum dimers through pentamers are the most reactive, as noted in Fig. 11. A factor of four drop in reactivity is noted going from the five atom cluster to that with six. The other small clusters containing 6–24 atoms are less reactive,

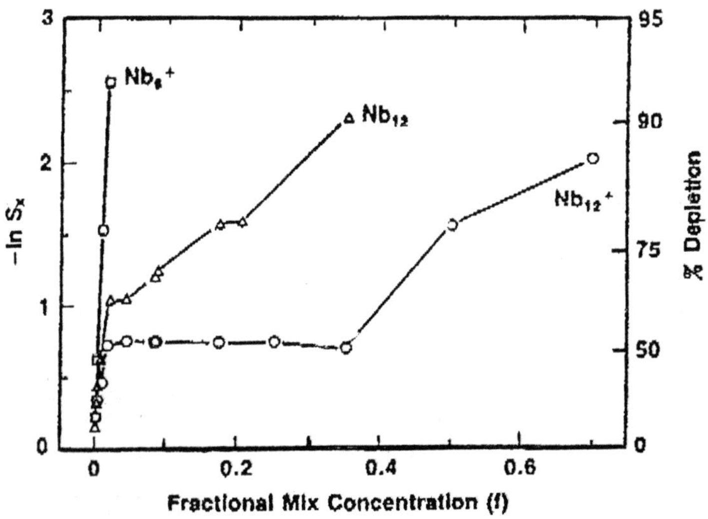

Fig. 9. Reactivity of Nb_6, Nb_{12}, and Nb_{12}^+ with D_2. S_x = survival fraction of injected metal cluster as a function of D_2/He fraction injected into the reactor.

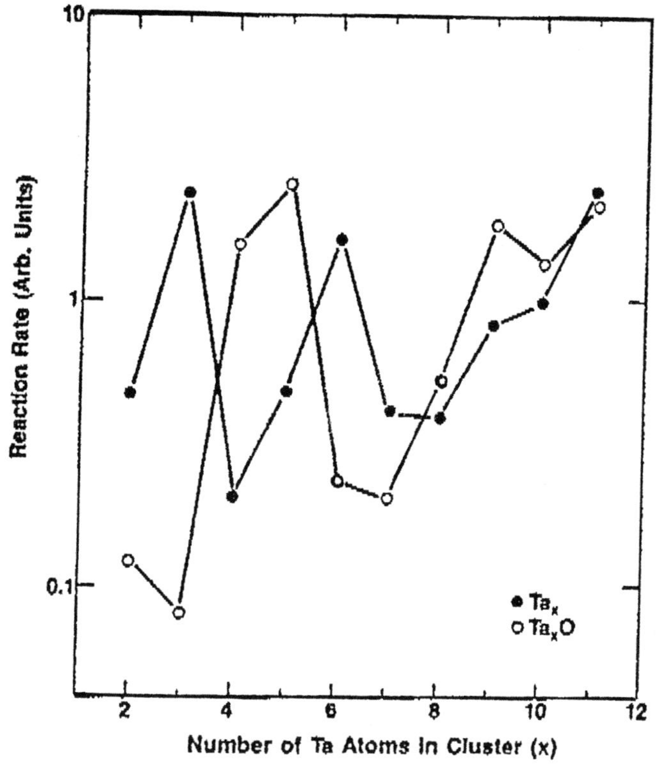

Fig. 10. Reaction rate Ta_x and Ta_xO(o) with H_2.

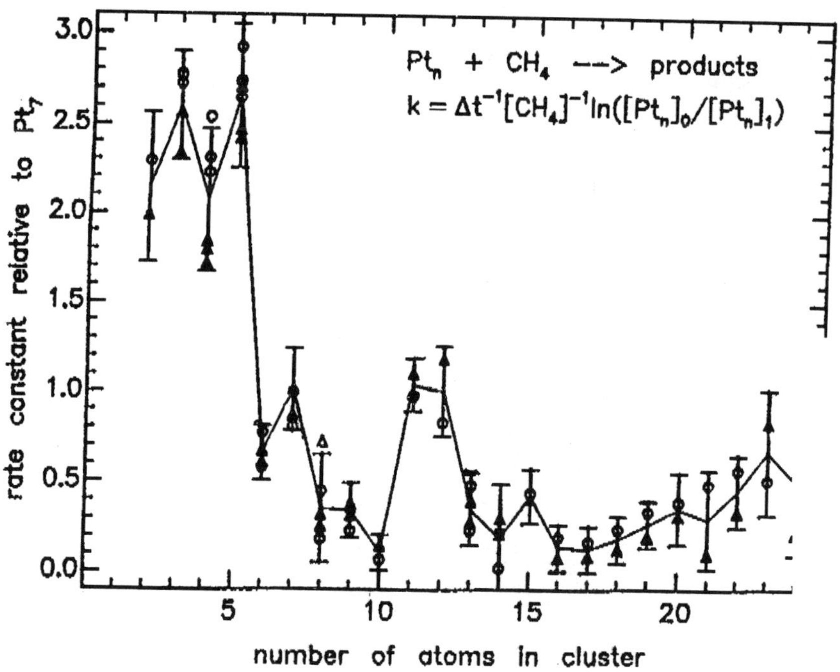

Fig. 11. Reaction rate of Pt_x with CH_4 normalized to Pt_7.

though there is a small increase in reactivity at 11- and 12-atom clusters (*16*). It is interesting to note that the primary effect of the positive charge on Pt clusters is to extend the larger reaction rate to the seven atom cluster ion; the slight reaction rate increase at 11- and 12-atom clusters, noted above, is also observed for the ions. The dominant products observed are Pt(x)C(1,2)H(y), with the larger clusters retaining more hydrogen. The decrease in reactivity with increasing cluster size is unusual and in contrast to what we see for most dihydrogen systems. A correlation is noted between reactivity and the ability for clusters to form compact structures with three or less coordinated surface atoms in a close packed structure. This is consistent with the organometallic model that coordinative unsaturation is critical for methane activation. Yet, these strongly bound products may be analogs for poisoned sites for catalysis. Palladium clusters activate methane, but follow a size selective pattern quite different observed for platinum (*14*). The maximum of reactivity is observed for the 10-atom cluster, and the overall pattern of size dependence is similar to dihydrogen. The deuterated methane reaction rate is significantly larger than for methane, similar to what is observed for hydrogen isotopes.

Benzene

Chemisorption of benzene on niobium and vanadium clusters shows no size dependence, other than a function of change in collision parameter with size, but an analysis of the reaction products is much more informative. Clusters of niobium containing 5, 6, and 11 atoms not only chemisorb benzene, but there is also some dissociation and loss of hydrogen. Kinetic isotope effects have shown that this loss is subsequent to the chemisorption step, and is related to hydrogen activation. In this case, as can be seen in Fig. 6, there is anticorrelation to the measured IPs of the clusters, similar to observations made for dihydrogen activation by niobium (*17,18*). Experiments with cluster ions confirmed these observations (*12*).

Figure 12 shows the data for the dehydrogenation of benzene by vanadium cluster ions. Sequential dehydrogenation channels are observed for several cluster ions, in contrast to niobium ions where primarily only two species coexist, Nb_xC_6 and $Nb_xC_6H_6$, we find dehydrogenation products with vanadium ions, such as $V_xC_6H_m^+$ with m = 2, 4, and 6. Only V_{5-7}^+ produce a completely hydrogen-free species, m = 0. There is a possibility that multiple oxidation states account for these products (*19*).

DISCUSSION

The model that does explain a significant number of the observations discussed above invokes the donor/acceptor model of chemical reactions. This model has been used by a number of workers to explain surface

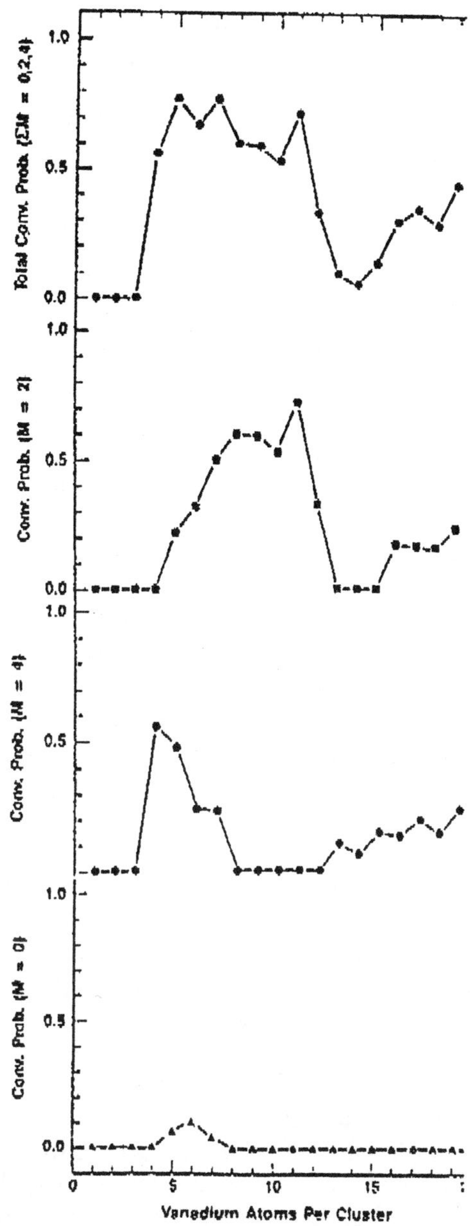

Fig. 12. Vanadium cluster ion size dependence for dehydrogenation of benzene; M = hydrogen content of vanadium benzene adduct $V_xC_6H_M$. Lower panels show that the degree of dehydrogenation is size dependent.

chemical reactions, as well as reactions of metal complexes. Cluster reactivity is governed by a number of different factors, ranging from electronic and structural effects to thermodynamic and kinetic factors.

From a simple frontier orbital point of view, the key features of the activation of oxidative coupling, as amplified by dihydrogen, are (1) attractive metal cluster donor interactions between the high-lying unoccupied hydrogen sigma antibonding orbital and the highest occupied molecular orbital of the cluster, (2) attractive metal acceptor interactions between the occupied hydrogen sigma bonding orbital and the lowest unoccupied cluster orbital, and (3) Pauli repulsion between the filled hydrogen sigma orbital and the filled cluster valence orbital. Interactions (1) and (2) are attractive, leading to H–H bond weakening and M–H bond formation. In (1), charge density is transferred into the antibonding sigma orbital, and in (2), charge density is transferred out of the sigma bonding orbital. The repulsive interaction, (3), is directly responsible for the activation barrier to dissociative hydrogen chemisorption and dominates the dynamics in the approach to the transition state. The size selectivity of the reaction rate arises primarily from (3) and (1). the energetic cost in overcoming the activation barrier is expected to be the smallest when the binding energy of the electron involved in the charge transfer is the lowest, assuming that the metal cluster orbitals are of the appropriate symmetry to interact with the hydrogen sigma and sigma star orbitals, respectively.

The same charge transfer model has been invoked for metal surfaces, but if the reaction is site specific, the effect is going to be magnified for metal clusters. For small metal clusters with high IPs, the donation from hydrogen sigma orbitals to the metal cluster will start to dominate. Similarly, symmetry considerations are also expected to play an increasingly important role since there is an increasing sparsity of states near the Fermi level as the cluster gets smaller. Indeed, for iron clusters with fewer than 8 atoms, a direct correlation of reaction rate with ionization potential is observed. Similar situation is found for vanadium, niobium, and tantalum.

The chemisorption rate levels off at some maximum value for clusters with more than 25 atoms. This is what one would expect for a nonactivated process. For all metal clusters examined so far, hydrogen chemisorption leads to an increased ionization potential. This follows from interactions (1) and (2) directly since the creation of the new metal hydride bonds removes two electrons from the cluster valence manifold. This oxidative addition process suggests that, as the cluster grows, the shift in IP diminishes. The addition of other reagents, such as ammonia, is a reductive addition process with respect to the metal cluster, resulting in an increase in the Fermi level and a reduction in IP.

We have explored other tests of the reactivity model by considering the role of charge on reaction rates and find that if one accounts for the induced polarizability of the reagent the model holds. Also, if one includes in the model reduction, as well as oxidation, the model is predictive for small clusters as well. We are still in search of a model that is consistent with the nonreactive clusters though, as discussed in the introduction, structural stability associated with high degree of symmetry may be an overriding factor. Also, the saturation level coverages are surprisingly large and require further insight. The transition metal cluster reactivity area is quite fertile and there is every reason to expect significant continued progress in the future.

REFERENCES

1. Kaldor, A., Cox, D. M., and Zakin, M. R. (1988), *Evolution of Size Effects in Chemical Dynamics*, part 2 Advances in Chemical Physics, vol. XLXX, Prigogine, I. and Rice, S. A., eds., Wiley, New York.
2. Spencer, M. S. (1986), *Nature* **323,** 685.
3. Bond, G. C. and Yide, X. (1984), *JCS Faraday Trans. I,* **80,** 969.
4. *JMRS report.*
5. Albert, M. R. and Yates, J. T. (1987), *The Surface Scientist's Guide to Organometallic Chemistry,* American Chemical Society.
6. Gates, B. C., Guczi, L., and Knozinger, H., eds., (1986), *Studies in Surface Science and Catalysis,* Vol. 29, Elsevier, New York.
7. Moskovits, M., ed. (1986), *Metal Clusters,* Wiley, New York.
8. Fayet, P., McGlinchey, M. J., and Woste, L. (1987), *J. Am. Chem. Soc.* **109,** 1733.
9. Zakin, M. R., Brickman, R. O., Cox, D. M., and Kaldor, A. (1988), *J. Chem. Phys.* **88,** 3555.
10. Huttig, G. F., *Arch. Metallk, 2,* 93, (1948)
11. Zakin, M. R., Brickman, R. O., Cox, D. M., and Kaldor, A. (1988), *J. Chem. Phys.* **88,** 6605.
12. Zakin, M. R., Brickman, R. O., Cox, D. M., and Kaldor, A. (1988), *J. Chem. Phys.* **88,** 5943.
13. Upton, T. H., Cox, D. M., and Kaldor, A. (1987), *Physics and Chemistry of Small Clusters,* Jena, P., Rao, B. K., and Khanna, S. N., eds., Plenum, New York.
14. Fayet, P., Kaldor, A., and Cox, D. M., *J. Chem. Phys.* in press.
15. Work in progress.
16. Trevor, D. J., Cox, D. M., and Kaldor, A., *J. Am. Chem. Soc.,* in press.
17. Zakin, M. R., Cox, D. M., and Kaldor, A., (1987), *J. Phys. Chem.* **91,** 5224.
18. St. Pierre, R. J., Chronister, E. L., and El-Sayed, M. A., (1987), *J. Phys. Chem.,* **91,** 5228.
19. Zakin, M. R., Cox, D. M., Brickman, R. O., and Kaldor, A., *J. Chem. Phys.,* in press.

Structure and Infrared Spectroscopy of the C_6 and C_8 Carbon Clusters

Martin Vala,* T. M. Chandrasekhar, Jan Szczepanski, and Robert Pellow

University of Florida, Department of Chemistry, Gainesville, FL 32611

ABSTRACT

Carbon clusters have been generated by Nd/YAG laser ablation of graphitic material and trapped in argon matrices for study via infrared (IR) spectroscopy. The most prominent bands observed (2164, 2039, 1998, and 1952 cm^{-1}) have been observed previously from a sample prepared by thermal vaporization followed by strong matrix annealing. The first two bands have previously been assigned to linear C_5 (2164 cm^{-1}) and linear C_3 (2039 cm^{-1}). The latter two, the focus of this study, were investigated via annealing behavior and isotopic substitution (1:1 mixtures of ^{12}C:^{13}C powder). Arguments based on IR band densities (no. of bands/cm^{-1}) and isotopomer band intensity diminution upon isotopic substitution are used to assign the 1952 cm^{-1} band to cumulenic chain-like C_6 and the 1998 cm^{-1} band to linear C_8. Normal coordinate calculations have been performed on C_6 and confirm the present assignment.

Index Entries: Carbon clusters; infrared spectroscopy; C_6; C_8.

INTRODUCTION

The structure, stability, and spectroscopy of small carbon clusters is the focus of considerable current attention. In addition to their relevance to interstellar space (*1*) and combustion chemistry (*2*), carbon clusters are intrinsically interesting since the structures of most are not yet precisely known. The thermal vaporization of graphite (*3,4*) yields predominantly carbon atoms, plus C_2 and C_3 clusters; C_2 and C_3 have been much studied and are now well-characterized (*5–7*).

*Author to whom all correspondence and reprint requests should be addressed.

Early theoretical work (*8*) predicted that all clusters containing up to 10 carbons should be linear. More refined recent calculations (*9–14*) have indicated that some of the ground state cluster geometries with four or more carbons may not be linear. Raghavachari and Binkley (*11*) calculate that odd carbon clusters have lowest energy linear structures, whereas even carbon clusters should be cyclic. Other recent theoretical work by Bernholc and Phillips (*15*) predicts linear structures for all clusters containing between three and about nine carbons.

To date, experimental spectroscopic studies on the small neutral C_n clusters ($3 < n \leq 10$) have been limited. Work has been reported on C_4 (ESR and optical spectra (*16*)), C_6, C_8, and C_{10} (ESR (*17*)), and several other neutral small clusters (infrared (*18,19*) and optical spectra (*20*)). We have recently reported (*21*) on the identification and IR spectral assignment of the linear C_5 cluster. In the present paper, we report on a Fourier transform infrared spectroscopic study of several matrix-isolated carbon clusters produced by the laser vaporization of graphite. Mixed isotopic ($^{12}C:^{13}C$) studies have led to the location and identification of the IR frequencies associated with the C_6 and C_8 clusters. Both C_6 and C_8 are chain-like structures and are probably linear or slightly bent.

EXPERIMENTAL

The ^{12}C clusters were prepared by laser vaporization of carbon from a rotating, spectroscopic-grade graphite rod held in close proximity to a CsI window mounted to the cold finger of a closed-cycle helium refrigerator (Displex Model DE202, Air Products). The isolant gas, argon, was inlet continuously for periods usually ranging from 40 to 60 min. The laser (frequency-doubled Spectra Physics DCR-11 Nd-YAG), which was operated at 10 Hz and ~0.1mJ/pulse, was slightly focused. It is known (*17*) from ESR experiments that sizable quantities of C_6 and C_8 are produced under similar conditions.

For the mixed isotope experiments, equal parts of ^{12}C (Union Carbide, spectroscopic purity) and ^{13}C (Isotec, 99 at. % enriched) powders were mixed, degassed in a Knudsen cell to 1500°C for 1 h, compressed to form a pellet, and placed in a small metal cup. The cup was mounted to the inside of the cryostat shroud. The laser beam focus was displaced to a new position on the pellet about every 10 min during the deposition to ensure a fresh vaporization surface.

Infrared data were recorded on a Nicolet 7199 Fourier transform spectrometer using a Hg-Cd-Te detector (0.25 cm^{-1} resolution, 200 scans). The moderately high resolution was found to be necessary to resolve close-lying peaks.

BACKGROUND

Recent theoretical studies, which include electron correlation effects, are now sufficiently powerful to predict, with high accuracy, the ground state vibrational frequencies of unknown molecules, such as the carbon clusters. These calculations generally predict frequencies that are about 10% too high. For the infrared active ungerade modes, Raghavachari and Binkley (*11*) calculate 2311 cm^{-1} for linear C_3 (observed: 2039 cm^{-1}) and 2344 cm^{-1}, 1632 cm^{-1} for linear C_5 (observed: 2164 and 1544 cm^{-1}). Raghavachari, Whiteside, and Pople (*12*) find that, for C_6, the distorted cyclic (D_{3h} symmetry) structure is the most stable form. Its IR-active carbon stretching frequencies are calculated as 1726 and 1365 cm^{-1}. These authors also compute the IR frequencies for cumulenic linear C_6 as 2184 and 1289 cm^{-1} and for acetylenic linear C_6 and 1173 and 1056 cm^{-1}. For C_8, Raghavachari and Binkley (*11*) conclude that, after correction for spin contamination effects in the triplet states of the linear form, the cyclic (C_{4h}) and linear structures are almost isoenergetic, with the cyclic form slightly more stable. The calculated IR frequencies for the cyclic cluster are 2007 and 1097 cm^{-1}. Frequencies for the linear C_8 structure were not reported.

The prediction of greater stability for the even numbered cyclic structures has, to the present time, not been substantiated by experiment. Using electron spin resonance spectroscopy of carbon species isolated in rare gas matrices, Weltner and coworkers (*16,17*) find that C_4, C_6, and C_8 are linear species. These authors do point out, however, that if the cyclic species were present and trapped out together with the linear species, they would not be observable via ESR since the ring structures are predicted to have singlet ground states. The application of an optical spectroscopic technique, such as infrared, would therefore be useful in the resolution of this problem.

An infrared study of matrix isolated carbon clusters has been performed by Thompson et al. (*18*). In this work, carbon was vaporized from a tantalum furnace (yielding C, C_2, and C_3), and the larger clusters were produced by annealing the matrix. Infrared bands were correlated by their intensity behavior upon annealing and structures assigned from partial isotopic studies using 95%-enriched ^{13}C in conjunction with normal coordinate calculations. Using a similar approach, but with complete isotopic information (1:1 mixture of $^{12}C:^{13}C$), a recent study (*21*) has determined the infrared active frequencies and geometry (linear) of the C_5 cluster.

In the present investigation, a similar approach has been adopted. First, we identify, via intensity behavior with matrix annealing, those bands that correlate with each other. Second, we use the mixed isotope effect (1:1 mixtures of $^{12}C:^{13}C$) both to assign a particular IR band to a specific carbon cluster and determine the cluster geometry. This is done by comparing the number of isotopomeric structures expected for a certain geometry with the number of isotopomeric peaks observed. We show here that C_6 and, less firmly, C_8 are chain-like clusters in the matrix whose most intense infrared active frequencies lie at 1952 and 1998 cm^{-1}, respectively. These IR attributions differ from previous assignments (*18*). We further show that the C_8 cluster peak grows dramatically upon matrix annealing, whereas the C_3 and C_5 cluster peaks diminish. This strongly indicates that C_3 and C_5 react in the matrix to form C_8.

RESULTS

The infrared spectrum of the ^{12}C products obtained via laser ablation of graphite is similar to that obtained by thermal vaporization followed by strong annealing. The prominent bands observed here (2164, 2039, 1998, and 1952 cm^{-1}) has been seen previously by Thompson, DeKock, and Weltner (*18*). Upon annealing, it is observed that the 2164 and 2039 cm^{-1} bands decrease in intensity, whereas the 1998 cm^{-1} band increases sharply. Figure 1 shows the difference spectrum (annealed minus unannealed) over the 2200–1800 cm^{-1} region after strong annealing. The negative peaks decrease in intensity upon annealing, whereas the positive ones increase. The 1952 cm^{-1} band grows slightly with annealing and then decreases.

Isotopic substitution is a powerful method for elucidating the structures of small carbon clusters. Using a 1:1 mixture of ^{12}C:^{13}C powder, *all* the isotopomeric structures possible for a particular cluster geometry can be expected if, during the laser vaporization process, the carbon is atomized and the atoms react to form clusters in the vapor before deposition. In Table 1, the important cyclic and linear cluster geometries for species containing between three and eight carbons and the expected number of isotopomers for each are listed. For linear C_3 and C_5, there are six and 20 isotopomers possible, respectively. Each isotopmer will give rise to an IR band whose frequency is shifted from its parent (containing only ^{12}C). As has been shown previously (*21–23*) and can be seen in Figs. 2(C_3) and 3 (C_5), this expectation is met. These spectra also illustrate the differences in band densities seen for different size clusters. The isotopomer band density (i.e., no. of bands/cm^{-1}) is significantly smaller for C_3 than C_5. As Table 1 shows, the maximum possible number of isotopomers for a given cluster size occurs for the linear species.

For linear carbon clusters it can be shown (*18*) from the Teller-Redlich rule (*24*) that the frequencies for the two completely substituted isotopomers will obey the relation

$$^{13}\nu_i/^{12}\nu_i = 0.9606 \qquad (1)$$

This relation dictates that for the ^{12}C cluster frequencies in the 2000 cm^{-1} (C-C stretch) range, its ^{13}C partner will absorb ~80 cm^{-1} to lower energy. The mixed isotopomers will, of course, fall within these limits. For the smaller linear clusters, the density of bands will be rather sparse but will increase nonlinearly with cluster size. Table 1 gives the expected average densities for linear clusters from C_3 to C_8. For linear C_6, a band is predicted about every 2 cm^{-1}, whereas for linear C_8, a band is expected about every 0.5 cm^{-1}.

Figure 2 gives the unannealed spectra for ^{12}C products and ^{12}C:^{13}C mixed products in the 2070–1930 cm^{-1} range. The three major ^{12}C prod-

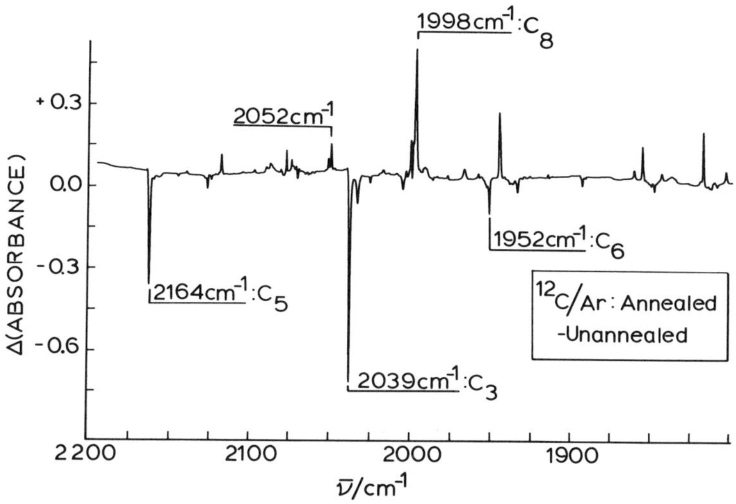

Fig. 1. Difference spectrum (annealed 35 K)-unannealed) of ^{12}C laser-ablation products in an argon matrix at 12 K. Resolution: 0.5 cm^{-1}.

uct peaks appear at 2039, 1998, and 1952 cm^{-1}. In the mixed isotope spectrum, the six isotopomer peaks owing to C_3 built on the 2039 cm^{-1} parent band are readily seen. Strikingly, in the mixed isotope spectrum, the 1998 and 1952 cm^{-1} ^{12}C parent peaks and their isotopomer partners appear to have disappeared. In fact, they have just diminished greatly in intensity. The 1952 cm^{-1} band is just barely visible in the mixed isotope run, but becomes easily discernable upon absorbance scale expansion (Fig. 4). Its isotopomer partners stretch down to 1877 cm^{-1}, 75 cm^{-1} away. However, the isotopomeric peaks based on the 1998 cm^{-1} band are lost in the baseline in an unannealed matrix. With strong annealing, they emerge from the baseline and are readily recognized (Fig. 5). Thompson et al. (*18*) have shown that the ^{12}C partner to the species responsible for the 1998 cm^{-1} band absorbs at 1920 cm^{-1}, a spread of 78 cm^{-1}. By tracking the behavior of the bands in the overlap region (1952–1920 cm^{-1}) with progressive annealing, it was possible to decide which bands belong to the 1998 cm^{-1} set and which to the 1952 cm^{-1} set. The 1952 cm^{-1} set is listed in Table 2; detailed analysis of the 1998 cm^{-1} set will be presented at a future date. The band density for the 1998 cm^{-1} set is markedly higher (~1.4 bands/cm^{-1}) than the 1952 cm^{-1} set (~0.47 bands/cm^{-1}). From the band density criterion alone, it is clear that the 1998 cm^{-1} band cannot be ascribed to a linear cluster smaller than C_8. Furthermore, the 1952 cm^{-1} band could be assigned to linear C_6. We propose here that the 1998 cm^{-1} band is owing to linear C_8 and the 1952 cm^{-1} to linear C_6.

This attribution nicely explains the apparent disappearance of the 1998 cm^{-1} band and the great diminution of the 1952 cm^{-1} band upon isotopic substitution. For linear C_8, the intensity originally concentrated in the $^{12}C_8$ band must be spread among 136 isotopomer peaks, whereas

Table 1
Expected Number of Isotopomers for Cyclic and Linear Carbon Clusters

Cluster	Cyclic Symm.	Cyclic	Linear	Band Density,[a] bands/cm^{-1}
C_3	D_{3h}	4	6	0.075
C_4	D_{2h}	9	10	0.125
	D_{4h}	6		
C_5	D_{5h}	8	20	0.25
C_6	D_{3h}	20	36	0.45
	D_{6h}	13		
C_7	D_{7h}	22	72	0.90
C_8	C_{4h}	45	136	1.725

[a]For linear clusters only.

for linear C_6, it must be spread among only 36 isotopomer peaks. So, even though the ^{12}C run shows similar intensities for the 2039, 1998, and 1952 cm^{-1} peaks, complete isotopic substitution would reduce the individual isotopomeric peak intensities to ~1/6 (C_3), 1/136 (C_8), and 1/36 (C_6) the intensity of the respective ^{12}C parent.

It is possible to use the isotopic intensity diminution in a semiquantitative way to determine the number of carbons (n) in a linear cluster. We use the following relation with C_3 as our internal standard

$$I(^{12}C_3)/I(^{12}C_n) = (N_3/N_n) \bullet (I^i(C_3)/I^i(C_n)) \tag{2}$$

where $I(^{12}C_3)$ and $I(^{12}C_n)$ are the intensities of the ^{12}C parents for the major IR bands owing to C_3 (i.e., 2039 cm^{-1}) and the unknown C_n; N_3 and N_n are the number of possible isotopomers for C_3 and C_n, respectively; and $I^i(C_3)$ and $I^i(C_n)$ are the intensities of the ^{12}C parent peaks in the mixed isotope spectrum. Expressions giving the possible number of isotopomers for a even numbered linear cluster containing n atoms are (*21*)

$$N_n = 2(2N_{n-2} - 2^{\frac{n}{2}-2}) \tag{3}$$

and $N_{n'} = 2N_n$ where n′ is odd and $= n + 1$. Obviously, $N_1 = 2$ and $N_2 = 3$. Thus, $N_3 = 6$ possible isotopomers. By solving Eq. (2) for N_n and substituting it into Eq. (3), it is possible to determine n. To test this approach, the data for C_5 (Fig. 3) were used: $I(^{12}C_5)/I(^{12}C_3) = 0.49$, $I^i(C_3)/I^i(C_5) \cong 6.2$ from which $N_n \cong 18.2 \approx 20$. From Eq. (3), this gives $n = 5$. For the 1952 cm^{-1} band intensity data, we find $I(^{12}C_n)/I(^{12}C_3) = 0.51$, $I^i(C_3)/I^1/(C_n) \cong 11.3$, so that $N_n \cong 34.6 \approx 36$. Thus $n = 6$. This is in agreement with the conclusion drawn above from the band density data.

Finally, we present the results of the normal coordinate analysis for chain-like C_6. Thirty-five isotopomeric peaks lying between 1952 and

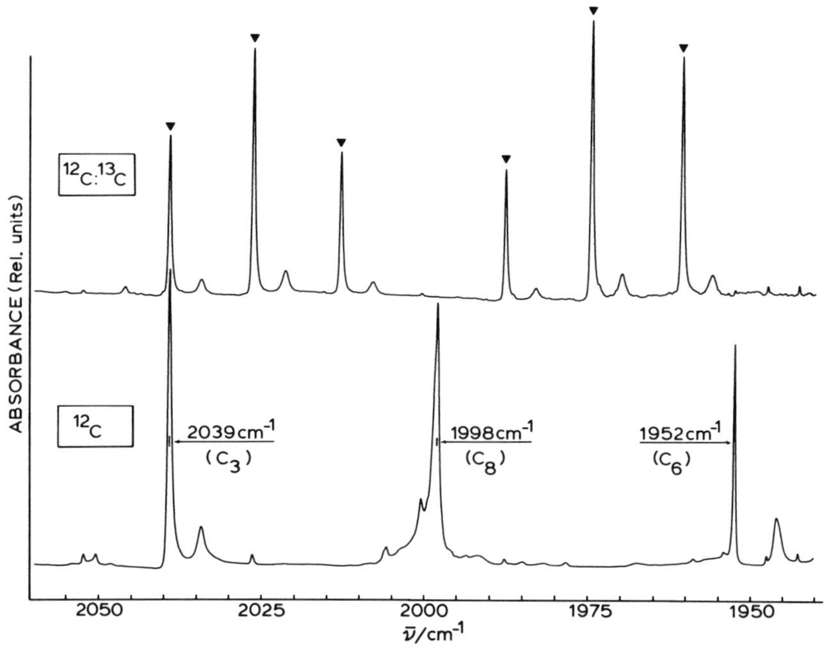

Fig. 2. Infrared spectra over the 2070–1930 cm^{-1} range of ^{12}C products (lower spectrum) and $^{12}C:^{13}C$ isotopically-mixed products (upper spectrum) isolated in unannealed argon matrices at 12 K showing the six C_3 isotopomeric peaks. Resolution: 0.25 cm^{-1}.

1877 cm^{-1} are observed (36 are expected for chain-like C_6) and are listed in Table 2. The normal coordinate analysis was done by fitting five force constants to 16 of the bands (marked in Table 2), 10 from the ^{12}C end, and six from the ^{13}C end of the isotopomer list. The force constants varied to effect the fit were: f_R, outer bond stretching force constant; f_r, next inner bond stretch; f_m, middle bond stretch; f_{rm}, interaction between r and m bond stretches; and f_{Rr}, interaction between R and r bond stretches. The fit described here is for a slightly bent C_6 (hence, use of the term "chain-like" instead of linear). The fit to a linear C_6 geometry was considerably worse. To describe the "best fit" geometry, consider the C atoms numbered in sequence: C_1–C_2–C_3–C_4–C_5–C_6. Keeping the C_1–C_2, C_3–C_4, and C_5–C_6 bonds parallel, the C_2–C_3 and C_4–C_5 bonds are bent 15° away from linear in opposite directions (i.e., trans-like) and further twisted 7° out-of-plane in an opposite sense. We are of the opinion that this distortion from linear geometry is the result of packing stresses induced by the argon matrix. It is interesting to note in this regard that C_5 which is known to be linear, fits easily into a vacancy created by the removal of two adjacent argon atoms, but linear C_6 will not fit into this cavity. No normal coordinate analysis has been attempted for C_8 since the large number of isotopomeric bands and their density probably precludes any meaningful fit.

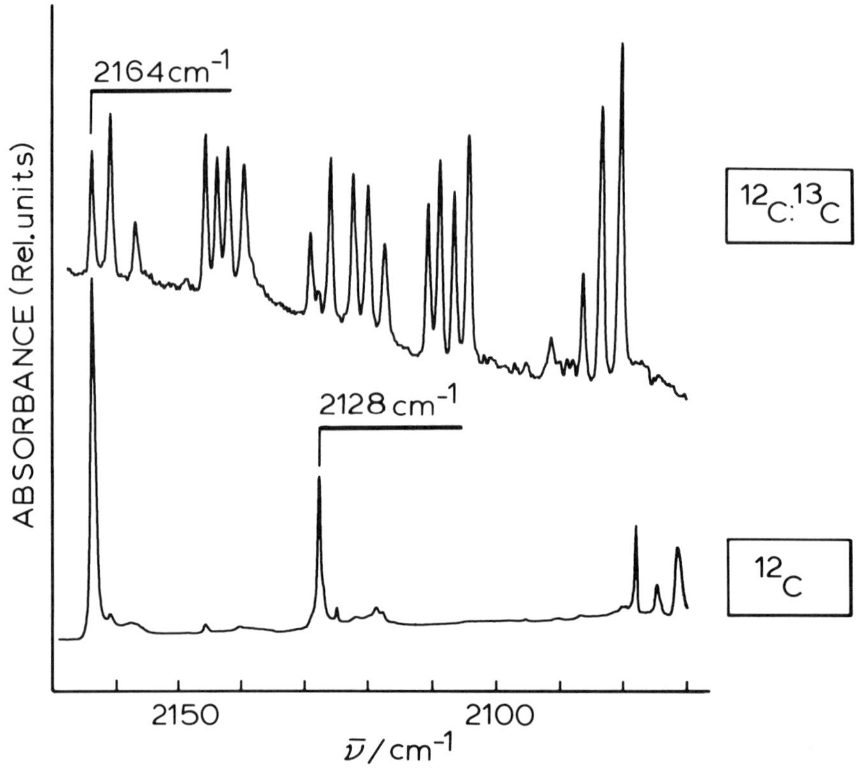

Fig. 3. Infrared spectra over the 2170–2070 cm^{-1} range of ^{12}C products (lower spectrum and $^{12}C:^{13}C$ isotopically-mixed products (upper spectrum) isolated in unannealed argon matrices at 12 K showing the 20 C_5 isotopomeric peaks. Resolution: 0.25 cm^{-1}.

DISCUSSION

In this paper, evidence has been presented that supports the reassignment of the prominent 1952 and 1998 cm^{-1} infrared bands to linear C_6 and C_8 species, respectively. These assignments differ from those proposed by Thompson et al. (*18*). Using intensity changes upon thermal annealing and partial isotopic substitution (5% ^{12}C and 95% ^{13}C), these authors had attributed the 1998 cm^{-1} band to C_6 and the 1952 cm^{-1} band to C_5. Because the 1998 cm^{-1} band increases with annealing, whereas the 2039 cm^{-1} (C_3) band decreases, these authors suggested that two C_3 clusters react in the matrix to form C_6. Given our assignments above, it seems likely that C_3 reacts with C_5 (both of whose bands decrease upon annealing) to form C_8. Whether two C_3's react to form C_6 is more problematic. We have found that the 1952 cm^{-1} (C_6) band increases slightly upon annealing and then (with stronger annealing) decreases. Perhaps two C_3's react to form C_6, which, if the matrix becomes soft enough at higher annealing temperatures, can react further to form higher clusters (C_7, C_8, C_9, and so on). This possibility needs to be pursued further.

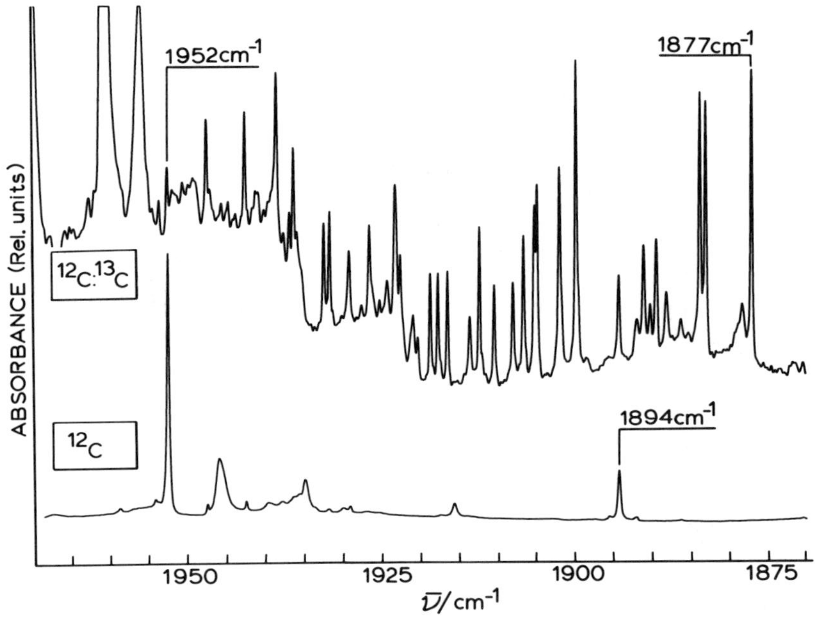

Fig. 4. Infrared spectra over the 1980–1870 cm^{-1} range of ^{12}C products (lower spectrum) and ^{12}C:^{13}C isotopically-mixed products (upper spectrum) isolated in unannealed Ar matrices at 12 K showing the C_6 isotopomeric peaks. Resolution: 0.25 cm^{-1}.

The assignment proposed here for C_6 is in good accord with recent theoretical predictions. Raghavachari et al. (*12*) calculated IR-active frequencies of 2184 and 1289 cm^{-1} for cumulene-like linear C_6 and 1173 and 1056 cm^{-1} for acetylene-like linear C_6. The experimental frequencies of 1952 and 1197 cm^{-1} are within 12 and 8%, respectively, of the predicted cumulenic-C_6 values. They are sufficiently distant from the acetylenic-C_6 values to conclude that the cumulenic form is the structure favored by linear C_6. This conclusion is further supported by the theoretical stability calculations. They show that the cumulenic form is 33.3 kcal/mol more stable than the acetylenic structure. It is worth recalling that the theory finds cyclic C_6 structures considerably more stable than the linear forms: the D_{3h} distorted ring lies lowest with the symmetric D_{6h} ring 3.5 kcal/mol higher. The cumulenic linear structure lies at 19.3 kcal/mol. The natural question then is: why do we observe linear C_6 rather than cyclic C_6?

The answer has been provided by Van Zee et al. (*17*), who estimate that for equilibrium concentrations in a laser vaporization plume at 3000 or 4000 K, with a $\Delta H^o_o = +10$ kcal/mol (for the conversion of the D_{3h} ring to the cumulenic linear form), the K_p ratio ($= P_{linear}/P_{ring}$) will be 50 at 3000 K and 200 at 4000 K. Thus, the concentration of the linear form in the vapor before deposition is expected to be far greater than of the cyclic form, and our observation of the linear form is understandable. It will be

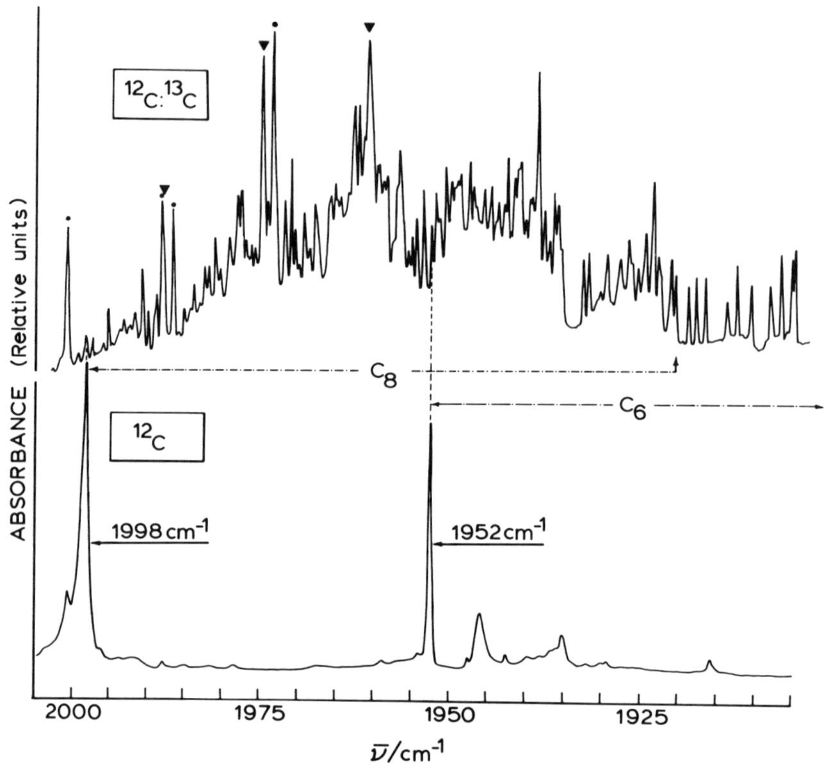

Fig. 5. Infrared spectra over the 2000–1920 cm^{-1} range of ^{12}C products (lower) in an unannealed Ar matrix (12 K). Upper spectrum shows the IR of the $^{12}C:^{13}C$ mixed products in Ar at 12 K after annealing to 35 K for 15 min. Resolution: 0.25 cm^{-1}. Note the increased density of bands compared to Fig. 2–4.

interesting to determine whether the more stable cyclic structure may be formed from the linear upon vigorous matrix annealing.

No comparison of the experimental C_8 frequency with recent theoretical values is possible since none has been reported for linear C_8. However, comparison to a simple model is instructive. Halford (25) has given an expression for the stretching frequencies of a model linear chain that has equal force constants (k) for all bonds. For N number of atoms of mass m in the chain, the n stretching frequencies are given by

$$\nu_n = (k/\pi^2 m)^{1/2} \sin(n\pi/2N) \tag{4}$$

where $n = 1, 2 \ldots N-1$ and only the even n frequencies will be IR-active. Choosing the C_3 frequency (2039 cm^{-1}) to calibrate this expression yields $(k/\pi^2 m)^{1/2} = 2354.5$ cm^{-1}. Then, the expression gives for C_5: 2239 cm^{-1} (2164 cm^{-1}) and 1384 cm^{-1} (1544 cm^{-1}); experimental frequencies are given in parentheses. For C_6: 2039 cm^{-1} (1952 cm^{-1}) and 1177 cm^{-1}

Table 2
Observed and Calculated Isotopomer Stretching Frequencies for Chain-Like C_6 in Argon Matrix[a]

Observed $\bar{\nu}/cm^{-1}$	Calc'd $\bar{\nu}/cm^{-1}$	Diff. $\Delta\bar{\nu}/cm^{-1}$	Observed $\bar{\nu}/cm^{-1}$	Calc'd $\bar{\nu}/cm^{-1}$	Diff. $\Delta\bar{\nu}/cm^{-1}$
1952.5[b]	1950.4	2.1	1910.3	1911.7	−1.4
1947.5[b]	1945.1	2.4	1908.	1907.3	0.7
1942.5[b]	1941.5	1.0	1906.5	1906.5	0
1938.5[b]	1942.4	−3.9	1905.2	1905.6	−0.4
1937[b]	1938.5	−1.5	1904.7	1905.3	−0.6
1936.4[b]	1934.9	1.5	1901.9	1901.8 , 1901.6	0.1, 0.3
1932.4[b]	1932.8	−0.4	1899.7	1900.5	−0.8
1931.5[b]	1934.3	−2.8	1894.3	1895.9	−1.6
1929.2[b]	1928.5	0.7	1892	1895.3	−3.3
1926.5[b]	1924.6	1.9	1891	1895.1	−4.1
1923.3	1923.3	0	1890.2	1892.4	−2.2
1922.5	1922.0	0.5	1899.5	1891.5	−2.0
1918.2	1915.7	2.5	1888.2[b]	1888.3	−0.1
1917.5	1915.2	2.3	1886.3[b]	1885.1	1.2
1916.5	1914.4	2.1	1883.7[b]	1882.7	1.0
1913.5	1913.2	0.3	1883[b]	1883.8	−0.8
1912.3	1912.1	0.2	1878.3[b]	1877.9	0.4
			1877.2[b]	1873.9	3.3

[a]Force constants found: f_R = 10.9367 md/Å, f_r = 9.4468 md/Å, f_m = 9.7160 md/Å, f_{Rr} = 1.1401 md/Å, and f_{rm} = 0.9306 md/Å (R = outer bond, r = inner bond, m = middle bond).

[b]Bands were used in the initial fitting to determine force constant parameters.

(1197 cm^{-1}); for C_8: 2175 (1998 cm^{-1}), 1665, and 902 cm^{-1}. Thus, even this simple expression mimics the placement of the highest IR band of C_8 between the C_5 and C_6 bands, as we find. Although further work is needed for a firm conclusion, it appears that C_8 is also present in the cumulenic form.

In summary, evidence from infrared spectroscopic studies has led to the assignment of the 1952 and 1998 cm^{-1} bands to C_6 and C_8 clusters, respectively. The structures of these species are found to be probably linear or possibly slightly bent with cumulenic-type bonding.

ACKNOWLEDGMENTS

We wish to acknowledge William Weltner for his generous loan of the Nd/YAG pulsed laser without which this work would not have been possible. We also gratefully acknowledge the Petroleum Research Fund, American Chemical Society (17896-ACH) for partial support, and the National Science Foundation (grant nos. CHE-8601705 and CHE-8514585) for its support of this research.

REFERENCES

1. Ballick, E. A. and Ramsey, D. A., *J. Chem. Phys.* **29,** 1418 (1958); and *Astrophys. J.* **137,** 84 (1963).
2. Edwards, J. B., *Combustion: Formation and Emission of Trace Species,* Ann Arbor Science, Ann Arbor, 1974.
3. Drowart, J., Burns, R. P., De Maria, G., and Inghram, M. G., *J. Chem. Phys.* **31,** 1131 (1959).
4. Skell, P. S., Havel, J. J., and McGlinchey, M. J., *Acc. Chem. Res.* **6,** 97 (1973).
5. Gausset, L., Herzberg, G., Lagerqvist, A., and Rosen, B., *Astrophys. J.* **142,** 45 (1965).
6. Altman, R. L., *J. Chem. Phys.* **32,** 615 (1960).
7. Weltner, W., Jr., Walsh, P. N., and Angell, C. L., *J. Chem. Phys.* **40,** 1299 (1964); Weltner, W., Jr., and McLeod, D., Jr., *J. Chem. Phys.* **40,** 1305 (1964); and Weltner, W., Jr. and McLeod, D., *J. Chem. Phys.* **45,** 3096 (1966).
8. Pitzer, K. S. and Clementi, E., *J. Am. Chem. Soc.* **81,** 4477 (1959); Clementi, E., *J. Am. Chem. Soc.* **83,** 450 (1961); Clementi, E. and Clementi, H., *J. Chem. Phys.* **36,** 2824 (1962); Clementi, E. and McLean, A. C., *J. Chem. Phys.* **36,** 45 (1962); and Hoffman, R., *Tetrahedron* **22,** 521 (1966).
9. Slanina, Z. and Zahradnik, R., *J. Phys. Chem.* **81,** 2252 (1977); Whiteside, R. A., Krishnan, R., DeFries, D. J., Pople, J. A., and Schleyer, P. v. R., *Chem. Phys. Letts.* **78,** 538 (1981); and Koutecky, J. and Pacchioni, G., *Ber. Bunsenges, Physik, Chem.* **88,** 233 (1984).
10. Magers, D. H., Harrison, R. J., and Bartlett, R. J., *J. Chem. Phys.* **84,** 3284 (1986).
11. Raghavachari, K. and Binkley, J. S., *J. Chem. Phys.* **87,** 2191 (1987).
12. Raghavachari, K., Whiteside, R. A., and Pople, J. A., *J. Chem. Phys.* **85,** 6623 (1986).
13. Ritchie, J. P., Ing, H. F., and Young, W. S., *J. Chem. Phys.* **85,** 5175 (1986).
14. Bernholdt, D. E., Magers, D. H., and Bartlett, D. J., *J. Chem. Phys.,* **89,** 3612 (1988).
15. Bernholc, J. and Phillips, J. C., *Phys. Rev.* **B33,** 7395 (1986); and *J. Chem. Phys.* **85,** 3266 (1986).
16. Graham, W. R. M., Dismuke, K. I., and Weltner, W., Jr., *Astrophys. J.* **204,** 301 (1976).
17. Van Zee, R. J., Ferrante, R. F., Zeringue, K. J., and Weltner, W., Jr., *J. Chem. Phys.* **88,** 3465 (1988).
18. Thompson, K. R., DeKock, R. L., and Weltner, W., Jr., *J. Am. Chem. Soc.* **93,** 4688 (1971).
19. Krätschmer, W. and Natchigal, K., *Polycyclic Aromatic Hydrocarbons and Astrophysics,* editors: Léger, A., d'Hendecourt, L., and Boccara, N., Reidel, Dordrecht, Holland, 1987.
20. Krätschmer, W., Sorg, N., and Huffman, D. R., *Surf. Sci.* **156,** 814 (1985).
21. Vala, M., Chandrasekhar, T. M., Szczepanski, J., Van Zee. R., and Weltner, W., Jr., *J. Chem. Phys.* **90,** 595, 1989.
22. Douglas, A. E., *Astrophys. J.* **114,** 466 (1951).
23. Weltner, W., Jr., and McLeod, D., Jr., *J. Chem. Phys.* **40,** 1305 (1964).
24. Herzberg, G., *Infrared and Raman Spectra of Polyatomic Molecules,* Van Nostrand, Princeton, NJ, 1945.
25. Halford, J. O., *J. Chem. Phys.* **19,** 1375 (1951).

Mechanism of Cluster Formation During Evaporation of Alloys

ANDREAS HARTMANN AND KONRAD G. WEIL*

Institut für Physikalische Chemie, Technische Hochschule Darmstadt Petersenstrasse 20, 6100 Darmstadt FRG

ABSTRACT

We used a mass spectrometer, which allowed us to evaporate alloys either from a Knudsen cell or directly from a heated source (Langmuir evaporation) in order to study the mechanism of formation of heteronuclear compounds. The systems cesium with antimony and bismuth, in some cases with additions of indium or tin, seemed promising owing to the wide variety of intermetallic species. Three types can be discerned: highly polar compounds like Cs_2Sb_2, Cs_2Sb_4; clusterlike particles like Cs_6Sb_4, Cs_6Sb_3Bi, or $Cs_6Sb_2Bi_2$; Zintl-type molecules like CsSnSb, $CsSnSb_3$, $Cs_2Sn_2Sb_2$, and Cs_2InSb_3. Nevertheless, the main constituents of the gas phase are homonuclear particles like Cs atoms and Sb_2 and Sb_4 molecules.

All the intermetallic particles, mentioned above, are found by Knudsen evaporation of a molten alloy and by Langmuir evaporation from the solid. We conclude that they are formed in an adsorbed layer during the evaporation process. The relative abundances of these particles are nearly the same under both evaporation conditions. This indicates the possibility that in neither case is an equilibrium established.

Index Entries: Evaporation of alloys; cluster formation; intermetallic particles; evaporation equilibrium; high temperature mass spectrometry.

INTRODUCTION

In the majority of cases, the vapors over alloys consist of a mixture of the constituents, which are also present in the vapor over the pure components. Often heteroatomic dimers can be found as minority spe-

*Author to whom all correspondence and reprint orders should be addressed.

cies (*1,2*). Triatomic molecules are also observable, but at even lower abundances (*3–5*). Large and rather complex species are found in the vapors over alloys containing antimony with alkali metals (*6–8*). Quite recently, we detected ternary compounds in the vapor over cesium/antimony/indium and cesium/antimony/tin alloys (*9*). Quaternary compounds were detected in the system cesium/antimony/bismuth/tin. In these particles, we assume an electron transfer from the cesium atoms to the central complexes $(InSb_3)^{2-}$, $(SnSb_3)^-$, $(Sn_2Sb_2)^{2-}$, $(SnSb_2Bi)^-$, and $(SnSbBi_2)^-$, which are all isoelectronic with Sb_4, which is known to be a stable tetrahedron in the gas phase. A similar electron transfer can stabilize the $(SnSb)^-$ unit, isoelectronic to Sb_2, in the molecule CsSnSb.

The crystal structures of Cs_3Sb and of CsSb are known: The first is a superstructure of the NaTl-type (*10*), the latter consists of infinite antimony chains (*11*). One need not go into the details of these structures in order to observe, that the species seen in the gas phase are not building units of the crystals. Therefore, two questions can be raised: Are the intermetallic species formed in the gas phase or are they formed in an interlayer between condensed phase and gas? Is there a solid/gas equilibrium established under the conditions of Knudsen evaporation? The situation is similar to that of the evaporation of As_4O_6, which occurs as arsenolite, containing As_4O_6 building units and evaporates with an evaporation coefficient of nearly unity, and of claudite, which does not contain these units and has an evaporation coefficient much lower than unity (*12*).

METHODS

We used a magnetic sector field mass spectrometer MM 30 from VG. In order to minimize mass discrimination, a Daly converter connected to an ion counting system was employed. Microprocessor control of the Knudsen cell temperature and of the magnetic field allowed automatic recording of ion intensities during variation of the temperature.

The Knudsen cells were machined from stainless steel. They had orifice diameters of 0.5, 0.8, and 1.0 mm. No substantial dependence of the mass spectra from the orifice diameter could be detected, the results presented in the following section all are obtained with Knudsen cells with 0.8 mm orifice diameter. The inner diameter of the cell, which determines the surface of the evaporating phase was 12 mm. The cells were mounted within an electrical heating unit; the temperature was measured with a thermocouple, which was fastened in a bore-hole in the bottom of the cell. In previous experiments with the same system, we calibrated the temperature measurement with the melting points of silver and of antimony. The maximum deviation between the temperature reading of the thermocouple and the standard melting temperatures of these materials never exceeded 2 K.

In the case of Knudsen cell experiments, the molecular beam from the cell entered the ionization chamber through a grid. Ionization occurred by electron impact. The electron energy was calibrated using the known ionization energies of Cs and of Sb_4. In order to be able to perform experiments with Langmuir evaporation, we fixed an evaporation crucible directly within the ionization chamber, *see* Fig. 1. It was made from ceramics and was mounted within a heating coil. A thermocouple was placed in the crucible. Owing to the special geometry of our ionization chamber, it had to be mounted horizontally. Prior to each experiment, the crucible was loaded in a glove box with the desired alloy. We used a small, single piece of the alloy that was positioned close to the open end of the crucible in order to maintain well defined evaporation conditions. This arrangement allowed us to perform Langmuir evaporation experiments up to temperatures just below the melting temperature of the alloy. As a consequence of the direct evaporation into the ionization space, the sensitivity of the system was high enough to observe even the largest intermetallic species at temperatures below the melting point.

In those cases, when absolute partial pressures within the Knudsen cell should be determined, the calibration constant of the mass spectrometer was determined from the equilibrium constant for the reaction

$$2\ Sb_2 = Sb_4\ (13)$$

In these cases, the ion intensities Sb_2^+ and Sb_4^+ were measured at an electron energy of 10 eV. At this energy, no fragmentation of Sb_4 is to be expected. The following values were used for the ionization cross sections at 10 eV: $\sigma(Sb_2) = 1.883 \cdot 10^{-20}\ m^2$; $\sigma(Sb_4) = 3.765 \cdot 10^{-20}\ m^2$. These values were estimated from the ionization cross section of antimony atoms σ(Sb) (*14*) with the equation $\sigma(Sb_i) = 0.75 \cdot i \cdot \sigma(Sb)$.

The alloys were prepared from the elements in a glove box. Weighed amounts of the elements were mixed in a quartz ampoule under an argon atmosphere. The ampoule was sealed and subsequently annealed at a temperature higher than that of the highest melting compound in the respective system. In the case of Cs_2InSb_3, crystals could be grown during extended annealing at 1073 K. The crystals are extremely soft and ductile; a structure determination has not yet been performed.

DISCUSSION

First we compared the mass spectra, obtained by either Knudsen effusion or by Langmuir evaporation of solids with the following compositions: Cs_2Sb_3, CsSb, and Cs_3Sb. For both modes of evaporation, the following intermetallic particles were detected: $CsSb_2$, Cs_2Sb, Cs_2Sb_2, Cs_2Sb_4, and Cs_6Sb_4. The first four of them were identified as parent molecules by Knudsen effusion mass spectrometry in a previous study (*6*); on the existence of Cs_6Sb_4 we reported recently (*8*). The only significant difference in the spectra obtained by the different techniques is the appearance of the molecule Cs_3Sb during Langmuir evaporation.

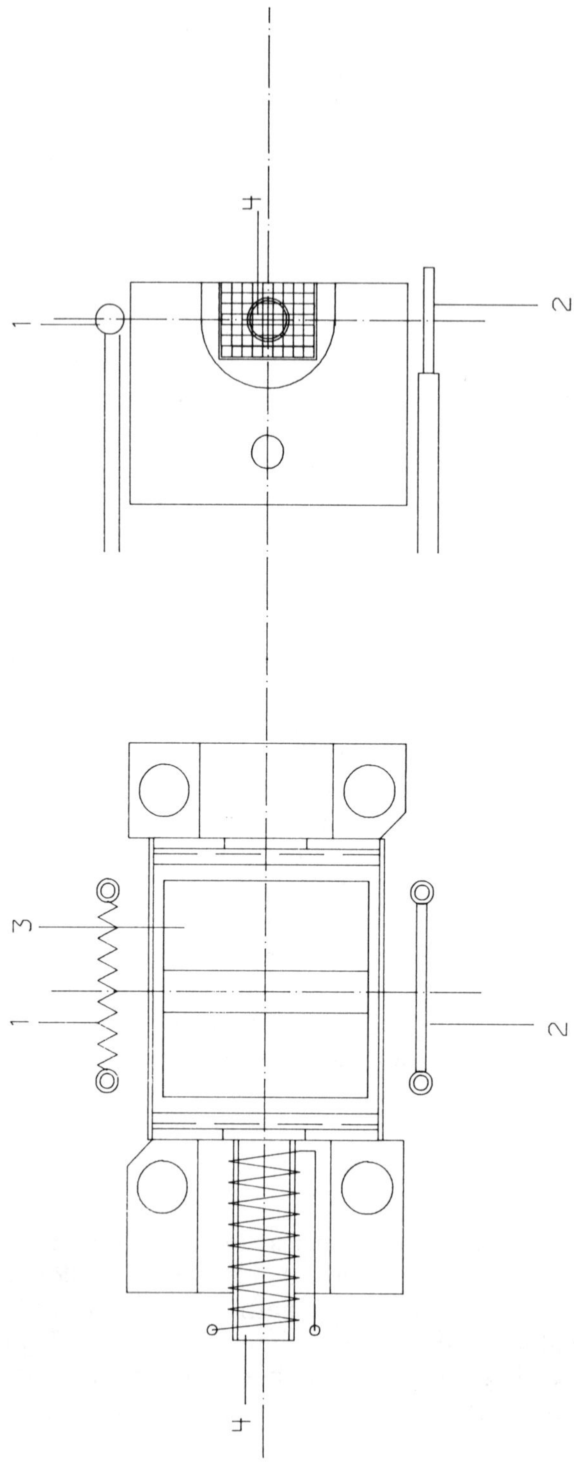

Fig. 1. Frontview and sideview of the ion source, used for Langmuir evaporation. 1: Emitter filament. 2: Electron collector. 3: Repeller plate. 4: Ceramic Langmuir source. The ion extraction lenses are not incorporated in the figure.

The same correspondence between the spectra is observed, when the molecular beams, emanating from Cs_2InSb_3 are analyzed: Not only the same small molecules are detected with both evaporation techniques, but also Cs_2Sb_4, Cs_6Sb_4, and Cs_2InSb_3 are observed in the different molecular beams. Analogously, Knudsen effusion and Langmuir evaporation of a solid with the composition $Cs_2Sn_2Sb_2$ both yield, in addition to the small molecules the same particles, namely Cs_2Sb_4, Cs_6Sb_4, $CsSnSb_3$, Cs_2Sn_2 Sb_2, and CsSnSb. One should remember, that all the intermetallic species mentioned in this paragraph are minority constituents of beams, which predominantly contain the homonuclear species Cs, Sb_2, and Sb_4.

This qualitative comparison demonstrated that in all investigated cases the mass spectra, concerning the intermetallic minority species are virtually independent of the evaporation mode. We therefore tried to compare the relative abundances of these species quantitatively. This was done for a solid of the composition CsSb. In Table 1a, we present the measured intensities, corrected for the isotopic abundances, of 8 mass signals at different temperatures. These data are collected during an experiment with Knudsen effusion. In Table 1b, similar data are collected, obtained by Langmuir evaporation. Note, that the temperature ranges of both experiments are different.

In Tables 2a and 2b, we present relative abundances, derived from the data in Tables 1a and 1b. To obtain these figures, we first added the intensities of parent and fragment signals, then divided the intensities by the respective ionization cross sections and finally normalized the intensities relative to the sum of the intensities of all antimony species.

A comparison of Tables 2a and 2b clearly indicates that the relative abundances of the intermetallic species are only slightly temperature dependent and that they do not depend on the mode of evaporation.

In order to present a more detailed description of the gas phase within the Knudsen cell, we calibrated the partial pressures of the parent molecules over a solid with the composition CsSb at 900.1 K. Calibration of the mass spectrometer occurred as described in the experimental section. The results are given in Table 3.

CONCLUSIONS

The results presented in the previous sections can be summarized as follows: In the case of cesium/antimony mixtures with and without additional indium, tin, or bismuth, the molecular beams emanating from a Knudsen cell, contain the same molecules as are formed by Langmuir evaporation of the solids, even when the Knudsen cell contains the material in the molten state. This observation clearly demonstrates, that the intermetallic particles, present in these systems, are formed during the evaporation step. More puzzling is the finding, that even the relative abundances of the intermetallic molecules are virtually independent of the mode of evaporation. A possible interpretation can be found, when

Table 1a
Signal Intensities in Counts/Second for Different Temperatures[a]

T/K	Sb	Cs	Sb_2	Sb_3	Sb_4	Cs_2Sb_2	Cs_2Sb_4	Cs_6Sb_4
898.1	$576 \cdot 10^3$	$634 \cdot 10^4$	$139 \cdot 10^4$	$153 \cdot 10^4$	$692 \cdot 10^4$	$213 \cdot 10^2$	$515 \cdot 10^2$	877
896.6	$580 \cdot 10^3$	$638 \cdot 10^4$	$139 \cdot 10^4$	$153 \cdot 10^4$	$681 \cdot 10^4$	$213 \cdot 10^2$	$507 \cdot 10^2$	829
907.6	$725 \cdot 10^3$	$712 \cdot 10^4$	$177 \cdot 10^4$	$197 \cdot 10^4$	$874 \cdot 10^4$	$275 \cdot 10^2$	$646 \cdot 10^2$	934
908.1	$761 \cdot 10^3$	$732 \cdot 10^4$	$181 \cdot 10^4$	$199 \cdot 10^4$	$863 \cdot 10^4$	$278 \cdot 10^2$	$685 \cdot 10^2$	$101 \cdot 10^1$
918.1	$924 \cdot 10^3$	$840 \cdot 10^4$	$214 \cdot 10^4$	$238 \cdot 10^4$	$104 \cdot 10^5$	$352 \cdot 10^2$	$820 \cdot 10^2$	$109 \cdot 10^1$
919.1	$931 \cdot 10^3$	$845 \cdot 10^4$	$219 \cdot 10^4$	$238 \cdot 10^4$	$104 \cdot 10^5$	$355 \cdot 10^2$	$824 \cdot 10^2$	$114 \cdot 10^1$

[a]Knudsen effusion, orifice: 0.8 mm. Liquid CsSb. Electron energy: 14 eV; emission current: 100 μA.

Table 1b
Signal Intensities in Counts/Second for Different Temperatures[a]

T/K	Sb	Cs	Sb_2	Sb_3	Sb_4	Cs_2Sb_2	Cs_2Sb_4	Cs_6Sb_4
709.6	$192 \cdot 10^2$	$189 \cdot 10^3$	$131 \cdot 10^3$	$221 \cdot 10^3$	$942 \cdot 10^3$	$181 \cdot 10^1$	$337 \cdot 10^1$	3
710.6	$281 \cdot 10^2$	$228 \cdot 10^3$	$177 \cdot 10^3$	$240 \cdot 10^3$	$858 \cdot 10^3$	$137 \cdot 10^1$	$280 \cdot 10^1$	60
711.6	$353 \cdot 10^2$	$252 \cdot 10^3$	$183 \cdot 10^3$	$242 \cdot 10^3$	$898 \cdot 10^3$	$173 \cdot 10^1$	$337 \cdot 10^1$	48
713.1	$407 \cdot 10^2$	$294 \cdot 10^3$	$213 \cdot 10^3$	$306 \cdot 10^3$	$114 \cdot 10^4$	$195 \cdot 10^1$	$521 \cdot 10^1$	55
715.8	$460 \cdot 10^2$	$362 \cdot 10^3$	$258 \cdot 10^3$	$335 \cdot 10^3$	$121 \cdot 10^4$	$241 \cdot 10^1$	$538 \cdot 10^1$	55
733.6	$103 \cdot 10^3$	$716 \cdot 10^3$	$552 \cdot 10^3$	$687 \cdot 10^3$	$259 \cdot 10^4$	$552 \cdot 10^1$	$109 \cdot 10^2$	124
734.7	$934 \cdot 10^2$	$888 \cdot 10^3$	$667 \cdot 10^3$	$107 \cdot 10^3$	$417 \cdot 10^4$	$969 \cdot 10^1$	$220 \cdot 10^2$	212
751.3	$142 \cdot 10^3$	$120 \cdot 10^4$	$118 \cdot 10^4$	$200 \cdot 10^4$	$776 \cdot 10^4$	$185 \cdot 10^2$	$406 \cdot 10^2$	593
751.6	$172 \cdot 10^3$	$136 \cdot 10^4$	$132 \cdot 10^4$	$209 \cdot 10^4$	$819 \cdot 10^4$	$197 \cdot 10^2$	$456 \cdot 10^2$	816
755.1	$197 \cdot 10^3$	$154 \cdot 10^4$	$159 \cdot 10^4$	$199 \cdot 10^4$	$104 \cdot 10^5$	$223 \cdot 10^2$	$576 \cdot 10^2$	937
756.3	$175 \cdot 10^3$	$122 \cdot 10^4$	$165 \cdot 10^4$	$276 \cdot 10^4$	$993 \cdot 10^4$	$211 \cdot 10^2$	$570 \cdot 10^2$	910
771.3	$538 \cdot 10^3$	$393 \cdot 10^4$	$416 \cdot 10^4$	$666 \cdot 10^4$	$229 \cdot 10^5$	$677 \cdot 10^2$	$143 \cdot 10^3$	1690

[a]Langmuir evaporation. Solid CsSb. Electron energy: 14 eV; emission current: 100 μA.

Table 2a
Abundances of Cesium Atoms and Cesium/Antimony Species, Relative to the Total Abundance of Antimony Species[a]

T/K	Σ Sb_i	Cs	Cs_2Sb_2	Cs_2Sb_4	Cs_6Sb_4
898.1	1	$5.7 \cdot 10^{-1}$	$9.5 \cdot 10^{-4}$	$1.8 \cdot 10^{-3}$	$1.4 \cdot 10^{-5}$
896.6	1	$5.8 \cdot 10^{-1}$	$9.6 \cdot 10^{-4}$	$1.8 \cdot 10^{-3}$	$1.3 \cdot 10^{-5}$
907.6	1	$5.1 \cdot 10^{-1}$	$9.7 \cdot 10^{-4}$	$1.8 \cdot 10^{-3}$	$1.2 \cdot 10^{-5}$
908.1	1	$5.2 \cdot 10^{-1}$	$9.8 \cdot 10^{-4}$	$1.9 \cdot 10^{-3}$	$1.3 \cdot 10^{-5}$
918.1	1	$5.0 \cdot 10^{-1}$	$1.0 \cdot 10^{-3}$	$1.9 \cdot 10^{-3}$	$1.1 \cdot 10^{-5}$
919.1	1	$5.0 \cdot 10^{-1}$	$1.0 \cdot 10^{-3}$	$1.9 \cdot 10^{-3}$	$1.2 \cdot 10^{-5}$

[a]Knudsen effusion. The figures were calculated using the intensities given in Table 1a.

Table 2b
Abundances of Cesium Atoms and Cesium/Antimony Species, Relative to the Total Abundance of Antimony Species[a]

T/K	Σ Sb_i	Cs	Cs_2Sb_2	Cs_2Sb_4	Cs_6Sb_4
709.6	1	$1.4 \cdot 10^{-1}$	$6.6 \cdot 10^{-4}$	$9.7 \cdot 10^{-4}$	$4.4 \cdot 10^{-7}$
710.6	1	$1.6 \cdot 10^{-1}$	$4.8 \cdot 10^{-4}$	$7.9 \cdot 10^{-4}$	$7.9 \cdot 10^{-6}$
711.6	1	$1.7 \cdot 10^{-1}$	$5.9 \cdot 10^{-4}$	$9.1 \cdot 10^{-4}$	$6.1 \cdot 10^{-6}$
713.6	1	$1.6 \cdot 10^{-1}$	$5.3 \cdot 10^{-4}$	$1.1 \cdot 10^{-3}$	$5.5 \cdot 10^{-6}$
715.8	1	$1.8 \cdot 10^{-1}$	$5.9 \cdot 10^{-4}$	$1.0 \cdot 10^{-3}$	$5.0 \cdot 10^{-6}$
733.6	1	$1.7 \cdot 10^{-1}$	$6.5 \cdot 10^{-4}$	$1.0 \cdot 10^{-3}$	$5.3 \cdot 10^{-6}$
734.7	1	$1.4 \cdot 10^{-1}$	$7.6 \cdot 10^{-4}$	$1.3 \cdot 10^{-3}$	$6.1 \cdot 10^{-6}$
751.3	1	$1.0 \cdot 10^{-1}$	$7.9 \cdot 10^{-4}$	$1.3 \cdot 10^{-3}$	$9.3 \cdot 10^{-6}$
751.6	1	$1.1 \cdot 10^{-1}$	$7.9 \cdot 10^{-4}$	$1.4 \cdot 10^{-3}$	$1.2 \cdot 10^{-5}$
755.1	1	$1.0 \cdot 10^{-1}$	$7.4 \cdot 10^{-4}$	$1.5 \cdot 10^{-3}$	$1.1 \cdot 10^{-5}$
756.3	1	$0.8 \cdot 10^{-1}$	$6.7 \cdot 10^{-4}$	$1.4 \cdot 10^{-3}$	$1.0 \cdot 10^{-5}$
771.3	1	$1.1 \cdot 10^{-1}$	$9.3 \cdot 10^{-4}$	$1.5 \cdot 10^{-3}$	$8.5 \cdot 10^{-6}$

[a]Langmuir evaporation. The figures were calculated using the intensities given in Table 1b.

extremely small evaporation coefficients for the intermetallic molecules are assumed: In order to maintain equilibrium within the Knudsen cell, the cross section of its orifice must be small, compared to the product of the liquid/gas interface, multiplied by the evaporation coefficient. In our cases, the cross section of the orifice is $3.14 \cdot (0.4\ \text{mm})^2 = 0.50\ \text{mm}^2$. The interface between condensed phase and gas is $3.14 \cdot (6\ \text{mm})^2 = 113\ \text{mm}^2$. When the evaporation coefficient is $\leqslant 4 \cdot 10^{-3}$, this condition is not longer fulfilled. Then even in the Knudsen cell, the evaporation is more of the Langmuir type and no differences between Knudsen effusion and Langmuir evaporation can be expected. An evaporation coefficient of this magnitude is not extremely unlikely in cases, where species are formed during the evaporation, which are not contained in the condensed phase. The example of the different evaporation behavior of arsenolite and claudite were already mentioned in the introduction. A similar case is the evaporation of $(AgI)_3$ from molten AgI. Here, we had to assume

Table 3
Partial Pressures of the Parent Molecules in the Vapor over Liquid CsSb[a]

	Cs	Sb_2	Sb_4	Cs_2Sb_2	Cs_2Sb_4	Cs_6Sb_4
p/Pa	2.39	0.352	6.13	$3.15 \cdot 10^{-3}$	$7.19 \cdot 10^{-3}$	$4.18 \cdot 10^{-5}$

[a] $T = 900.1$ K.

evaporation coefficients $\leq 10^{-2}$, in order to devise experimental conditions that lead to reliable values for the thermochemistry of $(AgI)_3$ evaporation (*15*). Further experiments to determine the evaporation mechanism of intermetallics from cesium-antimony condensed phases are in progress.

ACKNOWLEDGMENT

This paper is part of the Dissertation of Andreas Hartmann, Darmstadt D17. It was generously sponsored by the Deutsche Forschungsgemeinschaft within the Schwerpunktprogramm Physik anorganischer Cluster. Further assistance by the Fonds der Chemischen Industrie is greatly acknowledged.

REFERENCES

1. V. Piacente and A. Desideri, *J. Chem. Phys.* **57,** 2213 (1972).
2. G. DeMaria, L. Malaspina, and V. Piacente, *J. Chem. Phys.* **56,** 1978 (1972).
3. K. A. Gingerich, D. L. Cocke, H. C. Finkbeiner, and C.-A. Chang, *Chem. Phys. Letts.* **18,** 102 (1973).
4. G. DeMaria, J. Drowart, and M. G. Inghram, *J. Chem. Phys.* **31,** 1076 (1959).
5. G. Riekert, G. Rainer-Harbach, P. Lamparter, and S. Steeb, *Z. Metallkde* **72,** 406 (1981).
6. B. Busse and K. G. Weil, *Ber. Bunsenges. Phys. Chem.* **86,** 93 (1982).
7. T. Scheuring and K. G. Weil, *Ber. Bunsenges. Phys. Chem.* **89,** 811 (1985).
8. A. Hartmann and K. G. Weil, *Z. Phys. D,* in press.
9. A. Hartmann and K. G. Weil, *Angew. Chem. Int. Ed. Engl.* **27,** 1091 (1988).
10. K. H. Jack and M. M. Wachtel, *Nature* **178,** 1408 (1956).
11. H. G. von Schnering, W. Hönle, and G. Krogull, *Z. Naturforsch.* **34b,** 1678 (1979).
12. I. N. Stranski and G. Wolff, *Research* **4,** 15 (1951).
13. R. Hultgren, P. D. Desai, D. T. Hawkins, M. Gleiser, K. K. Kelley, and D. D. Wagman, *Selected Values of the Thermodynamic Properties of the Elements,* Am. Soc. Met. 1973.
14. J. B. Mann, *J. Chem. Phys.* **46,** 1646 (1967).
15. U. Pittermann and K. G. Weil, *Ber. Bunsenges. Phys. Chem.* **84,** 542 (1980).

Investigation of Small Lithium and Potassium Clusters by Knudsen Effusion Mass Spectrometry Using Broad Band Photoionization

K. HILPERT* AND D. KATH

Institute for Reactor Materials, Nuclear Research Centre (KFA) Jülich, P.O. Box 1913, D-5170 Jülich, FRG

ABSTRACT

A novel Knudsen cell-mass spectrometer system using broadband photoionization has been developed and used for the study of the equilibrium vaporization of lithium and potassium in the temperature ranges of 540 to 830 and 350 to 460 K, respectively. The ions M^+, M_2^+, M_3^+ (M = Li, K), and Li_4^+ were detected and their intensities determined except that of Li_4^+. Partial pressures of M, M_2, and M_3 (M = Li, K) were tentatively evaluated from the ion intensities and enthalpies of dissociation of Li_2 and K_2 determined according to the second- and third-law methods as well as of Li_3 and K_3 according to the third-law method. The data obtained for the dimer K_2 indicate an unusually high photoionization cross section for the ion K_2^+ relative to that of K^+. This effect increases the K_2^+/K^+ ion intensity ratio in alkali metal cluster beams, also generally analyzed using broad-band photo-ionization, by a factor of 11. Enthalpies of atomization for Li_3, and K_3, ΔH°_0, were obtained as 152 ± 3 and 90 ± 2 kJ mol^{-1}, respectively, neglecting possible errors of the molecular parameters of these species and assuming no unusual photoionization cross sections of Li_3^+ and K_3^+ relative to that of Li^+ or K^+, respectively.

Index Entries: Knudsen effusion mass spectrometry; photoionization; vaporization; dissociation energies; bond energies; ionization cross section; alkali metal clusters; Li_2; Li_3; K_2; K_3.

*Author to whom all correspondence and reprint orders should be addressed.

INTRODUCTION

Alkali metal clusters are of great interest, as shown by the recent review articles of Schumacher et al. (*1*), Kappes (*2*), Kappes and Schumacher (*3*), Koutecký (*4*), and Gole (*5*). Most of the investigations were carried out on homo nuclear sodium and potassium clusters as well as on hetero nuclear clusters composed of these two metals. Abundances, ionization potentials, and electrical polarizabilities resulted by the various studies (*3*). They were all carried out by generating the clusters under nonequilibrium conditions by either neat vapor or seeded beam expansion. The relative abundance of the different clusters obtained by the expansion is related to their stability (*2*). The cluster abundances resulted by mass spectrometric analysis of the expanded beam using broadband photoionization. Multimodal abundance distributions of alkali clusters were obtained in this way by Schumacher's (*6*) and Knight's (*7*) groups after the pioneering work on sodium clusters by Robbins and coworkers (*8*).

We report for the first time on the abundance distribution of gaseous alkali metal species generated under equilibrium conditions by vaporization of lithium and potassium, and determined by a novel Knudsen cell-mass spectrometer system using broadband photoionization. Equilibrium partial pressures and dissociation enthalpies were tentatively evaluated for the gaseous species identified, thereby also giving information on photoionization cross sections.

EXPERIMENTAL

The vacuum chamber of the apparatus developed (Fig. 1) is made of stainless steel. *Conflat* sealings are generally used except the two *Viton* O-ring seals for the entrance and exit windows made of *Suprasil* quartz glass.

Two different Knudsen cells, cell A and cell B, were used for the vaporization studies of lithium and potassium, respectively. Cell B was developed by us earlier for the study of alkali metals and is described in (*9*). It is made of molybdenum and can be adjusted from outside. The cylindrical effusion orifice had a diameter of 0.8 mm and a length of 2 mm for the study of lithium as well as a diameter of 0.5 mm and a length of 1 mm for the vaporization of potassium. Chromel/Alumel thermocouples are used for the temperature measurement. Cell A used for the study of lithium resembles cell B. It can be used at higher temperatures than the latter and does not allow the change of reactive samples without avoiding completely chemical interactions with the air or moisture unlike cell B.

A quadrupole mass spectrometer (type Q 156, RIBER S.A., Rueil Malmaison, France) is used for the analysis of the molecular beam collimated by a water cooled diaphragm of 2 mm diameter and originating in the Knudsen cell. The molecular beam can be interrupted by a shutter

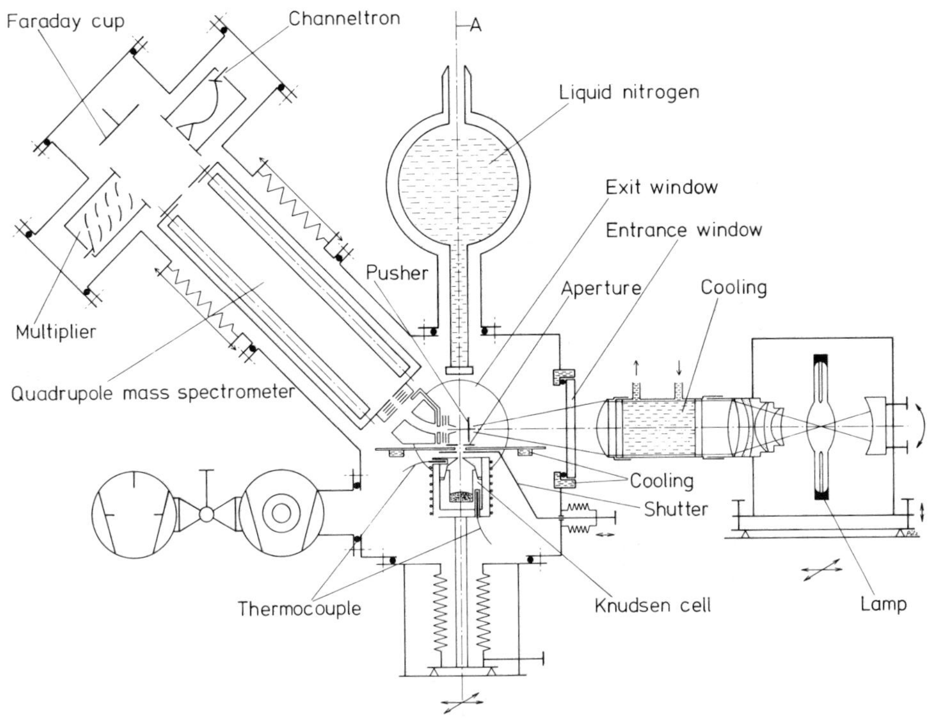

Fig. 1. Schematic representation of a Knudsen cell-quadrupole mass spectrometer system using broad-band photoionization. (The entrance window, the optical system, and the lamp have to be rotated on the axis A by 90°).

and is condensed by a cryogenic pump filled with liquid nitrogen. A 45° ion deviation system was used to diminish the distance between the ionizing region in the ion source and the effusion orifice. The ion source was modified for photoion collection. Practically no light touches the metal parts near the ion source to prevent photoelectron formation.

Photoionization of atoms and molecules in the beam is accomplished by a 1 kW Hg/Xe lamp (model 977B0010, Hanovia) equipped with a 68 mm diameter four element f 10.7 UV condensor (model ASPHERAB, Oriel, Stratford, CT). Thereafter, the light is focussed at right angles onto the molecular beam using a 100 mm focal length lens. The light is passed through the two windows on either side of the ion source.

RESULTS

The ions Li_n^+, K_n^+ (n = 1,2,3), and Li_4^+ were detected in the mass spectrum upon vaporizing lithium or potassium (nominal purity 99.95 mass%) supplied by Ventron, Karlsruhe, FRG. These ions were completely generated by photoionization. This was demonstrated by vaporizing sodium. The intensity of the ions Na_i^+ (i = 1–4) measured by using different UV cutoff filters agrees well with the known photoionization efficiency curves (6a,b) of these ions.

Figures 2 and 3 show the intensities of the ions $^{7}Li^{+}$, $^{14}Li_2^{+}$, $^{39}K^{+}$, and $^{78}K_2^{+}$ measured at different temperatures in the course of various runs. Those of Li_3^{+} and K_3^{+} have so far only been determined at temperatures of 800 and 450 K, respectively. Typical relative ion intensities determined for the monomer, dimer, and trimer ions and corrected for isotopic distribution are, therefore, given at these temperatures in Table 1. The detection of the tetramer ion Li_4^{+} was not possible at the temperature given in the table due to its low relative abundance. This ion was observed at temperatures of 920 and 1000 K.

Enthalpies of sublimation (Table 2) were obtained according to the second-law method by the use of the equation

$$\Delta_{sub}H^{o}_{T} = -R d \ln (I_i T)/d (1/T) \tag{1}$$

where I_i is the intensity of the ion i, and the product $I_i T$ is proportional to p_i (*see* Eq. (2)). R is the gas constant. The enthalpies of sublimation at 0 K (Table 2) were computed by using the enthalpy increments in the JANAF tables (*10*). Dissociation enthalpies of the molecules M_2 (M = Li, K) (Table 2) result by the relation: $\Delta_{dis}H^{o}_{0}(M_2) = 2\,\Delta_{sub}H^{o}_{0}(M) - \Delta_{sub}H^{o}_{0}(M_2)$.

The tentative evaluation of partial pressures for Li_i and K_i (i = 1,2,3) and of dissociation enthalpies of dimer and trimer species according to the third-law method is reported later, owing to possible uncertainties of the photoionization cross sections.

DISCUSSION

Partial Pressures

Partial pressures $p(M_i)$ of the species M_i at the Knudsen cell temperature T may be obtained by the equation

$$p(M_i) = k \cdot 1/N(M_i^{+}) \cdot 1/\sigma(M_i^{+}) \cdot I(M_i^{+})\, T \tag{2}$$

where k, $\sigma(M_i^{+})$, and $I(M_i^{+})$ are the pressure calibration factor, the ionization cross section, and the intensity of the ion M_i^{+} corrected for isotope abundances, respectively. Constant ionization cross sections are assumed for photon energies larger than the ionization potential of the ion considered. $N(M_i^{+})$ is the number of the effective incident photons for M_i^{+}. Relative values result by integrating the lamp spectrum using selected ionization potentials as integration limits. Equation (2) is obtained by analogy to the corresponding relation for conventional Knudsen effusion mass spectrometry using electron impact ionization (*11*) and by analogy to the estimation of neutral cluster abundances from ion intensities (2b) obtained mass spectrometrically at nonequilibrium expanded beam experiments by the use of broad-band photoionization.

Relative partial pressures $p(M)/p(M_i)$ (M = Li, K) (i = 1,2,3) (Table 3) result tentatively from the ion intensity ratios in Table 1 by the relation

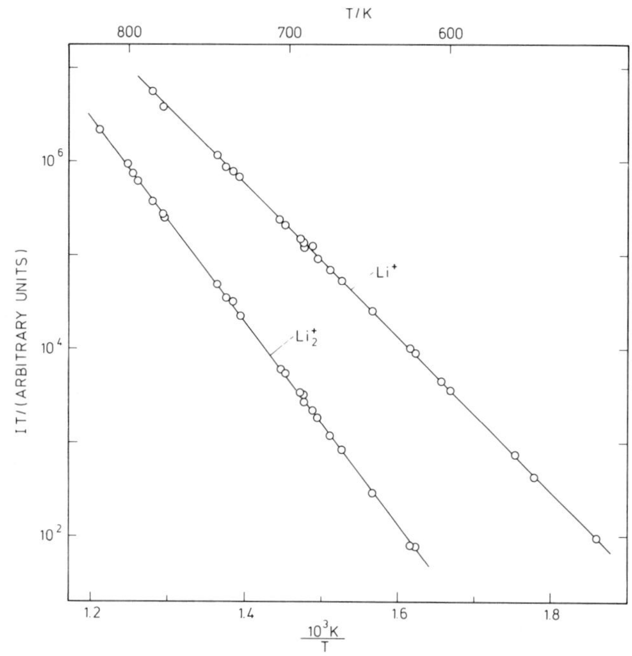

Fig. 2. Temperature dependence of the intensities of the ions Li^+ and Li_2^+ observed on vaporizing lithium.

$$p(M_i)/p(M) = \sigma(M^+)/\sigma(M_i^+) \cdot N(M^+)/N(M_i^+) \cdot I(M_i^+)/I(M^+) \quad (3)$$

as follows from Eq. (2). The ionization cross section ratios were estimated as $\sigma(M)/\sigma(M_2) = 0.667$ and $\sigma(M)/\sigma(M_3) = 0.444$ according to the rule $\sigma(M_i) = 0.75\ i\ \sigma(M)$ (i = 2,3). It is generally used for the estimation of ionization cross sections of homo nuclear molecules, $\sigma(M_i)$, from that of the atom, $\sigma(M)$. The following ratios of incident photons were employed for the computation of the relative partial pressures: $N(Li_2)/N(Li) = 8.83 \times 10^{-2}$, $N(Li_3)/N(Li) = (8.50^{+1.04}_{-0.90}) \times 10^{-3}$, $N(K_2)/N(K) = 4.14 \times 10^{-1}$, and $N(K_3)/N(K) = 1.60 \times 10^{-1}$. They were determined as aforementioned by using the following ionization potentials in eV for the ions in parentheses: 5.392 (Li^+) [Ref. (*13*)], 5.145 (Li_2^+) [Ref. (*14*)], 4.17 ± 0.04 (Li_3^+) [Ref. (*15*)], 4.341 (K^+) [Ref. (*13*)], 4.064 (K_2^+) [Ref. (*14*)], and 3.33 ± 0.01 (K_3^+) [Ref. (*16*)]. The lamp spectrum was redetermined for the integration procedure (*17*). It agrees well with that given by the supplier of the lamp.

The errors given in Table 3 for the relative partial pressures are probable uncertainties obtained by the error propagation if Eq. (3) is used. Errors of 10 and 20% were assumed for the ion intensity ratios $I(M_2^+)/I(M^+)$ and $I(M_3^+)/I(M^+)$, respectively. The uncertainty of the estimated relative ionization cross sections $\sigma(M_i)$ (i = 2,3) was estimated as 30% neglecting possible unusual cross sections.

The partial pressure ratio $p(Li_2)/p(Li)$ obtained in this work is roughly the average of those given by the literature (cf., Table 3). The deviation

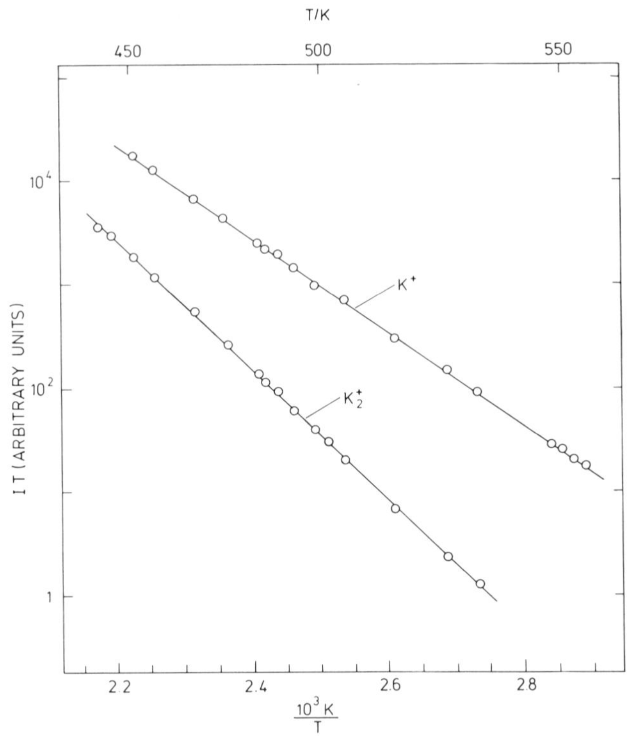

Fig. 3. Temperature dependence of the intensities of the ions K^+ and K_2^+ observed on vaporizing potassium.

of the two literature values from that of this work may be explained by possible uncertainties. The partial pressure ratio $p(K_2)/p(K)$ obtained in this work exceeds those of the literature by about one order of magnitude. The high value of this work might be explained by autoionizing Rydberg states in K_2, thereby increasing the ionization probability compared to the direct photoionization process. Such states in K_2 were detected and investigated by one-, two-, and three-photon ionization experiments carried out by Wöste's group (*18,19*) using laser radiation.

A further explanation for the stated difference of the $p(K_2)/p(K)$ pressure ratios in Table 3 might be erroneous literature data. A change of the ratio given by the literature to the value of this work corresponds to an increase of the dissociation enthalpy by more than 10 kJ mol^{-1} since the partial pressures from the tables in Refs. (*11*) and (*12*) depend on the dissociation enthalpies adopted there. Such an error of the adopted dissociation enthalpy is improbable.

There is no experimental evidence for an unusually high ionization cross section for K_3^+. The laser one-photon ionization efficiency curve for K_3^+ (*16*) shows no autoionization structure as observed for K_2^+ (*19*). The same has also been observed for Na_3^+ (*20*).

Third-Law Evaluation of Enthalpies of Dissociation

Dissociation enthalpies of the dimer and trimer molecules for the reactions

Table 1
Relative Ion Intensities $I(M_i^+)/I(M^+)$ Measured upon Vaporizing Different Alkali Metals M (M = Li, K) at the Temperature T and at the Saturation Pressure $p_M(T)$ in the Knudsen Cell.

M	T/K	$p_M(T)$/Pa	M_i^+	$I(M_i^+)/I(M^+)$
Li	800	1.1	Li^+	1
			Li_2^+	9.23×10^{-2}
			Li_3^+	1.19×10^{-5}
K	450	0.34	K^+	1
			K_2^+	1.29×10^{-1}
			K_3^+	1.08×10^{-6}

Table 2
Enthalpies of Sublimation and of Dissociation Evaluated According to the Second-Law Method for Different Gaseous Species, M_i, for the Mean Temperature of the Measurements, $\tilde{T}$, and 0 K[a].

M_i	$\tilde{T}$/K	$\Delta_{sub}H^\circ_{\tilde{T}}$ kJ mol^{-1}	$\Delta_{sub}H^\circ_o$ kJ mol^{-1}	$\Delta_{dis}H^\circ_o$ kJ mol^{-1}
Li	650	156.17 ± 0.63	160.31 ± 0.63	
Li_2	700	207.35 ± 0.88	220.87 ± 0.88	99.75 ± 1.54
K	400	86.21 ± 0.38	90.51 ± 0.38	
K_2	400	119.02 ± 0.63	129.63 ± 0.63	51.39 ± 0.99

[a]Errors are standard deviations.

$$M_2(g) \rightleftarrows 2\,M(g) \text{ and} \tag{4}$$

$$M_3(g) \rightleftarrows 3M(g)\ (M = Li,\ K) \tag{5}$$

were evaluated according to the third-law method (Table 4) by using the equation

$$\Delta_{dis}H^\circ_o = -T\,[R \ln K_p + \Delta(G^\circ_T - H^\circ_o)/T] \tag{6}$$

R, K_p, and $\Delta(G^\circ_T - H^\circ_o)/T$ are the gas constant as well as the equilibrium constant and the change of the Gibbs energy function of the reaction considered.

The equilibrium constants for the reactions Eqs. (4) and (5) (Table 4) are obtained from the partial pressure ratios in Table 3 and the monomer partial pressures from the JANAF tables (*10*) by the relations $K_p(4) = p(M)\,p(M)/p(M_2)$ and $K_p(5) = p(M)^2p(M)/p(M_3)$. The Gibbs energy function of the monomer and dimer species, M = Li, K, was obtained from the JANAF tables. That of the trimer species was computed from the molecular parameters by the use of the rigid rotator-harmonic oscillator approximation (*11*) as $\Delta(G^\circ_{800} - H^\circ_o)/800\ K = -274.90\ J\ mol^{-1}$ for Li_3 and $\Delta(G^\circ_{450} - H^\circ_o)/450\ K = -324.035\ J\ mol^{-1}$ for K_3. No experimental values for the molecular parameters are available for Li_3 and K_3. We,

Table 3
Relative Equilibrium Partial Pressures $p(M_i)/p(M)$ Evaluated in this Work and from the Literature at the Temperature T for the Vaporization of the Alkali Metals M (M = Li, K)

M	T/K	M_i	This work	Literature	
			$p(M_i)/p(M)^a$	$p(M_i)/p(M)^b$	$p(M_i)/p(M)^c$
Li	800	Li	1	1	1
		Li_2	$(5.43 \pm 2.77) \times 10^{-3}$	2.90×10^{-3}	8.02×10^{-3}
		Li_3	$(4.50\ ^{+2.63}_{-2.47}) \times 10^{-8}$		
K	450	K	1	1	1
		K_2	$(3.55 \pm 1.12) \times 10^{-2}$	2.92×10^{-3}	3.64×10^{-3}
		K_3	$(7.68 \pm 2.77) \times 10^{-8}$		

[a] Errors see text.
[b] Ref. (*11*).
[c] Ref. (*12*).

therefore, used those computed recently by Thompson et al. (*21*) on the basis of data from various ab initio computations. These authors report the structure of an isosceles triangle for Li_3 and K_3 with side lengths of 2.78, 2.78, and 3.25 Å for Li_3 as well as 3.96, 3.96, and 4.92 Å for K_3 if the most stable configuration is considered. The vibrational frequencies belonging to these configurations were computed as 78, 163, and 327 cm^{-1} for Li_3 as well as 56, 36, and 89 cm^{-1} for K_3. A $^2B_2$ electronic ground state was assumed for these molecules.

The enthalpy of dissociation of Li_2 evaluated according to the second-law method, $\Delta_{dis}H^\circ_o = 99.75 \pm 1.54$ kJ mol^{-1} (Table 2), agrees well with the third-law value of this work and the dissociation enthalpy selected by the JANAF group (Table 4). In contrast to this, the second-law dissociation enthalpy of K_2 evaluated as $\Delta_{dis}H^\circ_o = 51.39 \pm 0.99$ kJ mol^{-1} (Table 2) and the dissociation enthalpies selected by the literature are by more than 10 kJ mol^{-1} smaller than that obtained according to the third-law method (Table 4). This is caused by the neglect of the unusual relative photoionization cross section of K_2^+ that leads to substantially increased values for the relative K_2 partial pressure as discussed earlier.

The atomization enthalpies of Li_3 and K_3 evaluated tentatively according to the third-law method (Table 4) are smaller than the experimental literature values given in the same table. The value for Li_3 was obtained on the basis of a second- and third-law evaluation by Knudsen effusion mass spectrometry using electron impact ionization (*22*). The upper bound given for the K_3 atomization enthalpy was obtained from the enthalpy of dissociation of K_2 selected by the JANAF group (*11*) and the upper bound for the K_2-K bond energy determined by laser-induced atomic fluorescence (LIAF) by Gole's group (*23*) as 43.19 kJ mol^{-1}. Moreover, the latter group assumed that the onset of the K_3 LIAF spectrum appear to provide the lower bound for the K_2-K bond energy as 24.16 kJ mol^{-1}. The lower bound for the enthalpy of atomization of K_3 results by

Table 4
Equilibrium Constant, Change of Gibbs Energy Function, and Dissociation Enthalpy for the Reactions $M_i(g) \rightleftarrows iM(g)$ (M = Li, K; i = 2,3) Evaluated in this Work According to the Third-Law Method and Experimental Dissociation Enthalpies from the Literature

M_i	T/K	K_p at T	$\Delta\ [(G^o_T - H^o_o)/T]$ J mol^{-1} K^{-1}	This work $\Delta_{dis}\ H^o_o$ kJ mol^{-1}	Literature $\Delta_{dis}\ H^o_o$ kJ mol^{-1}
Li_2	800	$(2{,}029\ ^{+0.938}_{-0.488}) \times 10^{-3}$	−78.780	104.3 ± 3	99.8 ± 0.1[a], 107.8 ± 1.3[b]
Li_3	800	$(2.702\ ^{+1.626}_{-0.807}) \times 10^{-3}$	−140.64	151.8 ± 3	173.6 ± 16.7[c]
K_2	450	$(9.520\ ^{+4.396}_{-2.285}) \times 10^{-5}$	−67.483	65.0 ± 2	53.5 ± 1.2[a], 53.8 ± 0.9[b]
K_3	450	$(1.484\ ^{+0.838}_{-0.394}) \times 10^{-4}$	−126.18	89.8 ± 2	<96.67[d]

[a]Ref. (11).
[b]Ref. (12).
[c]Ref. (22).
[d]Ref. (23).

the use of this value as 77.64 kJ mol^{-1}. The assumption made by Gole and coworkers is, therefore, supported by our value (cf., Table 4).

ACKNOWLEDGMENT

The author is indebted to H. Nickel for the support of this work and his kind advice. He wishes to thank K. G. Weil for the helpful discussions.

REFERENCES

1. Schumacher, E., Kappes, M. M., Marti, K., Radi, P., Schär, M., and Schmidhalter, B., *Ber. Bunsenges. Phys. Chem.* **88,** 220 (1984).
2. Kappes, M. M., *Chem. Rev.* **88,** 369 (1988).
3. Kappes, M. M. and Schumacher, E., *Z. Phys. Chem. Neue Folge* **156,** 23 (1988).
4. Koutecký, J., *Chem. Rev.* **6,** 539 (1986).
5. Gole, J. L., *Metal Clusters*, Moskovits, M., ed., Wiley, New York, 1986, p. 131.
6. (a) Herrmann, A., Schumacher, E., and Wöste, L., *Helvetica Chim. Acta* **61,** 453 (1978); (b) Herrmann, A., Schumacher, E., and Wöste, L., *J. Chem. Phys.* **68,** 2327 (1978); (c) Kappes, M. M., Kunz, R. W., and Schumacher, E., *Chem. Phys. Lett.* **91,** 413 (1982); (d) Kappes, M. M., Radi, P., Schär, M., and Schumacher, E., *Chem. Phys. Lett.* **119,** 11 (1985); (e) Kappes, M. M., Schär, M., Radi, P., and Schumacher, E., *J. Chem. Phys.* **84,** 1863 (1986).
7. Knight, W. D., Clemenger, K., de Heer, W. A., Saunders, W. A., Chou, M. Y., and Cohen, M. L., *Phys. Rev. Lett.* **52,** 2141 (1984); Knight, W. D., de Heer, W. A., Clemenger, K., and Saunders, W. A., *Solid State Commun.* **53,** 445 (1985).

8. Robbins, E. J., Leckenby, R. E., and Willis, P., *Adv. Phys.* **16,** 739 (1967); Foster, P. J., Leckenby, R. E., and Robbins, E. J., *J. Phys. B.* **2,** 478 (1969).
9. Hilpert, K., *Ber. Bunsenges. Phys. Chem.* **88,** 260 (1984).
10. Chase, M. W., Jr., Davies, C. A., Downey, J. R., Jr., Frurip, D. J., McDonald, R. A., and Syverud, A. N., *JANAF Thermochemical Tables*, Third Edition, Part II, American Institute of Physics, New York, 1986.
11. Drowart, J. and Goldfinger, P., *Angew. Chem.* **79,** 589 (1967); Gingerich, K. A., *Current Topics of Materials Science*, vol. 6, Kaldis, E., ed., North Holland, Amsterdam, 1980, p. 345; Hilpert, K., *Structure and Bonding*, Springer, Heidelberg, in press.
12. Vargaftik, N. B., *Tables on the thermophysical properties of liquids and gases*, 2nd ed., Wiley, New York, 1975.
13. Moore, C. E., *Ionization Potentials and Ionization Limits Derived from the Analysis of Optical Spectra*, Nat. Bur. Stand. US, NSRDS-NBS 34 (1970).
14. Kappes, M. M., Schär, M., and Schumacher, E., *J. Phys. Chem.* **89,** 1499 (1985).
15. Kappes, M. M. and Schumacher, E., *Surf. Sci.* **156,** 1 (1985).
16. Brechignac, C. and Cahuzac, Ph., *Surf. Sci.* **156,** 183 (1985).
17. Hilpert, K. and Kath, D., in preparation.
18. Broyer, M., Chevaleyre, J., Delacrétaz, G., Martin, S., and Wöste, L., *Chem. Phys. Lett.* **99,** 206 (1983).
19. Broyer, M., Chevaleyre, J., Delacrétaz, G., Fayet, P., and Wöste, L., *Chem. Phys. Lett.* **114,** 477 (1985).
20. Broyer, M., Chevaleyre, J., Delacrétaz, G., Fayet, P., and Wöste, L., *Surf. Sci.* **156,** 342 (1985).
21. Thompson, T. C., Izmirlian, G., Lemon, S. J., Truhlar, D. G., and Mead, C. A., *J. Chem. Phys.* **82,** 5597 (1985).
22. Wu, C. H., *J. Chem. Phys.* **65,** 3181 (1976).
23. Hayden, J. S., Woodward, R., and Gole, J. L., *J. Phys. Chem.* **90,** 1799 (1986).

Oxidation of Small Metal Clusters

T. C. DeVore*[,1] and J. L. Gole[2]

[1]Department of Chemistry, James Madison University, Harrisonburg, VA 22807; and [2]School of Physics, Georgia Institute of Technology, Atlanta, GA 30010

ABSTRACT

A specially designed furnace system has been built that enables the creation of a large flux of molecular clusters containing up to six atoms. By varying the conditions in this furnace assembly, the cluster of interest can be optimized.

We have used this source to investigate the oxidation of small clusters of boron, chromium, manganese, copper, and silver with oxidants such as fluorine, ozone, N_2O and NO_2 under a variety of experimental conditions. The results of these investigations indicate that cluster reactions can be more energetic than the corresponding atomic reactions; and that the initial products of metal cluster reactions are kinetically controlled.

The evidence supporting these conclusions and the possible implications that these processes have to metal oxidation processes in general will be discussed.

Index Entries: Ag_x, oxidation by ozone; Cu_x, oxidation by ozone; Mn_x, oxidation by ozone; Cr_x, oxidation by ozone; Ag_xO; Cu_xO; Mn_xO; CrF.

INTRODUCTION

Most heterogeneous chemical conversion processes occur on or near the surface of a condensed phase. The widespread applicability of heterogeneous chemical processes in a variety of diverse practical applications encompassing metal combustion, chemical vapor deposition, plasma etching, and general industrial catalysis has fueled significant theoretical and experimental research into surface processes. A novel,

*Author to whom all correspondence and reprint orders should be addressed.

but increasingly popular, technique now used to develop this research involves modeling the surface as a small group of interacting atoms (cluster). Since clusters represent a unique intermediate state of matter occupying the middle ground between small inorganic molecules and the bulk solid, several elegant studies have emerged to explore their structures, their molecular bonding, and the chemical reactivity of these unique species (*1*). However, this work has largely neglected the internal mode structure and the dynamics associated with the products of cluster oxidations. The limited information available indicates that clusters often undergo unique and totally unexpected reactive branching. The study of this reactive branching is fundamental to the development of rules required to extrapolate from simple A + BC reactions. Also, the study of the quantum level structures of the products of metal cluster oxidation provides detailed information pertinent to the understanding of molecular structure and bonding. In correlation with quantum chemistry, this data allows the generation of reliable potentials that can be used to describe highly exothermic combustion processes; and local cooperative atomic phenomena and mobilities whose description can be useful in modeling at the interface of surfaces.

We have developed several techniques that can be used to generate a substantial and usable continuous flux of metal clusters. Cluster sources have been used to investigate the electronically excited products of the reactive systems $B_x + NO_2$ (*2,3*), $Cu_x + O_3$ (*2,4*), $Ag_x + O_3$ (*2,5,6*), $Mn_x + O_3$ (*2,7*), and $Cr_x + F_2$ (*8*). This work had led to three general observations about cluster reactions:

1. Cluster reactions often produce small molecules in higher energy states than do the corresponding atomic reactions.
2. The products of the reactions of larger metal clusters often contain smaller metal rich molecules. In many cases, metal-metal bonds are present in these product molecules.
3. The kinetics of the reaction often dictate the initially formed reaction products.

The data supporting these conclusions are described more completely in this paper. Since the kinetic data has been published (*2*), we largely concentrate on a description of the quantal structure of the products of metal cluster oxidation.

EXPERIMENTAL

Chemiluminescent probes have been used to explore the dynamics of metal cluster oxidation. The versatile entrainment apparatus used in these investigations has been described in detail elsewhere (*2–8*). The heart of the system is the large capacity furnace assembly shown in Fig. 1. The metal of interest is placed in a crucible (usually machined from high grade graphite). This crucible is placed into a resistively heated tungsten basket heater and the assembly is surrounded by tantalum or

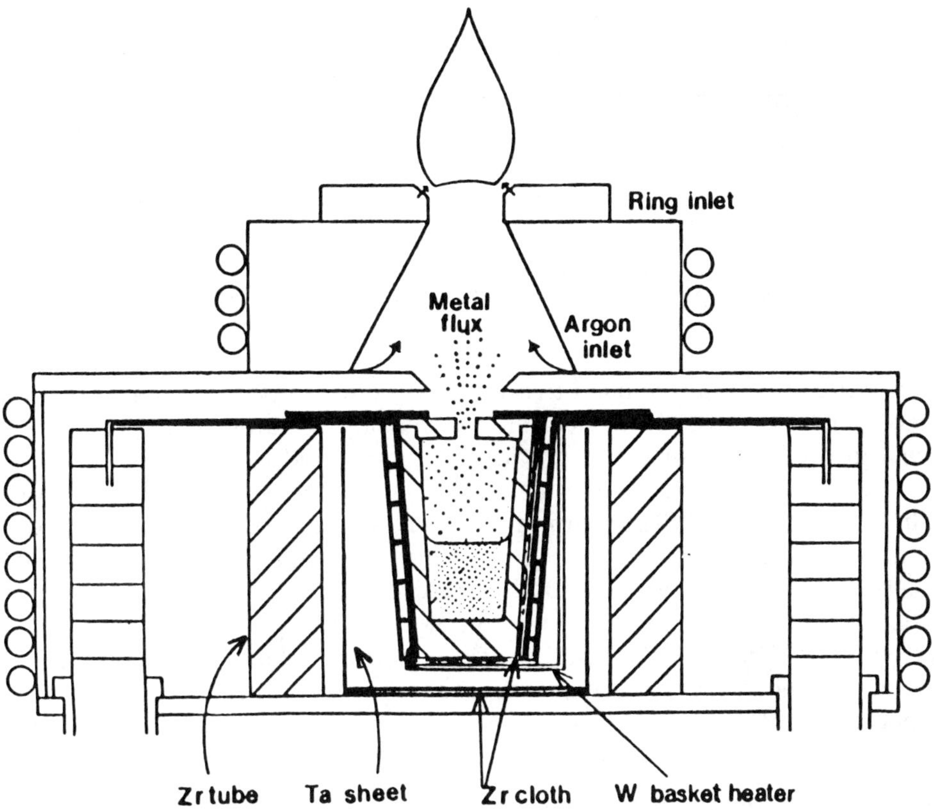

Fig. 1. Schematic of the vaporization-entrainment apparatus used to generate small metal clusters.

tungsten heat shields and is heavily insulated with zirconia cloth (Zircar Products, Florida, *NY*). This effective heat containment allows vaporization temperatures in excess of 2300 K to be easily obtained.

A high flux continuous metal vapor flow is used to create a large concentration of small metal clusters. By operating the metal source at temperatures such that the existing metal flow is characterized by a Knudson number less than one, we form the seed for the initial phases of a cluster generating environment. This high metal flux is further agglomerated to small clusters by an entraining gas flow (usually Ar, He, N_2, or CO), which can be cooled to obtain a range of conditions from room temperature to near liquid nitrogen temperature. The outer walls of the furnace are cooled to the same temperature as the rare gas entrainment flow, thus maintaining cooling. The oxidant enters the flow after its agglomeration through a ring injector or through a nozzle positioned approximately 5 cm above the cooled furnace assembly. Highly exothermic, primarily multicentered reactions lead to emission from a number of product molecules. Several factors, including metal flux, entrainment gas temperature and pressure, pumping speed, and oxidant configuration, can be adjusted to maximize the process being studied. Changes in

Table 1
Summary of the Reaction Systems and Conditions Used to Investigate the Oxidations of Metal Clusters

Metal	Oxidant[a]	Rare gas	T(Metal)	P(Rare gas)
Ag	O_3	He/Ar	1400–1700 K	10– 65 Pa
Cu	O_3	He/Ar	1650–1960 K	60–200 Pa
Mn	O_3	Ar	1350–1600 K	60–130 Pa
Cr	F_2	He/Co	1700–1900 K	200–300 Pa

[a]Oxidant pressures were adjusted to maximize the desired emission under study.

pumping speed can be used to adjust the conditions in the reaction zone across a wide range from near single collision conditions ($p \sim 1 \times 10^{-3}$ Pa.) to multiple collision conditions ($P_{max} \sim 1000$ Pa).

The chemiluminescent emission is usually focused onto the entrance slit of a 1 m spex 1704 scanning monochrometer operated in first order with a Bausch and Lomb 1200 groove/mm grating blazed at 500 nm. The dispersed fluorescence is usually detected using a dry ice cooled EMI 9808 photomultiplier tube. The signal from the phototube is fed to a Keithley 417 fast picoammeter whose partially damped output signal drives a Leeds and Northrup strip chart recorder used to record the spectra of interest.

RESULTS AND DISCUSSION

Reactions between metal atoms and oxidants have long been used to produce metal oxides and halides for spectroscopic and kinetic investigation (*9*). The energy released in these reactions of the form

$$M + XY \longrightarrow MX^* + Y \tag{1}$$

is given by

$$E_{RXN} \sim D_o^o\,(MX) - D_o^o\,(XY) + \Delta E_{INT} \tag{2}$$

where D_o^o (A) is the bond dissociation energy of A and ΔE_{INT} is the contribution from reactant translational energy. For a typical oxidant (O_3, N_2O, F_2) reacting with a transition metal, the process (1) gives a maximum exothermicity between approximately 2 and 4 ev. (*10*). For example, the estimated reaction exothermicity for the reaction

$$Cu + O_3 \longrightarrow CuO^* + O_2 \tag{3}$$

is 1.76 ± .05 eV. This process is energetic enough to populate the $v' = 2$ level of the $\delta^2\Sigma^+$ state of CuO. As the data obtained for copper atom and cluster oxidation depicted in Fig. 2 demonstrates, the available energy for the atomic oxidation process is in excellent agreement with the chemiluminscent spectrum observed for reaction (3) under near single collision conditions.

There is, however, growing evidence that the oxidation of metal dimers (or larger clusters) can result in significantly more energetic processes which populate higher lying energy states of those produce molecules formed in reaction. Four center dimer reactions of the type

$$M_2 + XY \longrightarrow MX^* + MY \tag{4}$$

$$E_{RXN} \sim D_o^o\,(MX) + D_o^o\,(MY) - D_o^o\,(M_2) - D_o^o\,(MY) + \Delta E_{INT} \tag{5}$$

can release 1–2 eV more energy than do the corresponding atomic reactions (*10*). If this energy is largely collected in one of the product molecules, then higher lying energy states of this product can be populated. As indicated in Fig. 2, the oxidation of copper clusters with ozone under multiple collision conditions produces emission that originates from the $D^2\Delta$, $C^2\Pi$, $A^2\Sigma$, $B^2\Delta$, and $A'^2\Sigma$ states of CuO. The population of these states requires up to 1eV more energy than is available from the copper atom reaction. This increased exothermicity could be provided by the reaction of vibrationally hot copper dimer, or more likely by ground state copper trimer via the process

$$Cu_3 + O_3 \longrightarrow CuO^* + Cu_2O_2 \tag{6}$$

The increased exothermicity inherent to the multicentered capability of metal cluster reactions provides an opportunity to produce and explore new energy states in previously studied product molecules. For example, the reaction between chromium atoms and dimers, and molecular fluorine has recently been used to study three previously unknown electronic states of CrF (*8*). Prior to this investigation, only the $A^6\Sigma - X\,^6\Sigma$ transition of CrF had been analyzed (*11*). The metal atom reaction

$$Cr + F_2 \longrightarrow CrF^* + F \tag{7}$$

which has a reaction exothermicity close to 3 eV., produces two additional product CrF band systems. Vibrational analysis indicates that both terminate in the ground state. The reaction of chromium dimers with molecular fluorine, viz.

$$Cr_2 + F_2 \longrightarrow CrF^* + CrF \tag{8}$$

nearly doubles the reaction exothermicity. This increase in reaction exothermicity is sufficient to populate an additional electronic state of CrF. The molecular constants determined for each of the newly observed states of CrF are summarized in Table 2.

The oxidation of silver clusters with ozone under multiple collision conditions produces emission from the $A\,^2\Pi$ (400–420 nm) and the $B^2\Pi$ (320–370 nm) states of AgO (*2,5,6*). Since neither the reaction of silver atoms nor the reaction of silver dimers can be expected to populate these AgO product states, this emission must result from the oxidation of Ag_3 or larger silver clusters.

At higher silver fluxes, the AgO emission is quenched and the spectrum is characterized by a combination of $Ag_xO(x \geq 2)$ and Ag_2 band

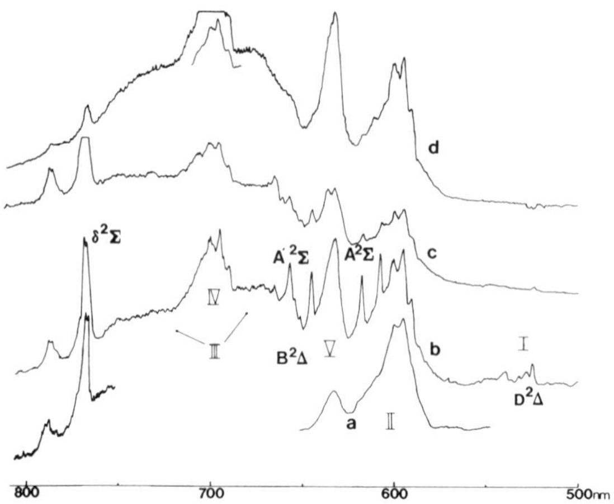

Fig. 2. Chemiluminescent spectra observed for the oxidation of small copper clusters, (a) Single collision chemiluminescent spectrum; (b) Multiple collision chemiluminescent spectrum obtained with room temperature water cooling; (c) Chemiluminescent spectrum obtained using dry ice/methanol cooling; and (d) Chemiluminescent spectrum obtained using liquid nitrogen cooling. Systems II, III, IV, and V are attributed to Cu_xO species.

systems, being largely dominated by features arising from Ag_2. The reactions or energy transfer producing the Ag_2 emission have not been clearly established. However, multicentered processes such as

$$Ag_4 + O_3 \longrightarrow Ag_2^* + AgO + AgO_2 \tag{9}$$

$$Ag_5 + O_3 \longrightarrow Ag_2^* + AgO + Ag_2O_2 \tag{10}$$

represent the most likely candidate reactive encounters, whereas the energy transfer

$$AgO^* + Ag_2 \longrightarrow Ag_2^* + AgO \tag{11}$$

or the reaction–dissociation sequence

$$\begin{gathered} Ag_3 + O_3 \longrightarrow Ag_2O^* + AgO_2^* \\ Ag_2O^* \longrightarrow Ag_2^* + O \end{gathered} \tag{12}$$

are also possible sources of the dimer excited electronic state (Fig. 3).

There is some evidence that this result can be generalized and the reactions of large metal clusters can be used as a source of electronically excited smaller clusters. The oxidation of alkali metal trimers by halogen atoms

$$\begin{gathered} M_3 + X \longrightarrow M_2^* + MX \\ M = Na, K; X = Cl, Br \end{gathered} \tag{13}$$

Table 2
Molecular Constants Determined for CrF Formed in the Chemiluminescent Reactions of Chromium Atoms and Dimers with Molecular Fluorine

State	T_e	ω_e	$\omega_e\chi_e$
$X^6\Sigma$	0	622.3 (*23*)	3.4 (*9*)
$A^6\Sigma$	9957.6	577.0 (*13*)	0.6 (*4*)
$B^6\Sigma$	16893 (*10*)	605.2 (*20*)	1.8 (*20*)
$C^6\Pi$	~20600	~580 (*10*)	—
$D^6\Sigma$	31705 (*15*)	692 (*6*)	1.8 (*20*)

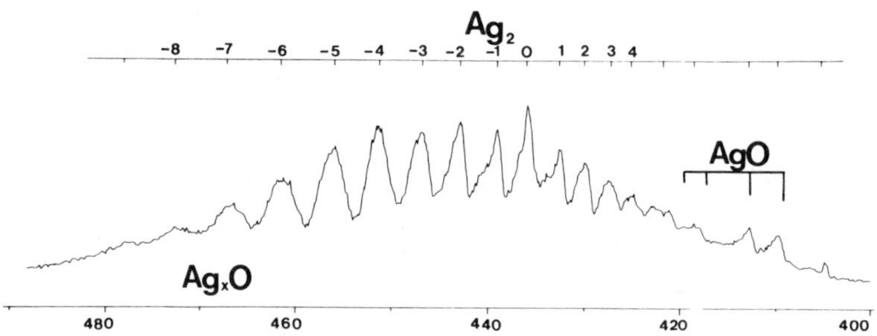

Fig. 3. Portion of the chemiluminescent spectrum observed for the oxidation of Ag_x clusters with ozone.

has now been well investigated (*12,13*). The oxidation of copper clusters at relatively high pressures (~700 Pa) with ozone produces emission which has been correlated with the A-X and B-X band systems of Cu_2. However, no middle transition metal dimer emissions have yet been identified.

The oxidation of moderately sized metal clusters also produces novel metal rich molecules. Experimental conditions that generate emission spectra for these molecules have been established for the reaction systems $B_x + NO_2$, (*2,3*), $Mn_x + O_3$ (*2,7*), $Cu_x + O_3$ (*2,4*), $Cr_x + F_2$ (*8*), and $Ag_x + O_3$ (*2,5,6*). An analysis of product emission spectra has provided the first quantal information on the energy levels and optical signatures of these metal atom clustered molecules.

When a moderate flux of boron vapor entrained in argon is cooled to dry ice temperatures and reacted with NO_2, a new electronic transition is observed (*2,3*). A comparison of the vibrational frequencies obtained from the analysis of this spectrum to the vibrational frequencies calculated for B_2O and B_2O_2 using SCF-CI calculations (*14–16*) clearly shows that this molecule is BBO, and not the thermodynamically more stable BOB isomer. This result is significant for two reasons. First, the products of the reaction (at least B_2O) are kinetically controlled. Through their study, we begin to probe the dynamics of boron agglomerate reactions.

Second, the product contains a metal–metal bond. This suggests that metal cluster reactions may provide a way to study thermodynamically less stable, but kinetically important isomers that are not easily produced using currently established laboratory techniques. It is reasonable to suggest that if large enough molecules of this type are produced, their properties can be used to provide a reasonable model for oxidant surface interactions.

Although the data is not as straightforward, there is evidence that other systems behave in a manner analogous to the boron system. The oxidation of manganese clusters by O_3 also produces a molecule that contains a metal–metal bond (*7*). The vibrational analysis of this spectrum clearly shows ground state vibrational frequencies of 534 cm^{-1} and 306 cm^{-1}. The 306 cm^{-1} frequency is consistent with a frequency estimated for Mn_2^+. Since the 534 cm^{-1} frequency is similar to that metal oxygen stretch associated with the alkali metal oxides (*17*), this molecule is assigned as Mn_2O bonded as $(Mn_2)^+ O^-$. The geometry of this molecule has not been firmly established, but the presence of the Mn–Mn bond seems certain (*see* Fig. 4).

The $Ag_x + O_3$ reaction also produces a product molecule emitter that has tentatively been assigned as Ag_2O. Three vibrational frequencies (440 cm^{-1}, 160 cm^{-1}, and 250 cm^{-1}) established for this molecule indicate that it contains an Ag–Ag bond. The 440 cm^{-1} frequency is readily assigned to the Ag-O stretch. Since the geometry of the molecule emitter has not been established with certainty, definite assignments for the two low frequency modes have not been established. However, either the 160 cm^{-1} or the 250 cm^{-1} frequency could reasonably be assigned as an Ag-Ag stretch. Since only one bending mode is expected, the two frequencies in this region are strongly suggestive of a silver-silver bond in the molecule.

The $Cu_x + O_3$ reaction provides evidence that emission from the metal clustered oxides can be observed. Emission features associated with Cu_2O, Cu_3O, and/or Cu_4O have been identified (*4*). Unfortunately, the emission spectra do not contain the detail needed to determine the structure of these molecules. However, the experiments do indicate that larger metal clustered oxides can be produced using the techniques which we have outlined herein.

CONCLUSIONS AND FUTURE EXTENSIONS

The characteristics of the metal cluster oxidation reactions summarized in this outline lead us to four generalized observations.

The oxidation of larger metal clusters can be more energetic than the corresponding reactions of a smaller cluster or metal atom. This trend holds well for reactions of atoms, dimers, and trimers, but it is expected to level off considerably as one considers the reactions of larger clusters. Extensions of this work to larger clusters will be needed before more specific rules can be formulated.

Table 3
Comparison of Observed and Calculated Vibrational Frequencies for the Isomers of B_2O

Observed	B-O-B	B-B-O
Frequencies	(CALC)	(CALC)
1970 (?)	1510[a]/1430[b]	1935[a]/1856[b]
—	77/—	282/—
~582	1005/926	581/579

[a]Scaled values from Ginn, H., Jones, Li, and Shillady, D., private communication.

[b]Scaled values from Zyubina, T. S., Charkin, O. P., Zyubin, A. S., and Zakzhevokii, A. S., *Zh. Neorg. Khim* **27**, 558 (1982).

Table 4
Observed Emission Systems for Ag_xO and Cu_xO

Ag_xO				
Transition	T_o	$\omega_1''(cm^{-1})$	$\omega_2''(cm^{-1})$	$\omega_3''(cm^{-1})$
A – X	~15670		~163	
B – X	~19766	442	165	(256)

Cu_xO		
Assignment	T_o	Frequency observed (cm^{-1})
Cu_2O	17092	~121
Cu_3O or Cu_4O	14525	~102

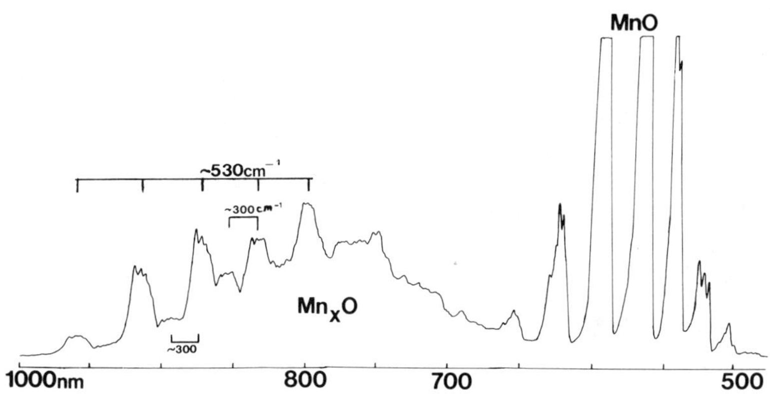

Fig. 4. The chemiluminescent spectrum observed for the oxidation of Mn_x clusters with ozone. The chemiluminescence from the MnO band system results from the oxidation of Mn atoms.

The energy of a cluster reaction is often largely channelled into only one of the reaction products. The emission from electronically excited molecules produced in this manner can be used to obtain previously inaccessible optical signatures for the excited states of these molecules. Thus, these studies serve to better define the molecular electronic structure of the product molecules.

The products of the oxidation of larger clusters contain cluster agglomerate bonds. As the cluster reacts, it usually fragments. However, as the reacting clusters become larger, their reaction should facilitate the formation of a greater metal agglomeration in those product molecules formed. The studies thus far have not established if there is a direct relationship between the relative sizes of the clusters that undergo reaction and the size of the oxidized cluster products. The current investigations need to be extended to larger clusters to determine whether the observed trends will continue. However, the current observations hold the promise that larger cluster oxide species can be made and explored using this technique. In most cases, the products of these cluster oxidations contain metal–metal bonds. Thus, the distinct possibility exists that the internal mode structure associated with the reactions of cluster molecules and their products may aid in the modeling of surface behavior.

Reaction kinetics is a primary determinant of the products that are formed in these systems. The characterization of the products contributes information useful to the exploration of the mechanisms for cluster oxidation. Together with recent elegant mass spectrometric kinetic investigations of cluster reactions and recent advances in theory, this information is useful for the understanding of reaction pathways and the energetics of cluster reactions.

The study of the spectroscopy of metal cluster compounds formed as the products of metal cluster oxidation reactions is still in its infancy. The work discussed here establishes the usefulness of these investigations. Much future work, particularly involving high resolution techniques, will be needed before the bonding, molecular geometries, and reactivity of these compounds is totally established. Also, it is desirable to continue these investigations using larger clusters to determine if the observations given above hold in general. We also wish to discern whether larger cluster compounds can be synthesized and characterized in order to learn more about those reactive systems which more closely approach the bulk surface. This will require a modification of the furnace source to permit the synthesis of larger clusters. These modifications are currently in progress.

ACKNOWLEDGMENTS

TCD would like to thank the NSF-RUI program (CWE-8511710) for supporting this work. JLG would like to thank the National Science Foundation (CHE-8604471) and the ACS-PRF (PRF 17322-A6) for support. Also, without the capable assistance of J. R. Woodward, S. H. Cobb, and M. McQuaid from Georgia Tech and T. R. Burkholder, C. Pettyjohn, K. Morris, M. Cuttino, B. Graham, and R. Gordon from JMU, much of the work outlined here could have not have been accomplished.

REFERENCES

1. For example, *see* (a) *Metal Clusters,* Moskovits, M. ed., Wiley, New York (1986), (b) J. Phys. Chem. Symposium Issue, Vol. 91, No. 1 (1987). (c) *The Physics and Chemistry of Small Clusters,* Jena, P., Rao, B. K., Khana, S. N., eds., NATO ASI series B: Physics 158, Plenum, New York (1987).
2. Woodward, R., Le, P. N., Temmen, M., and Gole, J. L., *J. Phys. Chem.* **91,** 2637 (1987).
3. DeVore, T. C., Woodward, J. R., and Gole, J. L., *J. Phys. Chem.* **92,** 6919 (1988).
4. DeVore, T. C., Burkholder, T., Woodward, J. R., and Gole, J. L., *High Temp. Sci.,* submitted.
5. Gole, J. L., Woodward, J. R., Hayden, J. S., and Dixon, D. A., *J. Phys. Chem.* **89,** 4905 (1985).
6. DeVore, T. C., Woodward, J. R., Le, P. N., and Gole, J. L., *J. Phys. Chem.* submitted for publication.
7. DeVore, T. C., Woodward, J. R., and Gole, J. L., *J. Phys. Chem,* submitted for publication.
8. DeVore, T. C., McQuaid, M., and Gole, J. L., *J. Phys. Chem.,* submitted for publication.
9. For example, *see* (a) *Proceedings of the Symposium on High Temperature Metal Halide Chemistry,* Hildenbrand, D. L., Cubicciotti, D. D., eds., The Electrochemical Society, Princeton (1978), (b) Cheung, A. S-C., Zyrnicki, W., and Merer, A. *J. J. Molec. Spectrosc.* **104,**315 (1984); (c) *Gas Phase Chemiluminescent and Chemiionization,* Fortijn, A., ed., Elseuien Science Publishers, 1985.
10. Based on Data from Suchard, S. N., *Spectroscopic Data,* Vols. 1 and 2; IFI/ Plenum, New York (1975).
11. Dubov, V. M. and Shenyauskaya, E. A., *Opt. Spectrosc. (USSR)* **62,** 195 (1987).
12. Cobb, S. H., *Gas Phase Characterization of the Molecular Electronic Structure of Metal Clusters and Metal Cluster Oxidation,* Ph.D. dissertation, Georgia Institute of Technology, 1988.
13. Cobb, S. H., Woodward, R., and Gole, J. L., *Chem. Phys. Letters* **143,** 205 (1988).
14. DeKock, R. L., and Barbachyn, M. R., *J. Inorg. Nucl. Chem.* **43,** 2645 (1981).
15. Zyubine, T. S., Charkin, O. P., Zyubin, A. S., and Zakzhevokii, A. S., *Zh. Neorg. Khim* **27,** 558 (1982).
16. Ginn, H., Jones, L., and Shillady, D., private communication.

Freezing and Melting of Metallic and Salt-Like Clusters

R. STEPHEN BERRY,* HAI-PING CHENG, AND JOHN ROSE

Department of Chemistry and the James Franck Institute, The University of Chicago, Chicago, IL 60637

ABSTRACT

Simulations have been carried out of small clusters of KCl and of Cu, by means of classical, isoergic molecular dynamics, in order to study the stability of coexistence of phase-like forms. To study Cu clusters, it was necessary to generate relatively reliable but efficient potentials, which was done by iterating efficient embedded-atom potentials against comparisons with more reliable but much less efficient density functional potentials. Born-Mayer-Rittner potentials without polarization were used for KCl. Both solid–liquid and solid–soft solid transitions were found. No sharp bimodal temperature distributions were found for coexisting Cu_{13} solid-like and liquid-like clusters, but the highly asymmetric temperature distributions actually found would not preclude such coexistence.

INTRODUCTION

Clusters offer a tantalizing potential for creating new materials and new reactions, as recognized by this session of the Conference. But to realize that potential, we shall have to understand and control the species we make. This means that one of our important tasks is learning the forms clusters may assume, and then finding how these forms differ in their behavior. Some examples illustrate the issues. One is the prediction from several early simulation studies that clusters at least of rare gas atoms, may exhibit solid-like and liquid-like forms (*1*). Another is the evidence that clusters of a given size may have several locally stable geometric forms, and that what these forms are depends on the number N of particles comprising the cluster (*1–3*). A third is the observed differences (*4*) in reactivity of clusters of similar but not identical size,

*Author to whom all correspondence and reprint orders should be addressed.

presumably reflecting how the forms of clusters depend on *N*. Here, we focus on the nature of the solid-like and liquid-like forms of clusters, and of the equilibrium between them. To do this, we emphasize results originating in general arguments regarding densities of states, but which are derived by simulations, largely by molecular dynamics (MD) methods. The discussion following is a progress report on the extension of our own earlier work (*1,2*) on rare gas clusters to high-temperature materials, notably to alkali halides and metals.

To carry out simulations by molecular dynamics, one must have an effective potential energy function, a surface, for the system one is describing. In the case of alkali halides, rather good potentials are available (*5,6*), based on Born-Mayer-like interactions, representing polarizable ions with rather hard-core, short-range repulsions in addition to the long-range attractive forces. Representing clusters of metal atoms is a different kind of problem. It has been very difficult to obtain accurate representations of their potentials in terms of pair potentials, even when supplemented by single-particle, density-dependent terms. Most attempts to construct effective potentials for metal clusters have fallen back onto quantum mechanical solutions of the electronic Schroedinger equation at some level of approximation. This might be a simple spherical shell model, or an elaborate multiconfiguration calculation, but the approaches have been largely through the electronic structure. The consequence has been that only a very few molecular dynamics calculations have been done for metal clusters (*7,8*); most of the calculations have been restricted to the regions around the (often presumed) lowest minima on the potential surface. We have taken a pragmatic approach, not so ambitious as the elaborate scheme of Car and Parinello (*9*) nor as simple and economical as that of Sawada and Sugano (*8*), but an intermediate way that passes back and forth in a series of iterations between dynamics done with simple potentials and more accurate calculations done in very restricted regions of the surface. The next section describes how the potential for the copper cluster was derived and also what potentials were chosen for the alkali halides. The third section describes the results of our preliminary simulations of alkali halide clusters and compares these results with the earlier results of others. The fourth section describes the results of our simulations of copper clusters of several sizes and compares those with analogous simulations of argon clusters. The final section summarizes the results and makes some general inferences.

THE POTENTIALS

For the alkali halides, we have used the potentials of Welch et al. (*6*). These were chosen in preference to those of Brumer and Karplus (*5*) in large part because we wish to compare our results with those of other simulations, most of which have used those of (*6*).

For the copper clusters, we used Cu_6 as the basis for selecting parameters of the potential and larger clusters for verifying those parameters. To fix the potential, we iterated between two methods, the embedded atom (*10,11*) (EA) and the discrete variational (DVM) version (*12,13*) of the Xα method, a local-density formalism that has been applied previously to the determination of structures and binding energies of copper clusters (*14*). The EA potential is a sum of two-body terms and effective many-body terms that are represented by charge-density-dependent one-body terms. The first trial values of the parameters of the EA potential were taken from a prior computation of bulk copper (*11*) and molecular dynamics calculations were carried out with that EA potential. The molecular dynamics simulation was used at that stage to find first approximations to all the potential minima and saddles, the latter by the method of "slowest slides" introduced recently (*15*). Then DVM calculations were carried out in the vicinity of the lowest potential minimum, a regular octahedron, particularly along the direction of the symmetric stretching mode of the cluster. The results of these calculations were next used as the energy curve to which a second, refined set of EA parameters was fit. The Cu-Cu potential based on these parameters was then used to determine the minima of the potentials for Cu_7, Cu_8, and Cu_9. Two minima and a saddle were found for Cu_6, with energies, respectively, of −3.00, −2.970 and −2.969 eV/atom. Previously, dynamics of 6 and 7-atom metal clusters had been simulated (*8*) with the Gupta potential, a sum of exponentials.

The results were structures very close in geometry to the minimum-energy structures for Cu_8 and Cu_9 found by the DVM (no DVM calculation was done for Cu_7), and the energies around the minima were also as well represented for the larger clusters by the refined EA potential as was the shape of the Cu_6 surface. Further refinements are planned, based on geometries at the saddles found with EA functions and the energies at those saddles found by the DVM.

THE SIMULATED ALKALI HALIDE CLUSTERS

Previously, extensive simulations have been done to determine structures of alkali halide clusters (*3,16*). These gave results showing strong variation of equilibrium geometry with the number of alkali halide ion pairs in the cluster, and decided exceptional stability for those numbers from which cubic or rectangular parallepiped structures can be made. Simulations done to probe dynamics were done for NaCl (*3*); these showed that some sizes exhibit transitions like solid-solid, softening transitions, and at higher energies, melting like that seen (*1*) for Ar_{13}. One example, $(NaCl)_{108}$ at an energy equivalent to 811 K, was quoted as showing liquid behavior on one side and solid behavior on the other. This last situation is especially interesting for us because it appears to exemplify a liquid that does not wet its corresponding solid.

We felt that the dynamics of alkali halides other than NaCl needed to be examined, partly to get a sense of the generality of the conclusions drawn in (*3*), partly to probe further into the wetting-nonwetting question and partly to lay the groundwork for other studies that deal with sintering. Here we report some of our first findings for KCl clusters, specifically $(KCl)_4$ and $(KCl)_5$. These were chosen because the former has as a lowest-energy form a "cube" with one ion at each corner, whereas the latter has no such magic-number geometry.

The easiest way to see the outcomes of the simulations is to look first at the caloric curves of $\langle T \rangle$ vs E, then at the distribution functions of short-time (500-step) averages of $\langle T \rangle$, and then at the properties of the cluster that account for the effective temperatures and their distributions. The caloric curve for $(KCl)_4$, Fig. 1, based on the effective temperature average over an entire long run appears at first sight to be qualitatively like that of the argon trimer (*17*). There is a low-temperature region in which the slope, the heat capacity at a constant pressure of zero, is a straight line and the cluster is solid-like. There is a high-temperature region, also with a nearly constant slope but displaced to higher energies than the solid-like region. Between them is a region of negative slope connecting the two extreme regions. The intermediate region of E and $\langle T \rangle$ for some clusters, such as Ar_7 and Ar_{13}, is the domain of coexisting solid-like and liquid-like clusters (*1*). (In those two examples, the slope of the caloric curve is everywhere positive, but for Ar_3 it is negative there.)

For $(KCl)_4$, the region linking the lowest and next-lowest branches of the caloric curve is a region of two coexisting solid-like forms of this cluster. This becomes clear when one constructs the curves of the mean square displacement of an ion as function of time, as shown in Fig. 2. The diffusion process simply does not carry an ion everywhere in the cluster when the energy of the cluster is of order -2.92 eV per ion or -23.4 eV; on a time scale of about 10^{-12} s, the mean square displacement is only about 3 $Å^2$ and at longer times, this curve oscillates around a value that seems constant on the picosecond time scale. At an energy slightly higher, -2.83 eV per ion, the mean square displacement reaches a level value of about 20 $bohr^2$ in a picosecond; this corresponds to traversal of the entire cluster and to liquid-like behavior. The process associated with the "transition" at about -2.95 eV/ion is the opening of an isomerism between the lowest-energy structure, Fig. 3a, and other structures, including the rectangular form, Fig. 3b, and the sort of octagon, Fig. 3c. The cohesive energies of these structures are -24.97, -24.50, and -24.53 eV, respectively. A fourth structure was reported (*3*) for $(NaCl)_4$ but has not been seen for $(KCl)_4$ in this work. These structures are, of course, the geometric forms of the $(KCl)_4$ cluster at the minimum energies of the potential wells on the effective energy surface; the clusters vibrate within these wells and pass among them, spending virtually no time in the geometries of the minima.

It seems that the saddles separating the "octagon" structure and the other two, the "cube" and the "rectangle," are all relatively low. If one of the minima were separated from the others by high saddles, isoergic simulations based on increasing the energy in steps from the lowest-energy structure could reach regions in which some basins allowed on

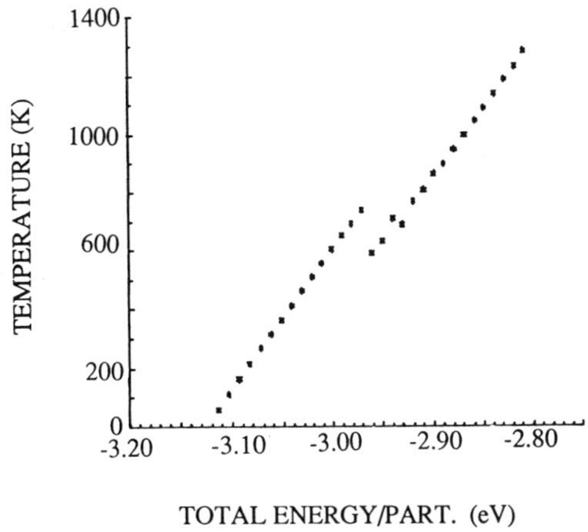

Fig. 1. The caloric curve for $(KCl)_4$ based on molecular dynamics (MD) simulations for clusters at constant energy. The smooth, continuous curve is based on long-time averaging of the kinetic energy to fix the mean temperature. The two terminating branches are based on the short-time average temperatures within each peak of the temperature distribution in the region of energy where it is bimodal.

the basis of total energy would be unreachable by any classical path. For such cases, caloric curves based on isothermal simulations would be expected to differ from those based on isoergic simulations carried out as ours were. However, whether this occurs with any alkali halide cluster will be answered only when the corresponding saddles have been identified. If there are "puddles" on the potential surface, one might think it possible to use several different initial points for the isoergic simulations in order to develop caloric curves from microcanonical ensembles or Hamiltonian systems that are equivalent to the curves developed from canonical ensembles. However, there is a difficulty in doing this: One does not know without a detailed analysis of the available volumes of the phase space in the unconnected regions how to weight their contributions.

The distributions of mean kinetic energies, that is, of mean temperatures, for $(KCl)_4$ for several energies are shown in Fig. 4. At low energies, these are unimodal and at very low energies, not at all like Gaussian distributions (a), but become Gaussian (b), and then unsymmetrical and then bimodal (c) at temperatures in the region of coexisting isomers., At still higher temperatures, the curves become Gaussian again (d). In contrast to some of the argon clusters, this cluster does not show bimodality between solid and liquid forms, but between regular and less regular, soft solid structures. It seems likely that the clusters are so nonrigid in the energy range of coexisting solid-like isomers that the small increment required to make such clusters liquid-like is enough to facilitate frequent passages among all the wells so that only the liquid form is seen at the higher temperatures.

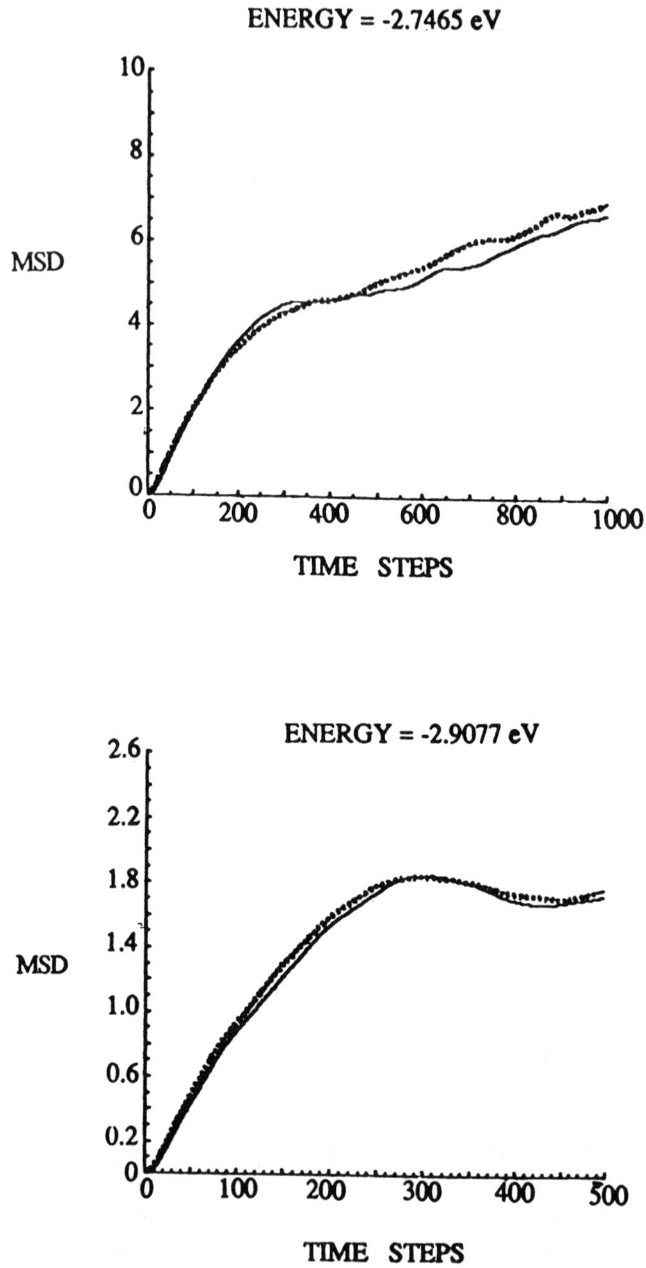

Fig. 2. The mean square displacements of the K^+ (solid curve) and Cl^- (dotted curve) ions of $(KCl)_4$ at energies of -2.92 and -2.73 eV per ion or total energies of -23.36 and -21.84 eV, corresponding to temperatures of approximately 800 and 1200K, respectively. One unit on the time scale is one time step, 3×10^{-15} s.

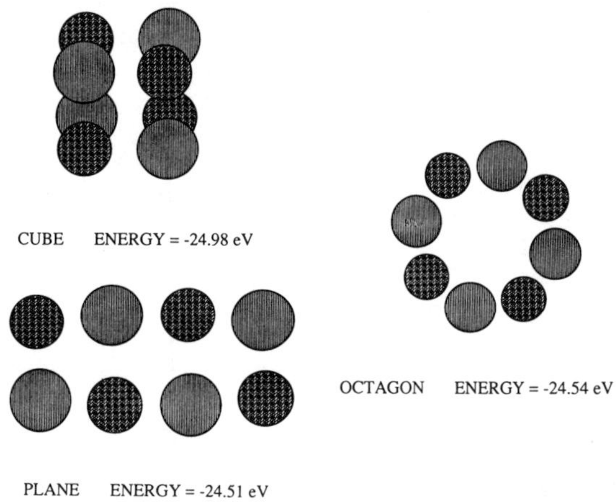

Fig. 3. Structures of the stable forms of $(KCl)_4$ found in these simulations. The energies of the cluster in these forms are −24.97, −24.53, and −24.50 eV for the "cube," "octagon," and "rectangle," (a), (b), and (c).

One diagnostic used for clusters is δ, the root-mean-square nearest neighbor distance. As Fig. 5 shows, this quantity seems to satisfy the Lindemann criterion very nicely, staying well under 10% of the interionic spacing for low energies at which the cluster is a "cube," and rising rapidly to about 10% in the transition region. However, the rise starts well below the energy at which the liquid appears, at the energy of appearance of the "rectangular" isomer. In other words, this criterion may sometimes be weaker or more ambiguous in its meaning than has been supposed.

The $(KCl)_5$ cluster is rather different from its $(KCl)_4$ homologue. The lowest-energy structures are shown in Fig. 6; their energies are, respectively, −31.25 and −31.12 eV, considerably closer than the lowest two minima of the tetramer. The caloric curve of $(KCl)_5$ is shown in Fig. 7; this shows two regions of nonconstant slope, the lower around −2.96 eV/ion and the upper, around −2.71 eV/ion. Even as low as −3.0 eV/ion, the distribution of temperatures is bimodal and remains so for most of the range up to about −2.85, about halfway between the two wiggly parts of the caloric curve. The curves of mean square displacement of an ion as a function of time, Fig. 8, shows how this cluster exhibits solid-like rigidity at an energy of −3.01 eV/ion, softness and slow diffusion at −2.95, and, like $(KCl)_4$ illustrated by the higher-energy curves of Fig. 2, two distinct time scales at −2.91 and −2.85 eV/ion. For the pentamer, these occur with fast motion on a distance scale of about 2.2Å and time scale of about 400 time steps of 3×10^{-15} s, and slow diffusive motion on longer distance and time scales. At about −2.85 eV/ion or even lower energies, the cluster seems to be liquid-like. The interruption in the caloric curve at about −2.71 eV/ion is associated with vaporization.

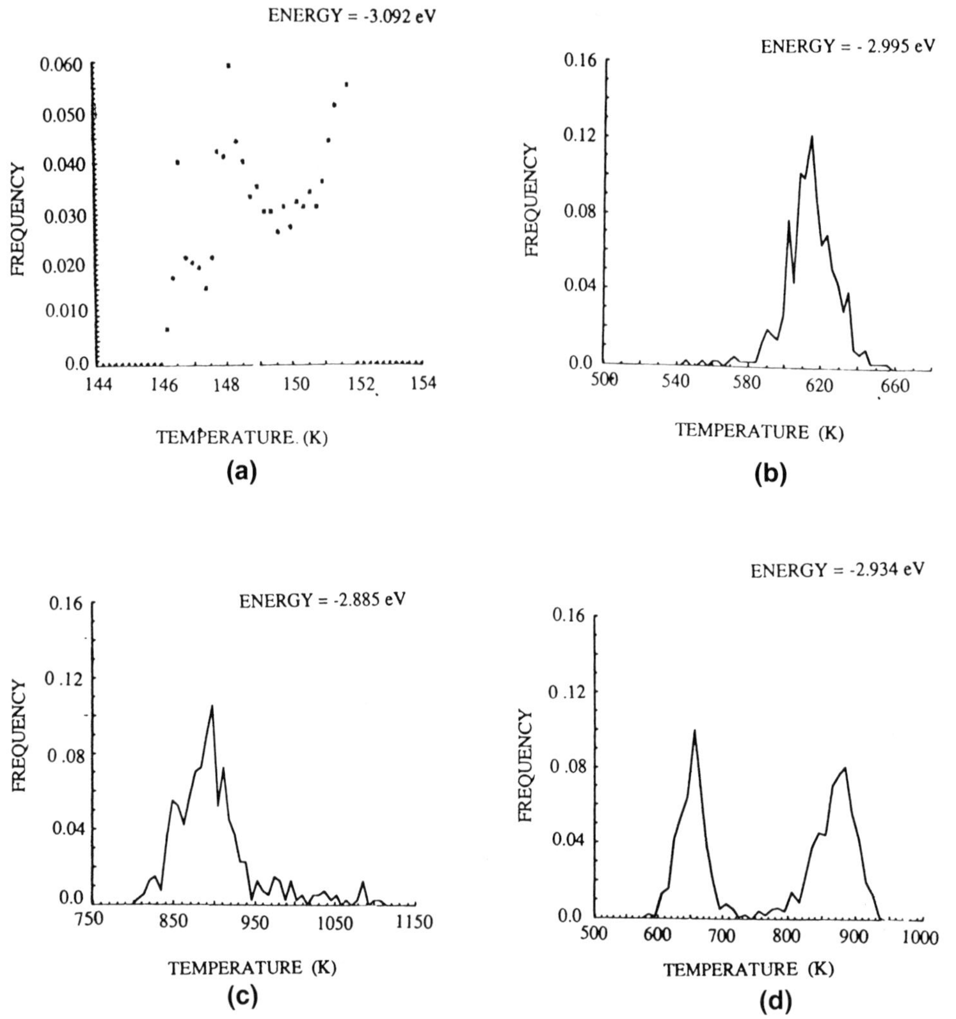

Fig. 4. Distributions of mean kinetic energies for $(KCl)_4$ at energies of -3.09, -2.99, -2.94, and -2.884 eV per ion, or totals of 24.7, 23.9, 23.5, and 23.07 eV for (a), (b), (c), and (d), respectively.

COPPER CLUSTERS

We report here on simulations of Cu_7 and Cu_{13}. The results from simulations based on the once-iterated embedded atom potentials developed for Cu_6 are strikingly like those of the corresponding argon clusters based on Lennard Jones 6-12 potentials. The reason is that the seven-particle cluster is the smallest that seems to show unambiguous liquid-like behavior, as least for Lennard-Jones clusters, and the thirteen-particle cluster is the paradigm and most-studied small cluster, therefore most suited for comparison of different potentials.

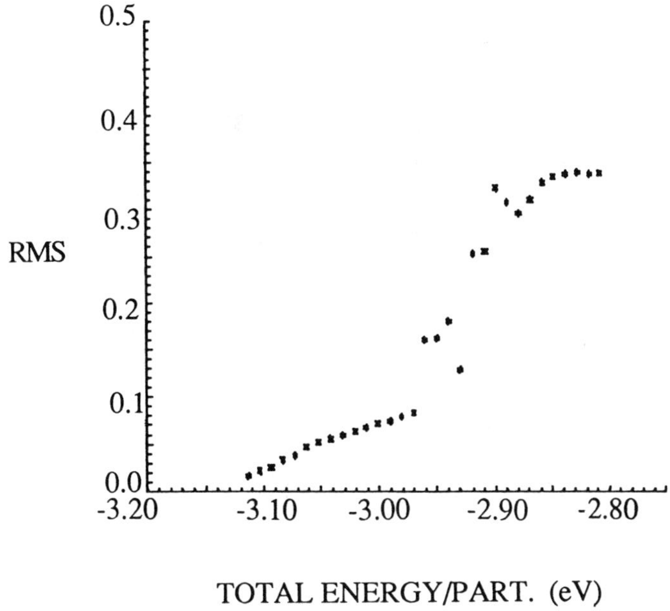

Fig. 5. The mean displacement δ from equilibrium of the ions in $(KCl)_4$ as functions of temperature. The step corresponds closely to illustrating the Lindemann criterion of 10% as the value of δ at the solid–liquid transition—except that the new phase is not liquid-like at the energy of the threshold.

At this stage, Cu_7 was studied for its structures and potential surface; Cu_{13} was studied primarily for dynamics. Like Ar_7 and the hypothetical metal simulated by Sawada and Sugano (*8*) the Cu_7 cluster has four kinds of potential minima, of which the lowest has the D_{5h} symmetry and structure of a pentagonal bipyramid. The order of the next states is also that of Ar_7 and the Gupta-potential 7-cluster. The next-to-lowest equilibrium geometry is that of an octahedron with one face capped, giving the structure C_{3v} symmetry. The third structure has symmetry C_{3v}, a tricapped tetrahedron, and the highest has only C_2 symmetry. These are all shown in Fig. 9. The energies of these structures are −3.05, −3.03, −3.014, and 3.011 eV/atom.

Dynamics were carried out for the Cu_{13} cluster over a range of temperatures. The geometry of minimum energy is, like the Lennard-Jones cluster, an icosahedron. The caloric curve for this cluster is shown in Fig. 10. Like Ar_{13}, the curve has a long, solid-like region at low energies and a liquid-like region at higher energies, and the two are separated by a transition region that is flatter but still monotonically increasing. The melting transition occurs in the region of 650 K, about half the bulk melting temperature of 1351 K.

The geometry of the lowest-energy form of Cu_{13} is the regular icosahedron shown in Fig. 11a. Its energy is −3.285 eV/atom or −42.71 eV total. The next-lowest structures are those of Fig. 11b, c, and d, with one atom removed from the octahedral shell and placed on a face of the

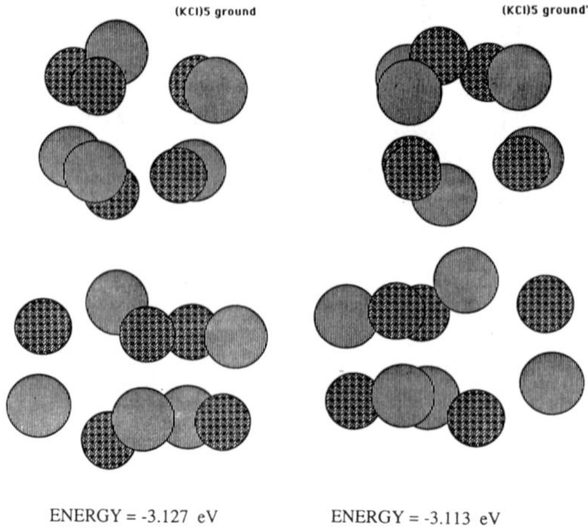

Fig. 6. Two views of each of the two lowest-energy structures of $(KCl)_5$.

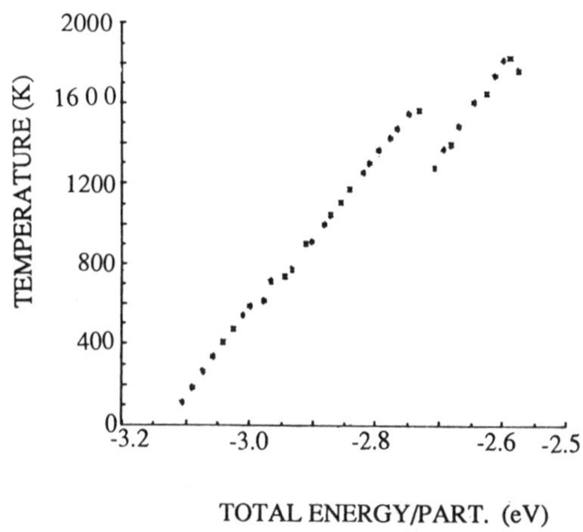

Fig. 7. The caloric curve for the $(KCl)_5$ cluster from about -3.0 to -2.6 eV/ion, showing two "unstable" regions. The one at higher energy is associated with vaporization; that at lower energy is associated with the onset of a solid–solid interconversion.

icosahedron. There are three geometrically distinct kinds of such faces and all three are shown. The energies are almost identical: the lowest of these, Fig. 11b, is -3.233 eV/atom or -42.03 eV total; the energies of the next two are equal, -3.2326 eV/atom in our simulation. (The significance of all the figures is only in how close the energies are.) These are the same as the lowest-energy geometries (2) of Ar_{13}.

The solid-like and liquid-like character of the low-energy and high-energy forms of the cluster are exhibited by the mean square displacements of the atoms. Two examples are shown in Fig. 12, for energies corresponding to 627 and 711K. The curve for 627 K is essentially a

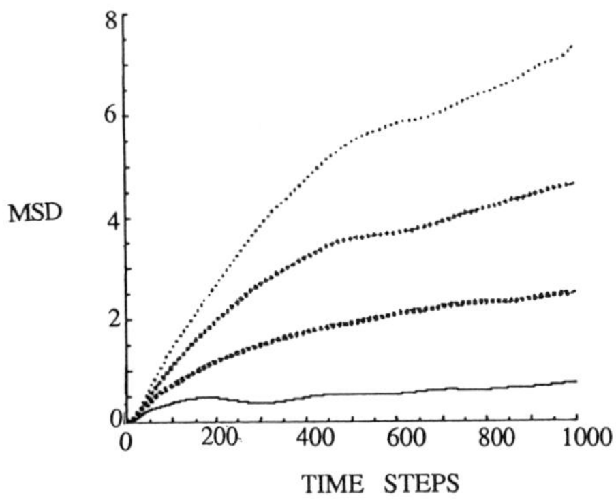

Fig. 8. Four curves of the mean square displacement of an ion in $(KCl)_5$ at four different energies; from bottom to top, −3.02, −2.96, −2.92, and −2.86 eV/ion. Note the two distinct slopes in the two upper curves, corresponding to fast local processes and slow diffusive processes.

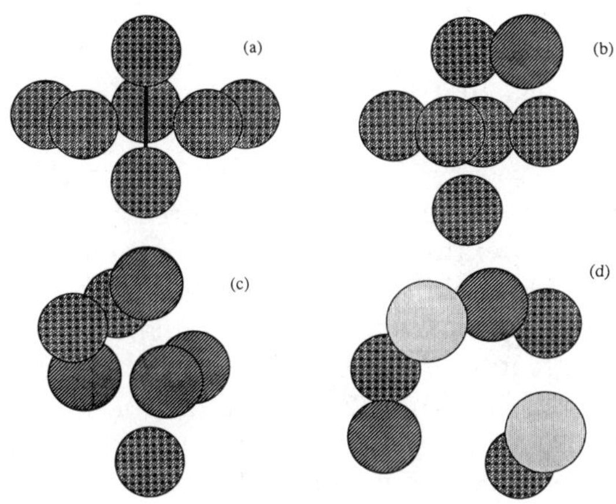

Fig. 9. The structures of the four locally stable forms of the Cu_7 cluster based on the once-iterated embedded-atom potential. The lowest-energy structure is a pentagonal bipyramid. The next has C_{3v} symmetry, corresponding to an octahedron with one capped face, the cap being the top atom in (b); the third also has C_{3v} symmetry and the highest, C_2. The next-highest corresponds to a tetrahedron with three capped faces, or a trigonal bipyramid with two capped faces.

horizontal line with small fluctuations around a mean of approximately 0.005 Å (2). The corresponding diffusion coefficient is zero. At the higher temperature of 711 K, the mean square displacement slopes, in the mean, upward, exhibiting an average slope of 6.4×10^{-6} cm²/s. The root-mean-square nearest neighbor distance δ is shown in Fig. 13. This function rises somewhat more gradually with mean temperature than its counterpart for Ar_{13}, but liquid-like behavior does set in where δ reaches about 10%, the Lindemann criterion value.

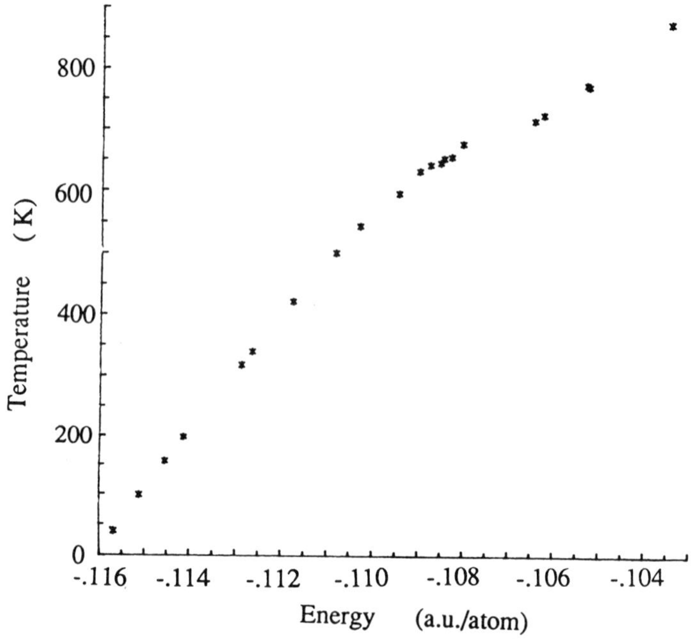

Fig. 10. The caloric curve for Cu_{13}.

The final set of data we present here consists of the distributions of short-term-mean temperatures for Cu_{13} at various energies, Fig. 14. At low temperatures, we see a phenomenon that should not be surprising but has apparently not been reported for simulations previously. This is a sharp cutoff at high temperatures, i.e., at high kinetic energy, that is presumably associated with the restricted vibrations available to the cluster at low energies. The cluster's vibrations can carry it up only to a maximum kinetic energy equal to its total energy, and the dynamics in this cluster seem to concentrate the clusters probable position into the region not far from that deepest part of the potential surface. Furthermore, whereas the distributions for higher energies, in particular in the transition region, are very asymmetric, the Cu_{13} cluster as we have simulated it has not displayed a bimodal distribution of temperatures. This result is sufficiently different from what was expected, based on the example of Ar_{13}, that it will be investigated further.

SUMMARY AND FUTURE ISSUES

Simulations of salt-like and metal clusters show softening, solid-like to solid-like changes with two or more forms coexisting over ranges of temperature, like chemical isomers, in addition to solid–liquid transformations. The changes seem to be less sharp for these substances than for

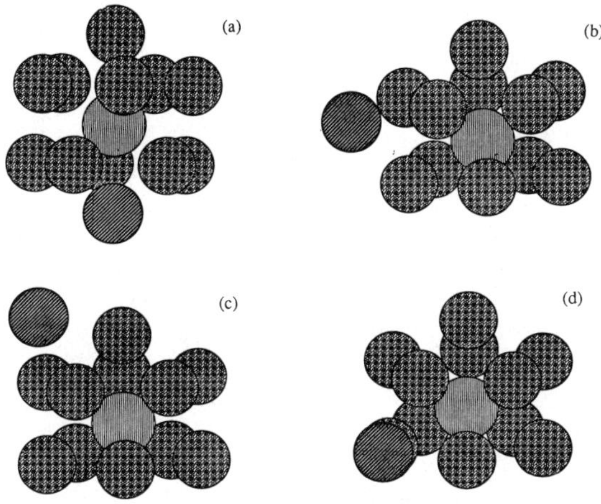

Fig. 11. The four lowest-energy geometries of Cu_{13}. The atoms are shown as inequivalent only to clarify the geometric relationships. The central atom of the icosahedron of (a) is shaded there and its counterpart is similarly shaded in the other three figures. The lowest atom of (a) is the most darkly shaded; the other three figures are arranged so that the lowest icosahedral site in these figures is vacant. In (b), (c), and (d), the atom which occupied that lowest site in the most stable structure occupies each of the three kinds of surface faces available to the promoted atom.

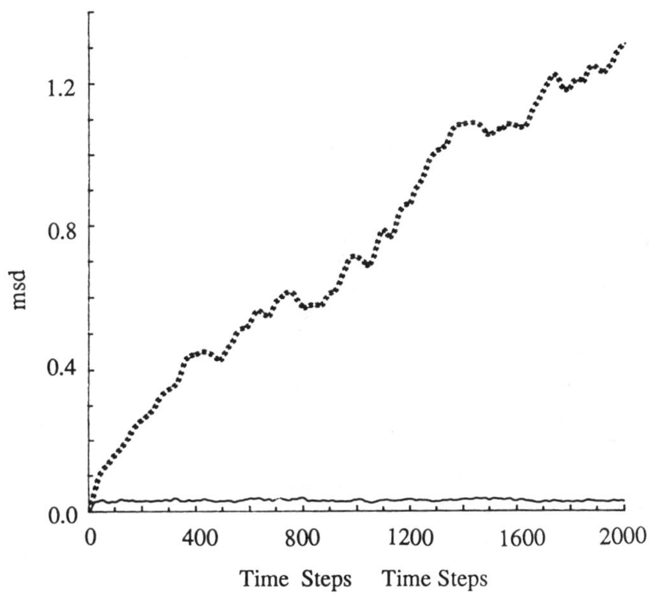

Fig. 12. Two curves of mean square displacement of a Cu atom in Cu_{13}. The lower, nearly horizontal curve is drawn for an energy corresponding to a mean temperature of 627 K; the upper, dotted curve is drawn for 711 K.

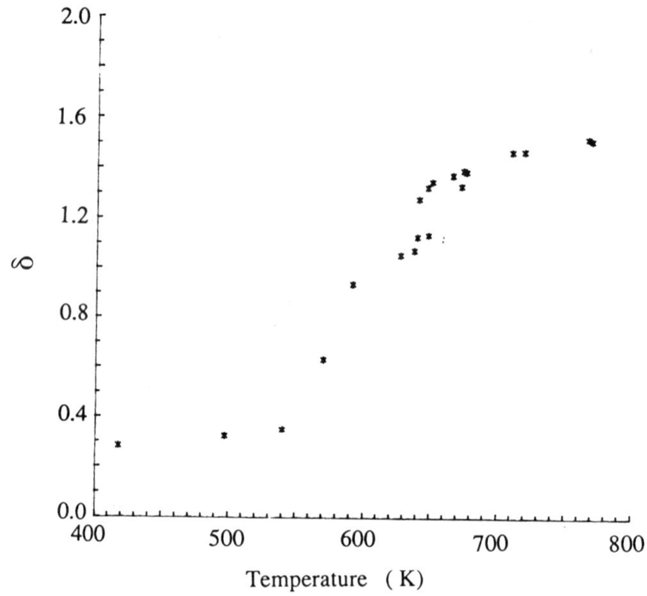

Fig. 13. The rms fluctuation of the nearest-neighbor distance of "bond length" for Cu_{13}, as a function of temperature.

rare-gas (Lennard-Jones) clusters, perhaps because of the longer ranges and softer character of the potentials of salts and metals.

Questions that remain to be studied even for small clusters of metals and salts are the establishment of the saddles as well as of the minima on their potential surfaces and the connection of that information to the "phase coexistence" phenomenon in such clusters. How sharp are the limits of stability for any one form? What are the time scales for passage from one phase-like region of the potential surface to another?

Going beyond these questions of phase stability and coexistence, we can turn to questions of whether clusters of metals or salts can be used to construct phases of bulk matter unavailable by conventional mass processing, and of whether cluster reactivity can be distinguished from and related to bulk heterogeneous reactions. In particular, the study of reactions of clusters will have to confront the question of whether heats of reaction can be trapped in clusters and transform clusters from one phase to another during a sequence of reactions. If this can occur, and we see no reason why it should not, the reactivity of clusters becomes a more complicated matter that has sometimes been supposed, but offers a rich new range of possibilities for control of reactions with clusters.

ACKNOWLEDGMENTS

The research described here was supported by a contract from the Office of Naval Research.

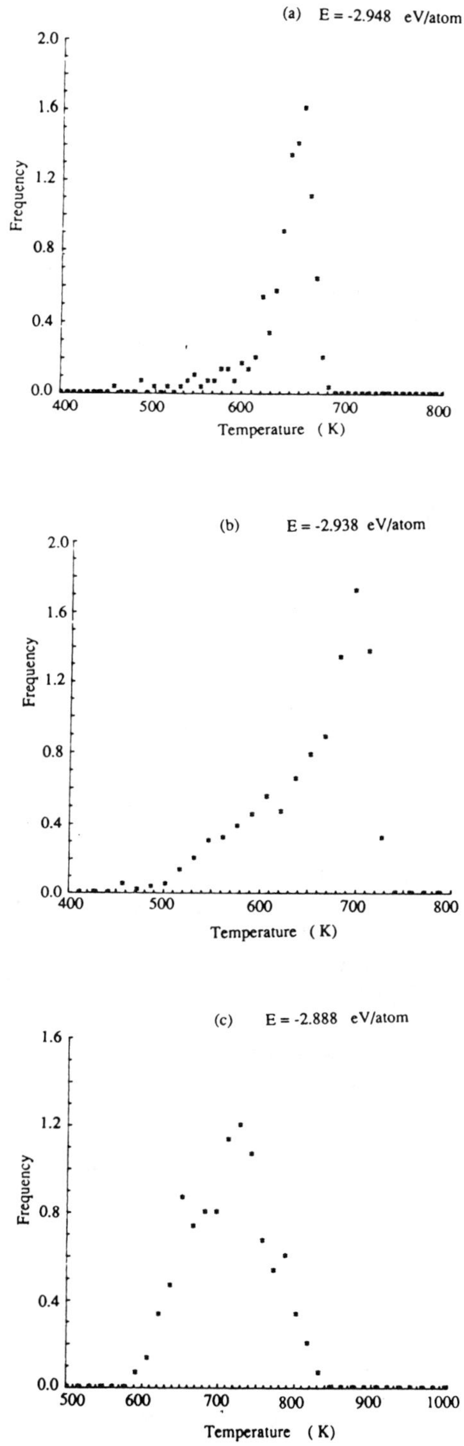

Fig. 14. Distributions of short-time-average temperatures for Cu_{13} at energies of (a) -2.948 eV/atom, (b) -2.938 eV/atom, and (c) -2.888 eV/atom. The first is essentially solid, the third liquid and the second, a mixture of the two.

REFERENCES

1. For a review of this subject, *see* Berry, R. S., Beck, T. L., Davis, H. L., and Jellinek, J., *Evolution of Size Effects in Chemical Dynamics, Part 2*, Advances in Chemical Physics, Vol. LXX, Prigogine, I. and Rice S. A., eds., Wiley, New York, 1988, p. 75.
2. Beck, T. L., Jellinek, J., and Berry, R. S., *J. Chem, Phys.* **87**, 545 (1987); Beck, T. L. and Berry, R. S., *J. Chem. Phys.* **88**, 3910 (1988).
3. Luo, J., Landman, U., and Jortner, J., *Physics and Chemistry of Small Clusters*, Jena, P., Rao, B. K., and Khanna, S. N., eds., Plenum, New York, 1987, p. 201.
4. Parks, E. K., Weiler, B. H., Bechtold, P. S., Hoffman III,W. F., Neiman, G. C., Pobo, L. G., and Riley, S. J., *J. Chem. Phys.* **88**, 1622 (1988); Parks, E. K., Neiman, G. C., Pobo, L. G., and Riley, S. J., *J. Chem. Phys.* **88**, 6260 (1988).
5. Burmer, P. and Karplus, M., *J. Chem. Phys.* **58**, 3903 (1973).
6. Welch, D. O., Lazareth, O. W., Dienes, G. J., and Hatcher, R. D., *J. Chem. Phys.* **64**, 835 (1976).
7. Holian, B. L., Straub, G. K., Swanson, R. E., and Wallace D. C., *Phys, Rev.* **B27**, 2873 (1983).
8. Sawada, S., *Microclusters*, Sugano, S., Nishina, Y., and Ohnishi, S., eds., Springer, New York, 1987, p. 211; Sawada, S. and Sugano, S., *Z. Physik D*, in press, 1989.
9. Car, R. and Parinello, M., *Phys. Rev. Lett.* **55**, 2471 (1985); Car, R., Parinello, M., and Andreoni, W., *Microclusters*, Sugano, S., Nishina, Y., and Ohnishi, S., eds. Springer, New York, 1987, p. 134.
10. Nørskov, J. K., *Phys. Rev.* B **26**, 2875 (1982); Daw, M. S. and Baskes, M. I., *Phys. Rev.* **B29**, 6443 (1984).
11. Foiles, S. M., Baskes, M. I., and Daw, M. S., *Phys. Rev.* **B33**, 7983 (1986).
12. Baerends, E. J., Ellis, D. E., and Ros, P., *Chem. Phys.* **2**, 41 (1973); Rosén, A., Ellis, D. E., Adachi, H., and Averill, F. W., *J. Chem. Phys.* **65**, 3629 (1976); Rosén, A. and Ellis, D. E., *J. Chem. Phys.* **62**, 3039 (1975).
13. Delley, B. and Ellis D. E., *J. Chem. Phys.* **76**, 1949 (1982); Delley, B., Ellis, D. E., Freeman, A. J., Baerends, E. J., and Post, D., *Phys. Rev.* **B27**, 2132 (1983).
14. Delley, B., Ellis, D. E., Freeman, A. J., Baerends, E. J., and Post, D., *Phys. Rev.* **B27**, 2132 (1983).
15. Berry, R. S., Davis, H. L., and Beck, T. L., *Chem. Phys. Lett.* **147**, 13 (1988).
16. Martin, T. P., *Phys. Repts.* **95**, 167 (1983).
17. Beck, T. L., Leitner, D. M., and Berry R. S., *J. Chem. Phys.* **89**, 1681 (1988).

Laser CVP of Ultrafine Ceramic Powders of Si and Si_3N_4

A Study of the Flow Pattern of the Laser Flame

R. A. BAUER,[1] F. E. KRUIS,[2] J. G. M. BECHT,[1] B. SCARLETT,*[,2] AND J. SCHOONMAN[1]

[1]Laboratory for Inorganic Chemistry; and [2]Laboratory of Powder Technology, Delft University of Technology, PO Box 5045, 2600 GA Delft, The Netherlands

ABSTRACT

The flow patterns of flames in a Laser Chemical Vapor Precipitation Reactor (L-CVP) for the production of Si and Si_3N_4 submicron powders have been studied using interferometry and the shadow method. In the case for Si_3N_4 production, the reactant gases, SiH_2Cl_2 and NH_3 were mixed in the laser beam, thus preventing low temperature reactions. The methods used give detailed information about flows in and out of the flame. This information is important for controling the uniformity of the physical and chemical properties of the synthesized powder and to maintain long production times. Interferometry also provides information on the temperature distribution in the flame.

Index Entries: Si; Si_3N_4; SiH_2Cl_2; NH_3 flames, flow patterns of.

INTRODUCTION

Recent interest in high temperature structural ceramics has led to the development of chemical vapor phase synthesis techniques for the for-

*Author to whom all correspondence and reprint orders should be addressed.

mation of a variety of high quality oxidic and nonoxidic ceramic powders. Regarding the gas phase reaction process, we have described Chemical Vapor Precipitation (CVP) (*1*) of submicron Si_3N_4 powders using thermal (*1*) and laser (*2,3*) activation.

The primary characteristic of the laser-driven gas phase synthesis process is a well defined reaction zone that facilitates a considerable degree of control over composition, size, and size distribution of the product powders. In the laser-CVP of Si, SiC, and Si_3N_4, SiH_4 is invariably used as the silicon reactant (*4–8*), whereas in the traditional thermally-activated CVP of powders and thin films of Si_3N_4 chlorinated silanes, i.e., SiH_2Cl_2 and $SiCl_4$, are frequently employed (*1,9*). It has been shown that ultrafine spherical silicon powder can be produced using SiH_2Cl_2 as a reactant in L-CVP (*2,3*). In the production of Si_3N_4, NH_3 reacts with chlorinated silanes, such as SiH_2Cl_2 and $SiCl_4$, to form silicon diīmides at ambient temperatures. To avoid this reaction, a laser reactor was equipped with a specially designed nozzle to inject the reactants separately into the laser beam.

Earlier experiments (*2,3*) have shown that the silicon nitride powders contain hard agglomerates and some imide contamination. Normally, the silicon nitride powders are collected at a high temperature to separate silicon nitride from the waste product, NH_4Cl, which dissociates above 610 K. However, the collection at ambient temperature also leads to powders that are agglomerated. Therefore, the agglomeration is probably caused by collisions of the particles in the flame. Although the particles leave the laser beam within a millisecond, the gases and particles take tens of milliseconds to be cooled down. To lower the agglomeration rate, the cooling down must be more rapid. By entering NH_3 in the flame above the laser beam, silicon powder is first formed and subsequently nitrided in the upper part of the flame. Using this method, silicon nitride will be formed at a lower temperature; hence, the cooling down is expected to occur faster because of mixing with surrounding nitrogen. This method of injection will also avoid imide formation.

When NH_3 is injected directly into the laser flame, the flow pattern shows disturbance, illustrated by the fact that the powder does not leave the reactor through the exhaust funnel, but spreads through the reactor. The collection of powder on the KCl windows causes a breakdown because of laser heating. Furthermore, spreading of the powder lowers the transmission of the CO_2 laser beam (*2,3*).

Various techniques, such as the shadow method, interferometry, and visual observation, are being used to study the gas flow pattern. The shadow method shows the contour of the gas flows and flow pattern of clouds of powder. With interferometry, the density distribution of the gases in place and time can be determined. With these techniques, the NH_3 flow into the flame and the emission of the powder and exhaust gas out of the reactor can be optimized; these methods also give an indication of the temperature distribution in the flame.

EXPERIMENTAL PROCEDURE

Reactants

The reactants for silicon and silicon nitride synthesis by Laser Chemical Vapor Precipitation were dichlorosilane (Air Products and Chemicals, Inc., Pacoima, CA, 97% SiH_2Cl_2, 3% other chlorosilanes), and NH_3 (Air Products, 99.999%) were utilized. The carrier gases, N_2 and H_2, were purified with copper (BASF Wyandotte Corp., Parsippany, NJ, R 3-11) and palladium (BASF, R 0-20) catalyst, respectively, and a zeolite (Dow Chemical Co., Midland, MI, A4).

Laser CVP

The selection of the present reactants for the synthesis of Si_3N_4 implies the formation of silicon diīmide, $Si(NH)_2$, at room temperature unless precautions are taken in the introduction of the reactant gases into the laser beam. This can be achieved by a special nozzle design. In addition ammonium chloride is a waste product, requiring adequate separation of Si_3N_4.

A schematic drawing of the stainless steel reactor is presented in Fig. 1. The two KCl windows were flushed with N_2. The reactor was operated at atmospheric pressure. An industrial untuned 150 W cw CO_2-laser (Electrox, PB 2000) radiating at 10.6 μm was partially focused with a ZnSe lens (focal length, 30 cm) in an orthogonal stream of reactants. A conic and straight exhaust funnel were explored to optimize the gas and powder emission from the laser flame.

The nozzle was built up of four separated concentric outlets for NH_3 pointed toward the middle and a centered outlet for the chlorinated silanes, sensitizer, and carrier gas. The outlets for NH_3 can be altered in horizontal, as well as in a vertical direction. The gases were mixed in or above the laser beam, thus avoiding low temperature reactions.

The reaction products were confined to a small well-defined reaction zone by a concentric N_2 gas stream. This gas stream was also used to feed the products into a horizontal electrostatic precipitator.

Interferometry and the Shadow Method

To study the flow and temperature profiles, interferometry and shadow measurements have been made. Schematics of the interferometry and shadow measurement setups are shown in Fig. 2 top and bottom, respectively. A Mach-Zehnder interferometer was used, consisting of a HeNe laser beam (632.8 nm) expanded by two lenses. This expanded beam was split with a beam splitter and one beam was passed through the reactor and the other was guided around the reactor. Finally, the two were combined and directed to a photographic plate without the use of

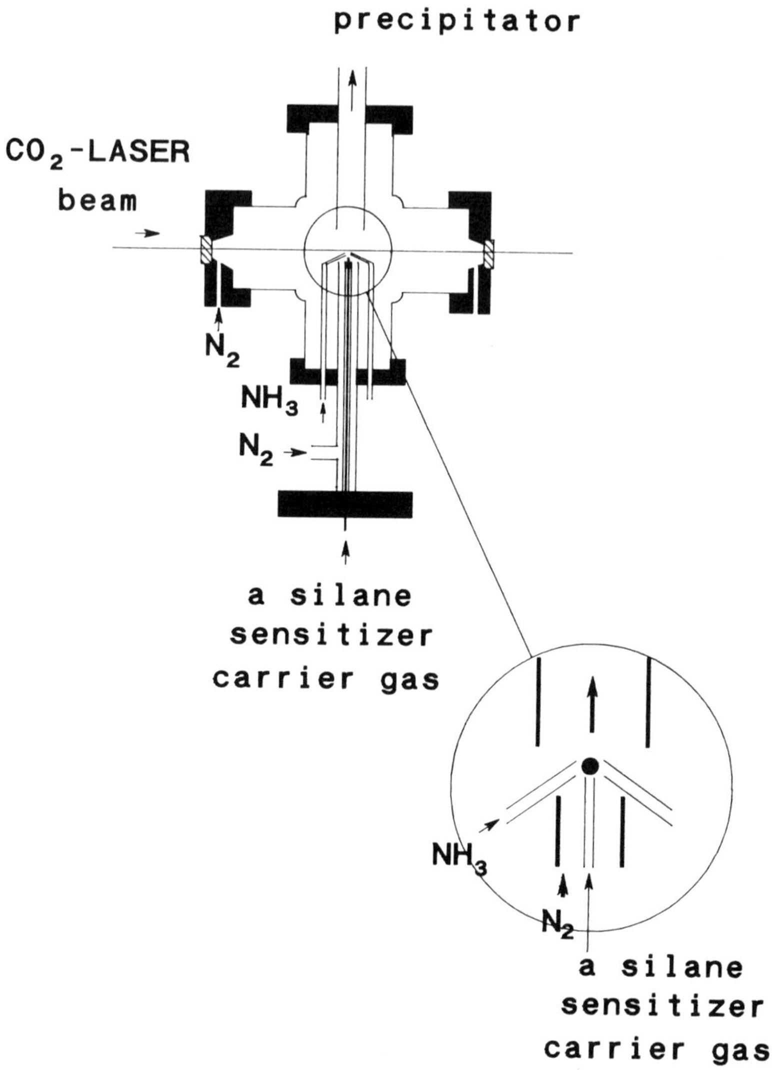

Fig. 1. Schematic of the reactor and the nozzle with the gases used.

lenses. By adjusting one of the mirrors, the difference in optical path length could be changed, and consequently, the size and direction of the fringes could be altered. For the shadow measurements, the reference beam was not used.

Both methods rely on the difference in refractive index. Since the density is strongly a function of the temperature, the influence of the variation in the composition of the gases can often be neglected compared to the influence of a difference in temperature (*12*). For ideal gases, the relation between the refractive index, n, and the temperature, T, is

$$(n-1)T = \text{constant} \quad (1)$$

The fringes seen with interferometry represent areas where the difference in optical length is a multiple of the wavelength

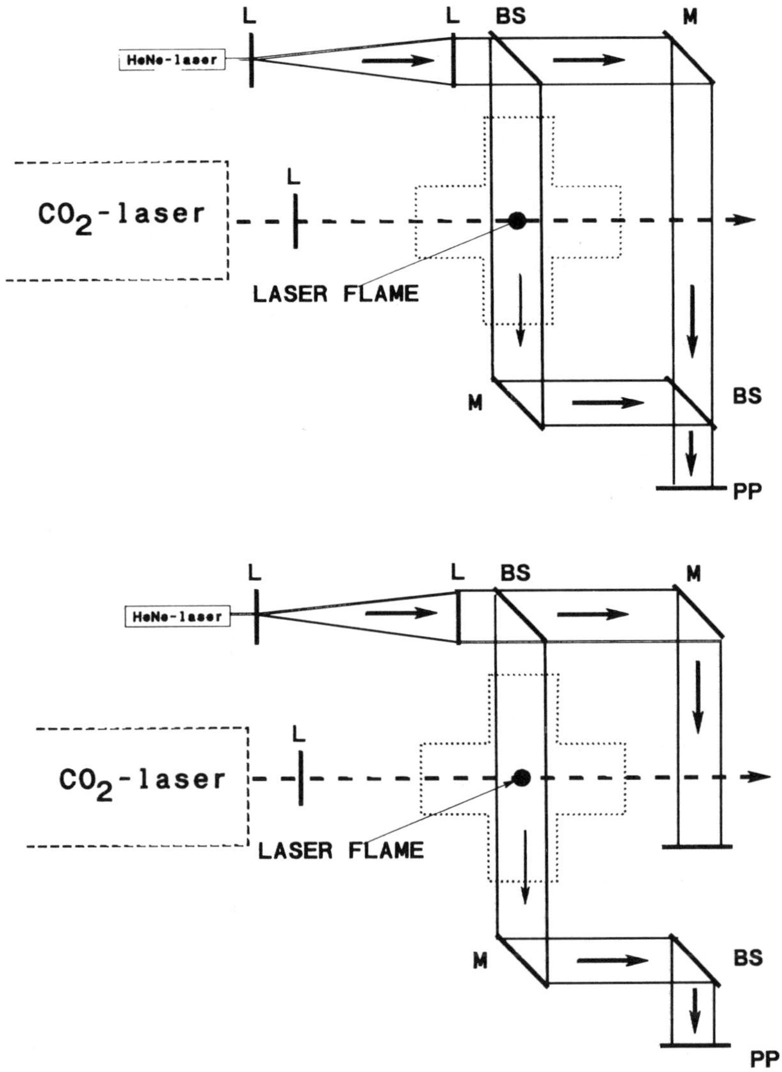

Fig. 2. Schematics of the setup for interferometry (top) and the shadow method (bottom): L = lens, BS = beam splitter, M = mirror, and PP = photographic plate.

$$\int_{flame} n_{flame}dx - \int_{ref.} n_{air}dx = p\lambda \quad (2)$$

If the wave fronts of the two separated beams are perfectly parallel, the photographic plate is uniformly illuminated or completely black, i.e., the fringe is infinite. In other cases, the fringes are finite and fringes will be seen at a distance q from one another. If there is a small disturbance, the fringes will be shifted by δq, and can be related to the refractive index as follows

$$\int_{flame} (n_{flame} - n_{nitrogen})dx = (\delta q/q)\lambda \quad (3)$$

The variation in refractive index can be calculated and from which the relative temperature can also be calculated. To calculate the absolute temperature, the temperature dependence of the refractive index has to be known. By measuring a reference point, the constant can be calculated. The exact interpretation of complex flow patterns is often difficult with interferometry.

Using the shadow method, the dark lines represent areas where $d^2(n - l)/dx^2$ is maximal. This means that regions in the flame with strong variations are visible, so the contour lines of flows within the flame are shown. Line broadening of the shadow line gives an indication of the turbulence.

RESULTS AND DISCUSSION

The flow pattern in the flame has been studied for a variety of process conditions using interferometry and the shadow method.

Figure 3a shows the shadowgraph of an unheated SiH_2Cl_2 stream, which hardly expands upon leaving the nozzle. If SiH_2Cl_2 is heated with the CO_2 laser radiation, the gas stream expands strongly, as is shown in Fig. 3b, and a long yellow flame was seen. When N_2 was injected, the flame flattened and small flames were observed in the center and between the centric and concentric outlets. This was corroborated by interferometry and the shadow method. Injection of reactive NH_3 resulted in the formation of additional, small flames associated with the NH_3 outlets. The overall flame became whitish and brighter. The shadowgraph revealed only slight broadening of the flow. However, the interferogram showed a larger shift of the fringes at the concentric part of the flame.

When the concentric outlets were increased in height in order to change the reaction mechanism, growth of a deposit was observed on the concentric outlets, as shown in Fig. 4. Although the outlets were outside the luminescent part of the flame, this shadowgraph shows that in reality the outlets were inside an outer, non-luminescent, part of the flame. The particle formation clearly occurs in a larger region than the luminescent area, as can be seen with shadowgraphs. The outlets had to be altered in a horizontal direction in order to avoid growth on the nozzle.

Using interferometry, flames with low N_2 flow as well as those with low NH_3 flow, had flow patterns that were only slightly perturbed, compared to the flame with only SiH_2Cl_2. Both flames were yellow but shorter than the SiH_2Cl_2 flame. Visual observation showed that the flame with a nitrogen injection showed a sharp decrease of luminescence and a narrowing of the luminescent area above the place where the N_2 was injected. The flame with NH_3 was separated into two parts, one with only SiH_2Cl_2 below the injection point of NH_3 and the upper part including NH_3. The upper part of the flame seemed to be even broader than the lower part. Both parts produced about the same luminescence. The reaction

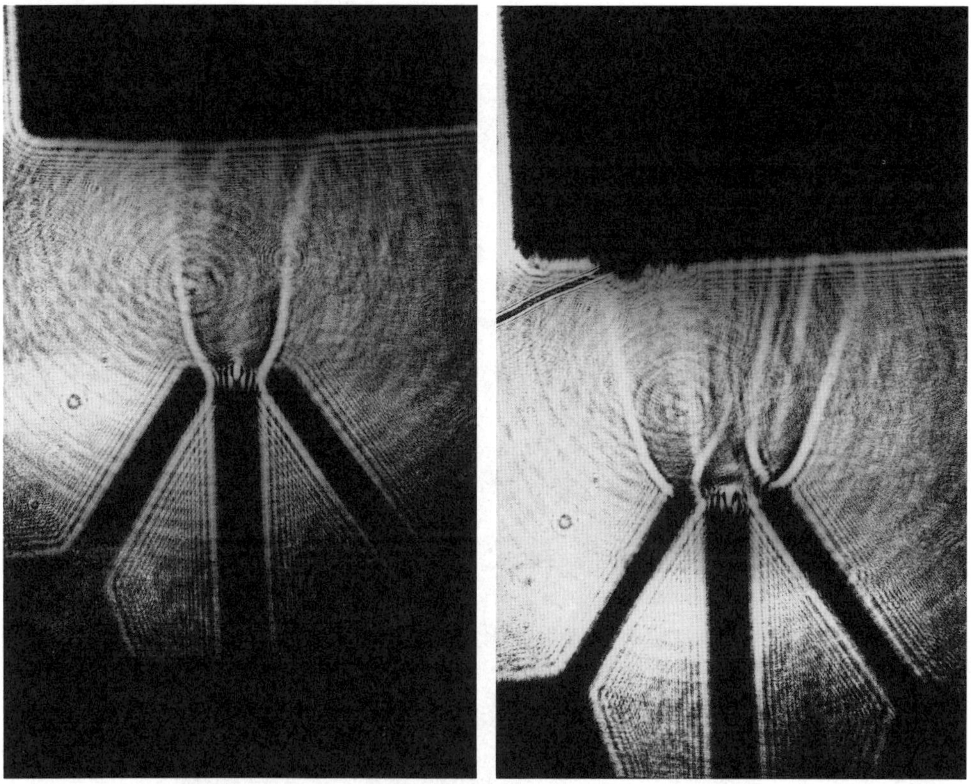

Fig. 3. Shadowgraphs of laser flames (a) with only dichlorosilane and (b) with ammonia and dichlorosilane.

$$6\,Si + 8\,NH_3 \longrightarrow 2\,Si_3N_4 + 12\,H_2\,(\Delta H(298K) = -607\,kJ/mol\,Si_3N_4)$$

is clearly exothermic.

Fig. 5 shows interferograms of different flames with a large conic exhaust funnel. With only SiH_2Cl_2 used as a reactant (Fig. 5a), the shift of the fringes was fairly constant over the whole flame and did not change noticeably in time, which indicates that the turbulence in the flame must be small. The injection of up to 7.5 L/h N_2 into the laser beam causes the occurrence of turbulence as indicated by a less constant and time dependent shift of the fringes (Fig. 5b). The injection of more than 4.5 l/h NH_3 into the laser beam creates a totally disturbed flow (Fig. 5c), and the flame could not be observed any more with the employed methods, because a mist of powder occurred in the reactor. The interferogram in Fig. 5c also illustrates this turbulence by the vagueness of the fringes.

Different chimneys are used to observe the stabilization of the flame, as shown in Fig. 6. When a small tube is used to guide the gas and the powder from the flame into the exhaust funnel, only little improvement was observed. The shadowgraph shows that the powder circulates back from the conical funnel into the reactor. When a tube with an increased

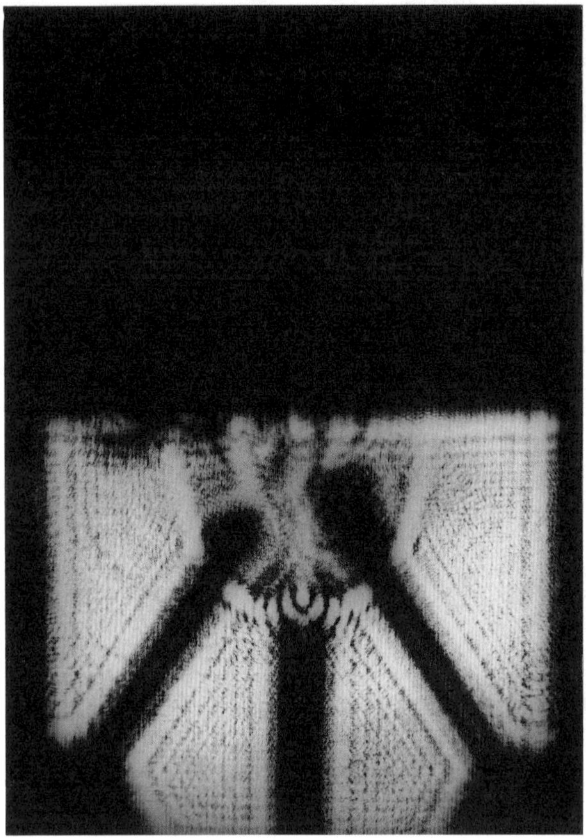

Fig. 4. Shadowgraph of the flame and nozzle with growths on the outlets for ammonia.

length was used, thus forming a straight tube to the precipitator, the disturbance was minimal, as shown in Figs. 3 and 7.

These experiments demonstrate that using interferometry and the shadow method, information can be obtained on the flame and particle formation. This information is necessary in order to improve the performance of the reactor. An example of these improvements is the replacement of the exhaust funnel.

Preliminary results of Si_3N_4 powder synthesis show that when using the modified exhaust funnel and injecting NH_3 above the laser beam, considerably smaller agglomerates of Si_3N_4 particles are formed than in the preceding experiments.

CONCLUSION

The employed methods, i.e., interferometry and the shadow method, provide detailed information on the flow pattern of the laser-activated flame, which cannot be determined with visual observation.

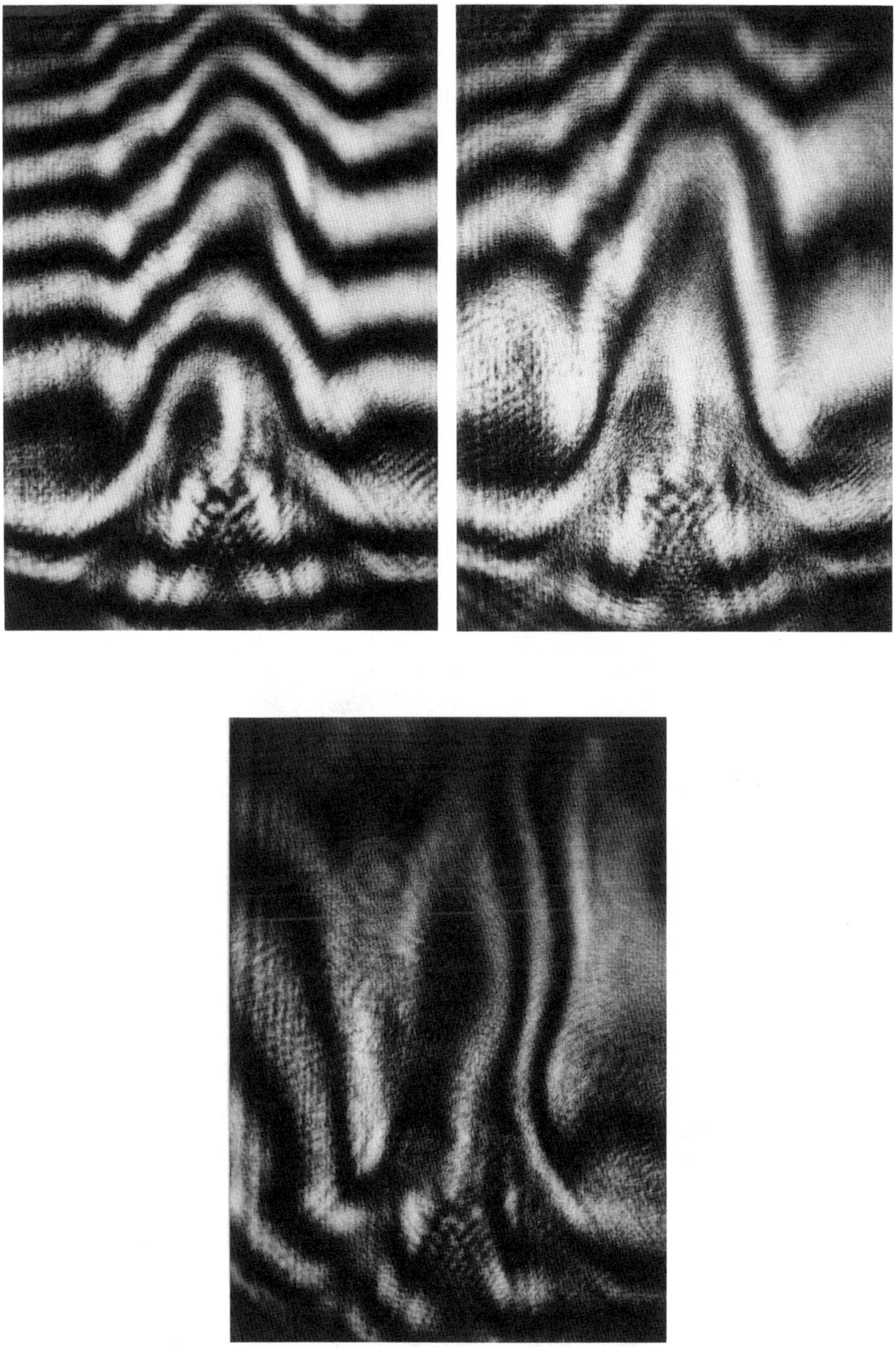

Fig. 5. Interferograms of flames without a chimney to stabilize the flame with reactants: (a) dichlorosilane, (b) dichlorosilane and nitrogen, and (c) dichlorosilane and ammonia.

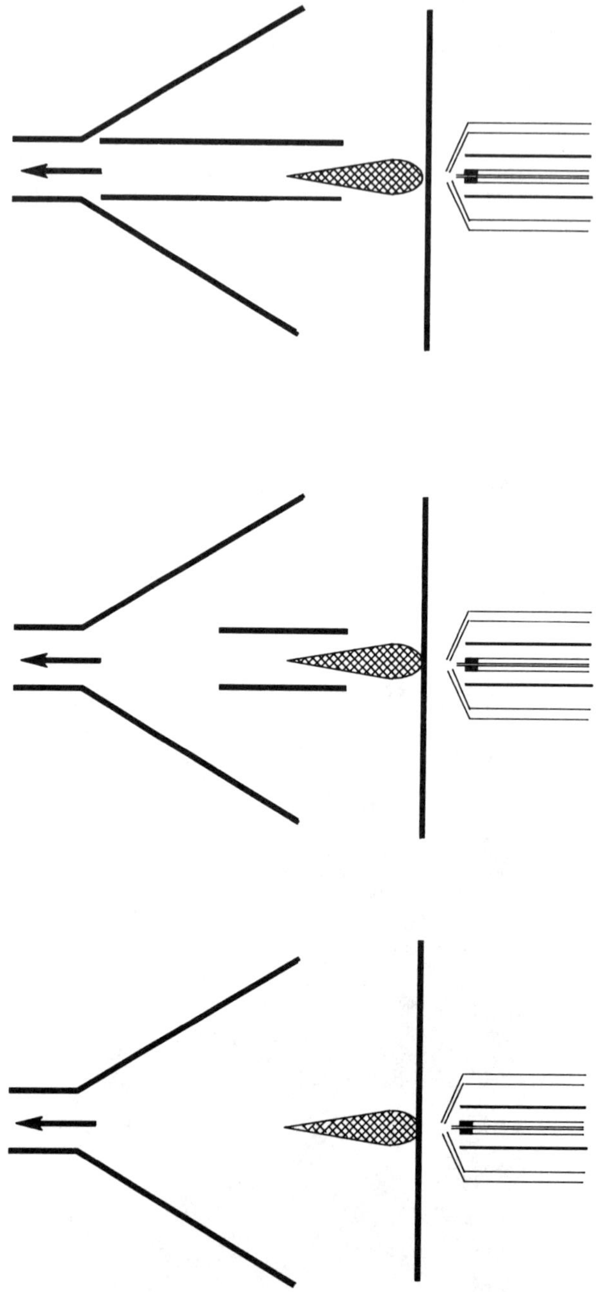

Fig. 6. Schematics of the nozzle and exhaust funnels with different tubes used as a chimney to stabilize the flame. From left to right, the flame is increasingly stable.

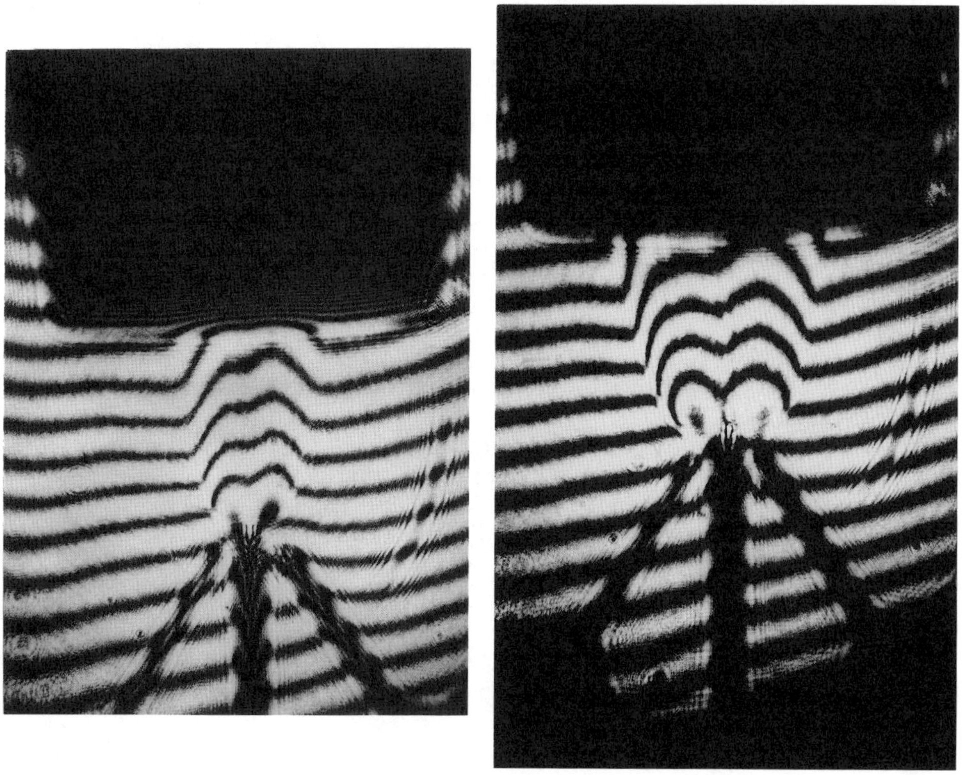

Fig. 7. Interferograms of flame stabilized by a modified exhaust funnel with the reactants: (a) only dichlorosilane and (b) dichlorosilane and ammonia.

With this information, the position of the reactant inlets can be optimized, the flame stabilized, and the flow of the powder out of the reactor improved. In addition the degree of powder agglomeration can be reduced. These aspects of the flow pattern are essential for the control of the uniformity of the physical and chemical characteristics of the powders and for achieving long production times.

ACKNOWLEDGMENT

This research is financed by the Netherlands Ministry of Economic Affairs in the framework of the Innovation Directed Research Programs (IOP).

REFERENCES

1. Jaquemijns, E. J., van der Put, P. J., and Schoonman, J., *High Temperatures-High Pressures* **20,** 31–34 (1988).
2. Bauer, R. A., Smulders, R., Becht, J. G. M., van der Put, P. J., and Schoonman, J., *J. Am. Cer. Soc.* **72,** 1301–1304 (1989).

3. Bauer, R. A., Smulders, R., Geus, E., van der Put, P. J., Becht, J. G. M., and Schoonman, J., *Ceram. Eng. Sci. Proc.* **9,** 949–956 (1988).
4. Cannon, W. R., Danforth, S. C., Flint, J. H., Haggerty, J. S., and Marra, R. A., *J. Am. Ceram. Soc.* **65,** 324–330 (1982).
5. Cannon, W. R., Danforth, S. C., Haggerty, J. S., and Marra, R. A., *J. Am. Ceram. Soc.* **65,** 330–335 (1982).
6. Sawano, K., Haggerty, J. S., and Bowen, H. K., *J. of Ceram. Soc. Japan* **95,** 64–69 (1987).
7. Lihrmann, J. M., Haggerty, J. S., Luce, M., Croix, O., and Cauchetier, M., *Materiaux Mèchanique Electricitè* **422,** 32–36 (1987).
8. Symons, W. and Danforth, S. C., *Advances in Ceramics*, vol. 21, Messing, G. L., Mazdiyasni, K. S., McCauley, J. W., and Haber, R. A., eds., American Ceramic Society, Westerville, OH, 1987, pp. 249–256.
9. Grieco, M. J., Worthing, F. L., and Schwartz, B., *J. Electrochem. Soc.* **115,** 525–531 (1968).
10. Haggerty, J. S. and Cannon, W. R., *Laser-Induced Chemical Processes*, Steinfeld, J. I., ed., Plenum, New York, 1981, pp. 165–241.
11. Sausa, R. C. and Ronn, A. M., *Chem. Phys.* **96,** 183–189 (1985).
12. Weinberg, F. J., *Optics of Flames,* Butterworths, London, 1963, p. 37.

Laser Production of Metal Fine Particles from Organometallic Compounds

NOBUO SHIMO*,[1] AND KEITARO YOSHIHARA[2]

[1]Idemitsu Kosan Co., Ltd., Central Research Laboratories Kami-izumi, Sodegaura, Chiba, 299-02, Japan; and [2]Institute for Molecular Science, Myodaiji, Okazaki, 444, Japan

ABSTRACT

A new type of laser-induced decomposition reaction of organometallic compounds and the production of metal fine particles have been discovered in which a single laser pulse is used only for ignition of a thermal chain reaction. Two key processes of this reaction, the formation of high density active species and the propagation of the reaction caused by these active species, are discussed in detail. An orange emission from some reaction intermediates are also discussed.

Index Entries: Laser chemistry; explosive reaction; organometallic compound; thermal chain reaction; metal fine particles.

INTRODUCTION

Recently, chemical vapor deposition has been attracted much attention for the preparation of thin films and fine particles. For example, amorphous silicon film is deposited from silane or disilane; silicon carbide is deposited from an organosilicon compound, and various metals are deposited from organometallic compounds.

In these processes, laser beams are used for the decomposition of starting gaseous materials. However, a photon of laser beam is effective, on an average only, to produce one or less product, and the quantum efficiency is usually low.

We have found a new process with an extremely high efficiency for producing metal fine particles from organometallic compounds, in which

*Author to whom all correspondence and reprint requests should be addressed.

laser light is used only for igniting a decomposition reaction. The quantum efficiency exceeds 2×10^5. In this article, we report some features of a unique decomposition reaction of organometallic compounds and the spectra of orange emission that is observed during the decomposition reaction.

Chain reactions, initiated by a CO_2 laser were reported with a reaction of chromyl chloride with hydrogen to produce chromium oxide, water, and hydrogen chloride (*1*) and a reaction of vanadium oxychloride with hydrogen and oxygen to produce vanadium oxides (*2*).

EXPERIMENTAL DATA

An excimer laser (Lumonics HE-460 or Lambda Physik EMG-101E, ArF (193 nm) or KrF (248 nm)), having a fluence of 5–90 mJ/cm^2 and a pulse width of ca. 10 ns, was used to decompose organometallic compounds. In some cases of laser irradiation, the laser beam was focused with a convex lens of f = 100 mm, in order to make active organometallic compounds at high concentrations.

Tetramethyllead (TML), tetraethyllead (TEL), and trimethylbimuths (TMBi) were purchased from TRI Chemical Laboratory Inc., and their stated purities were 99.9999% based on metal analysis. Dimethylzinc (DMZ), diethylzinc (DEZ), and dimethylcadmium (DMC) were from TOYO Stauffer Chemical. These organometallic compounds were purified by at least five freeze-pump-thaw cycles followed by trap-to-trap distillation in vacuum before transferring them to the reaction cell. Nitrogen, helium, and hydrogen (all from Nippon Sanso, stated purity of 99.9%) were used without further purification.

Gaseous products were analyzed by a gas chromatograph (Hewlett Packard, model 5890A) using a methyl silicon gum column (HP-1, 5 m $\times$ 0.53 μm) with a flame ionization detector.

Organometallic compounds were introduced into the reaction cell of about 100 mL vol. A reaction cell with a vol. of 10 L was used for the quantum efficiency measurement. Background pressure of 10^{-5} torr was achieved by using an oil diffusion pump. The shape and size of metal particles were measured by a scanning electron microscope (Hitachi, S-800). Total emission light was detected with an optical multichannel analyzer (Princeton Instruments, SMA).

RESULTS AND DISCUSSION

Description of the Explosive Reaction

After organometallic compounds, such as tetramethyllead (TML) and trimethylbimuths (TMBi), were introduced into the reaction cell, the sample was irradiated by a single laser pulse through a quartz window. Immediately after laser irradiation, most of the organometallic compounds decompose within a few tens of milliseconds accompanied by a strong orange emission from the whole volume of the sample cell. After that, black fine spherical metal particles having a size below 1 μm were deposited on the wall of the reaction cell. The size of metal particles depended on the reaction conditions.

There are two key points that make this unique reaction possible. One is the formation of high density active species by a single laser pulse irradiation, and the second is the propagation of a thermal chain reaction.

High density active species, such as metal atoms and methylated metal radicals, are formed in relatively confined volume because TML, for example, has a large molar extinction coefficient at the excited laser wavelength. These active species can cause the propagation of a thermal chain reaction. After the explosive reaction, most of the TML compound is decomposed to produce metal fine particles and ethane as the gaseous main product.

Now let us consider the efficiency of the reaction. The irradiated laser light, for example ArF 193nm, 25mJ, contains 2.4×10^{16} photons/pulse. On the other hand, 2700 Pa of TML in the 10 L vol. reaction cell at room temperature (22°C) correspond to 6.5×10^{21} molecules. Since more than 80% of TML is decomposed, the reaction quantum efficiency becomes at least 2.2×10^5, which is limited by the volume of sample container.

Laser Fluence and Laser Wavelength

Figure 1 shows the laser fluence dependence and laser wavelength dependence for the explosive reaction. In this figure, the vertical axis is the ratio of pressure after laser irradiation to that before irradiation. If the decomposition reaction of TML proceeds completely according to scheme 1, the ratio should be 2, because main gaseous products are 2 molecules of ethane, which is obtained by the recombination reaction of methyl radicals.

$$(CH_3)_4\ Pb \rightarrow Pb + 2\ C_2H_6 \qquad \text{(Scheme 1)}$$

In fact, most of the TML is decomposed and 85% of ethane, several percent of methane and ethylene, and a few percent of propane are detected by gas chromatography as gaseous products.

As is clearly shown in this figure, there exists a threshold laser fluence, about 5 mJ/cm^2 in the case of 193 nm irradiation and 60 mJ/cm in the case of 248 nm irradiation without focusing the laser beam. When the KrF excimer laser beam is focused with a convex lens, having 100 mm focal length, at a distance of 80 mm from the window, the threshold laser fluence goes down to 10 mJ/cm^2 at 248 nm irradiation. The difference in the threshold laser fluence between 193 and 248nm irradiation can be interpreted by the difference in absorption coefficient of TML. The molar extinction coefficient at 193 nm is 1.03×10^4 *M*/cm, whereas at 248 nm it is 120 *M*/cm. Accordingly, the one photon absorbance at 193 nm is about two orders of magnitude greater than the one at 248 nm. These results of

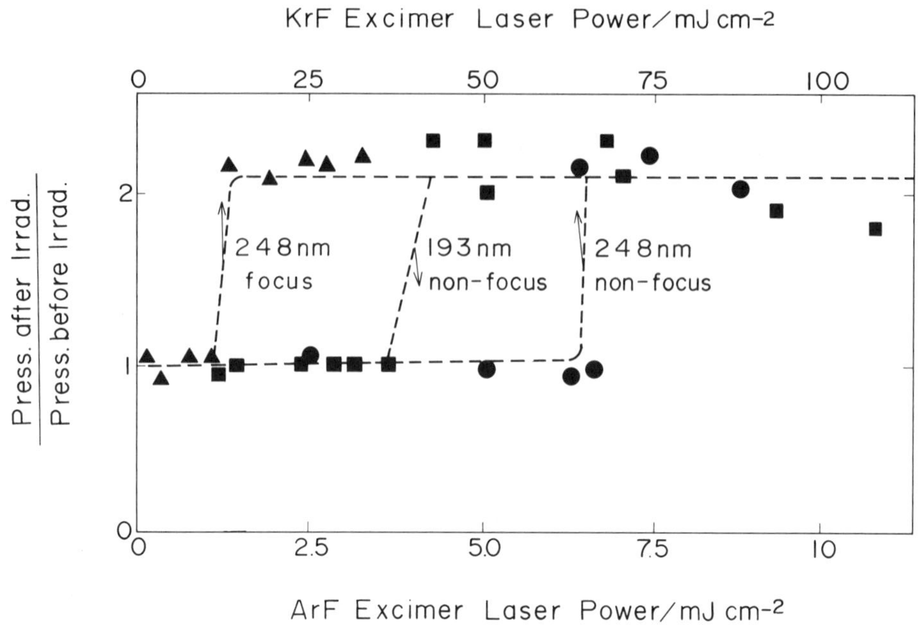

Fig. 1. Laser power dependence and laser excitation wavelength dependence for the explosive chain reaction. Vertical axis is a ratio of the pressure after laser irradiation and the pressure before laser irradiation. This value of about 2 indicates the occurrence of the chain reaction, and 1 is for the absence of the chain reaction (*see text*). Horizontal axis is ArF excimer laser power (indicated below) and KrF excimer laser power (indicated above). KrF excimer laser was focused with a convex lens (f = 100 mm).

laser power dependence and laser wavelength dependence clearly show that it is very important for this reaction to produce the high density active species.

Organometallic Compounds Possible to Induce a Chain Reaction

Among the organometallic compounds we have investigated at the present stage, TML and TMBi decomposed with the explosive reaction. These two compounds have low metal-carbon bond dissociation energies. As the bond dissociation energy decreases, organometallic compounds can be expected to decompose to smaller fragments, such as monomethylated metal radical species and metal atoms species.

Among the various types of organometallic compounds, alkylmetal, metalcarbonyl, halogenated metal, and metallocene compounds have weak bond dissociation energies. Decomposition of some of the alkylmetal compounds can be by an exothermic reaction because the recombination of an alkyl–alkyl radical is highly exothermic. Table 1 shows the average bond dissociation energy (D_n) and energy balance (ΔH) of some alkylmetal compounds for the reaction of scheme 2.

Table 1

Alkylmetal compounds	kJ mol^{-1}	
	D_n	ΔH
Bi $(CH_3)_3$	428	−124
Pb $(CH_3)_4$	632	−104
Pb $(C_2H_5)_4$	512	−224
Sn $(CH_3)_4$	896	160
Ge $(CH_3)_4$	1040	304
Si $(CH_3)_4$	1276	540
Zn $(CH_3)_2$	351	−17
Zn $(C_2H_5)_2$	290	−78
Cd $(CH_3)_2$	279	−88

$$MR_n \rightarrow M + n/2\ (R\text{-}R) \qquad \text{(Scheme 2)}$$

where *M* is metal, *R* is methyl or ethyl, and *n* is the number of alkyl groups. ΔH is calculated as bond dissociation energy of alkylmetal (D_n) minus bond formation energy of an alkyl–alkyl radical; for instance, the bonding energy of methyl–methyl is 368 kJ/mol. The compound, which has a negative ΔH value, that is, the one preceding an exothermic reaction, has a possibility of producing a chain reaction. In fact, TML and TMBi can cause a chain reaction. However, some compounds, which have negative ΔH, did not cause a chain reaction. This is presumably because of the low vapor pressure at room temperature and other reasons.

Effect of Concentrations of Organometallic Compounds

The reaction does not take place below the concentration of ca. 1 torr in both cases of 193 nm and 248 nm irradiations. This pressure effect can be interpreted as follows: when the pressure of starting materials is low, active species formed by laser irradiation cannot collide with the other parent molecule and, perhaps, will be deactivated by the collision to the wall. On the other hand, with higher pressures, active species can collide with the other parent molecule and transfer the internal energies to it. The frequency of collision is attributed to the pressure of the parent molecule. The excited parent molecule then decomposes to form active species again, and thermal chain propagation proceeds. Metal fine particles and gaseous products are formed during the propagation of the chain reaction.

Effect of Foreign Gases

The effect of foreign gases for the explosive reaction is shown in Fig. 2. Under these conditions, the chain reaction is prohibited by the addition of foreign gases (He, N_2, and air) above 7 torr at the pressure of the parent molecule (TML) of 5 torr. This result can be interpreted as the

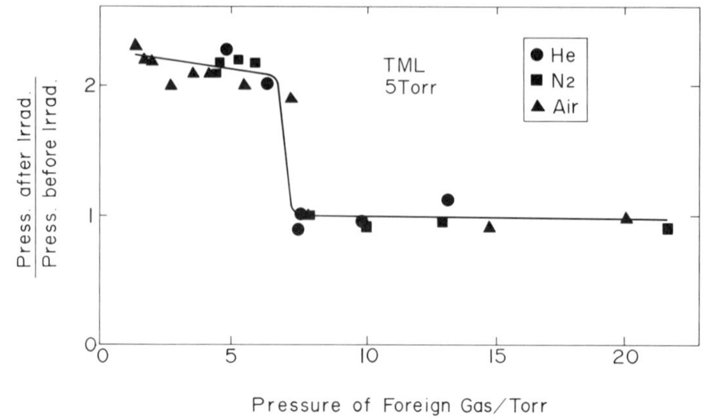

Fig. 2. Effect of foreign gases. Horizontal axis is the pressure of added foreign gases to 700 Pa (5 torr) of tetramethyllead (TML) under the same reaction condition.

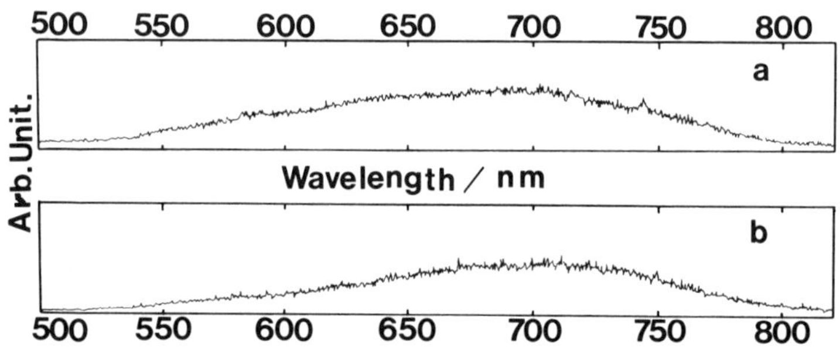

Fig. 3. Emission spectra obtained by the photolysis of organometallic compounds: (a) Emission spectra obtained by the ArF excimer laser irradiation of 3300 Pa of TML; and (b) Emission spectra obtained by the KrF excimer laser irradiation of 3300 Pa of TMBi.

deactivation of active species by added foreign gases. When the concentration of foreign gases is low, the active species can collide with the other parent molecules. However, upon increasing foreign gas pressure, the possibility of collision with foreign gases increases, and finally, the active species cannot propagate the thermal chain reaction.

Emission Spectra from Reaction Intermediates

A strong orange emission, which was observed during the decomposition reaction, was detected with an optical multichannel analyzer. Figure 3(a) shows the emission spectra, obtained by KrF excimer laser photolysis of 3300 Pa of TML, and Fig. 3(b) shows the one obtained by ArF excimer laser photolysis of 3300 Pa of TMBi. In both cases, the broad emission spectra with a peak at around 700 nm were detected and tentatively assigned to the one from black body radiation of metal clusters or metal fine particles. These types of emission spectra were also reported in the reaction of $SiH_4 + O_2$ (3). The emission spectra were not affected with the different laser wavelengths (193 and 248 nm) and laser power.

The emission from lead atoms can be observed within 12 mm from the irradiation surface. This emission can be also observed even when the thermal chain reaction does not take place. These results indicate that the excited lead atom is formed in the first step of laser irradiation.

The details of emission, such as pressure effect, time resolved spectra, position dependence, and so on, are now under investigation and will be reported in the near future.

In conclusion, we have discovered a unique laser-induced decomposition reaction of organometallic compounds. Since this reaction is initiated by only a single shot of laser irradiation and caused an explosive reaction, we would like to call this type of decomposition reaction "laser ignited mild explosive reaction" (LIMER). The key points of LIMER are the high density formation of active species and the propagation of a thermal chain reaction caused by these active species.

REFERENCES

1. Karny, Z., Gupta, A., Zare, R. N., Lin, S. T., Nieman, J., and Ronn, A. M., *Chem. Phys.* **37,** 15 (1979).
2. Sausa, R. C. and Ronn, A. M., *J. Chem. Phys.* **81,** 1716 (1984).
3. Koda, S. and Fujiwara, O., *21st Symp. (Int.) Combust.* (1988).

Nucleation with Simultaneous Chemical Reaction

JOSEPH L. KATZ* AND MARC D. DONOHUE

Department of Chemical Engineering, Johns Hopkins University, Baltimore, MD 21218

ABSTRACT

Phase transitions sometimes occur as a result of chemical reaction. This is common at high temperatures when gaseous reactants produce involatile products, e.g., combustion reactions resulting in the formation of SiO_2, TiO_2, Al_2O_3, and so on. It is fairly simple to describe such processes if the nucleation is independent of the reaction. However, such phase transitions can be self-catalytic, since subcritical-sized clusters of the condensing phase can serve as reaction sites. We show how to generalize nucleation theory so that it describes these processes. We find that when the arrival rate of condensing molecules and the rate of chemical reaction at these clusters are of comparable magnitude, the rate of nucleation will be greatly increased. This increase occurs even if the gas-phase reaction can be assumed to be infinitely fast. Results of calculations for the condensation of silica ($SiO + \frac{1}{2} O_2 \rightarrow SiO_2$) are presented.

Index Entries: Nucleation; chemical reaction; SiO_2, silica; surface reaction.

INTRODUCTION

At high temperatures, the chemical composition of a condensed phase is often different from the vapor phase. Consider, for example, the condensation of SiO_2 via the reaction of gaseous SiO and O_2. For condensation to occur, there has to be chemical reaction. Where does this reaction occur? Does it occur in the vapor phase, on the surface of the condensed phase, or perhaps on clusters of the condensing phase? In this paper, we show how to generalize nucleation theory to allow for these possibilities. This work is based on a very general and powerful

*Author to whom all correspondence and reprint requests should be addressed.

theory that was developed to describe nucleation of vacancies to form vacancy clusters (*1–4*) in the presence of their antinucleator, i.e., interstitial atoms. This theory is able to describe simultaneous multiple nucleation processes (*5–7*). It also has been applied to nucleation on a smooth surface where there can be two simultaneous processes occurring, direct condensation from the vapor to the cluster, and condensation to an adsorbed layer, followed by surface diffusion to the cluster (*8*). The application of interest here is condensation that occurs via chemical reaction when this chemical reaction can occur on or in clusters of the condensing phase, as well as in the vapor phase (*9*).

For simplicity of presentation, we will discuss only reactions of the type

$$\text{A(gas)} + \text{B(gas)} \longrightarrow \text{C(liquid or solid)} \qquad [1]$$

However, application of these results to a more complicated stoichiometry is straightforward.

One needs to consider whether the chemical reaction occurs homogeneously in the gas phase, or on the surfaces of clusters formed by homogeneous nucleation. If the chemical reaction occurs homogeneously in the gas phase only, then what occurs depends on whether the reaction rate is fast compared to the condensation rate. This will be discussed below. However, if the reaction can occur on or in the nuclei themselves, then the nucleation becomes self-catalytic.

There are a number of other factors that influence the actual condensation rate for a system in which chemical reaction is taking place. One factor is the presence or absence of foreign surfaces or heteronuclei. When large numbers of heteronuclei of sufficient effectiveness are present, their surface area completely controls the condensation process. However, since we are concerned with the effect of chemical reaction rate on the homogeneous nucleation rate, we shall confine our discussion to those situations in which there are few heteronuclei or the increase in supersaturation is sufficiently rapid that it cannot be relieved by condensation on the heteronuclei alone.

In high-temperature coal combustion, silica (i.e., silicon dioxide), which is the major component of coal slag, evaporates and dissociates to silicon monoxide and oxygen. As the combustion gases cool during subsequent passage through process equipment, these slag vapors condense, often with very deleterious effects (e.g., poisoning of catalysts, fouling of heat exchange tubes, and erosion of turbine blades). The condensation rate of silica depends on the concentration of silicon monoxide in the vapor phase, the number of type of any heteronuclei present, temperature, the cooling rate, and the oxygen partial pressure, via the chemical equilibrium

$$SiO + \tfrac{1}{2}O_2 \rightleftharpoons SiO_2 \qquad [2]$$

When the cooling rate is fast (or when there are few effective heteronuclei), homogeneous nucleation will occur. Depending on the temperature and oxygen partial pressure, this occurs either as the nucleation of pure silica or chemical nucleation. At very low temperature (and for very high oxygen partial pressures), the equilibrium favors SiO_2 over SiO in the vapor phase. Therefore, SiO_2 can condense via pure-component homogeneous nucleation. However, at high temperatures and reasonable oxygen partial pressures, the silicon in the vapor phase is predominantly silicon monoxide. Under these conditions, nucleation occurs by a second mechanism, nucleation with simultaneous chemical reaction. The higher the temperature or the lower the oxygen partial pressure, the more likely that nucleation with simultaneous reaction will dominate.

HOMOGENEOUS NUCLEATION

Before beginning the discussion of condensation processes that occur via chemical reaction, we first present a brief explanation of homogeneous nucleation. A snapshot of the molecules present in a supersaturated state would show small clusters of the new phase containing two or more molecules. There also would be occasional clusters containing a larger number of molecules; however, a motion picture of these larger clusters would show that most are very short-lived; they grow rapidly and then shrink rapidly. Nucleation occurs when a cluster of two, three, four, or some other small number of particles grows (fluctuates in size) to a size large enough that it will continue to grow rather than shrink.

A critical-sized nucleus is a cluster of size such that its rate of growth is equal to its rate of decay. Any cluster that fluctuates to a size larger than the critical size will probably continue to grow to macroscopic size, whereas clusters smaller than the critical size most likely will shrink. The nucleation rate is then the net number of clusters per unit time that grow larger than the critical size.

In dilute solution, whether gaseous or condensed, the concentrations of clusters (dimers, trimers, and so on) are much lower than the concentration of monomers. Because of these low concentrations, it is unlikely that two clusters will ever collide. Therefore, we assume that all clusters grow by the addition of single molecules. From the principle of microscopic reversibility (whereby at equilibrium every forward process has to be matched by its corresponding reverse process), it follows that evaporation also occurs one molecule at a time.

The net rate $J(i)$, at which clusters of size i grow to size $i+1$ is the difference between the rate at which clusters of size i add an additional molecule and the rate at which clusters of size $i+1$ lose a molecule, that is

$$J(i) = f(i)n(i) - b(i+1)n(i+1) \quad [3]$$

where $f(i)$ represents the forward rate—the rate of addition of molecules to a cluster $b(i+1)$ is the backward rate—the rate of loss of molecules from a cluster, and $n(i)$ is the number, or concentration, of clusters of size *i*. Each of these quantities, and hence $J(i)$, are functions not only of temperature, cluster size, and pressure, but also can be functions of time.

A convenient equation that contains all the nucleation information can be readily derived by defining a recursive relation

$$Z(i+1) = Z(i)b(i+1)/f(i+1) \quad \text{for } i > 1 \tag{4}$$

and

$$Z(1) = 1 \tag{5}$$

This recursive relation is easily solved yielding

$$Z(i) = \prod_{j=1}^{i} b(j)/f(j) \tag{6}$$

Multiplying Eq. [3] by $Z(i)$ and summing, one obtains

$$\sum_{i=1}^{\hat{i}} J(i)Z(i) = \sum_{i=1}^{\hat{i}} [f(i)n(i)Z(i) - f(i+1)n(i+1)Z(i+1)] \tag{7}$$

Note that the two terms on the right-hand side are identical except for their indices. Then, on summing, successive terms cancel and Eq. [7] simplifies to

$$\sum_{i=1}^{\hat{i}} J(i)Z(i) = f(1)n(1) - f(\hat{i}+1)n(\hat{i}+1)Z(\hat{i}+1) \tag{8}$$

On carrying out the summation to a sufficiently large value of $\hat{i}$, the last term on the right-hand side of this equation becomes negligible compared to the first because $Z(i+1)$ rapidly goes to zero as i becomes large (7). Furthermore, in most cases of interest, a steady state is rapidly established. Thus, J becomes a constant for all sizes, can be factored out of the summation, and one obtains

$$J = f(1)n(1)/\{1 + \sum_{i=2}^{\hat{i}} \prod_{j=2}^{i} b(j)/f(j)\} \tag{9}$$

To proceed further, one needs explicit expressions for the backward rate $b(i)$ and the forward rate $f(i)$.

Let's first consider the condensation of a one-component supersaturated gas (without any chemical reaction). The forward rate, that is the rate that impinging molecules condense, is the product of the surface area of the cluster, $a(i)$, the rate per unit area at which molecules impinge on the surface, β, and the condensation coefficient, α, i.e.,

$$f(i) = a(i)\beta\alpha \tag{10}$$

For an ideal gas, the impingement rate of molecules on a cluster is

$$\beta = P/(2\pi mkT)^{1/2} \tag{11}$$

The condensation coefficient, α, is a complicated and unknown function of cluster size. However, its effect on the nucleation rate is relatively small since, as will be shown later, the nucleation rate depends primarily on a ratio of forward rates and α cancels.

The backward rate is the rate at which molecules leave a cluster. This is not known in general. However, for dilute systems (i.e., systems in which the number density of solute molecules is much smaller than the number density of solvent molecules), the interactions between a molecule leaving a cluster and other solute molecules are negligible compared to the interactions with solvent molecules. Therefore, in dilute systems, the backward rate $b(i+1)$, although a complicated function of temperature and cluster size, is independent of the concentration of the nucleating species, i.e., the solute. We obtain $b(i+1)$ by realizing that, at true equilibrium (i.e., in a saturated vapor), the nucleation rate $J(i)$ is zero for *all* i and, therefore Eq. [3] can be solved to yield:

$$b(i+1) = f_e(i)n_e(i)/n_e(i+1) \qquad [12]$$

where the subscript e is used to remind the reader that these concentrations and forward rates refer to those found at saturation, not those in the actual nucleating vapor. Eqs [10], [11] and [12] can be substituted into Eq. [9]. However, a simplification occurs when $f(i)/f_e(i)$ is independent of the size of the cluster. This is usually the case and occurs when the forward rate is a product of terms that are either size-independent or whose size dependence is independent of supersaturation. Assuming so, one obtains

$$J = 1/\sum[a(i)\beta\alpha n_e(i)(P/P_e)^i]^{-1} \qquad [13]$$

Equation [13] can be written in the standard form of classical nucleation theory by making the additional assumption that the equilibrium cluster distribution is given by

$$n_e(i) = N\exp[-\sigma a(i)/kT] \qquad [14]$$

where N is a normalization constant with σ is the surface free energy of a flat liquid surface of the same composition and temperature as the nucleus.

HOMOGENEOUS GAS-PHASE REACTION

If the reaction occurs only in the gas phase, then the condensation can be represented as the two step process

$$A(gas) + B(gas) \rightarrow C(gas) \qquad [15]$$

$$C(gas) \rightarrow C(liquid\ or\ solid) \qquad [16]$$

where Eq. [16] represents the nucleation process.

The equation for the nucleation rate is then identical to the equation for a system where no reaction has occurred, i.e., Eq. [13] is valid. In order to determine β and J, the partial pressure of species C in the vapor phase is needed. When both the forward and the backward reactions

$$A(gas) + B(gas) \rightleftharpoons C(gas) \qquad [17]$$

are fast compared to the nucleation, the partial pressure of species C can be calculated from the equilibrium constant, $K_p = P_c/(P_a P_b)$.

SURFACE-ENHANCED CHEMICAL REACTION

When the reaction occurs both homogeneously in the gas phase and on the surface (or in the bulk) of clusters or droplets that are already present, the nucleation rate can be enhanced drastically. As was the case in homogeneous nucleation, the forward rate is the rate at which a cluster containing i molecules of species C grows to contain $i+1$ molecules. As before, clusters can grow by the condensation of C molecules that were formed previously by the gas-phase reaction. However, they also can form by the reaction on (or in) the cluster of adsorbed (or absorbed) A and B molecules. The forward rate then is given by

$$f_c(i) = \beta_c \theta_c \alpha a(i) + k[A][B]a(i) \qquad [18]$$

where the subscript C denotes the species. As before, the impingement rate β depends on the concentration or partial pressure of molecules of species C in the vapor phase and $a(i)$ is the surface area of a cluster containing i molecules of species C. The rate that clusters grow by the condensation of gas-phase molecules of species C differs from that for a pure vapor by the additional factor θ_c in the first term of the left-hand side of Eq. [18], i.e., the fraction of the surface of the cluster that has exposed C molecules. This implies that C molecules can not condense on that portion of the cluster surface that is covered by adsorbed A and B molecules. Although this may seem strange, it is necessary to maintain the principle of microscopic reversibility. Since it seems reasonable that a C molecule cannot evaporate from that portion of the surface that is covered by A and B molecules, it follows that C molecules cannot condense and evaporate through an adsorbed layer either. In reality, C molecules do condense and evaporate through an adsorbed layer, but at a rate that is much smaller than for exposed C molecules.

The second term on the right hand side of Eq. [18] represents the rate that the cluster increases its number of C molecules via the condensed-phase reaction. Here, k is the surface reaction rate constant and $[A]$ and $[B]$ represent the surface concentrations. This equation presumes that the heterogeneous chemical reaction occurs on the surface of clusters of C molecules. If the heterogeneous reaction occurs in the bulk of the cluster, then Eq. [18] would be

$$f(i) = \beta_c \alpha a(i) + k'[A]'[B]'v(i) \tag{18'}$$

where the primes on k, $[A]$, and $[B]$ indicate that this is a bulk liquid-phase reaction and $v(i)$ is the volume of a cluster of size i

A nucleus also can lose species C by two processes: direct evaporation, and the reverse reaction to form species A and B. The backward rate is thus given by

$$b_c(i) = \gamma_c(i)\theta_c a(i) + k'[C]a(i) \tag{19}$$

where γ_c is the rate of evaporation of C molecules/unit area, k′ is the rate constant for the reverse reaction, and [C] is the surface concentration of species C. The surface concentrations (molecules/unit area) and surface fraction are related by

$$\theta_c = s_c[C] \tag{20}$$

where s_c is the area on the surface occupied by one molecule of species C and

$$[A]s_a + [B]s_b + [C]s_c = 1 \tag{21}$$

The backward rate $b(i)$ then can be written as

$$b(i) = (\gamma_c s_c + k')[C]a(i) \tag{22}$$

Although the evaporation rate per unit area, γ_c, still can be assumed independent of supersaturation, Eq. [22] shows that the backward rate is supersaturation dependent if [C] depends on the supersaturation. Thus, one can not equate $b(i+1)$ and $b_e(i+1)$ as was done in deriving Eq. [12]. Nonetheless, it is still true that, at equilibrium, the forward and backward processes proceed at exactly equal rates; therefore

$$b_e(i+1) = f_e(i)n_e(i)/n_e(i+1) \tag{23}$$

Furthermore, the rate at which the individual processes (γ_c and k'), which lead to the loss of one C molecule from the cluster occur is unchanged. Thus, from Eq. [22], the backward rate can be related to the equilibrium backward rate, i.e.,

$$b(i+1) = b_e(i+1)[C]/[C]_e \tag{24}$$

Combining Eqs. [23], [24], and [18] one obtains

$$\frac{b(i)}{f(i)} = \frac{\beta_e^* a(i-1)n_e(i-1)}{\beta^* a(i)n_e(i)} \tag{25}$$

where

$$\beta^* \equiv \beta\theta_c\alpha + k[A][B]/[C]s_c \tag{26}$$

Substituting into Eq. [9] one obtains

$$J = \frac{\beta^* n(1)/n_e(1)}{\sum 1/[a(i)n_e(i)(\beta^*/\beta_e^*)^{i-1}]} \tag{27}$$

The summation can be converted to an integral, expanded around its maximum value, and integrated. This integration can be performed without any knowledge of $n_e(i)$ except that its i dependence is weaker than the *ith* power. However, to proceed further $n_e(i)$ has to specified. Equation [14], when used with a size-independent surface tension, is equivalent to the standard assumption used in classical nucleation theory. Since the purpose of this paper is to illustrate the generalization of nucleation theory to allow for surface chemical reaction, we used it and obtain

$$J = N_c a_0 [a_0 \sigma / kT]^{1/2} \beta^* \exp\left[\frac{-4a_0^3 \sigma^3}{27(kT)^3 (\ln \beta^* / \beta_e^*)^2}\right] \qquad [28]$$

where N_c is the number of molecules/unit volume of species C present as monomers in the vapor phase, $a_o \equiv 4\pi(3v/4\pi)^{2/3}$ and v is the vol./mol.

EXAMPLE CALCULATION

Silica Condensation

As discussed above, a condensed phase chemical reaction can dramatically affect the nucleation rate. Equations [28] and [26] show that the nucleation rate depends on the, thermodynamic driving force (the amount of undercooling), the rates of gas-phase and surface chemical reactions, and the extent to which the reactant molecules adsorb on the surfaces of nuclei of the condensing product. For the condensation of silica (under typical high-temperature combustion conditions), the rate of gas-phase reaction should be very fast. Therefore, we assume that the vapor-phase concentration of SiO_2 is controlled by its chemical equilibria with SiO and oxygen. But this does not mean that one can use standard homogeneous nucleation theory with the SiO_2 concentration determined from the $SiO \rightleftarrows SiO_2$ equilibrium.

Under most conditions, nucleation occurs by the formation of SiO_2 molecules on the surface of clusters of SiO_2, i.e., SiO molecules arrive and are adsorbed; O_2 molecules arrive and are adsorbed; and it is these adsorbed species (including atomic oxygen) that then surface migrate and react. This extra mechanism for the formation of SiO_2 can drastically enhance the nucleation rate. This enhancement is so drastic that a plot of nucleation rates as a function of the various parameters would not be meaningful. For example, at a total partial pressure of all silicon containing species of 5 Pa and a temperature of 2070 K, the various cases shown in Fig. 1 would predict rates of 10^{-10}, 10^{-5}, 10^3, 10^4, and 10^{13}, respectively. Therefore, a different way of presenting the data is needed.

In Fig.1, the temperature at which there is a nucleation rate of 1 nucleus/cm^3/sec is given as a function of total moles of all silicon containing species (i.e., Si, SiO, and SiO_2) for various values of the reaction rate constant, and various surface coverages. These calculations are for a stoichiometric amount of Si and O_2 at a Si concentration typical of a high-temperature coal combustion process. The calculations were made using a modified version of the NASA computer program "Computer Program

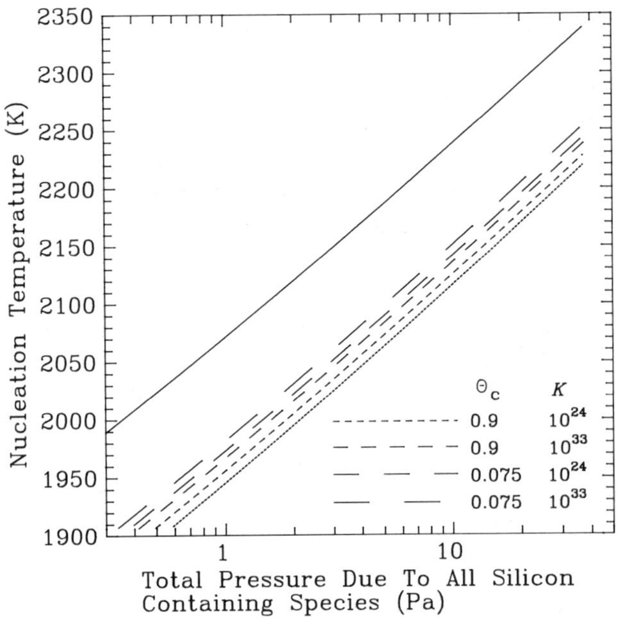

Fig. 1. Nucleation temperature for a nucleation rate of $1/cm^3/sec$ as a function of the total concentration of silicon present in the vapor phase. The solid line is the equilibrium vapor pressure of SiO_2 (l). The dotted line is the homogeneous nucleation temperature. θ_c is the fraction of the cluster surface covered by C molecules. $K \equiv k/(s_a s_b)^{1/2}$

for Calculation of Complex Chemical Equilibrium Compositions, Rocket Performance, Incident and Reflected Shocks, and Chapman-Jouguet Detonations" (*10*). The original computer program calculates chemical and phase equilibria for multicomponent systems. Our version makes two chemical equilibria calculations for each temperature and total pressure. In the first calculation, both chemical and phase equilibria are allowed. In the second calculation, the properties of a supersaturated state are calculated by forbidding the existence of all condensed phases. Using both sets of partial pressures and various values for the surface reaction rate constant and the extent of surface coverage, the nucleation rate with chemical reaction was then calculated using Eqs. [28] and [26]. (Note that for these examples calculations, $[O_2]^{1/2}$ power was used wherever [*B*] appears in these equations, as dictated by the stoichiometry of the reaction, Eq. [2].) The surface tensions needed to calculate n_c (*1*) were determined from the experimental data by Kingery (*11*). These calculations show that as the surface-reaction rate constant increases, or as the fraction of the surface covered by adsorbed reactant molecules increases, nucleation occurs at a higher temperature (i.e., lower supersaturation) for a given silicon concentration. An equilibrium calculation predicts that condensation would occur at 2181 K if the vapor phase consists of SiO, SiO_2, and stoichiometric oxygen at a sum of SiO and SiO_2 partial pressures of 5 Pa. But in a rapidly cooling gas stream, condensation will not occur at equilibrium. Nothing would occur until some nucleation mechanism was activated. Homogeneous nucleation of pure SiO_2 occurs at 2054 K; however, nucleation with simultaneous chemical reaction occurs more readily, probably between 2064 and 2090 K. The self-catalytic effect of chemical nucleation is clearly evident; condensation occurs more readily because the nuclei themselves serve as chemical reaction sites. This enhanced reaction rate results in nucleation sooner than expected. Of course, in any real coal combustion process, there would be numerous other species present. These species could themselves nucleate and can also result in multicomponent condensed phases. To treat the more complex real system will require a further major generalization of theory to include these multicomponent nucleation processes.

ACKNOWLEDGMENT

This research was supported by the Division of Materials Sciences, Office of Basic Energy Sciences, US Department of Energy.

REFERENCES

1. Katz, J. L. and Wiedersich, H., *J. Chem. Phys.* **55,** 1414 (1971).
2. Katz, J. L. and Wiedersich, H., *J. Nucl. Materials* **46,** 41 (1973).
3. Wiedersich, H., Burton, J. J., and Katz, J. L., *J. Nucl. Materials* **51,** 287 (1974).
4. Wiedersich, H. and Katz, J. L., *Adv. Colloid Interface Sci.* **10,** 33 (1979).
5. Katz, J. L. and Wiedersich, H., *J. Colloid Interface Sci.* **61,** 351 (1977).
6. Katz, J. L. and Spaepen, F., *Philosophical Magazine* **B 37,** 137 (1978).
7. Katz, J. L. and Donohue, M. D., *Adv. Chem. Phys.* **40,** 137 (1979).
8. Sheu, J. S., Maa, J. R., and Katz, J. L., *J. Statisical Phys.* **52,** 1143 (1988).
9. Katz, J. L. and Donohue, M. D., *J. Colloid Interface Sci.* **85,** 267 (1982).
10. Gordon, S. and McBride, B. J., *NASA #N71-37775,* Washington, DC (1971).
11. Kingery, W. D., *J. Am. Ceram. Soc.* **42,** 6 (1959).

Role of High Temperature Chemistry in CVD Processing

KARL E. SPEAR*,[1] AND RYAN R. DIRKX[2]

[1]*Materials Science and Engineering, Pennsylvania State University, University Park, PA 16802; and* [2]*M&T Chemicals, Rahway, NJ 07065*

ABSTRACT

High temperature chemistry principles are examined as important tools for developing an understanding of chemical vapor deposition (CVD) systems, and predicting deposition behavior. An attempt is made to qualitatively predict the complex interdependencies among controllable experimental parameters, process variables, and deposition properties. Examples of silicon boride deposition from SiH_4, BCl_3, and H_2 mixtures are used to illustrate high temperature thermochemical modeling techniques that utilize the concept of local equilibrium.

Index Entries: High temperature chemistry; chemical vapor deposition; modeling vapor deposition; thermochemical modeling; silicon borides.

INTRODUCTION

Chemical vapor deposition (CVD) has become an extremely important high temperature synthesis technique for coatings and powders. In a CVD process, gaseous species chemically react to form one or more condensed phases plus gaseous products. The chemistry involved in CVD is usually much more complex than in physical vapor deposition (PVD), a process in which gaseous species typically produced by evaporation or sputtering processes are condensed on a substrate. The present paper examines CVD processes, and particularly the role of high temperature chemistry in CVD as an aid in predicting the complex

*Author to whom all correspondence and reprint requests should be addressed.

chemistry and deposition behavior as a function of experimentally controllable CVD parameters.

An excellent overview of vapor deposition, a term collectively used to include all CVD and PVD methods, and its many applications is given in a book edited by Bunshah (*1*). The development of specific deposition techniques for densifying composites and other porous bodies, chemical vapor infiltration (CVI), is reviewed by Naslain (*2*), and those for achieving selective area or selective phase deposition are reviewed by Carlsson (*3*). These and every commercial coating process have a goal of achieving a particular set of coating properties. These sets of prioritized properties drive the CVD development studies, studies to determine the controllable parameters that will achieve the desired properties. For chemically-complex systems, using trial and error methods to achieve these properties is too costly in terms of both money and time. This is why research to develop predictive CVD models is so important. The ability to predict coating properties from a knowledge of the controllable experimental parameters, and the influence of these parameters on process variables, will typically impact a number of diverse applications for a particular coating.

Papers published over the past decade (*4–9*) have illustrated the value of high temperature thermochemical modeling techniques in CVD studies. The high temperature data needed for such modeling may be experimentally determined, or else predicted by calculational techniques and from established trends in high temperature behavior. More details on the uses of high temperature chemistry as a tool in developing an understanding of complex CVD processes is illustrated throughout this paper. The general areas of high temperature chemistry and modeling CVD processes are described first. This is followed by an outline of the complex interrelationships that exist among experimental CVD parameters, process variables, and deposit properties. Finally, examples of silicon boride deposition (*10*) are presented to illustrate the value of thermochemical modeling in analyzing CVD behavior.

HIGH TEMPERATURE CHEMISTRY

High temperature chemistry has been defined by Leo Brewer as *the chemistry at temperatures which are high enough to cause significant differences from the expected room temperature chemistry* (*11*). An effort to systematically understand the chemical behavior of materials at high temperatures was initiated by Brewer and coworkers more than 40 years ago during the Manhattan Project. These studies resulted in the birth of high temperature chemistry as a major area of chemical research, one that involves the chemical, thermodynamic, kinetic, and structural properties of species, systems, and processes at high temperatures.

Generalizations about the chemistry of reactions and compounds at room temperature often do not hold as the temperature of a system is increased (*11*). Chemical reactions that are not energetically favorable at room temperature may become important at high temperatures. Thermodynamic, rather than kinetic, limitations tend to determine high temperature reactivity. Vaporization processes and vapor species become increasingly important. Unusual compounds and vapor species that do not conform to the familiar oxidation states of the elements may form at high temperatures, and the complexity of the vapor phase increases. With increasing temperatures, ordered defect structures become disordered, and solid solution ranges increase significantly. Many of these effects result because reaction entropies gain in importance with respect to reaction enthalpies in determining the chemistry at high temperatures.

The above generalizations, along with experiences in the laboratory, led Searcy (*12,13*) to propose the first two "laws" of high temperature chemistry and Margrave (*14,15*) to add a third. These are (1) at high temperatures, everything reacts with everything else; (2) the higher the temperature, the more seriously everything reacts with everything else; and (3) the products might be anything. These laws humorously summarize the general essence of chemical trends in high temperature systems. Brewer (*16–19*) has shown that these laws are actually based on thermodynamic principles.

MODELING CVD PROCESSES

A CVD process involves gases flowing into a reactor where chemical reactions occur to produce a condensed phase plus gaseous products. These product gases, plus unreacted or partially reacted input gases, are then forced out of the reactor by the gas flow. A dynamic CVD process will achieve steady-state conditions unless fluctuations in temperature, total pressure, input gas composition, or flow rates or patterns dominate. The steady-state model plus local equilibrium concepts greatly simplify our ability to predict deposition behavior. Modeling deposition processes for a coating over a fixed geometric surface necessarily involves modeling not only the high temperature chemistry of the system, but also its mass transport and flow properties. This latter type of modeling has been reported for a few idealized systems with relatively "simple" chemistry (*20–25*), but the complexities of such chemically reactive flow models preclude their current use as a tool for most CVD studies.

The present discussions do not directly consider flow behavior and system geometry, but instead examine the local high temperature chemistry that can be predicted for a CVD system. The following discussion is intended to illustrate more clearly where our local modeling fits into a model for a complete CVD system.

Consider mentally placing a grid on a substrate to divide it into a finite number of cells, which include the gas phase above the substrate. The present paper concerns the steady-state chemistry in each of these substrate cells and the rates at which the mass transport and chemical processes occur within each cell. Coupling the residence time of the gas in each cell with the energy and mass transport from cell to cell would

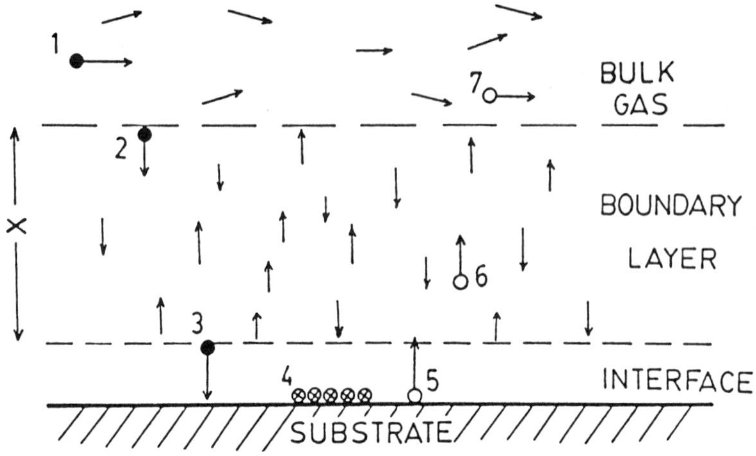

Fig. 1. Schematic diagram illustrating seven sequentially-linked mechanistic steps in a CVD process.

allow us to calculate the deposition behavior over an entire substrate. Of course, the system geometry and flow properties will determine the boundary conditions in each cell that are needed for calculating rates of deposition processes and the fraction of "input gas" that reacts within a given cell before transport to the next cell(s) occurs.

A simple physical model often used for describing a CVD system (*5,6,9*) is probably quite accurate for the substrate cell discussed above. As illustrated in Fig. 1, seven sequentially-linked mechanistic steps describe the deposition process

a) The first and seventh steps are the respective forced flow of reactant gases into the system and the forced exit of gases from the system.
b) The second and sixth steps are the respective diffusion and bulk (viscous) flow of reactants from the bulk gas through a gaseous concentration boundary layer to the substrate and the similar transport of product gases from the substrate out into the bulk gas.
c) The third and fifth steps are the respective adsorption of gases onto the substrate and desorption of adsorbed species from the substrate.
d) The fourth step includes the deposition reaction, as well as other chemical processes occurring on the surface.

High temperature research is needed for each of these steps, but is largely lacking for most CVD systems. For example, it is usually not known if the input gases react homogeneously before diffusing to the substrate surface, or if the rates of such reactions are slow in comparison to the transport rates. This information is required if we are to model the above second and third steps in the deposition process.

The adsorption and desorption steps typically have not been studied for the species and substrate or coating surfaces of interest. However, it is sometimes possible to deduce from experimental deposition results whether the desorption of important species such as HCl(g) could be rate-limiting, or if an abundant species like hydrogen is preferentially occupying (poisoning) the surface sites.

The fourth step, the surface reactions, may be a complex set of chemical reactions. The chemistry of the adsorbed species, relative species concentrations and activities, collision frequencies, lifetimes of adsorbed species, surface diffusion rates, reaction rates, and thermodynamic limits on the reactions are all examples of useful high temperature chemistry information for CVD modeling studies.

COMPLEXITIES OF CVD PROCESSES

The interrelationships in a CVD system among controllable experimental parameters, process variables, and coating properties are extremely complex, which means modeling a CVD system is extremely complex. This section summarizes a first attempt to qualitatively define these interrelationships to provide a framework for organizing experimental observations, and deducing experimental parameter changes that would lead to desirable coating properties. A more thorough discussion of these complex interrelationships has been given by Dirkx (*10*).

Generic listings are given in Table 1 for experimentally controllable parameters, process variables in a thermally-activated CVD reactor, and typically coating properties of interest. These lists, of course, could be modified for a specific system being used to produce a coating with specific properties.

The importance of understanding the relationships among these parameters, variables, and properties is illustrated by the simple schematic diagram shown in Fig. 2. Experimental parameters can be varied based on operator experience and "trial and error" methods until a coating with desired properties is achieved. Alternatively, models for the deposition system can be used to calculate the controllable parameters needed to achieve the desired deposition results. For complex chemical systems, the modeling techniques are needed to at least limit the parameter space most likely to yield the optimum deposition conditions.

Modeling requires understanding the influence of experimental parameters on process variables, and the influence of process variables on coating properties. Our attempts to qualitatively predict the generic interdependencies are shown in Tables 2 and 3.

The information in Tables 2 and 3 could be misleading in that the actual interrelationships are more complex than indicated. Many of the process variables are dependent upon other process variables. For example,the experimental parameter of substrate temperature influences the process variable of gas temperature, which influences the flow behavior, the gas chemistry, and, thus, the deposition rate. The flow behavior influences transport processes and deposition rate. The substrate tem-

Table 1
Listings of the Interrelated Experimentally Controllable Process Parameters, Process Variables, and Coating Properties

Controllable Parameters	Process Variables	Coating Properties
Reactor Geometry	Gas Temperature	Morphology
Substrate Temperature	Flow Behavior	Composition
Substrate Environment	Mean Gas Velocity	Thickness
Input Gas Composition	Gas Chemistry	Density
Total Flow Rate	Gas Homogeneity	Uniformity
Total Pressure	Deposition Rate	Adhesion
Substrate Surface	Deposition Surface	Crystallinity
Time of Deposition	Nucleation Density	Stress
Experimental Procedure		

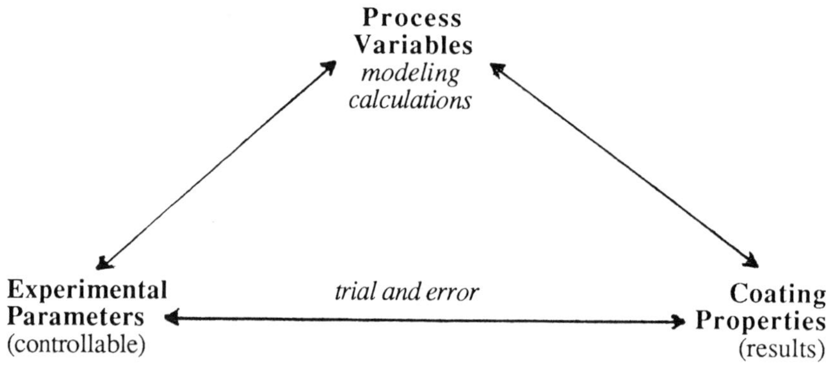

Fig. 2. Schematic relationship among experimental parameters, process variables, and coating properties.

Table 2
Experimental Parameters Influence on Process Variables[a]

Experimental Parameters	Influence on Process Variables[b]							
	GT	FB	MGV	GC	GH	DR	DS	ND
Reactor Geometry	X	X	X	X	X	X	X	X
Substrate Surface	–	X	–	–	–	X	X	X
Substrate Temperature	X	X	X	X	X	X	X	X
Substrate Environment	X	X	–	X	X	X	X	–
Input Gas Composition	X	X	–	X	X	X	X	–
Total Flow Rate	X	X	X	X	X	X	X	–
Total Pressure	X	X	X	X	X	X	X	–
Time of Deposition	–	X	–	–	–	–	X	–
Experimental Procedure	–	–	–	–	–	–	X	X

[a]The "X" qualitatively indicates a measurable dependence.

[b]Process Variables are represented by the following initials: GT, Gas Temperature; FB, Flow Behavior; MGV, Mean Gas Velocity; GC, Gas Chemistry; GH, Gas Homogeneity; DR, Deposition Rate; DS, Deposition Surface; and ND, Nucleation Density.

Table 3
Process Variables Influence on Coating Properties[a]

Process Variable	Influence on Coating Properties[b]							
	Mor	Com	Thi	Den	Uni	Adh	Cry	Str
Gas Temperature	X	S	–	X	X	–	X	–
Flow Behavior	X	X	X	–	S	–	–	–
Mean Gas Velocity	X	X	–	–	X	–	–	–
Gas Chemistry	X	S	X	X	X	X	–	–
Gas Homogeneity	X	S	–	–	S	X	–	–
Deposition Rate	X	–	S	X	S	–	S	S
Deposition Surface	S	X	–	X	S	S	X	S
Nucleation Density	X	–	–	S	S	S	X	–

[a]The "X" qualitatively indicates a measurable dependence, the "S" a strong dependence.

[b]Coating Properties are represented by the following initials: Mor, Morphology; Com, Composition; Thi, Thickness; Den, Density; Uni, Uniformity; Adh, Adhesion; Cry, Crystallinity; and Str, Stress.

perature directly influences the surface reaction rates and deposition rate.

Despite the complexities of the CVD parameter-variable-property interrelationships, they can often be qualitatively determined and can sometimes be quantitatively calculated. Quantitative limits on deposition behavior can often be predicted, as can the sensitivity of deposition to changes in experimental parameters. A knowledge of the high temperature chemistry and energetics of the chemical system is critical for all of these CVD process predictions.

PREDICTING CVD BEHAVIOR USING THERMOCHEMICAL MODELING

Equilibrium calculations provide a powerful tool for gaining insight into both equilibrium and nonequilibrium vapor deposition processes. Thermochemical modeling can be used to predict deposition behavior and explain experimental observations such as:

1. condensed phases that will deposit at equilibrium,
2. maximum rates or efficiencies of deposition,
3. partial pressures of all gaseous species in homogeneous equilibrium before any condensed phases are deposited,
4. partial pressures of all gaseous species in heterogeneous equilibrium with the condensed deposit,
5. sensitivity of the above predictions to changes in experimentally controllable parameters, and
6. mechanistic possibilities for explaining nonequilibrium behavior.

Examples of thermochemical modeling results for the Si–B–H–Cl system are used to illustrate the above types of predictions of limiting deposition behavior and molecular mechanisms responsible for the deposition. Both theoretical and experimental results on the deposition of silicon borides from precursor gases of SiH_4, BCl_3, and H_2 are discussed (*10*). However, before presenting these examples, the critical concept of local equilibrium must be discussed.

Local Equilibrium—A Critical Concept for Thermochemical Modeling

Many people question the value of thermochemical modeling of CVD systems since equilibrium calculations are being used and CVD is clearly a dynamic, nonequilibrium process. The reason thermochemical modeling is so useful as a tool in predicting and explaining CVD behavior is that local equilibrium will always be achieved by a portion of the chemical processes in a CVD system. That portion of the overall system can then be analyzed using equilibrium thermodynamic calculations. Such results often provide limiting values for deposition parameters for a system. For example, at temperatures high enough for mass transport of gaseous species through a concentration boundary layer to and/or from the substrate to be rate-limiting (steps 2 and/or 6 in Fig. 1), the chemical reactions at the substrate surface are in local equilibrium (step 4 in Fig. 1). Calculations of the surface equilibrium indicate which species are important in the deposition processes and provide partial pressures (activities) that are needed to calculate diffusion to and from the substrate. The extent of the changes in the equilibrium results with changing experimental parameters can also provide insight into the sensitivity of the deposition behavior to these experimental parameters.

It is common for the experimental composition of gas at the equilibrium surface to be different from that calculated by allowing the bulk input gas to reach heterogeneous equilibrium at the surface. This difference between the gas compositions for the experimental surface and calculated equilibrium surface depends on a number of factors, including two primary ones of: (1) the difference in the ratio of elements in the deposited phase and the bulk input gas, and (2) the difference in mass transport efficiencies of the elements to and from the surface (a variety of species may contribute significantly to the transport of a particular element). This topic has been discussed in more detail previously (*6*). Despite these differences between the bulk gas composition and that of the gas arriving at the surface to cause deposition, the surface reactions still attain equilibrium at high temperatures, and equilibrium calculations can provide a method for deducing these compositional differences. Reasonable changes in the input gas for the surface equilibrium calculations can be made until the calculated results agree with the experimental observations. The changes necessary to achieve this "forced agreement" between the experimental and modeling results can provide insight into mechanisms for a deposition process.

An example for this valuable use of thermochemical modeling (and local equilibrium) is for the case in which the rate-limiting step for deposition is either the adsorption or desorption of a particular gaseous species at the substrate. We considered such a case in the deposition of Nb_3Ge in the Nb–Ge–H–Cl system (*26*). If the product gas HCl removal from the surface is rate-limited by desorption, then its concentration will build up at the surface, and equilibrium calculations of deposition behavior must include increased quantities of HCl in the input to the calculations. Comparisons of modeling results with experimental observations, in general, can provide insight into the nature and extent of kinetic barriers to deposition.

Auxiliary Data and Procedures for the Si–B–H–Cl Deposition Modeling

The auxiliary high temperature chemistry data required for the above such predictive calculations must be published, measured, or predicted through correlations or calculations in order to perform even the simplest modeling studies. Thus, the vital role of high temperature chemistry is again apparent. Thermodynamic modeling calculations requires a knowledge of the chemical compositions and solid solution ranges of all possible gaseous species and condensed phases that can exist in the temperature and system composition ranges of interest. For our CVD calculations on the Si–B–H–Cl system, 33 gaseous species and five solid phases (no solid solutions) were used.

The required thermodynamic stabilities as a function of temperature for most of the gaseous species were available in the JANAF Tables (*27*). We developed a self-consistent set of data for the five solid phases in the Si–B system using free energy–phase diagram relationships (*28*). Each individual calculation required specifying the concentrations (relative flow rates) of starting components, and the equilibrium temperature and total pressure. Typical conditions chosen for both the calculations and the experimental studies were: temperature 1000–1800 K, total pressure 0.0263 atm (2,665 Pa), BCl_3 flow rate 30 sccm (standard cubic centimeters/min), SiH_4 flow rate 10 sccm, and H_2 flow rate 1000 sccm. Each calculation represents one point on the diagrams given below. The computer program used on microcomputers in the present study was a slightly modified version of SOLGASMIX written by Eriksson (*29–31*).

CVD Phase Diagrams

Generating CVD phase diagrams for a deposition system is one of the first types of thermodynamic modeling calculations that should be performed. These diagrams map the experimental parameters that will produce equilibrium deposition of a phase or phases of interest. Figure 3 is an example of a deposition temperature vs input gas composition CVD diagram for the Si–B–H–Cl system. The temperature range is from 1300 to 1800 K, whereas the input gas composition in terms of a ratio of [B/(B + Si)] ranges from 0.5 to 1.0 in this diagram. The diagram is for a constant total pressure of 0.0263 atm, and a constant $[H_2/(BCl_3 + SiH_4)]$ of 25.

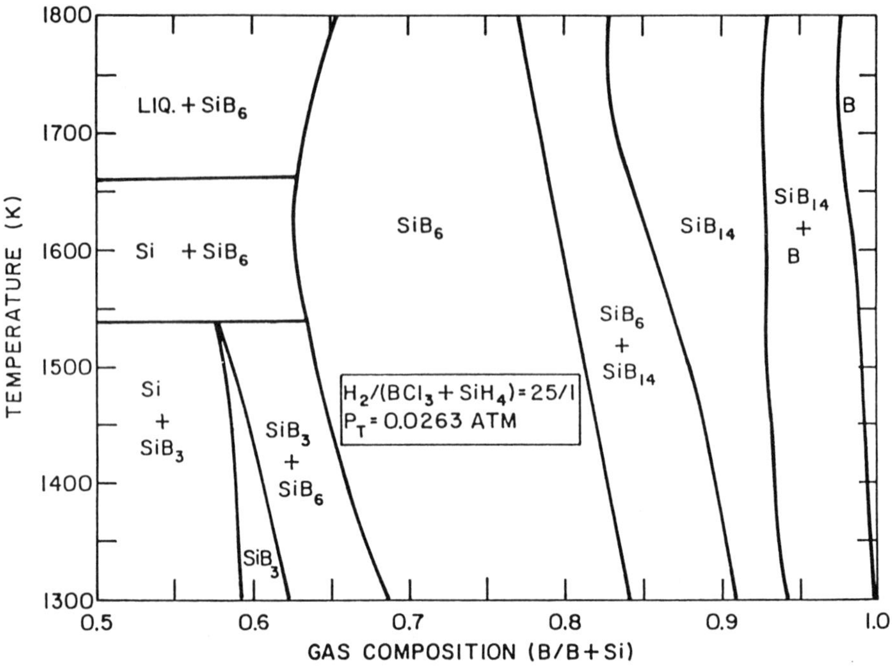

Fig. 3. CVD phase diagram for precursor gases SiH_4–BCl_3–H_2. The gas composition axis refers to the precursor gases and the temperature to the deposition temperature. This diagram predicts which condensed phase(s) deposit at equilibrium under the given experimental conditions.

These CVD phase diagrams provide a broad overview of the experimental parameter space in which the desired phase(s) can be deposited. The CVD diagram in Fig. 3 shows that small changes in experimental parameters will cause the system to shift from the single phase SiB_3 deposition region. The deposition of single phase SiB_3 may be experimentally difficult and quite sensitive to fluctuations in experimental parameters. The CVD diagram also shows that the temperature - composition range that will produce SiB_6 at equilibrium is large. However, possible deviations between calculated and experimental deposition may be expected for both SiB_3 and SiB_6 since the ratio of the elements in these condensed phases is significantly different from the ratio of these elements in the input gas required to deposit these phases. The diagram indicates that the surface gas composition will probably be richer in silicon than the bulk input gas composition.

The effects of doubling the hydrogen concentration on a portion of the above temperature vs composition CVD phase diagram is illustrated by Fig. 4. This figure shows the segment of the diagram in Fig. 3 that contains the single phase SiB_3 regions, but at two different hydrogen concentrations. The phase boundaries for single phase deposition are shifted considerably by hydrogen. Figure 5 shows the effect of changing hydrogen concentration by replacing the [B/(B + Si)] axis on the diagram in Fig. 4 with a changing [$H_2/BCl_3 + SiH_4$)] ratio. This diagram shows that the addition of more hydrogen to the reactant gas favors the deposi-

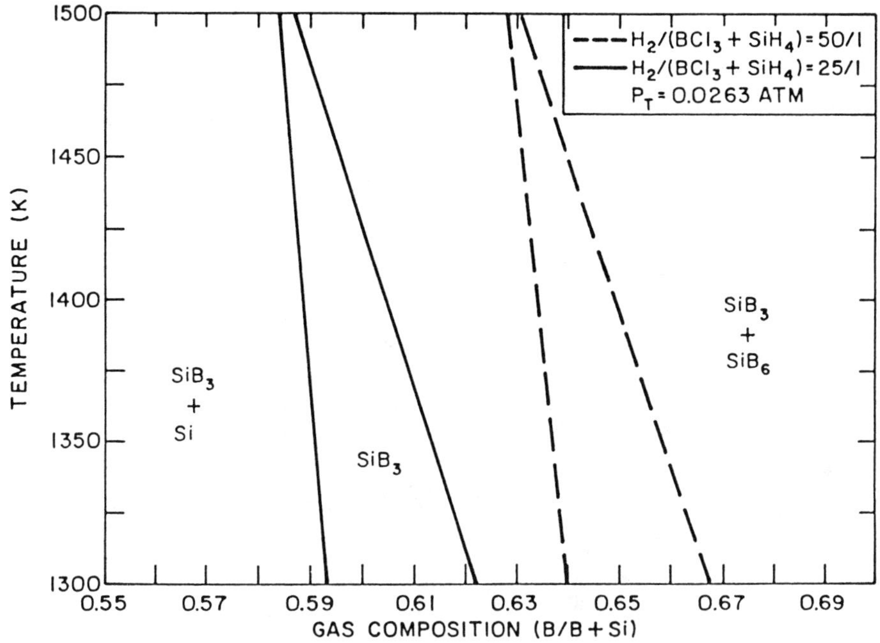

Fig. 4. CVD phase diagram illustrating the effect of doubling the H_2 concentration on the single phase triboride region.

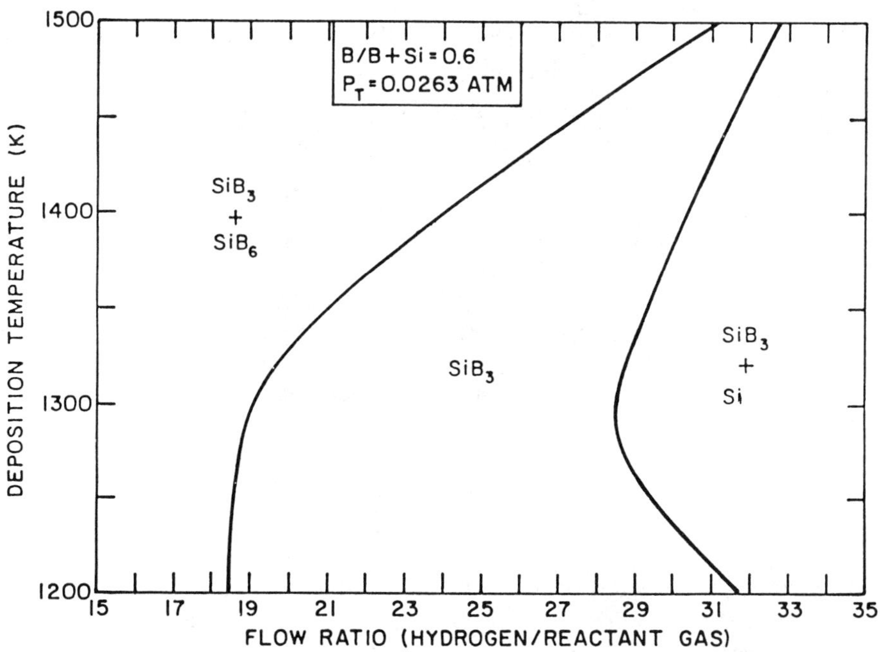

Fig. 5. CVD phase diagram illustrating the effect on the single phase triboride region of changing the input H_2 concentration while maintaining a constant input [B/(B + Si)] ratio.

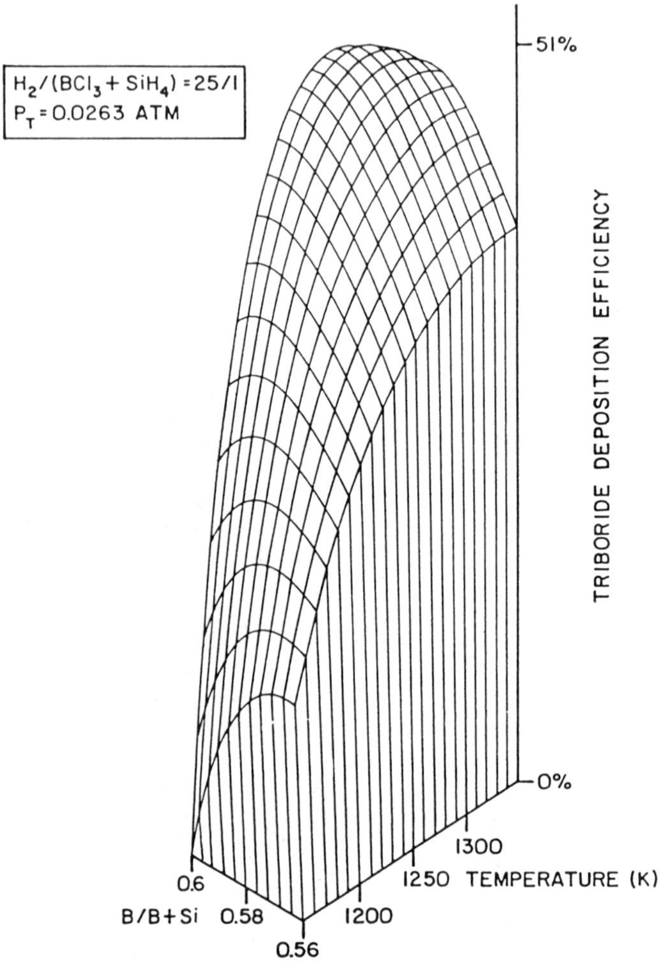

Fig. 6. Triboride deposition efficiency surface as a function of input gas composition and deposition temperature. Efficiency is defined as the equilibrium deposition rate divided by the rate if no kinetic or thermodynamic barriers to deposition existed.

tion of more silicon-rich phases. These diagrams in Figs. 4 and 5, depicting the effects of hydrogen on the deposition of SiB_3 could also be combined into a three dimensional representation showing the combined effects of changing two compositional parameters and temperature.

In Fig. 6, a temperature vs composition diagram that contains the single phase SiB_3 region is extended into the third dimension by an efficiency axis for the deposition of SiB_3. (Efficiency is defined as the equilibrium deposition rate divided by the rate if no kinetic or thermodynamic barriers existed). This figure clearly shows that uniform deposition rates are not expected if the gas composition or temperature changes over the surface on which SiB_3 is deposited.

At 1150 K in Fig 6, the SiB_3 deposition efficiency is practically zero at an [B/(B + Si)] ratio of 0.60, but then increases as this ratio decreases to 0.56. An explanation for this behavior is related to the stabilities of the chloride gaseous species. As the [B/(B + Si)] ratio increases, so does the chlorine content of the gas since its only source is BCl_3. The higher the concentration of chlorine in the bulk gas, the higher the fractions of silicon and boron that remain in the gas phase rather than depositing as solid SiB_3.

Also shown by Fig. 6 is that the efficiency of SiB_3 deposition increases rapidly with increasing temperature. As temperature is increased, the hydrogen reduction of the chlorides is more favorable and, thus, the deposition of solid SiB_3 increases. The rationalization of the results plotted in Fig. 6 are enhanced by considerations of partial pressure data presented in the next section.

Limits on Gas Phase Chemistry

CVD phase diagrams depict the condensed phases that can deposit at equilibrium for various sets of experimental parameters, but no information is given about which gaseous species may play important mechanistic roles in the deposition process. Detailed *in situ* gas phase diagnostics could provide the critical information, but these are difficult measurements that are not likely to be performed routinely on CVD reactors.

Modeling studies can provide predictive guidelines on important gaseous species in a CVD process, especially for high temperature operating conditions under which mass transport to and/or from the growing surface is rate-limiting. Gas diffusion to and from the deposition surface (steps 2 and 6 in Fig. 1) is driven by partial pressure gradients across the concentration boundary layer between the surface and the bulk gas. For mass transport limited deposition, the surface gases achieve equilibrium with the deposited solid. As a first approximation for the gas partial pressures (activities) at this surface, equilibrium values can be calculated for the input gas composition at the total pressure and temperature of the deposition process. Deviations from this approximation will be discussed later in the next section.

Two limiting values for the partial pressures in the bulk gas may be approximated. The first is that the input gas species do not react until reaching the deposition surface and that the product species are quickly swept out of the system. This approximation dictates that for diffusion calculations, all partial pressures in the bulk gas are zero except for the input species, whose partial pressures are calculated from their input mole fractions and the total system pressure.

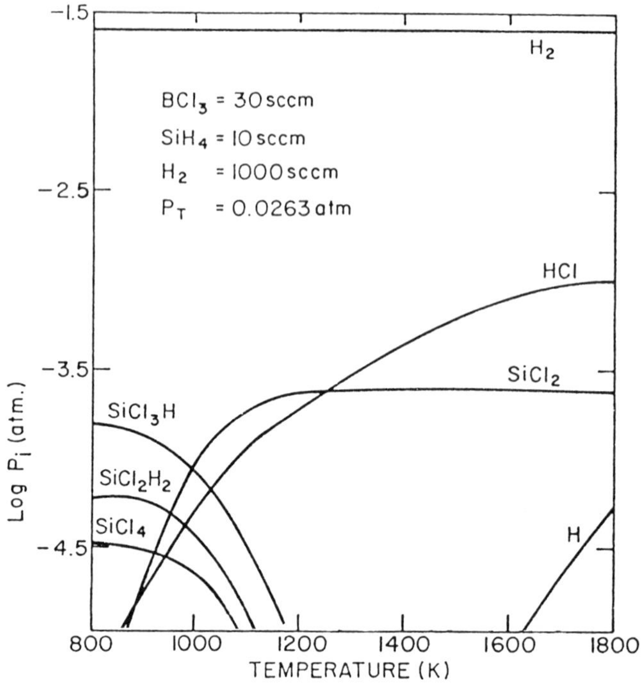

Fig. 7. Temperature profiles of the partial pressures of the predominant silicon-containing gaseous species in homogeneous equilibirum (no condensed phases are allowed). If the input gases did not react, they would have the following partial pressures [in $\log_{10}$(atm)]: SiH_4 (−3.60), BCl_3 (−3.12), and H_2 (−1.58).

The second approximation for the bulk gas partial pressures is that homogeneous equilibrium is reached in the input gas phase without the formation of a condensed phase. Whether the gas phase reaches homogeneous equilibrium depends not only on the basic kinetic properties of a chemical system, but also on the reaction design and operating conditions; for example, a hot-wall vs cold-wall reactor, the gas residence time in the reactor (flow rates, reactor dimensions), and the gas collision frequency (total pressure, temperature).

Figures 7 and 8 show the respective partial pressures of silicon- and boron-containing gaseous species calculated as a function of temperature for homogeneous gas phase equilibria (no solid phases were allowed to form). These figures also show the input gas partial pressures for the case in which no chemical reactions occur in the bulk gas. The experimental parameters utilized in these calculations were: total pressure (0.0263 atm), BCl_3 (30 sccm), SiH_4 (10 sccm), and H_2 (1000 sccm). Figures 9 and 10 show the results of similar calculations for the deposition surface when the gas has equilibrated with SiB_3(s), the phase experimentally deposited under these conditions.

In the limiting case of no reactions of the input species, the silicon and boron are transported to the deposition surface as SiH_4 and BCl_3, respectively. The equilibrium partial pressures at the deposition surface

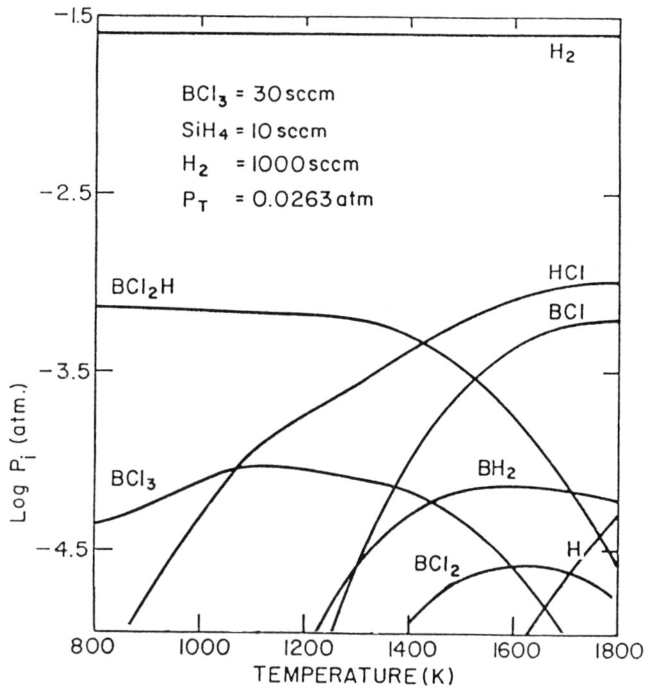

Fig. 8. Temperature profiles of the partial pressures of the predominant boron-containing gaseous species in homogeneous equilibrium (no condensed phases are allowed). If the input gases did not react, they would have the following partial pressures [in $\log_{10}$(atm)]: SiH_4 (−3.60), BCl_3 (−3.12), and H_2 (−1.58).

given in Fig. 9 and 10 show that the silane partial pressure is below the limit of this graph, and the boron trichloride partial pressure, although initially high at low temperatures, quickly decreases with increasing temperature.

In the limiting case of homogeneous equilibrium in the bulk gas, Fig. 7 shows that $SiCl_2$ is the most abundant silicon-containing species at temperatures greater than 1000 K, whereas $SiCl_3H$ is the most abundant at lower temperatures. At the equilibrium deposition surface, the $SiCl_4$ partial pressure depicted in Fig. 9 is relatively large at temperatures of about 1000 K and lower, but then quickly drops off. Above 1200 K, the dominant silicon-containing species at the surface is $SiCl_2$, but its partial pressure its more than an order of magnitude smaller than in the homogeneously equilibrated bulk gas.

If the input gases homogeneously equilibrate and the deposition is limited by mass transport processes (the deposition reactions reach equilibrium), the results in Figs. 7 and 9 predict the dominating net diffusion of silicon from the bulk gas to the surface occurs by $SiCl_3H$ at the lower temperatures and $SiCl_2$ at the higher temperatures. At lower temperatures, a significant net diffusion of $SiCl_4$ in the opposite direction from the surface to the bulk gas is predicted. As the temperature increases, this latter outward net diffusion rapidly decreases because of decreasing

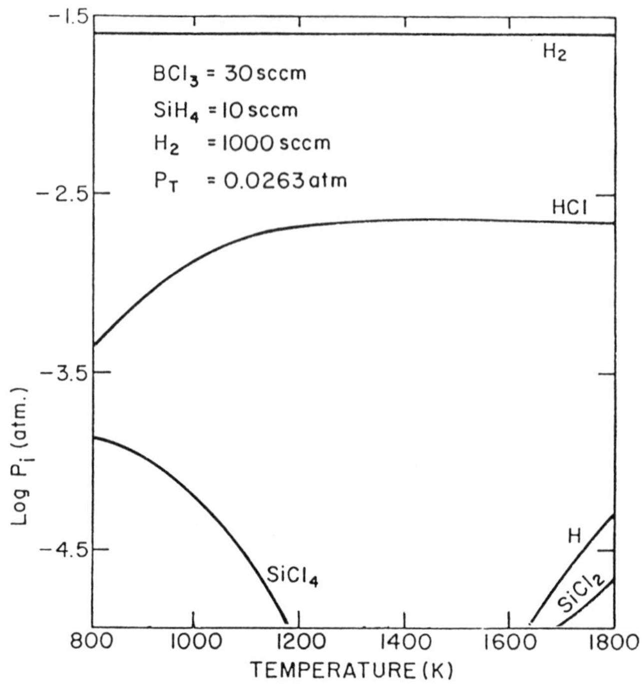

Fig. 9. Temperature profiles of the partial pressures of the predominant silicon-containing gaseous species in heterogeneous equilibrium with deposited SiB_3.

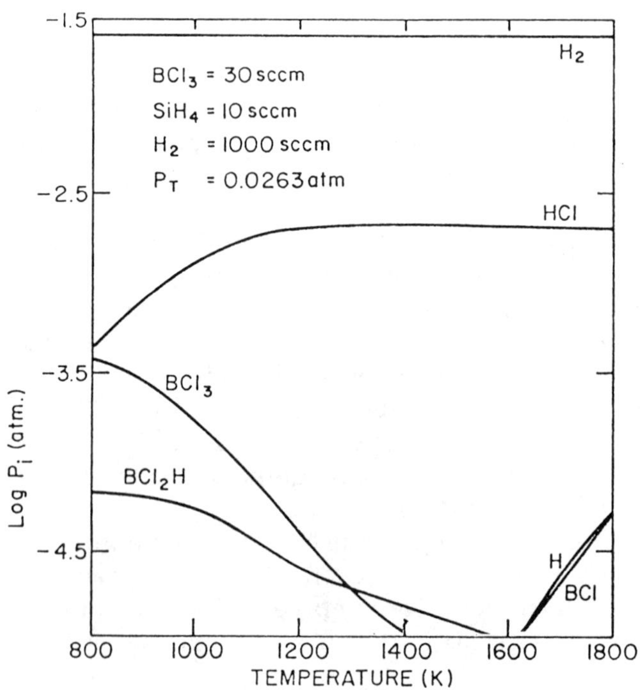

Fig. 10. Temperature profiles of the partial pressures of the predominant boron-containing gaseous species in heterogeneous equilibirum with deposited SiB_3.

partial pressures of $SiCl_4$ in the system. This conclusion helps to explain the low deposition efficiency of SiB_3 at low temperatures, which was indicated previously by the efficiency diagram in Fig. 6.

Figure 8 shows that BCl_2H is the dominant boron-containing gas up to about 1500 K in a homogeneously equilibrated bulk gas. Above this temperature, BCl dominates. At the equilibrium deposition surface, Fig. 10 shows the BCl_3 partial pressure to be dominating for boron-containing species at the deposition surface for temperatures up to about 1200 K, then BCl_2H and finally BCl take over, although the partial pressures of these latter species are always small in the temperature range depicted.

If the input gases homogeneously equilibrate at lower temperatures, the results in Figs. 8 and 10 predict the net diffusion of BCl_2H from the bulk gas to the surface, and a net diffusion of BCl_3 in the opposite direction from the surface to the bulk gas. As the temperature increases, this latter outward net diffusion decreases because the BCl_3 partial pressure decreases. Eventually, the net direction of BCl_3 diffusion reverses because this species partial pressure decreases more rapidly at the surface than in the bulk gas with increasing temperature.

Deviations from Predicted Deposition Behavior

Thermochemical modeling utilizes equilibrium calculations, but much can be learned about nonequilibrium behavior and other deviations from predicted "equilibrium behavior" through the use of local equilibrium concepts. The following examples illustrate such analyses.

Deposition studies as a function of temperature for input gas compositions of $[B/(B+Si)] = 0.75$ and $[H_2/(BCl_3+SiH_4)] = 25$ and a total pressure of 0.0263 atm produced SiB_3 at 1100–1500 K even though the CVD phase diagram calculations depicted in Fig. 3 predicted the formation of SiB_6. The first thought may be that the CVD phase diagram is not accurate because of uncertain system chemistry (for example, not including important gaseous species in the calculations) or uncertain thermodynamic data. The sensitivity of the CVD diagram to these uncertainties was considered, and such explanations for the differences between observed and calculated behavior were ruled out.

Two additional explanations resulted from analyses making use of thermochemical modeling. One partial explanation involves the deposition kinetics, the change from reaction kinetics limiting the deposition at lower temperatures to mass transport limiting the deposition rate at higher temperatures. The second explanation involves differences between the overall steady-state composition of the gas at the deposition surface and the bulk gas.

Deposition Rate-Limiting Processes

Experimental rate investigations and equilibrium rate calculations were performed for the deposition of the three solids SiB_3, Si, and B (*10*), and the data were analyzed in terms of the Arrhenius plots [logarithm (deposition rates) vs 1/T] shown in Figs. 11–13. Attempts were made to simulate the SiB_3 deposition conditions when performing the Si and B

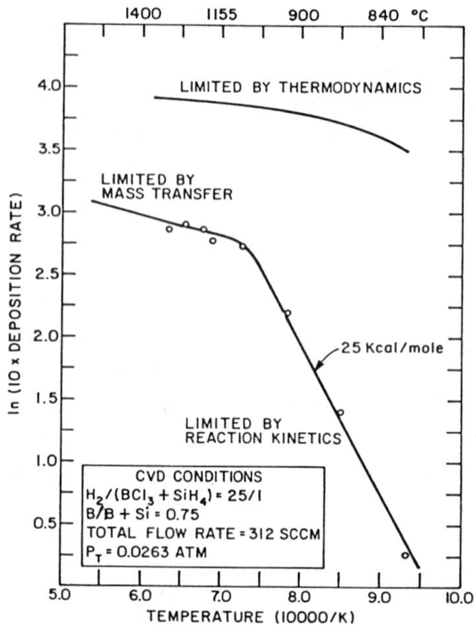

Fig. 11. Arrhenius plots of the logarithm (deposition rate) vs reciprocal temperature for SiB_3. Both experimental and calculated equilibrium data are given, as is the activation energy for the lower temperature rate-limiting deposition process.

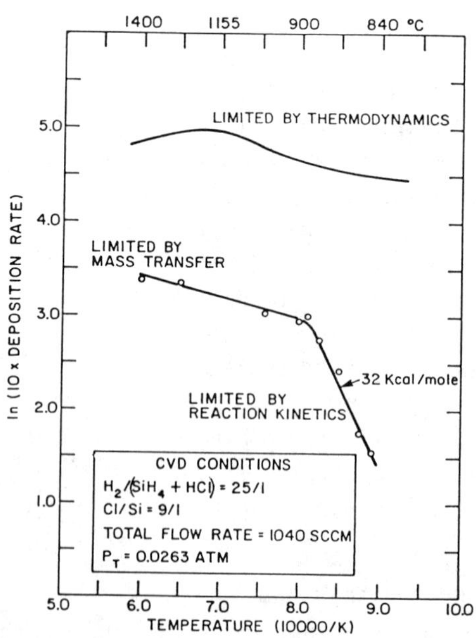

Fig. 12. Arrhenius plots of the logarithm (deposition rate) vs reciprocal temperature for silicon. Both experimental and calculated equilibrium data are given, as is the activation energy for the lower temperature rate-limiting deposition process.

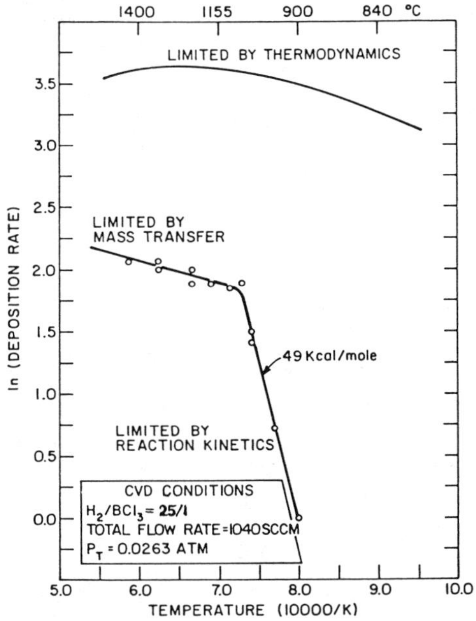

Fig. 13. Arrhenius plots of the logarithm (deposition rate) vs reciprocal temperature for boron. Both experimental and calculated equilibrium data are given, as is the activation energy for the lower temperature rate-limiting deposition process.

studies. For example, HCl was added to the SiH_4 and H_2 input gases in the Si deposition studies since chlorine appears to react with silicon in the SiB_3 deposition. The conditions experienced by the elements in the triboride deposition experiments, however, are not identical to those in the elemental deposition studies, so the following analysis does not compare the absolute rate data from these figures.

The slopes in each of the above figures for the experimental high temperature rate data are similar to those for the calculated equilibrium data. This is expected at high temperatures when mass transport through a gaseous concentration boundary layer is rate-limiting (*6*). The relatively large difference between experimental and equilibrium rates in this high temperature range is at least partially explained by the fact that a cold-wall reactor was used in the experimental studies (*10*).

The low temperature data are characterized by slopes related to activation energies for each rate-limiting chemical reaction. The break between the low and high temperature slopes occurs at about 1400 K for the SiB_3 and B deposition, but around 1150 K for the Si deposition. These data are consistent with a mechanism for SiB_3 deposition in which the Si and B deposit "independently" and then react to form the triboride. For such a mechanism, equilibrium at the SiB_3 deposition surface cannot occur until the temperature is high enough for equilibrium to be achieved in both the Si and B deposition processes; this is what is experimentally observed.

At temperatures from about 1150 K and higher, the Si deposition reactions reach equilibrium at the substrate, whereas the B deposition

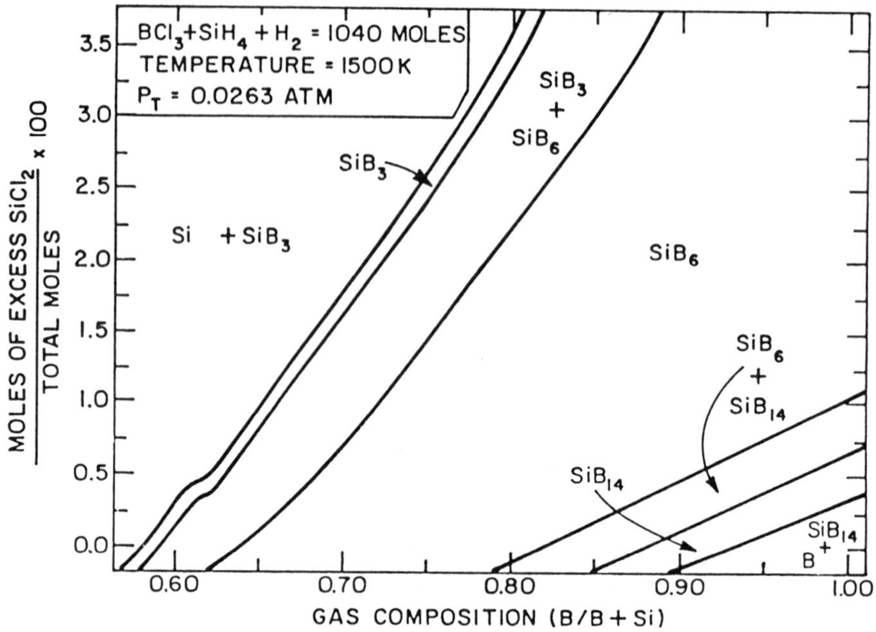

Fig. 14. CVD phase diagram at a constant 1500 K illustrating the effect of an excess concentration of $SiCl_2$ at the deposition surface. The input H_2 is 1000 moles, whereas the sum of the BCl_3 and SiH_4 input is 40 moles. Thus, every unit on the vertical axis represents 10.4 moles of excess $SiCl_2$ in addition to the input precursor gases.

processes are limited by reaction kinetics below 1400 K. At temperatures above 1400 K, the reactions should reach equilibrium for the deposition of SiB_3, Si, and B, and all three deposition systems should be limited by mass transport of gaseous species.

As we discussed at the beginning earlier, the CVD phase diagram in Fig. 3 predicts the equilibrium deposition of SiB_6 for the experimental parameters used, yet SiB_3 was observed at 1100–1500 K. The above rate data could explain this difference at low temperatures when the deposition of boron, but not silicon, was limited by reaction kinetics. However, above 1400 K, no chemical kinetic limitations are expected, so the observed deposition of SiB_3 at higher temperatures must be rationalized in other ways. This is done in the following section.

Steady-State Surface Concentrations

The observed high temperature deposition of SiB_3, when equilibrium calculations and the above kinetic analysis predicted SiB_6 deposition, is analyzed below in terms of the steady-state chemistry at the deposition surface.

The experimentally used input gas composition of $[B/(B+Si)] = 0.75$ is rapidly depleted in boron if SiB_6 is deposited since the $[B/(B + Si)]$ ratio is 0.86 in this solid. Comparisons of calculated gas phase chemistry in equilibrium with SiB_6 at 1500 K with that in equilibrium with SiB_3 showed a major difference to be that the SiB_6 containing system contained 10 times more $SiCl_2$ at equilibrium. What this means in terms of the deposition process is that if SiB_6 begins to form during the initial stages of deposition, as predicted by the CVD phase diagram in Fig. 3, $SiCl_2$ will begin to concentrate at the deposition surface. This would occur as long as SiB_6 is being deposited until a steady-state is reached in which the $SiCl_2$ rate of formation at the surface is equal to its rate of removal from the surface by mass transport through the gaseous boundary layer into the bulk gas stream.

The effects of a buildup of $SiCl_2$ at the deposition surface can be predicted using thermochemical modeling, as illustrated by Fig. 14. This figure is a CVD phase diagram at a constant temperature of 1500 K with the horizontal axis representing the input gas composition of SiH_4 and BCl_3 by the ratio [B/(B + Si)]. The vertical axis is a measure of the excess $SiCl_2$ build-up at the deposition surface by a ratio of the excess $SiCl_2$ moles over the total input gas moles. A value of 1.0 on that axis corresponds to 10.4 moles of excess $SiCl_2$ over the 10 moles of SiH_4, 30 moles of BCl_3, and 1000 moles of H_2 metered into the system at a composition of 0.75 for [B/(B + Si)].

At a value of 0.0 on the vertical axis in Fig. 14, which represents the behavior of the initial reactant gas before any dichloride build-up occurs, and at a [B/(B + Si)] input gas ratio of 0.75, the equilibrium deposition of SiB_6 is clearly predicted. Moving up the vertical axis at a constant [B/(B + Si)] input gas ratio of 0.75 shows that the build-up of $SiCl_2$ at the surface will eventually cause the deposition to shift from SiB_6 to SiB_3. This would occur at a ratio of about 25:10 for $SiCl_2$ (excess):SiH_4 (input), quite a large number. However, it is seen that such a large ratio is not necessary if the effect of the excess dichloride on the horizontal [B/(B + Si)] axis is taken into account. At an input gas composition of 0.75 for [B/(B + Si)] and a value of 0.5 on the vertical excess axis, the total ($SiCl_2$ + SiH_4) moles is 15, whereas the total BCl_3 moles is 30. This excess $SiCl_2$ causes a shift in the [B/(B + Si)] ratio from 0.75 to 0.67, which along with being at 0.5 on the vertical axis is almost enough to move into the single phase SiB_3 region of Fig. 14. Thus, instead of needing a ratio of 25:10 for $SiCl_2$ (excess):SiH_4 (input) at the surface to cause the equilibrium deposition of single phase SiB_3, a ratio of just over 5:10 is needed.

Calculations similar to those described above for the buildup of excess $SiCl_2$ at the deposition surface were performed to determine the possible influence of HCl, the major product gas in the deposition reactions. The results showed that a buildup of HCl at the surface would cause a shift from SiB_6 deposition toward more boron-rich phases, the opposite of what was experimentally observed. However, a depletion of HCl at the surface from the equilibrium amount would cause a shift in deposition behavior toward SiB_3. The fast removal of HCl from the deposition surface was discounted as a possible explanation for the observed deposition of SiB_3 on the basis of experiments from Mehalso and Diefendorf (32). They studied the deposition of boron by the hydrogen reduction of BCl_3, and concluded on the basis of kinetic results, that the rate-limiting process for the deposition was the slow removal of HCl from the deposition surface.

SUMMARY

The discussed complex interdependencies among controllable experimental parameters, process variables, and deposition properties in a CVD system could easily discourage any attempts to develop an understanding of such systems. Despite these complexities, many valuable analyses and predictions of deposition behavior can be made using the principles of high temperature chemistry, particularly high temperature thermochemistry. Many times the resulting predictions of deposition behavior are only numerically approximate or qualitative in nature, but they can still dramatically limit the range of conditions for deposition, or correctly give the trends in deposition behavior with changing experimental parameters.

Extremely efficient CVD research can be performed through a combination of a solid experimental program that is closely coupled with modeling studies. The combination of the two types of research yields much more than the sum of the two individual efforts, and either one by itself is of limited value if complex chemical systems are being examined. The value of this close coupling was illustrated by the high temperature thermochemical modeling that was used in explaining the experimental silicon–boride deposition behavior. Making use of local equilibrium concepts in setting up thermochemical modeling calculations for explaining experimental observations in a dynamic CVD system is critical. This is true for modeling any dynamic chemical process, whether it involves CVD, vaporization, corrosion, or reactions at interfaces in composite systems. The logic for understanding and modeling each of these types of systems is quite similar.

ACKNOWLEDGMENTS

The CVD investigations reported in this paper were largely supported by the Ceramics Program, Division of Materials Research, National Science Foundation.

REFERENCES

1. Bunshah, R. F., ed., *Deposition Technologies for Films and Coatings,* Noyes Publications, Park Ridge, NJ (1982).
2. Naslain, R. and Langlais, F., *High Temperature Science* (in press).
3. Carlsson, J. O., *High Temperature Science* in press.
4. Spear, K. E., *Proc. 7th International Conference on Chemical Vapor Deposition (CVD-VII),* Sedgwick, T. O. and Lydtin, H., eds., The Electrochemical Society, Pennington, NJ, 1979, pp. 1–16.
5. Spear, K. E., *Pure & Appl. Chem.* **54,** 1297–1311 (1982).
6. Spear, K. E., *Proc. 9th International Conference on Chemical Vapor Deposition (CVD-IX),* Robinson, McD., van den Breckel, C. H. J., Cullen, G. W., Blocher, J. M., and Rai-Choudhury, P., eds., The Electrochemical Society, Pennington, NJ, 1984, pp. 81–97.
7. Bernard, C., *High Temperature Science* in press.

8. Bernard, C., *Proc. 8th International Conference on Chemical Vapor Deposition (CVD-VIII)*, Blocher, J. M., Vuillard, G., and Wahl, G., eds., The Electrochemical Society, Pennington, NJ, 1981, pp. 3–16.
9. Besmann, T. M., *Surface Modifications Technologies*, Sudarshan, T. S. and Bhat, D. G., eds., The Metallurgical Society, Warrendale, PA, 1988, pp. 311–325.
10. Dirkx, R. R., *Phase Behavior in the Silicon-Boron System: Preparation of Silicon Borides by Chemical Vapor Deposition*, Ph.D. Thesis, Pennsylvania State University (1986).
11. Spear, K. E., *Treatise on Solid State Chemistry*, vol. 4, Hannay, N. B., ed., Plenum, New York, 1976, pp. 115–192.
12. Searcy, A. W., *International Symposium on High Temperature Technology*, McGraw-Hill, New York, 1960, p. 336.
13. Searcy A. W., *Chemical and Mechanical Behavior of Inorganic Materials*, Searcy, A. W., Ragone, D. V., and Colombo, U., eds., Wiley, New York, 1970, pp. 15–32.
14. Margrave, J. L., *High Temperature Materials and Technology*, Campbell, I. E. and Sherwood, E. M., eds., Wiley, New York, 1967, pp 21–55.
15. Margrave, J. L. and Mamantov, G., *High Temperature Materials and Technology*, Campbell, I. E. and Sherwood, E. M., eds., Wiley, New York, 1967, pp. 78–127.
16. Brewer, L., *Inorganic Chemistry*, Butterworths, London, 1958, pp. 559–570.
17. Brewer, L., *The Chemistry and Metallurgy of Miscellaneous Materials: Thermodynamics*, Quill, L. L., ed., McGraw-Hill, New York, 1950, pp. 193–275.
18. Brewer, L., *Topics in Modern Inorganic Chemistry*, Milligan, W. O., ed., Robert A. Welch Foundation, Houston, TX, 1963, pp. 47–92.
19. Brewer, L., *J. Chem. Educ.* **35,** 153–156 (1958).
20. Coltrin, M. E., Kee, R. J., and Miller, J. A., *J. Electrochem, Soc.* **131,** 425–434 (1984).
21. Coltrin, M. E., Kee, R. J., and Miller, J. A., *J. Electrochem, Soc.* **133,** 1206–1213 (1986).
22. Jensen, K. F., *Proc. 9th International Conference on Chemical Vapor Deposition (CVD-IX)*, Robinson, McD., van den Breckel, C. H. J., Cullen, G. W., and Blocher, J. M., and Rai-Choudhury, P. eds., The Electrochemical Society, Pennington, NJ (1984), pp. 3–17.
23. Rosner, D. E., Nagarajan, R., Gokoglu, S. A., and Kori, M., *Proc. 10th International Conference on Chemical Vapor Deposition (CVD-X)*, Cullen, G. W., and Blocher, J. M., eds., The Electrochemical Society, Pennington, NJ (1987), pp. 61–80.
24. Tsui, P., *Theoretical and Experimental Studies of Convective Effects on Chemical Vapor Deposition*, Ph.D. Thesis, Pennsylvania State University (1988).
25. Rosenberger, F., *Proc. 10th International Conference on Chemical Vapor Deposition (CVD-X)*, Cullen, G. W. and Blocher, J. M., eds., The Electrochemical Society, Pennington, NJ, (1987), pp 11–22.
26. Wan, C. F. and Spear, K. E., *Calphad* **7,** 149–155 (1983).
27. Chase, M. W., Davies, C. A., Downey, J. R., Frurip, D. J., McDonald, R. A., and Syverud, A. N., *JANAF Thermochemical Tables*, 3rd edition, American Institute of Physics, New York, (1986); and *J. Phys. Chem. Ref. Data* **14** (1985).
28. Dirkx, R. R. and Spear, K. E., *Calphad* **11,** 147–155 (1987).
29. Eriksson, G., *Acta Chem. Scand.* **25,** 2561–2568 (1971).
30. Eriksson, G. and Rosen, E., *Chemica Scripta* **4,** 193–194 (1973).
31. Eriksson, G., *Chemica Scripta* **8,** 100–103 (1975).
32. Mehalso, R. M. and Diefendorf, R. J., *Proc. 5th International Conference on Chemical Vapor Deposition (CVD-V)*, Blocher, J. M., Hintermann, H. E., and Hall, L. H., eds., The Electrochemical Society, Pennington, NJ, 1975, pp. 84–98.

Thermochemical Modeling of Vapor Deposition

CLAUDE BERNARD

Laboratoire de Thermodynamique et Physico-Chimie Métallurgiques, ENSEEG Domaine Universitaire, BP 75 - 38042 Saint-Martin-d'Hères Cedex, France

ABSTRACT

General modeling of the chemical vapor deposition (CVD) process should make allowance for the experimental parameters and the thermodynamic and transport properties in order to be able to discuss diffusion-controlled processes, and the kinetic properties in order to understand the deposition process controlled by surface reactions. A universal model has not yet been constructed; the real problem is not computer capacity but more the lack of data.

The various limitations that have to be overcome are emphasized and a current, more profitable approach is described. This approach starts with an a priori thermodynamical optimization to predict the range of experimental parameters for depositing a specific phase. A thermochemical mass transport model is then used for process optimization on the basis of the initial experimental results.

Impressive progress has been made in recent years in the design and upscaling of CVD reactors and especially in understanding CVD flow dynamics, as witnessed by the numerous publications on this subject over the past five years. Depending on the characteristics of the experiments, an assessment of what can reasonably be expected from these models will be made.

Various examples are presented to illustrate the general outlines of CVD modeling.

Index Entries: CVD; thermodynamics; hydrodynamics, models.

INTRODUCTION

The production of thin or thick films by vapor-phase chemical deposition is a process subject to an increasing number of specifications that have become increasingly demanding over the years. For example, for the production of integrated circuits, the homogeneity of the deposits must be improved while the area covered has increased. Similarly, for hard deposits, a well-defined stoichiometry may be required for a material with a wide solid solution range.

By using prior and then simultaneous simulation of the process, it is possible to avoid a number of errors in principle, to save considerable time in finalizing the experimental setup and to interpret certain phenomena observed during the experimental procedure.

PROCESS OPTIMIZATION CHRONOLOGY

A certain chronology must be respected in order to have the maximum chance of success when tackling a new problem.

1. The chemical system must be simulated as a whole, taking into account all agents involved: gas, reactor walls, crucibles, substrates, and assuming that the thermodynamic equilibrium is reached in a closed system with constant pressure and temperature. During this first step, which assumes good knowledge of the thermodynamic data available (which is rarely the case), a certain number of processes, gas vectors, crucible materials, etc. can be eliminated and, in certain cases, it is possible to assess the complexity of interrelations between the various deposition domains specific to the processes studied (*1,2*) Fig. 1 and 2.

   This type of approach is made by means of programs interfacing with data banks, some of which are available in software package form: for example Solgasmix (*4*) and its new version Sage by G. Erickson, the Phase program (*5*) by the National Physical Laboratory, Equi (*6*) or Mélange (*7*) by the L.T.P.C.M. These programs, which do not always agree on the method of mathematical treatment for minimization of the total Gibbs free energy of the system, nonetheless present similar facilities of use.
2. Once the operating conditions have been theoretically optimized, these results are used as a basis for designing a prototype reactor or a reactor of sufficient versatility so that it can be modified in the light of subsequent investigations.
3. A series of experimental manipulations are then performed in order to gather enough data to evaluate the kinetic parameters. There are two aims here:

   to reconstruct the performance of the pilot reactor in its initial configuration and to predict optimized performance by proposing changes to the basic design, to design a new type of reactor for the same chemical process.

This modeling stage is by far the most difficult since, in most cases, information on kinetic rate parameters of the reactions occurring in the process is nonexistent or, if such information is available, it is often

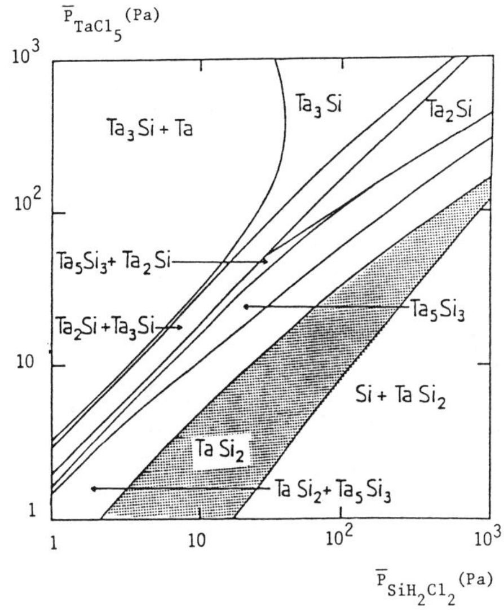

Fig. 1. Computed CVD phase diagram (*1*) giving the nature of the material deposited for the Ta-Si-Cl-H-Ar gas system at equilibrium. T = 1000 K, total pressure = 10^5 Pascal, Argon partial pressure 0.9×10^5 Pascal. The initial partial pressures of $TaCl_5$ and SiH_2Cl_2 are reported on the vertical and horizontal axes, respectively, and the initial partial pressure of hydrogen expressed in Pascal is given by $\bar{P}_{H_2} = 0.1 \times 10^5 - \bar{P}_{TaCl_5} - \bar{P}_{SiH_2Cl_2}$

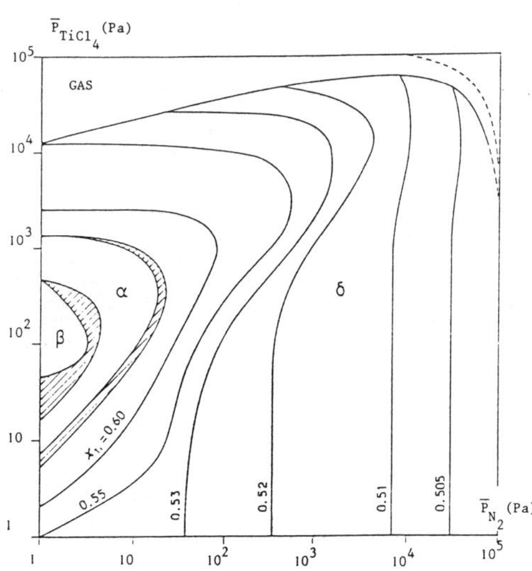

Fig. 2. Computed CVD phase diagram (3) giving the nature of the material deposited for the Ti-N-Cl-H gas system at equilibrium. α:hcp phase, β: bcc phase and δ: Ti_xN_{1-x} phase. T = 1900 K, total pressure = 10^5 Pascal. The initial partial pressure of $TiCl_4$ and N_2 are reported on the vertical and horizontal axes, respectively, and the initial partial pressure of hydrogen expressed in Pascal is given by $\bar{P}_{H_2} = 10^5 - \bar{P}_{TiCl_4} - \bar{P}_{N_2}$

Table 1
Reaction Mechanism for Silicon Chemical Vapor Deposition (*8*)

R1	$SiH_4 \rightarrow SiH_2 + H_2$
R2	$SiH_4 \rightarrow SiH_3 + H$
R3	$SiH_4 + SiH_2 \rightarrow Si_2H_6$
R4	$Si_2H_4 + H_2 \rightarrow SiH_4 + SiH_2$
R5	$SiH_4 + H \rightarrow SiH_3 + H_2$
R6	$SiH_4 + SiH_3 \rightarrow Si_2H_5 + H_2$
R7	$SiH_4 + SiH \rightarrow SiH_3 + SiH_2$
R8	$SiH_4 + SiH \rightarrow Si_2H_5$
R9	$SiH_4 + Si \rightarrow 2SiH_2$
R10	$Si + H_2 \rightarrow SiH_2$
R11	$SiH_2 + SiH \rightarrow Si_2H_3$
R12	$SiH_2 + Si \rightarrow Si_2H_2$
R13	$SiH_2 + Si_3 \rightarrow Si_2H_2 + Si_2$
R14	$H_2 + Si_2H_2 \rightarrow Si_2H_4$
R15	$H_2 + Si_2H_4 \rightarrow Si_2H_6$
R16	$H_2 + SiH \rightarrow SiH_3$
R17	$H_2 + Si_2 \rightarrow Si_2H_2$
R18	$H_2 + Si_2H_3 \rightarrow Si_2H_5$
R19	$Si_2H_2 + H \rightarrow Si_2H_3$
R20	$Si + Si_3 \rightarrow 2Si_2$

contradictory when gathered from different sources. A glance at the steps considered by Coltrin et al. (*8*) (Table 1) to represent CVD of Si by decomposition of SiH_4 suffices to illustrate this situation. In addition, it must be accepted that, just because a formula for deposition rate has been established that seems to be representative of experimental results, one cannot automatically assume that the corresponding kinetic mechanism is validated.

Faced with such difficulties, it would be easy to give up but, just as with post-modeling, where the molar production or destruction rates Rj of each species per chemical reaction have to be expressed, hypotheses must be formulated to simplify the phenomenon. Such hypotheses will occasionally be a far cry from the reality of the situation.

Graphs showing the logarithmic variation in desposition rate as an inverse function of temperature are generally available (Fig. 3). These graphs often have two distinct domains, one at low temperature generally accepted to be related to surface kinetics, and the other at high temperature, for which gas-phase diffusion plays a dominant role. For subsequent modeling of this second domain, the expression for the Rj values at the substrate surface can be obtained by an equilibrium calculation from the concentrations of species present at the surface. For the first domain, however, the slope of the experimental curve, Fig. 4, must be calculated to determine the apparent activation energy for each set of initial reagent concentrations. If sufficient information is available, it should be possible to deduce the value of kinetic parameters. Note that it is theoretically possible to estimate these parameters from the absolute

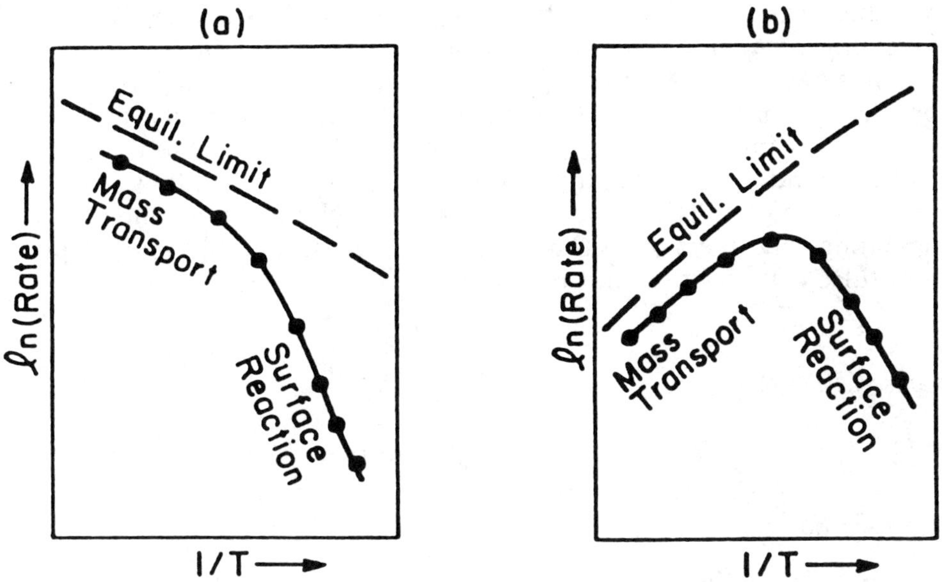

Fig. 3. Schematic Arrhenius Plots for the CVD Deposition Rate (*9*) as a function of Temperature. Both the equilibrium limiting rates and the experimental rates are shown. The CVD process is endothermic in (a) and exothermic in (b).

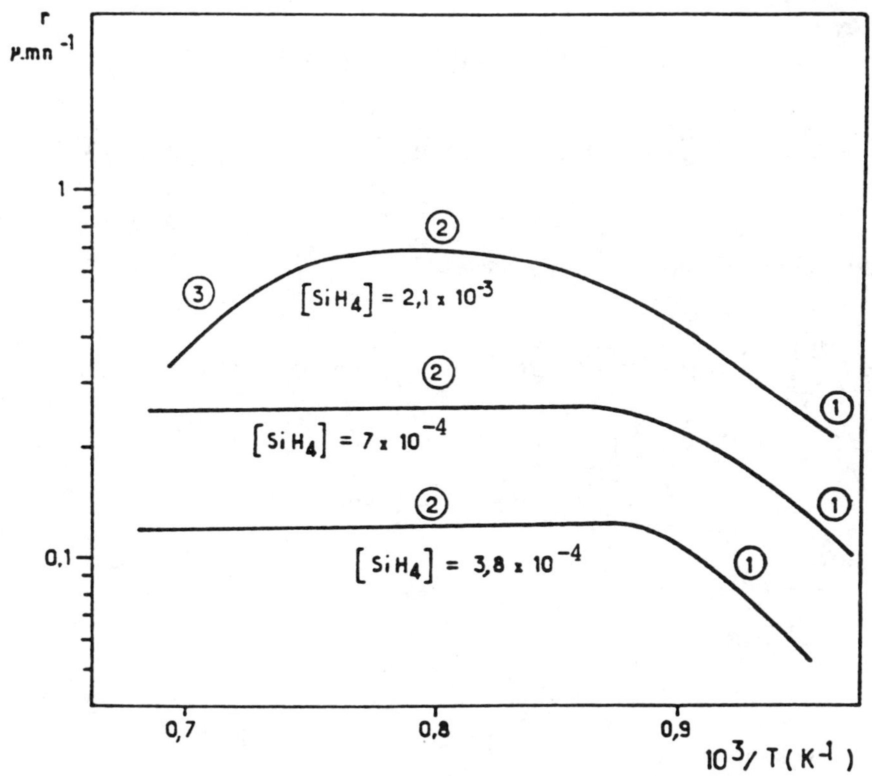

Fig. 4. Silicon growth rate as a function of the reciprocal temperature for different molar fraction of silane in a constant hydrogen flow (*10*). **1:** domain controlled by surface reaction. **2:** domain controled by mass transport. **3:** domain where the homogeneous decomposition takes place.

rate theory as already demonstrated, for example, by Cadoret (*11*). In the following paragraphs, it will be assumed that Rj is known.

It is now necessary to describe the system of equations to be used to predict reactor performance: deposition rate, film uniformity, and composition, by integrating the operating parameters specific to the experiment in hand: reactant mixture composition, temperature, flow patterns, reactor geometry. These equations are now well-known for the classical operating conditions of CVD and Low Pressure Chemical Vapor Deposition (LPCVD) reactors, and relate to:

Conservation of mass.

$$\frac{\partial \rho}{\partial t} = -\operatorname{div}(\overrightarrow{\rho v})$$

Conservation of momentum.

$$\frac{\partial(\rho\vec{v})}{\partial t} + \operatorname{div}(\rho\vec{v}) = -\overrightarrow{\text{grad}}\ P + \operatorname{div}\tau + \overrightarrow{\rho g}$$

Conservation of energy.

$$\rho\, Cp \frac{\partial T}{\partial t} + \rho\, Cp\, \vec{v}\, \overrightarrow{\text{grad}} T =$$

$$\operatorname{div}(k\, \overrightarrow{\text{grad}}\, T) + \frac{dP}{dt} + \vec{v}\, \overrightarrow{\text{grad}}\, P + \operatorname{div}\vec{v} - \sum_{j=1}^{J} \rho\, w_j \vec{v}_j Cp_j\, \overrightarrow{\text{grad}}\, T - \sum_{j=1}^{J} R_j h_j m_j$$

A conservation equation for each species.

$$\frac{\partial \rho j}{\partial t} + \operatorname{div}(\rho_j \vec{v}) = -\operatorname{div}(\overrightarrow{\text{grad}}\, \rho\, w_j v_j) + R_j m_j \quad \text{with } j = 1 \ldots J$$

and $$v_j = -\left[\frac{D_j}{x_j} \overrightarrow{\text{grad}}\, x_j - \frac{D_j^T}{x_j} \frac{1}{T} \overrightarrow{\text{grad}}\, T\right]$$

Nomenclature

ρ density
v velocity vector
t time
P pressure
τ momentum flux

g acceleration due to gravity
Cp heat capacity at constant pressure
T temperature
k thermal conductivity
Wj mass fraction of j
Vj velocity of diffusion of j
Rj molar rate of production or of disappearance of j per unit volume
kj enthalpy of j
mj molecular weight of j
Dj diffusion coefficient of j
xj molar fraction of j
D_j^T thermal diffusion coefficient of j

Once coupled with adequate boundary conditions, this absolutely general system should enable all cases to be solved. However, this assumes that a large number of parameters are known and that the actual mathematical resolution can be successfully completed.

State of the Art in Flow Modeling

For the various types of CVD and LPVCD reactors, with cold walls and hot walls, and depending on the complexity of the vapor phase used, numerous models of varying degrees of sophistication have been proposed. The most traditional, especially for the case of horizontal reactors, is the model based on the boundary layer theory (*12–16*). However, since this theory ignores significant hydrodynamic and thermal phenomena, it would appear today to be fairly restrictive for the purpose of designing a reactor and optimizing its operation. As Rosenberger quite rightly says (*17*), CVD processes "form some of the most complex fluid dynamics systems imaginable and it has become apparent that some of the crucial physics of CVD processes is lost in these simplifying models." Even so, it is perhaps rather severe to regret that these models have been able to give reasonable agreement with experimental results, purely by chance.

Indeed, the complexity of the conservation equations presented above was for many years sufficient justification for using numerous simplifications, some of which could, in certain cases, be fully justified. However, thanks to the much greater range of possibilities offered through the use of high-capacity computers, coupled with powerful numerical techniques such as the finite element method, allowance can now be made for far greater degrees of complexity in modeling CVD systems.

In order to solve these equations, the experimenter must have access to specific software, a sort of "black box" specialized in the resolution of complex differential equations. In the fluid mechanics field, a few software packages are available to deal with the problems involving coupled mass and heat transfer equations (*18*). Good examples are the Fidap finite element code (USA), which is suited to the resolution of aerothermal problems, the Phenix finite difference code (GB), which can be used to solve aerothermo-chemical problems, and the Flux expert software, written at the Institut Nationale Polytechnique de Grenoble and that is a general software package based on the finite element method, designed to solve all equations compatible with this formulation.

Simulations Performed or in Progress

The resolution of equations for the most general case, i.e., study of a given reactor under transient conditions by a three-dimensional approach, is still not for today. On the other hand, numerous studies have been carried out for cases with steady conditions, incorporating simplifications based on the reactor geometry, or the boundary conditions imposed by the experimenter, or again the specific nature of the gas mixture used. In view of the criticism occasionally leveled, in some cases justifiably, at such or such a simplifying hypothesis (boundary layer, for example), the examples presented will tend essentially to illustrate the potential of the resolution method.

The first example is derived from work by Pons et al. (*18*) who solved the conservation equations with the Flux expert system by decoupling the momentum and energy conservation equations from the species conservation equations. This is possible in a number of cases, especially in atmospheric reactors where a neutral carrier gas transports the reactive species which are generally highly diluted (electronics industry). The velocity and temperature fields can then be simulated independently of the concentration field. The influence of such experimental parameters as fluid velocity, shape factors (reactor walls, injection nozzle, substrate) can then be visualized. Thus, Fig. 5 shows how modifications to these factors can eliminate natural gravity convection of the hydrogen gas vector in a vertical CVD reactor. This type of approach should provide a means of revealing the flow and temperature profiles as already shown by Gilling (*19*) and subsequently by Bennet (*20*), thanks to interference holography, or visualised by means of TiO_2 smoke (*21–24*).

The second example concerns silicon deposition from SiH_4 diluted in hydrogen in a horizontal reactor. Coltrin's mechanism was simplified by Moffat and Jensen (*25*) who kept only SiH_4 and SiH_2 as primary reactive species. The advantage of the three-dimensional flow approach lies in the possibility of highlighting the secondary flows which gradually build up as the reactive gas mixture advances in the reactor. This model can be used to calculate a distribution profile for the SiH_4 above the substrate and according to the distance to the reactor axis, Fig. 6. Contrary to the two-dimensional model used by Coltrin, (*26*), this profile will depend on the choice of axis, midplane or not. Identical calculations have been carried out to study the deposition of GaAs by Metal Organic Chemical Vapor Deposition (MOCVD).

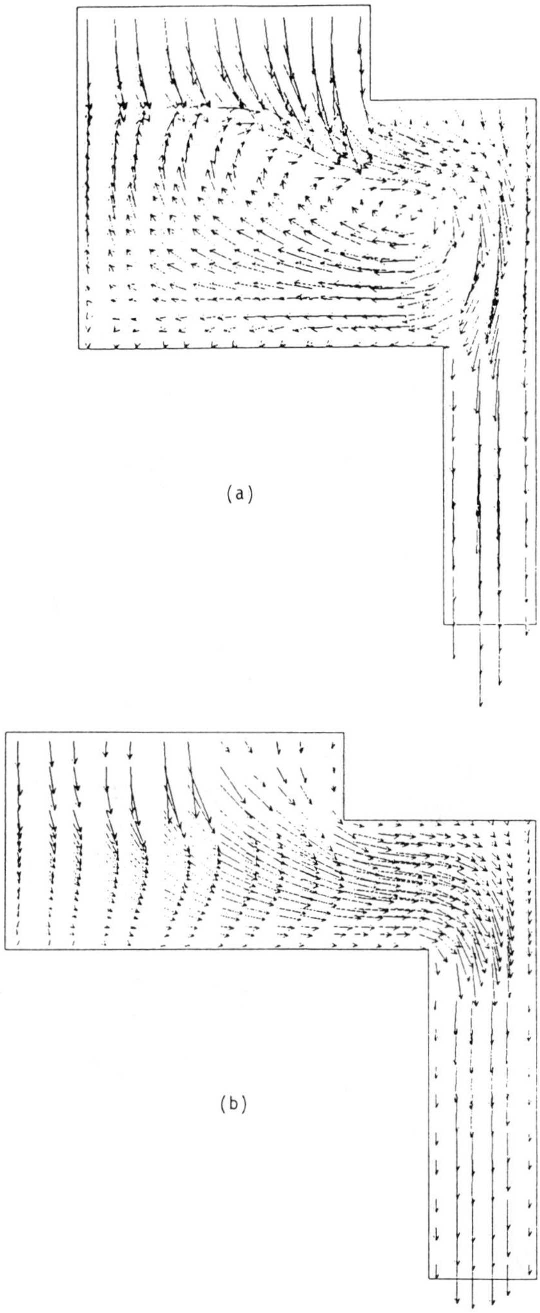

Fig. 5. Typical velocity field for the flow in a vertical reactor for 11 mn^{-1} at mass flow, with various dimensionless numbers: Reynolds number Re, Grashof number Gr, $A = \frac{Z}{R}$, $IA = \frac{L}{R}$, and $PA = \frac{l}{R}$. Z inlet-substrate length, R substrate radius, L inlet radius and l radial length of the outlet (*18*).

a) Re = 2; A = 0.65; IA = 0.8; PA = 0.25, Gr = 1200

b) solution of the thermal effect: Re = 2; A = 0.3; IA = 0.8; PA = 0.25, Gr = 1500

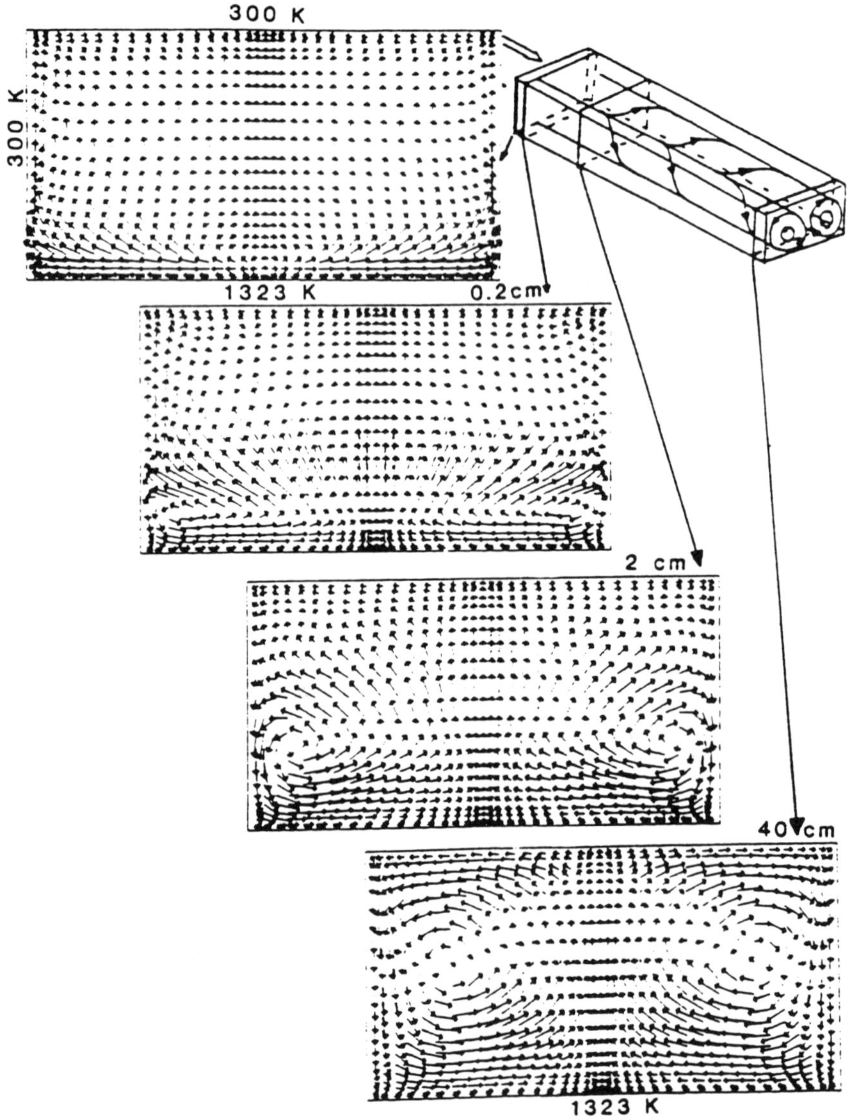

Fig. 6. Transverse velocities in a horizontal CVD reactor at four different axial positions (*25*).

CONCLUSIONS

This presentation does not pretend to be a review of current developments in CVD modeling, first because the literature on this subject is vast, and second because, as far as reactor modeling is concerned, excellent reviews have already been published on the subject (*25*) and (*27*). On the other hand, the possibilities offered by software currently available are worth mentioning.

For prior thermodynamic simulation of the process, the only possible limitation being the lack of thermodynamic data;

For reactor modeling, a field in which software packages capable of including most of the complexities of the phenomena involved are also available. In these calculations, two types of application can be considered;

Applications relating to the domain governed by reaction kinetics at the substrate surface: case of numerous LPCVD reactors designed for electronic applications, for which the crux of the problem lies in the cruel lack of kinetic data;

Applications relating to the domain governed by mass transfer: case of high-temperature deposition under atmospheric pressure, for which it is possible to consider coupling the fluid flow computations either with the substrate surface minimization computations or with the local thermochemical equilibrium computation at any point in the gas phase (*16*).

REFERENCES

1. Bernard, C., Vahlas, C., Million-Brodaz, J. F., and Madar R., *Proceedings of 10th International Conference on CVD*, Cullen, G. W., ed., The Electrochemical Society, Pennington, 1987, p. 700.
2. Spear, K., *Proceedings of 7th International Conference on CVD*, Sedgwick, T. O., Lydtin, H., eds., The Electrochemical Society, Pennington, 1979, p. 1.
3. Teyssandier, F., Bernard, C., and Ducarroir, M., *J. Mater. Sci.* **23,** 135 (1988).
4. Eriksson, G., *Chemica Scripta* **8,** 100 (1975).
5. Hodson, S. M., and Davies, R. H., *Proceedings of 15th Calphad Meeting*, Kaufman, L., ed., Calphad, 1986, p. 208.
6. Bernard, C., Deniel, Y., Jacquot, A., Vay, P., and Ducarroir, M., *J. Less Common Met.* **40,** 165 (1975).
7. Barbier, J. N., Bernard, C., *Proceedings of 15th Calphad Meeting*, Kaufman, L., ed., Calphad, 1986, p. 206.
8. Coltrin, M. E., Kee, R. J., and Miller, J. A., *Proceedings of 9th International Conference on CVD*, Robinson, Mc D., Van den Brekel, C. H. J., Cullen, G. W., Blocher, Jr, J. M., and Rai-Choudhury, P., eds, The Electrochemical Society, Pennington, 1984, p. 31.
9. Spear, K., *Proceedings of 9th Interantional Conference on CVD*, Robinson, Mc D., Van den Brekel, C. H. J., Cullen, G. W., Blocher, Jr., J. M., and Rai-Choudhury, P., eds., The Electrochemical Society, Pennington, 1984, p. 81.
10. Duchemin, J. P., 1976 Thesis, Universite de Caen, Caen.
11. Cadoret, R., *Current Topics in Materials Science*, Kaldis, E., ed., North Holland, 1980, p. 219.
12. Eversteijn, F. C., Severin, P. J. W., Van den Brekel, C. H. J., and Peck, H. L., *J. Electrochem. Soc.* **117,** 925 (1970).
13. Ban, V. S., and Gilbert, S. L., *J. Cryst. Growth* **31,** 284 (1975).
14. Ban, V. S., *J. Cryst. Growth* **45,** 97 (1978).
15. Spear, K. E., *Pure and Appl. Chem.* **54,** 1297 (1982).
16. Rosner, D. E., Nagarajan, R., Gokoglu, S. A., and Kori, M., *Proceedings of 10th International Conference on CVD*, Cullen, G. W., ed., The Electrochemical Society, Pennington, 1987, P. 61.
17. Rosemberger, F., *Proceedings of 10th International Conference on CVD*, Cullen, G. W., ed., The Electrochemical Society, Pennington, 1987, P. 11.

18. Pons, M., Klein, R., Arena, C., Mariaux, S., to be published in *Proceedings of 7th European Confence on CVD*, 1989.
19. Giling, L. J., *J. Electrochem. Soc.* **129,** 634 (1982).
20. Bennet, S., Berge, R., Brunet, S., Charar, S., Armas, B., and Combescure, C., *Proceedings of 8th International Conference on CVD*, Blocher, J. M., Vuillard, G. E., and Wahl, G., ed., The Electrochemical Society, Pennington, 1981, p. 188.
21. Takahasi, R., Koza, Y., and Sugawara, K., *J. Electrochem. Soc.* **119,** 1406 (1972).
22. Wahl, G., *Thin Solid Films* **40,** 13 (1977).
23. Ban, V. S., *J. Electrochem. Soc.* **125,** 317 (1978).
24. Wang, C. A., Groves, S. H., Palmateer, S. C., Weyburne, D. W., and Brown, R. A., *J. Crystal Growth* **77,** 136 (1986).
25. Jensen, K. F., *Chem. Eng. Sci.*, **42,** 923 (1987).
26. Coltrin, M. E., Kee, R. J., and Miller, J. A., *J. Electrochem. Soc.* **131,** 425 (1984).
27. Hess, D. W., Jensen, K. F., and Anderson, T. J., *Rev. Chem. Eng.* **3,** 97 (1985).

A Thermodynamic Analysis of Germanium Nitride CVD

F. DEFOORT, J. L. PONTHENIER, AND T. J. ANDERSON*

Department of Chemical Engineering, University of Florida, Gainesville, FL 32611

ABSTRACT

Complex chemical equilibrium calculations were performed for the Ge–N–Cl–H system to model chemical vapor deposition of stoichiometric Ge_3N_4 from $GeCl_4$ and NH_3 sources. Equilibrium deposition diagrams, gas phase compositions, and thermodynamic yields of reactants are computed as a function of deposition temperature, pressure, and feed composition for both H_2 and N_2 carrier gases. The choice of carrier gas has a strong influence on the range of operating conditions that produce single phase Ge_3N_4 deposition. Deposition of Ge_3N_4 is favored by operating at a lower temperature and excess NH_3 in the feed.

Index Entries: Ge_3N_4 chemical vapor deposition; thermodynamic yields; equilibrium.

INTRODUCTION

Germanium nitride, Ge_3N_4, has been suggested as a suitable surface-passivating material for compound semiconductors (*1,2*). In addition, Ge_3N_4 has been proposed as a gate dielectric and mask material in germanium MIS technology (*3,4*). There have been several reports of successful deposition of Ge_3N_4 films by chemical vapor deposition (CVD). Pande and coworkers (*1,2*) deposited Ge_3N_4 from germane and hydrazine sources in a nitrogen carrier gas by thermal decomposition at 673K. Gritsenko (*5*) used GeH_4 and NH_3 sources to deposit stoichiometric

*Author to whom all correspondence and reprint orders should be addressed.

Table 1
Summary of Thermochemical Values in the Ge–N–Cl–H System

Species	$\Delta H^\circ_{f,298}$, kcal/mol	Ref	$\Delta S^\circ_{f,298}$, cal/mol	Ref.	C_p, cal/mol K[a] a	b	c	d	Form	Ref.
N(g)	112.97	*(10)*	13.745	*(10)*	4.968	–	–	–	1	*(10)*
N_2(g)	0		0		7.024	−1.195	0.2985	1.132	2	*(10)*
N_3(g)	99.0	*(10)*	−14.555	*(10)*	6.072	15.29	−1.029	2.512	2	*(10)*
NH(g)	90.0	*(10)*	4.8215	*(10)*	7.265	−1.67	0.2758	−0.8403	2	*(10)*
NH_2(g)	45.5	*(10)*	−7.562	*(10)*	7.63	0.239	0.3676	−1.412	2	*(10)*
NH_3(g)	−10.97	*(10)*	−23.6355	*(10)*	5.99	8.60	−0.05614	−0.4986	2	*(10)*
N_2H_2(g)	50.9	*(10)*	−24.747	*(10)*	4.508	15.50	−0.608	0.766	2	*(10)*
N_2H_4(g)	22.79	*(10)*	−51.154	*(10)*	2.601	39.4	−2.536	0.6842	2	*(10)*
Cl(g)	28.992	*(10)*	12.809	*(10)*	5.779	−0.4083	−0.3869	–	1	*(10)*
Cl_2(g)	0		0		8.80	0.208	−0.670	–	1	*(10)*
H(g)	52.103	*(10)*	11.788	*(10)*	4.968	–	–	–	1	*(10)*
H_2(g)	0		0		15.256	2.12	−0.5906	−1.462	1	*(10)*
HCl(g)	−22.063	*(10)*	2.3945	*(10)*	6.224	1.29	0.3251	–	1	*(10)*
Ge(g)	89.5	*(11)*	32.673	*(11)*	29.096	1.418	−1.5164	−3.161	1	*(11)*
GeH_4(g)	21.70	*(12)*	−17.944	*(12)*	14.95	5.28	−5.17	–	1	*(12)*
GeH_3Cl(g)	−16.79	*(12)*	−17.906	*(12)*	−48.174	−4.768	1.059	10.795	1	*(12)*
GeH_2Cl_2(g)	−52.64	*(12)*	−18.157	*(12)*	1.57	−0.77	−3.04	3.24	1	*(12)*
$GeHCl_3$(g)	−85.87	*(12)*	−24.569	*(12)*	−14.03	−4.61	–	6.198	1	*(12)*
$GeCl_3$(g)	−76.68	*(13)*	−7.275	*(13)*	1.6066	10.082	−0.91754	2.7569	2	*(13)*
Ge_2N(g)	102.3	*(14)*	39.645	*(15)*	8.685	14.507	−1.2235	3.5024	2	*(10)*Si_2N
$GeCl_4$(g)	−118.6	*(12)*	−30.89	*(12)*	25.57	–	−2.28	–	1	*(12)*
Ge_2Cl_6(g)	−135.0	*(12)*	−66.73	*(12)*	45.90	1.03	−8.46	–	1	*(12)*
GeCl(g)	32.4	*(13)*	17.45	*(12)*	10.29	−0.62	−1.16	–	1	*(12)*
$GeCl_2$(g)	−40.76	*(12)*	10.12	*(12)*	11.263	−0.234	−1.105	0.4061	1	*(12)*
Ge(l)	8.48	*(11)*	6.59	*(11)*	6.6	–	–	–	1	*(11)*
N_2H_4(l)	12.1	*(10)*	−79.154	*(10)*	18.726	12.4	1.623	−8.807	2	*(10)*
NH_4Cl(sI)	−75.18	*(10)*	−89.284	*(10)*	9.29	38.3	–	–	1	*(10)*
NH_4Cl(sII)	−75.18	*(10)*	−89.284	*(10)*	6.297	31.5	−0.01508	−3.4058	2	*(10)*
Ge_3N_4(s)	−100	*(16,17,21)*	−73.87	*(15)*	34.293	16.45	–	–	1	*(15)*
Ge(s)	0		0		21.62	4.127	−1.3345	−2.7677	1	*(11)*

[a]C_p Form: (1) $C_p = a + b(10^{-3})T + c(10^5)/T^2 + d\ln T$; and (2) $C_p = a + b(10^{-3})T + c(10^{-5})T^2 + d(10^{-9})T^3$.

Ge_3N_4 by atmospheric pressure CVD at 823 K. Several investigators have used $GeCl_4$ and NH_3 sources at both atmospheric (*4,6–9*) and reduced pressure (*3*) to deposit Ge_3N_4 films. Using germanium tetrachloride as the Ge source, Ge_3N_4 has been deposited in both hot wall and cold wall processes and with both H_2 and N_2 carrier gases.

Though the deposition rate of Ge_3N_4 from $GeCl_4$ and NH_3 sources is reported to be limited by surface reaction at low temperatures (*3,6,9*), a thermodynamic analysis of the Ge–N–Cl–H system would be useful for establishing natural limits to the process. A complex chemical equilibrium analysis of the Ge–N–Cl–H system is presented in this paper and applied to CVD of Ge_3N_4 from $GeCl_4$ and NH_3 sources.

THERMODYNAMIC DATA

A total of 28 species were considered in these calculations. Values of the standard enthalpy of formation, $\Delta H^o_{f,298}$, standard entropy of formation, $\Delta S^o_{f,298}$, and heat capacity, Cp, used in the analysis are summarized in Table 1. A stoichiometric algorithm suggested by Cruise (*18*) and Smith and Missen (*19*) was adopted. Test of convergence for true equilibrium was made by assuring solution independence with respect to initial guess.

Of the major equilibrium species in this system, the thermodynamic property values of Ge_3N_4(s) have the most uncertainty. An excellent review of the thermodynamic data for Ge_3N_4(s) is presented by Pauleau et al. (*20*). The standard enthalpy of formation of the most stable structure of Ge_3N_4(s), $\Delta H^o_{f,298}$ ($\beta-Ge_3N_4$), has been measured by three different methods with a reported range in excess of 200 kcal/mol. The vapor pressure of N_2 above $\beta-Ge_3N_4$(s) was measured with a Knudsen effusion cell coupled to a mass spectrometer by Vorob'ev et al. (*16*) and coupled to a weighing balance by Ryklis et al. (*21*). Germanium nitride is most stable according to these measurements, $\Delta H^o_{f,298}$ ($\beta-Ge_3N_4$) = -99.9 kcal/mol (16) and -81.7 kcal/mol (21). The measured vapor pressure of these two studies differed by a factor of 2 and Vorob'ev et al. (*16*) had to correct for a measured pressure dependence on the effective effusion area. These facts suggest that a kinetic limitation may have been present during vaporization. The standard enthalpy of formation was also measured by combustion calorimetry by Hahn and Juza (22) and Pauleau (*23*). The results of these measurements indicate $\beta-Ge_3N_4$ is less stable, $\Delta H^o_{f,298}$ ($\beta-Ge_3N_4$) = -15.6 kcal/mol (22) and -29 kcal/mol (*23*). These measurements include uncertainties in the property values of reactants and products and errors owing to incomplete or competing reactions. Gas equilibrium studies in the Ge–NH_3–Ge_3N_4–H_2 system were performed by Morey and Johnson (*24*) and Pauleau (*23*) at a temperature near 900 K. Significantly different values were reported: $\Delta H^o_{f,298}$ ($\beta-Ge_3N_4$) = $+110.7$ kcal/mol (*24*) and -25.7 kcal/mol (*23*). These results may also be subject to reaction limitations. The homogeneous and heterogeneous decomposition rates of NH_3 are known to be relatively slow (*25*).

To better define a value for $\Delta H^{o}_{f,298}$ ($\beta-Ge_3N_4$), chemical equilibrium calculations were performed with different reported values of the enthalpy of formation at conditions known to deposit Ge_3N_4 by CVD(*3,4,9*). It was determined that only the more negative values of the standard enthalpy of formation produced deposition of solid Ge_3N_4 at typical CVD operating conditions. For example, the value of $\Delta H^{o}_{f,298}$ ($\beta-Ge_3N_4$) had to be changed to −70 kcal/mol before Ge_3N_4 was deposited when using the experimental conditions reported by Nagai and Niimi (*9*). Based on these calculations, a value close to that suggested by the mass spectrometric results (*16,17,21*) was adopted, $\Delta H^{o}_{f,298}$ ($\beta-Ge_3N_4$) = −100 kcal/mol. As further evidence, second-and third-law values of $\Delta H^{o}_{f,298}$ ($\beta-Ge_3N_4$) were calculated from the high temperature N_2 partial pressure measurements of Vorob'ev et al. (*16*). The values differed by about 1 kcal/mol; $\Delta H^{o}_{f,298}$ ($\beta-Ge_3N_4$, second law) = −99.86 kcal/mol and $\Delta H^{o}_{f,298}$ ($\beta-Ge_3N_4$, third law) = −98.74 kcal/mol.

The thermodynamic property values for species in the Ge–Cl–H system were taken as those suggested in the analysis of this system by Chang et al. (*12*). The agreement between experimental flow equilibrium studies of Hai et al. (*26*) and equilibrium calculations in the subsystem was shown to be very good (*12*). The heat capacity of Ge_2N was taken as identical for Si_2N.

RESULTS AND DISCUSSION

The system examined in this study contains species formed from the four elements Ge, N, Cl, and H. The only condensed phases found to be stable in the range of conditions studied were Ge and Ge_3N_4. Therefore, the system variance is equal to five when only the vapor phase is present and is reduced to three when Ge and Ge_3N_4 are codeposited. The independent variables were chosen from the controllable process variables pressure, temperature, and feed composition ($GeCl_4$, NH_3, H_2 or N_2).

The results are not necessarily dependent on this selection of gas sources. The results for use of other sources (e.g., hydrazine) can be obtained by transforming the inlet composition partial pressures to equivalent atom ratio. The Ge/Cl atom ratio, however, was fixed at the value 4.

Three types of information are presented in this analysis. The first type of calculation was a determination of deposition diagrams that specify the phases present as a function of a process variable. The equilibrium partial pressures of vapor phase species were also calculated to give insight into the process chemistry. Finally, thermodynamic yields of solid phases were determined to judge process deposition efficiencies. Calculations were performed using both H_2 and N_2 as the carrier gas, with the H_2 carrier gas results presented and discussed first.

Hydrogen Carrier Gas

Deposition domains for $GeCl_4$, NH_3, and H_2 sources are presented in Fig. 1 for three reactor pressures at a deposition temperature of 800 K. In

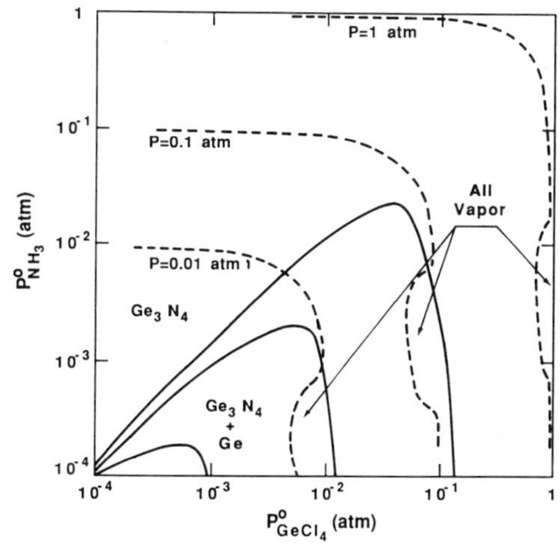

Fig. 1. Phase fields for solid species at equilibrium in the $GeCl_4$–NH_3–H_2 system at $T = 800$ K, $P = 0.01$, 0.1, and 1 atm.

this figure, the solid lines indicate the boundary between single-phase Ge_3N_4 deposition and two-phase $Ge_3N_4 + Ge$ deposition as a function of the feed partial pressure of $GeCl_4$, $P^o_{GeCl_4}$, and NH_3, $P^o_{NH_3}$. This boundary is seen to form a dome, with codeposition of Ge_3N_4 and Ge occurring inside the dome. With increasing reactant feed partial pressure, the boundary closely follows the equimolar feed composition line and then drops sharply at a value of $P^o_{GeCl_4}$ close to 10% of the total pressure. In order to achieve single phase Ge_3N_4 deposition in dilute feed mixtures, an excess of one reactant is required, the excess being significantly greater for $GeCl_4$. Figure 1 also shows that decreasing the reactor pressure scales the region of predicted two-phase deposition in a manner approximately proportional to the pressure. The dashed lines in Fig. 1 represent H_2-free inlet feed compositions and are bounds on possible operating conditions. A small region with no deposition (vapor phase only) is predicted.

The deposition phase diagram for the $GeCl_4$–NH_3–H_2 system is shown in Fig. 2 at a total pressure of 1 atm and a fixed value of $P^o_{GeCl_4} = 10^{-2}$ atm. This diagram contains three different phase fields: pure Ge, codeposited Ge_3N_4 + Ge, and pure Ge_3N_4. The boundary separating deposition of pure Ge_3N_4 and two-phase Ge + Ge_3N_4 is not sensitive to temperature below 1000 K and lies close to the equimolar feed composition. Ammonia-rich conditions are required to deposit single-phase Ge_3N_4 under these conditions. At elevated temperature, entropy effects become more important. This fact translates to deposition of pure Ge rather than Ge_3N_4 and formation of simpler gas phase molecular species. These conditions are similar to the experimental ones of Yashiro (*4*), who reported deposition of only single phase Ge_3N_4 in the temperature range 673–873 K at $P^o_{NH_3}/P^o_{GeCl_4} = 24$. This experimental observation is consistent with the results displayed in Fig. 2. Nagai and Niimi (*9*) reported the growth of epitaxial Ge above 873 K for $P^o_{GeCl_4} = 2 \times 10^{-3}$ atm and $P^o_{NH_3} = 3 \times 10^{-2}$ atm in H_2. Though their value of $P^o_{GeCl_4}$ was a

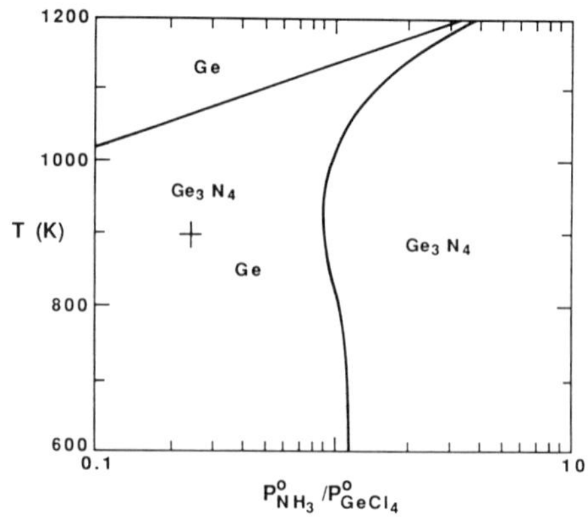

Fig. 2. Deposition phase diagram for the $GeCl_4$–NH_3–H_2 system as a function of temperature and the feed ratio $P^o_{NH_3}/P^o_{GeCl_4}$. $P^o_{GeCl_4} = 10^{-2}$ atm and $P = 1$ atm.

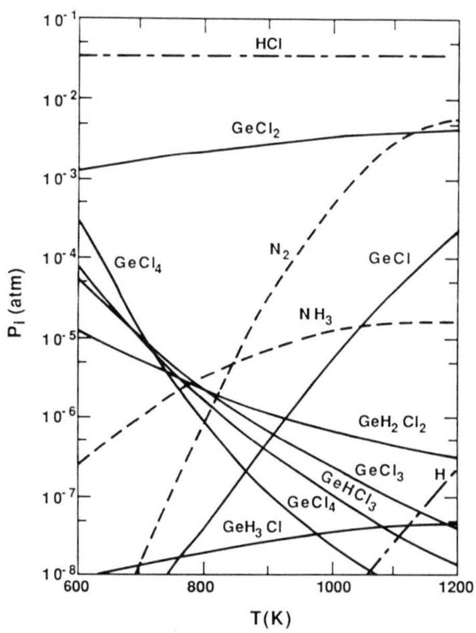

Fig. 3. Partial pressure of equilibrium species as a function of temperature in the system $GeCl_4$–NH_3–H_2 at 1 atm total pressure and a feed composition $P^o_{NH_3} = P^o_{GeCl_4} = 10^{-2}$ atm.

factor of 4 smaller than the value used to generate the results in Fig. 2, the calculated temperature required for deposition of pure Ge is higher than the observed one. This difference suggests that Ge_3N_4 is not quite as stable as assumed in these calculations.

The effect of changes in deposition temperature and inlet partial pressure of NH_3 or $GeCl_4$ on the equilibrium partial pressure of vapor species is shown in Figs. 3-5. The calculations were performed with the base operating conditions of $P = 1$ atm, $T = 800$ K, and $P^o_{GeCl_4} = P^o_{NH_3} = 10^{-2}$ atm. In these figures, Ge vapor species are represented by

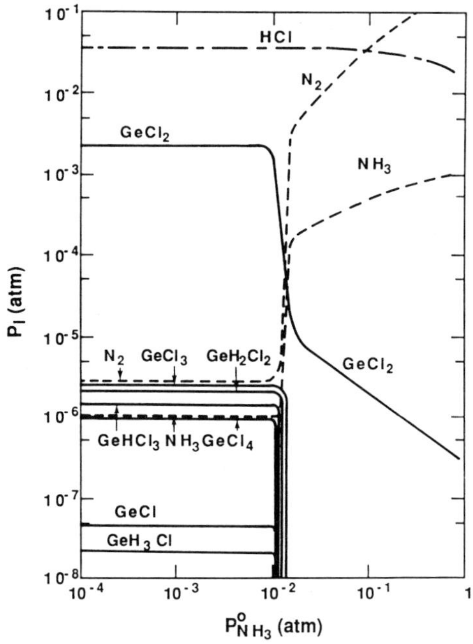

Fig. 4. Partial pressure of equilibrium species as a function of NH_3 partial pressure in the $GeCl_4$–NH_3–H_2 system at P = 1 atm, T = 800 K, and $P^o_{GeCl_4} = 10^{-2}$ atm.

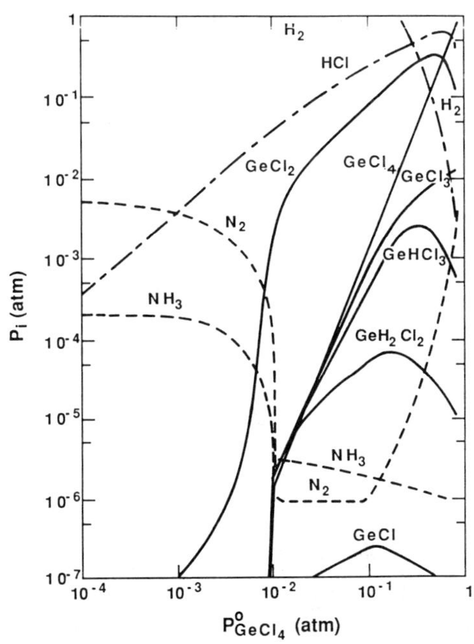

Fig. 5. Partial pressure of equilibrium species as a function of $GeCl_4$ partial pressure in the $GeCl_4$–NH_3–H_2 systems at P = 1 atm, T = 800 K, and $P^o_{NH_3} = 10^{-2}$ atm.

solid lines and the *N*-containing species by dashed lines. The major vapor reaction product is HCl. The major vapor Ge species is $GeCl_2$, though other chloride species and chlorogermanes can be significant at low temperature and high $P^o_{GeCl_4}$ values. In the range of process variables examined, N_2 and NH_3 are the only equilibrium nitrogen-containing vapor species with a concentration greater than 10 ppb. According to Fig. 1, Ge_3N_4 single-phase deposition occurs for inlet ammonia partial pressures greater than approximately 10^{-2} atm and two-phase Ge + Ge_3N_4 deposition at lower partial pressures. Referring to Fig. 4, the addition of NH_3 in the two-phase domain essentially converts solid Ge to solid Ge_3N_4, while consuming a small fraction of the available H_2 and slightly increasing the already significant HCl content. After the Ge feed is consumed, most of the additional NH_3 input decomposes to N_2 and H_2.

A similar behavior, with variations in the inlet $GeCl_4$ partial pressure, is shown in Fig. 5. As depicted in Fig. 1, single phase Ge_3N_4 deposition occurs, except in the partial pressure range 10^{-2}–10^{-1} atm, where Ge and Ge_3N_4 are codeposited. At $P^o_{GeCl_4}$ values less than 10^{-2} atm, additional $GeCl_4$ primarily react with available nitrogen to deposit Ge_3N_4. When a majority of the nitrogen has been consumed, additional $GeCl_4$ feed also produces solid Ge. A portion of the $GeCl_4$, however, disproportionates to $GeCl_2$ and competes with Ge deposition. At a value of $P^o_{GeCl_4} \approx 0.1$ atm, etching of Ge by disproportionation dominates, and the system returns to single-phase Ge_3N_4. Further increases in the inlet amount of $GeCl_4$ leads to a decrease in the amount of deposited Ge_3N_4 since chlorine, now a major element, retains Ge in the vapor phase. The net result is that nitrogen vapor species exhibit a minimum value of partial pressure, and most Ge species show a maximum value.

Equilibrium deposition efficiencies were also calculated for the two solid-phase species Ge and Ge_3N_4. The thermodynamic yield, *R*, is defined for a specific element in each solid phase in terms of the inlet mole numbers of $GeCl_4$ and NH_3, $N^o_{GeCl_4}$ and $N^o_{NH_3}$, and the deposited mole numbers of Ge and Ge_3N_4, N^{eq}_{Ge} and $N^{eq}_{Ge_3N_4}$, according to:

$$R_{Ge}(Ge) = N^{eq}_{Ge}/N^o_{GeCl_4}, \tag{1}$$

$$R_{Ge}(Ge_3N_4) = 3N^{eq}_{Ge_3N_4}/N^o_{GeCl_4}, \tag{2}$$

and

$$R_N(Ge_3N_4) = 4N^{eq}_{Ge_3N_4}/N^o_{NH_3} \tag{3}$$

The thermodynamic yields defined by Eq. (1–3) are plotted in Figs. 6–8 for the same deposition conditions used to generate Figs. 3–5. As shown in Fig. 6, nearly all the nitrogen is incorporated in Ge_3N_4 for the equimolar $GeCl_4$ and NH_3 feed mixture at deposition temperatures below 1000 K. The efficiency for incorporating Ge, however, does not exceed 75% under the same conditions, the balance of the Ge residing primarily in the vapor phase species $GeCl_2$. It is also concluded from Figs. 6–8 that nearly all of the Ge is incorporated in Ge_3N_4 when $P^o_{NH_3}/P^o_{GeCl_4} > 1$, and

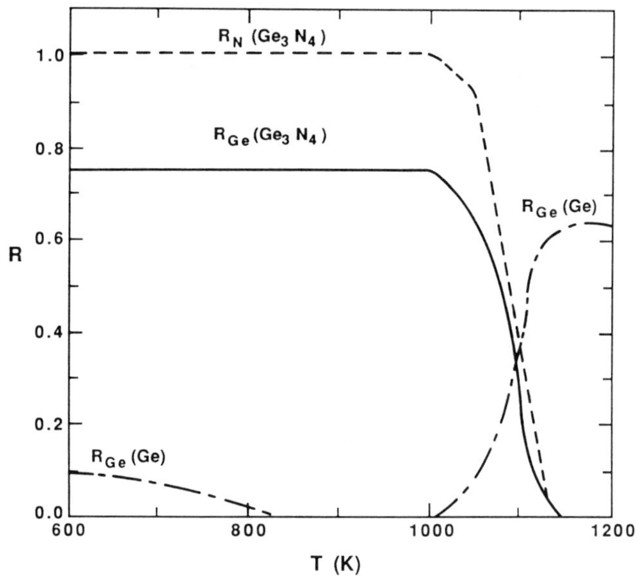

Fig. 6. Variation of the thermodynamic yields as a function of temperature for the $GeCl_4$–NH_3–H_2 system. $P^o_{NH_3} = 10^{-2}$ atm.

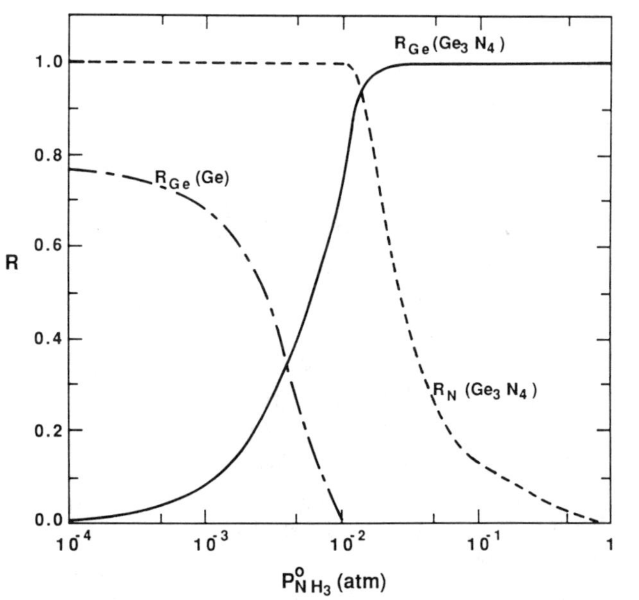

Fig. 7. Variation of the thermodynamic yields as function of NH_3 partial pressure for the $GeCl_4$–NH_3–H_2 system. $P = 1$ atm, $T = 800$ K, and $P^o_{GeCl_4} = 10^{-2}$ atm.

nearly all of the nitrogen is incorporated in Ge_3N_4 when $P^o_{NH_3}/P^o_{GeCl_4} < 1$, provided the deposition temperature is below 1000 K.

Nitrogen Carrier Gas

Similar calculations were performed with N_2 substituted for H_2 as the carrier gas. A N_2 carrier gas was used in the experimental studies of

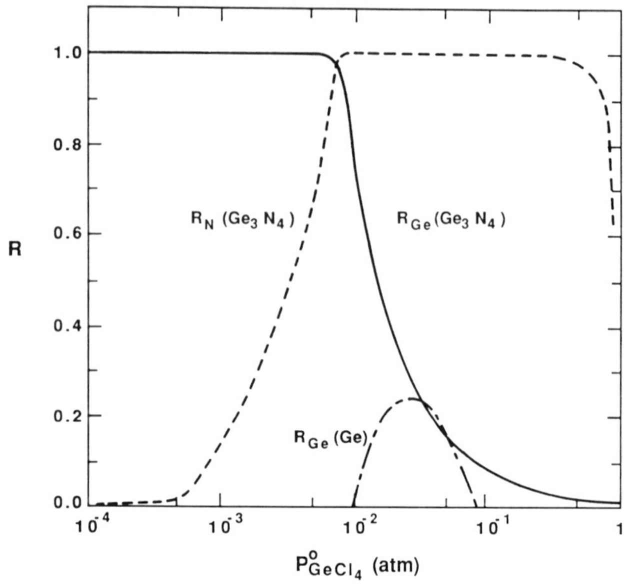

Fig. 8. Variation of the thermodynamic yields as function of $GeCl_4$ partial pressure for the $GeCl_4$–NH_3–N_2 system. $P = 1$ atm, $T = 800$ K, and $P^o_{NH_3} = 10^{-2}$ atm.

Zykov et al. (*6*), Pauleau and Remy (*7,8*), and Young et al. (*3*). Deposition domains for $GeCl_4$, NH_3, and N_2 sources are shown in Fig. 9 for three operating pressures at a temperature of 1000 K. The deposition phase diagram is shown in Fig. 10 at 1 atm total pressure. Single-phase Ge_3N_4 is deposited in a much larger domain when using N_2 as the carrier gas, even at a high temperature with the value of the ratio $P^o_{NH_3}/P^o_{GeCl_4}$ close to one. Unlike the H_2 carrier gas calculation, no elemental Ge deposition is predicted to occur in the range of conditions studied. Remy and Pauleau (*8*) observed Ge deposition at temperatures above 1073 K for values of $P^o_{NH_3}/P^o_{GeCl_4}$ close to unity. Zykov et al. (*6*) also reported the onset of Ge deposition at a slightly higher temperature (1123 K) using an NH_3-rich feed. These temperatures are higher than the value 873 K reported by Nagai and Niimi (*9*) for deposition of Ge in a H_2 carrier gas. Thus, the experimental trend is consistent with the equilibrium calculation results. Figure 9 also shows that decreasing the reaction pressure produces a smaller domain for deposition of single-phase Ge_3N_4. Deposition of single-phase Ge_3N_4 is favored by low temperature and large values of $P^o_{NH_3}/P^o_{GeCl_4}$.

The equilibrium partial pressures of vapor phase species are given in Figs. 11–13 as a function of deposition temperature, $P^o_{NH_3}$, and $P^o_{GeCl_4}$ with the same base conditions used to generate Figs. 3–5 (except N_2 replacing as H_2). The corresponding thermodynamic yields of Ge and N are represented in Figs. 14–16. With H_2 replaced by N_2 the deposition reaction is never limited by availability of nitrogen and the only source of hydrogen is decomposition of NH_3. Both Cl and Ge compete for hydrogen to form the deposition reaction product HCl and chlorogermanes.

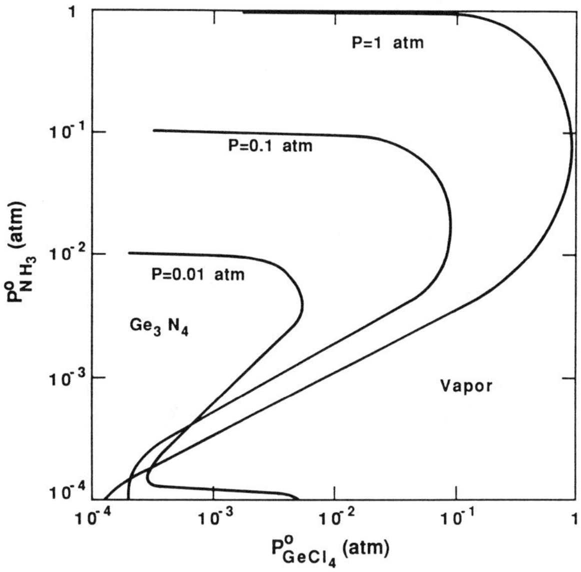

Fig. 9. Phase fields for deposition as a function of deposition temperature and the feed ratio $P^o_{NH_3}/P^o_{GeCl_4}$ for the $GeCl_4$–NH_3–N_2 system. P = 1 atm and $P^o_{GeCl_4} = 10^{-2}$ atm.

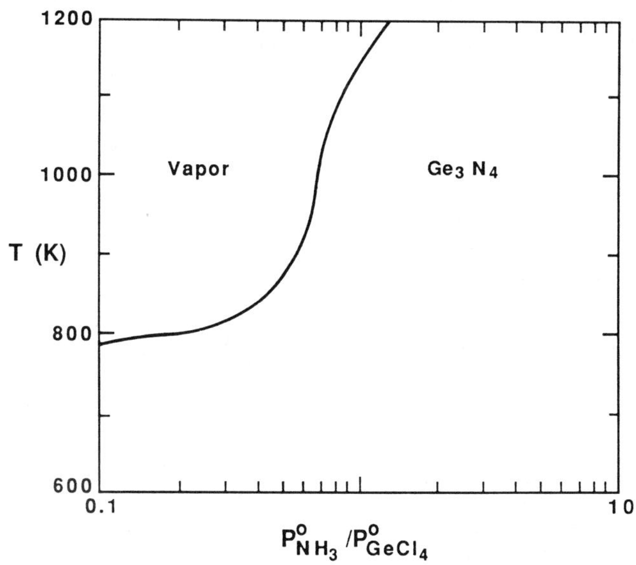

Fig. 10. Deposition phase diagram as a function of deposition temperature and the feed ratio $P^o_{NH_3}/P^o_{GeCl_4}$ for the $GeCl_4$–NH_3–N_2 system. P = 1 atm and $P^o_{GeCl_4} = 10^{-2}$ atm.

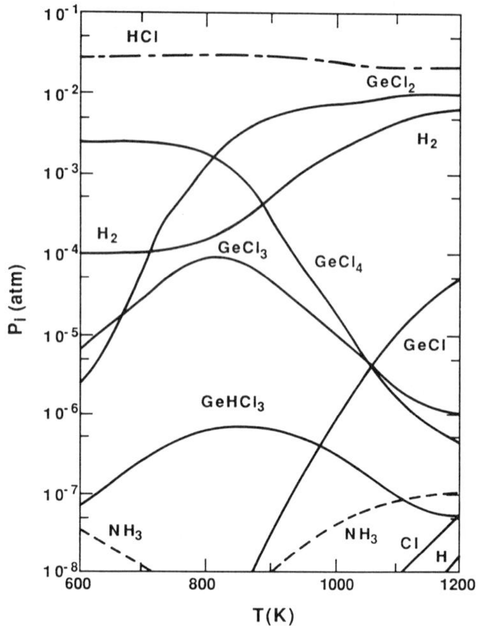

Fig. 11. Equilibrium composition of the gas phase in the $GeCl_4$–NH_3–N_2 system as a function of deposition temperature. P = 1 atm, and $P^o_{GeCl_4} = P^o_{NH_3} = 10^{-2}$ atm.

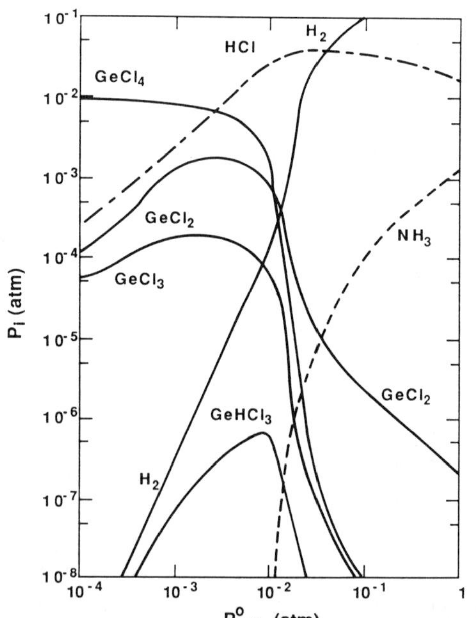

Fig. 12. Equilibrium composition of the gas phase in the $GeCl_4$–NH_3–N_2 system as a function of NH_3 partial pressure. P = 1 atm, T = 800 K, and $P^o_{GeCl_4} = 10^{-2}$ atm.

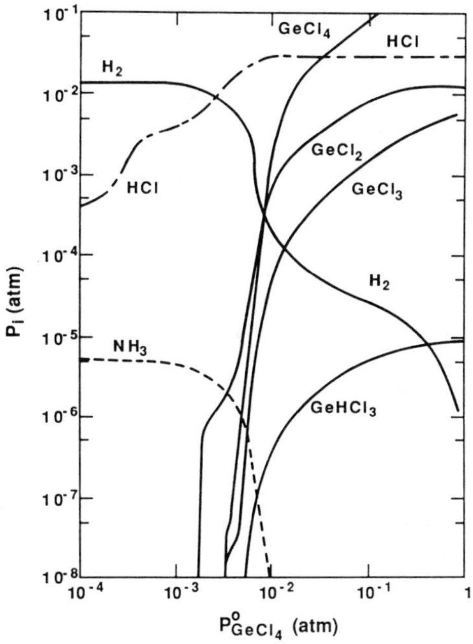

Fig. 13. Equilibrium composition of the gas phase as a function of inlet $GeCl_4$ partial pressure for the $GeCl_4$–NH_3–N_2 system. P = 1 atm, T = 800 K, and $P^o_{NH_3} = 10^{-2}$ atm.

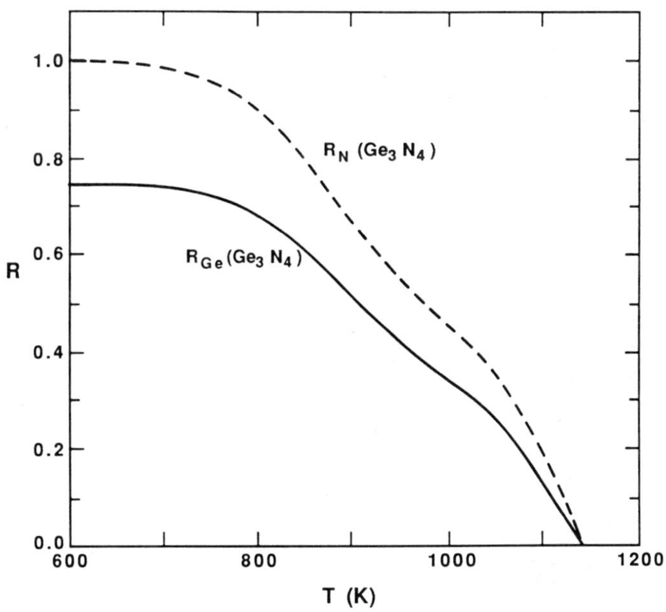

Fig. 14. Thermodynamic yields as a function of deposition temperature for the $GeCl_4$–NH_3–N_2 system. P = 1 atm and $P^o_{NH_3} = P^o_{GeCl_4} = 10^{-2}$ atm.

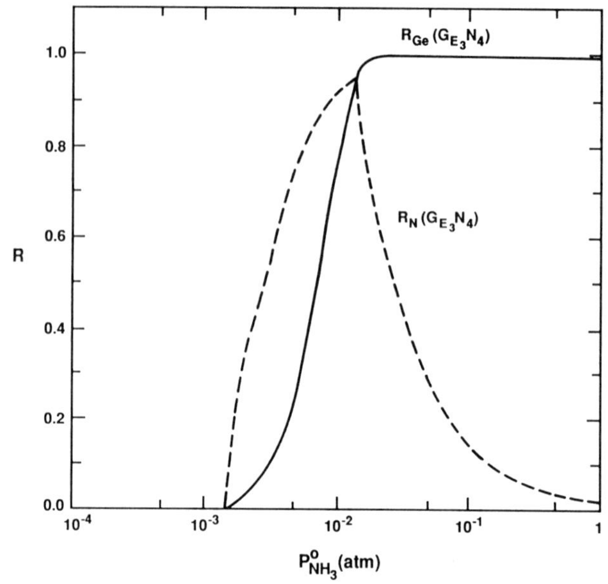

Fig. 15. Thermodynamic yields as a function of the feed NH_3 partial pressure for the $GeCl_4$–NH_3–N_2 system. P = 1 atm, T = 800 K, and $P^{o}_{GeCl_4} = 10^{-2}$ atm.

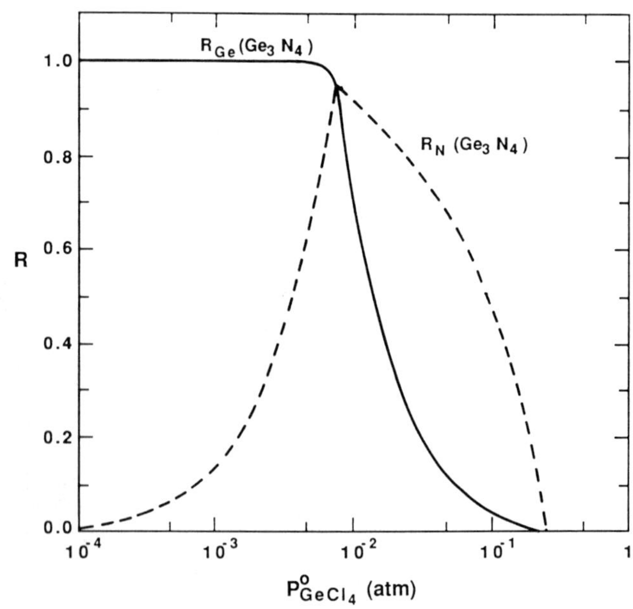

Fig. 16. Thermodynamic yields as a function of the inlet $GeCl_4$ partial pressure for the $GeCl_4$–NH_3–N_2 system. P = 1 atm, T = 800 K, and $P^{o}_{NH_3} = 10^{-2}$ atm.

Most of the hydrogen resides in the stable products HCl and only chlorine-rich Ge species are produced in the vapor phase. Comparing Figs. 6 and 13, the thermodynamic yields of N and Ge are nearly constant below 1000 K with a H_2 carrier gas, where as the yields steadily decrease as temperature increases with a N_2 carrier gas. This decrease in deposition efficiency is accompanied by disproportionation of $GeCl_4$ to $GeCl_2$ and, to a lesser extent at intermediate temperature, to $GeCl_3$ (Fig. 11). No deposition of any solid is predicted at temperatures greater than 1175 K. An experimental definition of the no deposition point would be useful in assessing the Gibbs energy of formation of Ge_3N_4 since this growth condition would be a close representation of true equilibrium (only gas phase reactions need occur and the temperature is elevated).

Thermodynamic yields for deposition of Ge_3N_4 are plotted as a function of $P^o_{NH_3}$ in Fig. 15, and the equilibrium vapor phase composition under the same conditions is shown in Fig. 12. At low values of $P^o_{NH_3}$, there is insufficient hydrogen available in the system to reduce $GeCl_4$, and most of the Ge remains in the vapor as $GeCl_4$. As the inlet NH_3 content approaches the inlet $GeCl_4$ content, the thermodynamic yield of Ge rapidly increases to nearly 100%. A similar behavior for variations in $P^o_{GeCl_4}$ is displayed in Figs. 14 and 16. At low values of $P^o_{GeCl_4}$, there is sufficient hydrogen available for reduction of $GeCl_4$ to deposit Ge_3N_4. For values of $P^o_{GeCl_4}$ larger than $P^o_{NH_3}$, the deposition efficiency of Ge decreases rapidly, and the concentration of the germanium chloride vapor species increases dramatically.

CONCLUSIONS

A complex reaction equilibrium analysis has been performed in the Ge–N–Cl–H system for deposition of stoichiometric Ge_3N_4. Though large differences exist in the reported values of the standard enthalpy of formation of Ge_3N_4, the qualitative trends should be valid. The range of operating parameters to achieve deposition of single-phase Ge_3N_4 were investigated. The choice of carrier gas has a strong influence on the domain extent for deposition of single-phase Ge_3N_4, with nitrogen carrier gas providing a larger range in processing variables. Deposition of Ge_3N_4 is favored by operating at lower temperature and values of $P^o_{NH_3}/P^o_{GeCl_4} > 1$, whereas the operating pressure has a small influence on the relative values of the operating parameters. Based on efficiency of converting reactants to single phase Ge_3N_4, the desirable operating conditions in a H_2 carrier gas are an NH_3 feed composition in slight excess of $GeCl_4$ and an operating temperature between 850 and 1000 K. With a nitrogen carrier gas, no deposition of elemental Ge is predicted, and the thermodynamic yields increase with decreasing deposition temperature.

ACKNOWLEDGMENT

This work was supported by the Office of Basic Energy Sciences, US Department of Energy, under Grant No. DE-FG05-86ER45276.

REFERENCES

1. Pande, K. P. and Shen, C. C., *Appl. Phys. A* **28,** 123 (1982).
2. Pande, K. P. and Pourdavoud, S. Sr., *IEEE Electron Device Lett.* **EDL-2,** 182 (1981).
3. Young, A. B., Rosenberg, J. J., and Szendro, I., *J. Electrochem. Soc.* **134,** 2867 (1987).
4. Yashiro, T., *J. Electrochem. Soc.* **119,** 780 (1972).
5. Gritsenko, V. A., *Phys. Status Solid A* **28,** 387 (1975).
6. Zykov, A. M., Borisov, O. P., Dubrovskii, G. P., and Ashkinadze, N. V., *Zh. Prikl. Khim.* (Leningrad) **44,** 2553 (1971).
7. Pauleau, Y. and Remy, J. C., *C.R. Acad. Sci.* (Paris) **280,** 1215 (1975).
8. Remy, J. C. and Pauleau, Y., *Inorg. Chem.* **15,** 2308 (1976).
9. Nagai, H. and Niimi, T., *J. Electrochem. Soc.* **115,** 671 (1968).
10. JANAF Thermochemical Tables, third edition, *J. Phys. Chem. Ref. Data* **14,** 1 (1985).
11. Hultgren, R., Desai, P. D., Hawkins, D. T., Gleiser, M., Kelley, K. K., and Wagman, D. D., *Selected Values of the Thermodynamic Properties of the Elements,* American Society of Metals (1973).
12. Chang, S., Unzicker, D., and Anderson, T. J., *Proc. Tenth Int. Conf. on Chemical Vapor Deposition,* **87-8,** 122 (1987).
13. Glushko, V. P., *Thermodynamic Properties of Elementary Substances,* Nauko (Moscow) **3,** 189 (1981).
14. Muenow, D. W., *J. Chem. Phys.* **60,** 3382 (1974).
15. Kubaschewski, O. and Alcock, C. B., *Metallurgical Thermochemistry,* 5th edition, Pergammon, Oxford, 1979.
16. Vorob'ev, A. M., Evseeva, G. V., and Zenkevich, L. V., *Russ. J. Phys. Chem.* **47,** 1691 (1973).
17. Fesenko, V. V., Bolgar, A. S., and Ryklis, E. A., *Proc. 1st Int. Conf. Calorimetry and Thermodynamics,* Warsaw (1969).
18. Cruise, D. R., *J. Phys. Chem.* **68,** 3797 (1964).
19. Smith, W. R. and Missen, R. W., *Can. J. Chem. Eng.* **46,** 269 (1968).
20. Pauleau, Y., Hantzpergue, J. J., and Remy, J. C., *Mat. Chem.* **5,** 371 (1980).
21. Ryklis E. A., Bolgar, A. S., Lyutaya, M. D., and Fesenko, V. V., *Porosh. Met.* **8,** 64 (1968).
22. Hahn, H. and Juza, R., *Z. anorg. allgem. Chem.* **244,** 111 (1940).
23. Pauleau, Y., *These de Doctrat es Sciences Physiques,* Angers, France, 130 (1975).
24. Morey, G. H. and Johnson, W. C., *J. Am. Chem. Soc.* **54,** 3603 (1932).
25. Homann, K. H. and Haas, A., *Comprehensive Chemical Kinetics,* Bamford, C. H. and Tipper, C. F. H., eds., Elsevier, Amsterdam, 1972, p. 1.
26. Hai, N. V., Hein, K., and Leipner, K., *Neue-Hutte* **16,** 711 (1971).

Modeling of Chemical Vapor Deposition or Etching in Closed Systems

Y. K. RAO* AND YANG DO

Department of Materials Science and Engineering, FB-10, University of Washington, Seattle, WA 98195

ABSTRACT

The sealed ampule chemical vapor transport reactor is commonly used for the growth of bulk crystals and thin films. The reactor consists of a high-temperature source zone, a lower temperature deposition zone, and a connecting zone with a temperature gradient. By assuming gas–solid equilibrium in the end zones, the rate of transport of matter, from the source to the seed or vice versa, by diffusion and Stefan flow mechanisms can be calculated. The method is illustrated for the growth and etching of nickel oxide in a closed reactor using chlorine gas as the transporting agent. The influences of zone temperatures, amount of chlorine, and temperature gradient on the diffusion rates were examined.

Index entries: Nickel oxide; vapor transport; diffusion; equilibrium; rate; deposition; etching; chlorine.

INTRODUCTION

Bulk crystals and thin films of high melting point oxides can be grown most advantageously by the chemical vapor transport technique. The preparation of NiO crystals by reacting $NiCl_2$ vapor with oxygen in a flow reactor was reported by Ito et al. (*1*). Crystals with an average size of 0.1 mm were obtained after 2 h at a growth temperature of 1000°C. The vapor-phase reaction method has also been used in the growth of epitaxial films of nickel ferrite ($NiFe_2O_4$) and cobalt ferrite ($CoFe_2O_4$) on <111>

*Author to whom all correspondence and reprint orders should be addressed.

MgO substrates (2). Emmenegger (*3*) analyzed the thermodynamic aspects of the chemical vapor deposition (CVD) of ternary metal oxides, such as crystalline spinels ($NiCr_2O_4$, $CoCr_2O_4$, and so on) grown from a stoichiometric mixture of the respective binary oxides sealed in a quartz ampule filled with chlorine gas. The transport of vapor species occurred over a temperature gradient of 100–150°C with the deposition zone held at 800°C.

Jeffes (*4*) examined a number of systems and developed a method of predicting the optimum thermodynamic conditions for the chemical vapor transport (CVT). The factors influencing efficient transfer of matter between the hot and cold zones in a sealed CVT reactor were considered in some detail by Schäfer (*5*). The sealed tube transport reactor may be visualized as consisting of three zones: a source zone at T_s, a connecting zone with a temperature gradient, and a deposition zone at T_d. In the two-zone model formulated by Schäfer (*6*), the vapor mass in the connecting zone of the reactor was not taken into consideration, and equilibrium was assumed to prevail in each of the two end zones, with the transport of species taking place by diffusion; a single "combined" diffusion coefficient was employed for all the species. Mandel (*7*) and Lever and Mandel (*8*) improved the model by the admitting values for the binary diffusion coefficients of the various pairs of gas species occurring in the system.

The diffusion-limited model for the growth of crystals in CVD systems was further developed by Faktor et al. (*9,10*); gas–solid equilibrium was assumed in both the source and deposition zones with rates of nucleation and surface reaction processes remaining very high and the deposition of crystals controlled by the gas-phase mass transport of species. Notable contributions to the theory of chemical vapor mass transport in closed ampule reactors were made by Bloem and coworkers (*11*) and Watanabe et al. (*12*); diffusion, laminar flow, thermal diffusion, and convection together determine the mass transfer rate.

The recent work of Nölang and Richardson (*13–15*) is particularly significant. They assume that the heterogeneous reactions proceed at very high rates so that equilibrium is maintained between the vapor phase and the transported solids; and the transfer of matter occurs by diffusion and bulk flow of fluid (vapor). Incorporating the temperature gradient into the diffusion controlled model, these authors (*13*) develop a flux function (ϕ) to express the rate of transport.

$$\phi = J/\bar{D}_o \frac{dT}{dZ} \quad , \text{ mole}\cdot\text{cm}^{-3}\cdot\text{K}^{-1} \tag{1}$$

where J(mole/cm²/s) is the molar flux, computed from the partial pressure gradient (dP_i/dZ), $\bar{D}_o$(cm²/s) is the average diffusion coefficient, and dT/dZ(K/cm) is the temperature gradient. In general ϕ is a function of temperature, total pressure, partial pressures of species (P_i), and temperature variation of partial pressure (dP_i/dT). This method was used to predict the rate of chemical transport of SnO_2(s) in SnO_2/HBr and SnO_2/Cl_2 systems (*14*).

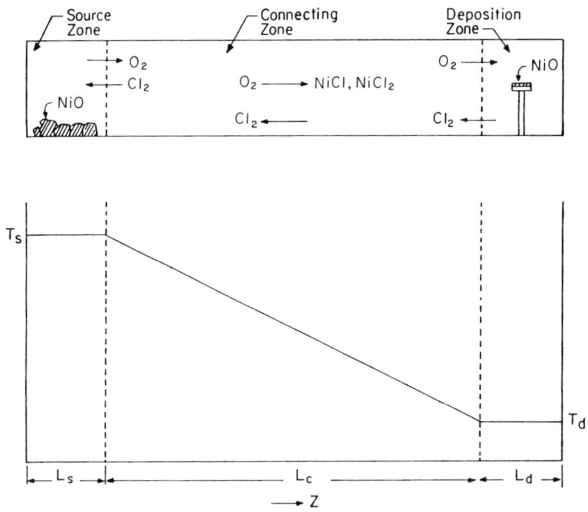

Fig. 1. Schematic diagram of the sealed chemical vapor transport reactor with a linear temperature gradient.

Despite the fact that the model of Nölang and Richardson (*13*) takes into consideration the existence of a temperature gradient between the end zones, the use of a single "average diffusion coefficient" for the vapor species constitutes an undue simplification that is hard to justify. The present work overcomes this debilitating assumption and analyzes the chemical vapor transport system in its full complexity. The sealed ampule reactor is regarded as a three-zone system with solid–gas equilibria prevailing in the source and deposition zones. The transport of species between zones occurs, under quasi steady-state conditions, by diffusion and laminar (Stefan) flow. Unlike the previous studies, where the equilibrium phase compositions in the end zones were computed by considering each zone (at a specified temperature) in isolation, full coupling is maintained between the three zones for the purpose of determining the equilibrium state. The mass-transport rates are computed using the Stefan-Maxwell diffusion formalism. The method is illustrated by the example of the chemical vapor transport of NiO(s) in a sealed tube with chlorine gas as the transporting agent. The influences of parameters, such as the end zone temperatures (T_s and T_d), temperature gradient, amount of halogen, and total pressure on the predicted transport rates, are investigated using the model.

CHEMICAL VAPOR DEPOSITION OF NiO(s)

The vapor phase transport of nickel oxide can be obtained through the use of a transport agent, such as Cl_2(g), HCl(g), or $NiCl_2$(g). In here, chlorine gas is selected as the "solvent" for the solid to be transported. Figure 1 shows the experimental set-up consisting of a sealed quartz ampule, heated in a horizontal furnace with the appropriate temperature gradient. The present system is prepared by placing NiO(s) in the source zone of the reactor, which is then evacuated and filled with chlorine gas in the desired amount.

The chlorination of the oxide is favored in the high temperature source zone; the gaseous products $NiCl_2$, NiCl, and O_2 diffuse through the connecting zone, and oxidation reaction produces a layer of nickel oxide on the substrate in the low temperature deposition zone. At elevated temperatures, in a short time interval, steady-state conditions are established in the sealed reactor. Equilibrium is assumed between the solid and gas phases in the two end zones, with the diffusion of species constituting a small perturbation of the equilibrium state. The gaseous species ($NiCl_2$, NiCl, and O_2) derived from the nickel oxide maintain stoichiometric composition in the source zone.

$$\frac{n^s_{Ni}}{n^s_{O}} = \frac{n^s_{NiCl_2} + n^s_{NiCl}}{2n^s_{O_2}} = 1 \qquad \text{(A.1)}$$

where n^s_i denotes the number of moles of the ith species in the source zone. Superscripts 's,' 'd,' and 'c,' are used to denote the quantities in the source, deposition, and connecting zones, respectively.

Source Zone

The species present in this zone include $NiCl_2$, NiCl, O_2, Cl_2, and NiO(s). As the number of the underlying elements (Ni, Cl, O), as well as the rank of the five species atom matrix, is three, it is clear that there can be two linearly independent chemical equilibria at the most, in accordance with the Gibbs Stoichiometric rule (*16*). These are chosen as follows

$$NiO(s) + Cl_2 = NiCl_2 + 0.5\ O_2 \qquad \text{(1s); } \varepsilon^s_1$$

$$NiCl_2 = NiCl + 0.5\ Cl_2 \qquad \text{(2s); } \varepsilon^s_2$$

where ε^s_1 and ε^s_2 are the extents of reactions for equilibria (1s) and (2s), respectively. These change with time as matter is transported from the source zone to the deposition zone. Under quasi steady-state conditions, the number of moles of each of the species can be expressed in terms of their respective initial amounts and the extents of reactions. For each of the four gas species, coupling is maintained between the three zones.

$$n^s_{NiCl_2} = \varepsilon^s_1 - \varepsilon^s_2 - n^{od}_{NiCl_2} - n^c_{NiCl_2} \qquad \text{(S.1)}$$

$$n^s_{Cl_2} = n^o_{Cl_2} - \varepsilon^s_1 + 0.5\varepsilon^s_2 - n^{od}_{Cl_2} - n^c_{Cl_2} \qquad \text{(S.2)}$$

$$n^s_{NiCl} = \varepsilon^s_2 - n^{od}_{NiCl} - n^c_{NiCl} \qquad \text{(S.3)}$$

$$n^s_{O_2} = 0.5\varepsilon^s_1 - n^{od}_{O_2} - n^c_{O_2} \quad (S.4)$$

$$n^s_{NiO} = n^o_{NiO} - \varepsilon^s_1 \quad (S.5)$$

where n^c_i is the number of mols of the ith species in the connecting zone and n^{od}_i denotes the number of mols of the ith species transferred to the deposition zone before the onset of NiO(s) formation. Note that $n^o_{Cl_2}$ and n^o_{NiO} are the initial amounts of Cl_2(g) and NiO(s) in the system.

Deposition Zone

The gaseous chlorides react with O_2-producing NiO(s) that deposits on the substrate. Equilibrium is assumed between the gas and solid phases. The gas composition in the vicinity of the solid does not necessarily conform to the stoichiometric constraint given by Eq. (A.1). The species present are five: $NiCl_2$, NiCl, Cl_2, O_2, and NiO(s). As noted earlier, only two independent chemical equilibria are possible.

$$NiCl_2 + 0.5O_2 = NiO(s) + Cl_2 \quad (1d);\ \varepsilon^d_1$$

$$NiCl_2 = NiCl + 0.5\,Cl_2 \quad (2d);\ \varepsilon^d_2$$

where ε^d_1 and ε^d_2 are the extents of reactions. For the amounts of each of these species, we can write the following equations:

$$n^d_{NiCl_2} = n^{od}_{NiCl_2} - \varepsilon^d_1 - \varepsilon^d_2 \quad (D.1)$$

$$n^d_{Cl_2} = n^{od}_{Cl_2} + \varepsilon^d_1 + 0.5\varepsilon^d_2 \quad (D.2)$$

$$n^d_{NiCl} = n^{od}_{NiCl} + \varepsilon^d_2 \quad (D.3)$$

$$n^d_{O_2} = n^{od}_{O_2} - 0.5\varepsilon^d_1 \quad (D.4)$$

$$n^d_{NiO} = \varepsilon^d_1 \quad (D.5)$$

On combining Eqs. (S.5) and (D.5), it is found that the difference $(\varepsilon^s_1 - \varepsilon^d_1)$ is equal to the number of mols of NiO(s) present as gas species ($NiCl_2$, NiCl, and O_2) in the closed CVT reactor; at steady state, this quantity remains unchanged. From Eqs. (S.1–S.4) and (D.1–D.4), it can be shown that

$$(n^s_{NiCl_2} + n^d_{NiCl_2} + n^c_{NiCl_2}) + (n^s_{NiCl} + n^d_{NiCl} + n^c_{NiCl}) \quad (A.2)$$
$$= 2(n^s_{O_2} + n^d_{O_2} + n^c_{O_2})$$

The chlorine atom balance equation derived using the foregoing relations is

$$[(n^{s}_{NiCl_2} + n^{d}_{NiCl_2} + n^{c}_{NiCl_2}) + (n^{s}_{Cl_2} + n^{d}_{Cl_2} + n^{c}_{Cl_2}) + 0.5(n^{s}_{NiCl} + n^{d}_{NiCl} + n^{c}_{NiCl})] = n^{\circ}_{Cl_2} \quad (A.3)$$

It is advantageous to express the constraining Eqs. (A.1–A.3) in terms of the partial pressures of the gaseous species. Let us suppose that V_s, V_c, and V_d denote the volumes of the source, connecting, and deposition zones, respectively. The temperature is considered to decline linearly from T_s to T_d, as shown in Fig. 1. It readily follows that

$$T = T_s - (T_s - T_d) \cdot \frac{(Z - L_s)}{L_c} \quad (2)$$

The concentrations of gas species can be assumed to vary linearly with distance Z in the connecting zone. The error in the total pressure resulting from such an assumption was found to be negligibly small in the case of chemical vapor transport of nickel in the $NiCl_2$–NiCl–Cl_2 system (*17*). For the ith gas species, the molar concentration (C^c_i) at Z = Z is given by

$$C^c_i = C^s_i - (C^s_i - C^d_i) \cdot \frac{(Z - L_s)}{L_c} \quad (3)$$

The number of mols of the ith species in the connecting zone can be found by integration.

$$n^c_i = S_c \int_{L_s}^{L_s + L_c} C^c_i dZ = 0.5\, L_c S_c (C^s_i + C^d_i) \quad (4)$$

Substituting $V_c = L_c S_c$ and using partial pressures in place of molar concentrations, we obtain

$$n^c_i = \frac{V_c}{2RT_s} (P^s_i + P^d_i \cdot \frac{T_s}{T_d}) \quad (5)$$

Assuming that the gaseous species behave ideally, the quantities n^s_i, n^d_i, and n^c_i in Eqs. (A.1–A.3) are replaced by the respective partial pressures. For the particular geometric configuation that ensures that $V_s = V_d$, the following relations are obtained:

$$P^s_{NiCl_2} + P^s_{NiCl} = 2P^s_{O_2} \quad (A.4)$$

$$P^s_{NiCl_2} + P^s_{NiCl} + (P^d_{NiCl_2} + P^d_{NiCl}) \frac{T_s}{T_d} = 2(P^s_{O_2} + P^d_{O_2} \cdot \frac{T_s}{T_d}) \quad (A.5)$$

$$P^s_{NiCl_2} + P^s_{Cl_2} + 0.5\, P^s_{NiCl} + (P^d_{NiCl_2} + P^d_{Cl_2} + 0.5\, P^d_{NiCl}) \cdot \frac{T_s}{T_d}$$

$$= n^{\circ}_{Cl_2} \cdot \frac{RT_s}{V_s} (1.0 + \frac{V_c}{2V_s})^{-1} \quad (A.6)$$

Under quasi steady-state conditions, the total pressure is the same in all the zones. Hence

$$P^s_{NiCl_2} + P^s_{NiCl} + P^s_{Cl_2} + P^s_{O_2} = P^d_{NiCl_2} + P^d_{NiCl} + P^d_{Cl_2} + P^d_{O_2} \quad (A.7)$$

The degrees of freedom (*f*) possessed by this closed system can be found by applying the extended form of the Gibbs phase rule (*18*).

$$f = f_s + f_d = (N_s - r_s - S_s) - p_s + 2 + (N_d - r_d - S_d) - p_d + 2 - (t_s + t_d) \quad (6)$$

where N_s is the number of species, r_s is the number of independent chemical equilibria, S_s is the number of stoichiometric constraints, p_s is the number of phases, and t_s is the number of special constraints in the source zone; N_d, r_d, S_d, p_d, and t_d have similar meanings for the deposition zone. It is seen that

$$N_s = 5 \; ; \; r_s = 2 \; ; \; S_s = 1 \; ; \; p_s = 2$$

$$N_d = 5 \; ; \; r_d = 2 \; ; \; S_d = 0 \; ; \; p_d = 2 \; ; \; t_s + t_d = 3$$

Substitutions yield

$$f = 2$$

These two degrees of freedom are satisfied by specifying the values for the temperatures of the source (T_s) and deposition (T_d) zones. Once these temperatures are known, the equilibrium constants for reactions (1s), (2s), (1d), and (2d) are defined.

$$K_{1s} = P^s_{NiCl_2} (P^s_{O_2})^{0.5} / P^s_{Cl_2} \quad (A.8)$$

$$K_{2s} = P^s_{NiCl} (P^s_{Cl_2})^{0.5} / P^s_{NiCl_2} \quad (A.9)$$

$$K_{1d} = P^d_{Cl_2} / [P^d_{NiCl_2} (P^d_{O_2})^{0.5}] \quad (A.10)$$

$$K_{2d} = P^d_{NiCl} (P^d_{Cl_2})^{0.5} / P^d_{NiCl_2} \quad (A.11)$$

Table 1
Data on Standard Free Energies of Formation (ΔG_f°) for Different Species

	ΔG_f°, J/mol,			Joules		
	$NiCl_2$	NiCl	NiO(s)			
T,K	*(19,20)*[a]	*(19,20)*	*(21)*	ΔG_{1s}°	ΔG_2°	ΔG_{1d}°
1000	−156,285	24,288	−149,365	−6,920	180,573	6,920
1100	−142,369	11,437	−140,750	−1,619	153,806	1,619
1200	−128,599	−1,282	−132,319	3,720	127,317	−3,720
1320	−122,938	−16,363	−122,019	−919	106,575	919

[a]Reference number.

The values of the four equilibrium constants are found from the standard free energy changes for the respective reactions. The latter, listed in Table 1, are deduced from published thermochemical data (*19–21*) for the species NiCl, $NiCl_2$, and NiO(s).

There are eight unknown equilibrium partial pressures and eight eqs.—(A.4) through (A.11). The solution is obtained by the iterative equilibrium constant method (*22*) for any given reactor size and initial amount of the transporting agent.

GAS-PHASE COMPOSITIONS

Using the coupled diffusion and virtual equilibrium model, the partial pressures of gas species were computed for different conditions. The sealed reactor dimensions employed in the present calculations are as follows:

$$L_s = L_d = 4 \text{ cm} \; ; \; L_c = 92 \text{ cm}$$

$$\text{area of cross-section, } S_c = \pi(2.5)^2 \text{ cm}^2$$

$$V_s = V_d = 78.54 \text{ cm}^3 \; ; \; V_c = 23V_s$$

For T_s(1320 K) → T_d(1200 K), with the initial amount of Cl_2(g) in the reactor is equal to 0.01 g, which corresponds to a pressure of about 7.43 × 10^{-3} atm at the mean temperature 1260 K for the transporting agent, the results are given in Table 2. The values of $P^s_{Cl_2}$, $P^d_{Cl_2}$, P^s_{NiCl}, and P^d_{NiCl}, not listed in the table, can be computed from the equilibrium constants K_{1s}, K_{1d}, K_{2s}, and K_{2d} using the eqs. (A.8)–(A.11). Transport of O_2, $NiCl_2$, and NiCl occurs left to right, whereas that of Cl_2 is in the opposite direction. Hence deposition of NiO(s) on the substrate is indicated.

On lowering the deposition zone temperature (T_d) to 1100 K, inspection of the data in Table 2 shows that $P^s_{O_2} < P^d_{O_2}$ and $P^s_{NiCl_2} < P^d_{NiCl_2}$ implying that etching is favored. For the T_s(1100 K) → T_d(1000 K) setting, a similar behavior obtains with the tendency for etching, judging from the value of $(P^s_{O_2}-P^d_{O_2})/P^s_{O_2}$, considerably enhanced.

A more extensive set of equilibrium partial pressures for the T_s(1320 K) → T_d(1200 K) temperature gradient, with different initial Cl_2(g) are shown in Table 3. In here, deposition is favored in all in-

Table 2
Diffusion-Limited Deposition or Etching Rates of NiO(s) in a Sealed CVT Reactor Filled with 10 mg $Cl_2(g)$

	$T_s \longrightarrow T_d$		
atm	1320 K ⟶ 1200 K	1320 K ⟶ 1100 K	1100 K ⟶ 1000 K
$P^s_{O_2}$	3.500×10^{-3}	3.363×10^{-3}	2.964×10^{-3}
$P^s_{NiCl_2}$	6.978×10^{-3}	6.704×10^{-3}	5.927×10^{-3}
$P^d_{O_2}$	3.432×10^{-3}	3.372×10^{-3}	3.006×10^{-3}
$P^d_{NiCl_2}$	6.864×10^{-3}	6.746×10^{-3}	6.012×10^{-3}
P	1.088×10^{-2}	1.045×10^{-2}	9.160×10^{-3}
$D^o_{O_2V}$ [a]	6.112×10^{-6}	6.115×10^{-6}	6.108×10^{-6}
N_{O_2} [b]	1.570×10^{-9}	-0.211×10^{-9}	-1.015×10^{-9}
R_{NiO}, mg/h	16.577	−2.225	−10.715

[a] $cm^2/atm/s/K^{1.75}$.
[b] mol $O_2/cm^2/s$.

stances. The possible condensation of $NiCl_2$ from the vapor phase was examined by comparing the equilibrium saturation pressure (23) for the phase transformation,

$$NiCl_2(g) \rightarrow NiCl_2(s,\ell)$$

with the P_{NiCl_2} values listed in Table 3. For the mass range of $Cl_2(g)$ studied (Table 3), $NiCl_2$ condensation is not feasible.

GROWTH OR ETCH RATES

The mass transport of $NiCl_2$, NiCl, O_2, and Cl_2 occurs by ordinary diffusion, bulk (or Stefan) flow, and thermal diffusion. The natural convection effects are not significant in a horizontal reactor. In addition, for T_s/T_d ratios of 1.1–1.2, the contribution of thermal diffusion to total mass transferred is about 2–6% (24); hence, it can be regarded as negligible. At quasi steady-state, in each of the three zones for the gaseous species,

$$(\partial C_i/\partial t) = 0 \tag{7}$$

Under these conditions, the rate of growth of nickel oxide on the substrate is equal to that of the corrosion of the starting oxide phase in the source zone. The transport rates of the elements Ni and O can be expressed in terms of the molar fluxes of the species.

$$N_{Ni} = N_{NiCl_2} + N_{NiCl}$$

$$N_O = 2N_{O_2}$$

The stoichiometry of NiO(s) requires that

$$N_{NiCl_2} + N_{NiCl} = 2N_{O_2} \tag{8}$$

Table 3
Diffusion-Limited Deposition Rates of NiO(s) in a Sealed CVT Reactor with T_s (1320 K) $\longrightarrow$ T_d (1200 K)

	Initial amount of Cl_2, g						
atm	0.02	0.05	0.1	0.2	0.5	0.8	1.0
$P^s_{O_2}$	6.846×10^{-3}	1.643×10^{-2}	3.150×10^{-2}	5.969×10^{-2}	0.1360	0.2055	0.2493
$P^s_{NiCl_2}$	1.367×10^{-2}	3.282×10^{-2}	6.296×10^{-2}	0.1193	0.2720	0.4109	0.4985
$P^d_{O_2}$	6.665×10^{-3}	1.580×10^{-2}	2.992×10^{-2}	5.586×10^{-2}	0.1244	0.1854	0.2234
$P^d_{NiCl_2}$	1.333×10^{-2}	3.160×10^{-2}	5.984×10^{-2}	0.1117	0.2487	0.3707	0.4467
P	0.02158	0.05316	0.1048	0.2059	0.5004	0.7877	0.9767
$D^o_{O_2V}$[a]	6.111×10^{-6}	6.110×10^{-6}	6.111×10^{-6}	6.111×10^{-6}	6.110×10^{-6}	6.109×10^{-6}	6.110×10^{-6}
N_{O_2}[b]	2.089×10^{-9}	2.910×10^{-9}	3.649×10^{-9}	4.424×10^{-9}	5.361×10^{-9}	5.805×10^{-9}	5.985×10^{-9}
R_{NiO}, mg/h	22.064	30.731	38.541	46.728	56.623	61.314	63.214

[a] $cm^2/atm/s/K^{1.75}$.
[b] $mol\ O_2/cm^2/s$.

Since there is no net transport of chlorine, the following relation is obtained by the atom balance.

$$2N_{NiCl_2} + N_{NiCl} + 2N_{Cl_2} = 0 \tag{9}$$

In the foregoing, N_{O_2}, N_{NiCl_2}, N_{NiCl}, and N_{Cl_2} are molar fluxes (in mol/cm^2/s) of species O_2, $NiCl_2$, NiCl and Cl_2 respectively.

The diffusion of a species in a multicomponent system is described most advantageously by the Stefan-Maxwell formalism. The letters A, B, C, and D can be used to denote the species O_2, $NiCl_2$, NiCl, and Cl_2 respectively. For the diffusion of A in a gas mixture composed of four species,

$$-dP_A = RT\Big[\frac{1}{D_{AB}}(X_B N_A - X_A N_B) + \frac{1}{D_{AC}}(X_C N_A - X_A N_C) + \frac{1}{D_{AD}}(X_D N_A - X_A N_D)\Big]dZ \tag{10}$$

where D_{AB}, D_{AC}, and D_{AD} are ordinary diffusion coefficients for the pairs A-B, A-C, and A-D, respectively, and X_i is the mol fraction of the ith species at Z = Z position in the connecting zone. Using the flux of O_2 as the basis, the following are defined:

$$\gamma_B = N_B/N_A \; ; \; \gamma_C = N_C/N_A \; ; \; \gamma_D = N_D/N_A \tag{11}$$

Substituting and solving for N_A, we obtain,

$$N_A = \frac{-(P_A/RT)(dX_A/dZ)}{\left(\frac{X_B}{D_{AB}} + \frac{X_C}{D_{AC}} + \frac{X_D}{D_{AD}}\right) - X_A\left(\frac{\gamma_B}{D_{AB}} + \frac{\gamma_C}{D_{AC}} + \frac{\gamma_D}{D_{AD}}\right)} \tag{12}$$

We can introduce an equivalent binary diffusivity D_{AV} for species A by the relation (*25*)

$$N_A = -D_{AV}(P/RT)(dX_A/dZ) + X_A N_A(1 + \gamma_B + \gamma_C + \gamma_D) \tag{13}$$

On combining Eqs. (*12*) and (13) and solving for D_{AV}

$$D_{AV} = \frac{(1 - \psi_A X_A)}{\left(\frac{X_B}{D_{AB}} + \frac{X_C}{D_{AC}} + \frac{X_D}{D_{AD}}\right) - X_A\left(\frac{\gamma_B}{D_{AB}} + \frac{\gamma_C}{D_{AC}} + \frac{\gamma_D}{D_{AD}}\right)} \tag{14}$$

where $\psi_A = 1 + \gamma_B + \gamma_C + \gamma_D$. Depending on the directions of the fluxes, γ_B, γ_C, and γ_D can be either positive or negative. For the mole fractions X_A, X_B, and so on, the averages computed from X_i^s and X_i^d with i = A,B, . . ., are substituted.

In general, the ordinary diffusivity D_{AB} for the O_2–$NiCl_2$ pair depends on temperature and total pressure. According to Fuller and coworkers (*26*),

$$D_{AB} = D^\circ_{AB} \cdot P^{-1} \cdot T^{1.75} \tag{15}$$

where P (in atm) is the total pressure and T is temperature at the position Z = Z. Furthermore

$$D^{\circ}_{AB} = \frac{10^{-3}[1/M_A + 1/M_B]^{1/2}}{(V_A^{1/3} + V_B^{1/3})^2} \tag{16}$$

in which M_A and M_B are the molecular weights and V_A and V_B are diffusion volumes for species A and B; the latter are 16.6 for O_2, 37.7 for Cl_2 (*26*), 51.6 for $NiCl_2$ and 32.1 for NiCl, respectively. An appropriate analog of equation (14) can be written for D°_{AV} using relations, such as Eq. (15). Since the quantity ψ_A is virtually independent of composition, Eq. (13) can be rearranged to yield

$$N_A = \frac{D_{AV}P}{RT\psi_A} \cdot \frac{d\ln(1 - \psi_A X_A)}{dZ} \tag{17}$$

Introducing the variable $u = T^{0.25}$ and expressing D_{AV}, in terms of D°_{AV}, total pressure, and temperature, we obtain

$$N_A = -\frac{D^{\circ}_{AV}w}{4R\psi_A} \cdot \frac{d\ln(1 - \psi_A X_A)}{du} \tag{18}$$

where $w = (T_s - T_d)/L_c$ is the temperature gradient along the connecting zone. At quasi steady-state, the molar flux N_A is independent of Z (or T) in the connecting-zone. Hence Eq. (18) can be integrated using the boundary conditions $u = T_s^{0.25}$ at $X_A = X_A^s$ and $u = T_d^{0.25}$ at $X_A = X_A^d$.

$$N_A = \frac{D^{\circ}_{AV}w}{4R\psi_A} \cdot \frac{\ln(1 - \psi_A X_A^d) - \ln(1 - \psi_A X_A^s)}{(T_s^{0.25} - T_d^{0.25})} \tag{19}$$

Using Eqs. (8) and (9) it can be shown that

$$\psi_A = 1.0 + 0.5(N_{NiCl}/N_{O_2}) \simeq 1.0 \tag{20.a}$$

Similarly, for D°_{AV}, we obtain

$$D^{\circ}_{AV} = \frac{(1 - \psi_A X_A)}{\left[\frac{X_B}{D_{AB}} + \frac{X_C}{D_{AC}} + \frac{X_D}{D_{AD}} - X_A(0.0715\text{x}10^6)\right]} \tag{20.b}$$

The procedure is illustrated by a calculation for the 1320 K → 1200 K setting in Table 2.

X_A^s = 0.321691; X_A^d = 0.315441; $\overline{X}_A$ = 0.318566 = X_A
X_B^s = 0.641360; X_B^d = 0.630882; $\overline{X}_B$ = 0.636121 = X_B
X_C^s = 0.001995; X_C^d = 0.000075; $\overline{X}_C$ = 0.001035 = X_C
X_D^s = 0.034899; X_D^d = 0.053658 $\overline{X}_D$ = 0.044279 = X_D

Substitutions yield $D^{\circ}_{AV} = 6.112 \times 10^{-6}$/cm^2/s/atm/K$^{1.75}$. On inserting the values for D°_{AV}, w, ψ_A, and the mole fractions in Eq. (19), we obtain,

$$N_A = N_{O_2} = 1.570 \times 10^{-9} \text{ mol } O_2 \cdot cm^{-2} \cdot s^{-1}$$

For the rate of growth of nickel oxide, R_{NiO}, it is found that

$$R_{NiO} = 2N_{O_2} S_c M_{NiO} (3600) \quad , \text{ g} \cdot hr^{-1} \tag{21}$$

where S_c(= 19.635 cm^2) is the area of cross-section of the connecting zone and M_{NiO}(= 74.71) is the molecular weight of nickel oxide. Substitutions give a value of 16.58 mg/h for the growth rate (Table 2). For the other sets of data in Tables 2 and 3, the growth or etch rates, as the case may be, were similarly calculated.

The relationship between the total pressure in the reactor and the growth rate of NiO(s) can be discerned from the results summarized in Table 3. The rate increases monotonically with the pressure in the range studied. The observed growth rate of NiO(s) is expected to be but a small fraction of the computed virtual maximum rate reported in Table 3; the value of 0.002 mg/h found for the growth rate of NiO(s) in a flow-reactor system by Ito et al. (*1*) lends credence to this viewpoint.

It is noteworthy that D°_{AV} is not very sensitive to the gas-phase composition; hence, the integration of equation with D°_{AV} treated as invariant is valid. The species NiCl occurs only in very small concentrations, ensuring that the flux N_{NiCl} is negligibly small compared to N_{O_2} in Eq. (20.a).

CONCLUDING REMARKS

The present method is advantageous in that it permits an *a priori* calculation of the maximum possible growth (or etch) rates for any combination of experimental variables. In the absence of such information, the experimentalist is compelled to proceed in a trial-and-error fashion and after considerable exploratory work establish the optimum conditions for the growth of the desired phase. With the present approach, the respective influences of the variables, such as the zone temperatures, the type of transporting agent and its amount, and the temperature gradient, can be investigated theoretically before embarking on an experimental study, thus saving considerable expense and time.

ACKNOWLEDGMENTS

The authors gratefully acknowledge the financial support of this work by the Washington State Mining and Mineral Resources Research Institute.

REFERENCES

1. Ito, S., Miyashita, H., and Yoneda, N., *J. Crystal Growth* **73,** 167 (1985).
2. Gibart, P., Robbins, M., and Kane, A. B., *J. Crystal Growth* **24/25,** 166 (1974).
3. Emmenegger, F., *J. Crystal Growth* **3/4,** 135 (1968).
4. Jeffes, J. H. E., *J. Crystal Growth* **3/4,** 13 (1968).
5. Schäfer, H., *Chemical Transport Reactions,* Academic, New York, 1964.
6. Schäfer, H., Jacob, H., and Etzel, K., *Z. Anorg. Allgem. Chem.* **286,** 27 (1956).
7. Mandel, G., *J. Phys. Chem. Solids* **23,** 587 (1962).
8. Lever, R. F., and Mandel, G., *J. Phys. Chem. Solids* **23,** 599 (1962).
9. Faktor, M. M., Garrett, I., and Heckingbottom, R., *J. Crystal Growth* **9,** 3 (1971).
10. Faktor, M. M. and Garrett, I., *J. Crystal Growth* **9,** 12 (1971).
11. Van der Putte, P., Giling, L. J., and Bloem, J., *J. Crystal Growth* **31,** 299 (1975).
12. Watanabe, H., Nishinaga, T., and Arizumi, T., *J. Crystal Growth* **31,** 179 (1975).
13. Nölang, B. I. and Richardson, M. W., *J. Crystal Growth* **34,** 198 (1976).
14. Nölang, B. I. and Richardson, M. W., *J. Crystal Growth* **34,** 205 (1976).
15. Richardson, M. W. and Nölang, B. I., *J. Crystal Growth* **42,** 90 (1977).
16. *The Collected Works of J. Willard Gibbs,* vol. I, Yale University Press, New Haven, CT, 1957, p. 62.
17. Rao, Y. K. and Kim, S. S., unpublished work, University of Washington, Seattle (1988).
18. Rao, Y. K., *Chem. Eng. Educ.* **19,** 40 (1985).
19. Barin, I. and Knacke, O., *Thermochemical Properties of Inorganic Substances,* Springer-Verlag, Berlin, 1973.
20. Barin, I., Knacke, O., and Kubaschewski, O., *Thermochemical Properties of Inorganic Substances,* Springer-Verlag, Berlin, 1977.
21. Pankratz, L. B., *Thermodynamic Properties of Elements and Oxides,* Bulletin 672, Bureau of Mines, US Department of the Interior, Washington, DC, 1982.
22. Rao, Y. K., *Stoichiometry and Thermodynamics of Metallurgical Processes,* Cambridge University Press, New York, 1985.
23. Rao, B. S. and Dadape, V. V., *High Temp. Sci.* **3,** 1 (1971).
24. Grew, K. E. and Ibbs, T. L., *Thermal Diffusion in Gases,* Cambridge University Press, Cambridge, 1952, p. 36.
25. Hsu, H.-W. and Bird, R. B., *A.I.Ch.E. Journal* **6,** 516 (1960).
26. Fuller, E. N., Schettler, P. D., and Giddings, J. C., *Indus. Engin. Chem.* **58(5),** 18 (1966).

Thermodynamics and High Temperature Chemistry of the Chemical Vapor Deposition of Diamond Films

M. SOMMER AND F. W. SMITH*

Department of Physics, City College of the City University of New York, New York, NY 10031

ABSTRACT

The chemical vapor deposition of carbon films is analyzed using the thermodynamic quasiequilibrium model and a phase diagram for the carbon–hydrogen system is obtained. When the enhanced etching of graphite by hydrogen is included in the model, a region in the phase diagram where diamond is the only stable solid phase of carbon is predicted, which is consistent with the experimental results of Matsumoto et al. (*3*). Details of the model and additional comparisons with experiment are presented.

Index Entries: Carbon; crystal growth; diamond; graphite; hydrocarbons, reactions, phase diagram.

INTRODUCTION

The growth of diamond films via chemical vapor deposition (CVD) at significant growth rates and low pressures where diamond is metastable has led to renewed interest in this technologically-unique material (*1–9*). A recent comprehensive review by Angus et al. (*10*) has provided a useful summary of the history and current developments in diamond CVD. A thermodynamic analysis of the carbon–hydrogen (C–H) system in relation to diamond CVD is of great interest since such an analysis can not only identify the predominant vapor species, H_x, C_xH_y, in the C–H system, which may be important for diamond growth, but can also

*Author to whom all correspondence and reprint orders should be addressed.

demonstrate the effect of varying thermodynamic parameters, such as temperature, pressure, and C/(C + H) ratio on the growth of solid carbon, i.e., diamond or graphite. Such a thermodynamic analysis of diamond CVD based on the quasiequilibrium model has been developed previously (*11*) and is presented in detail below. When the observed enhanced etching of graphite by hydrogen is included in the model, a region in the phase diagram for the C–H system is predicted where diamond is the only stable form of solid carbon present. Comparisons between the predictions of the model and relevant experiments will be presented.

MODEL

We will apply the quasiequilibrium (QE) model of Batty and Stickney (*12*) to the C–H system, the key assumption of the model being that thermochemical equilibrium exists between the solid carbon surface and the vapor species, H_x, C_xH_y, desorbed from it. The question of whether it is meaningful or useful to apply the equilibrium thermodynamics of the QE model to the nonequilibrium steady state deposition of solid carbon will ultimately be answered on the basis of comparisons between the predictions of the model and experimental results. Such comparisons will be presented below. Since the growth rates observed for diamond CVD can be very low, implying that the competing processes of deposition and etching of solid carbon may be nearly in local equilibrium, it seems reasonable to employ an equilibrium analysis at least as a starting point.

Kinetic theory is used from the outset in the QE model in order to express the rates at which vapor species are incident on and adsorbed species are evaporated or desorbed from the substrate surface in terms of the partial pressures of the species. These adsorption and desorption rates are given by

$$I(C_xH_y) = \frac{\eta\, P(C_xH_y)}{\sqrt{2\pi m k_B T}} \quad \text{and} \tag{1}$$

$$R_e(C_xH_y) = \frac{\eta'\, P_{eq}(C_xH_y)}{\sqrt{2\pi m k_B T}} \tag{2}$$

respectively. Here η and η' are the nonthermodynamic sticking (or equilibration) and desorption (or accomodation) coefficients, respectively. In addition, P (C_xH_y) is the partial pressure of the incident C_xH_y species (typically CH_4) and P_{eq} (C_xH_y) is the partial pressure of C_xH_y in equilibrium with the substrate surface at temperature *T*. Also, *m* is the molecular mass of C_xH_y (in grams), k_b is Boltzmann's constant, and the units of *I* and R_e, P, and *T* are molecules cm^2 sec, dynes/cm^2, and degrees K, respectively.

Under the assumption that thermochemical equilibrium exists, the composition (i.e., equilibrium partial pressures) of the mixture of vapor species in equilibrium with the surface can be obtained from the equilibrium constants *K* of the generalized reactions

$$xC\,(s) + (y/2)\,H_2\,(g) \leftrightarrow C_xH_y\,(g), \quad \text{or}$$
$$xC\,(s) + yH\,(g) \leftrightarrow C_xH_y\,(g). \qquad (3)$$

In addition, the required conservation of hydrogen arriving at and leaving the surface is expressed by

$$I\,(H) + 2I\,(H_2) + y\sum_{x,y} I\,(C_xH_y) = R_e\,(H) + 2R_e\,(H_2) + y\sum_{x,y} R_e\,(C_xH_y), \qquad (4)$$

where the $\sum_{x,y}$ are summations over all possible hydrocarbon species. The equilibrium constants K (C_xH_y) are given by

$$K\,(C_xH_y) = \exp\,[-\Delta G_o\,(C_xH_y)\,/RT] \qquad (5A)$$
$$= \frac{P_{eq}\,(C_xH_y)}{(P\,(H_2))^{y/2}}, \qquad (5B)$$

where $\Delta G_o(C_xH_y)$ is the standard free energy of formation of C_xH_y and *R* is the gas constant.

In addition to determining which are the important vapor species in the C–H system, the question of whether deposition or etching of solid carbon (diamond or graphite) occurs can be answered within the framework of the QE model by comparing the flux of carbon atoms reaching the substrate surface, given by

$$I\,(C) = x\sum_{x,y} I\,(C_xH_y), \qquad (6)$$

with the total carbon atom flux R_e (C) leaving the surface via the production of the volatile hydrocarbons C_xH_y, which is given by

$$R_e\,(C) = x\sum_{x,y} R_e\,(C_xH_y). \qquad (7)$$

The net rate of deposition of carbon, therefore, will be given by

$$R_d\,(C) = I\,(C) - R_e\,(C). \qquad (8)$$

Note that R_d (C) < 0 corresponds to etching of solid carbon, if any is present. The boundary in the C–H phase diagram between etching and deposition of solid carbon will correspond to the locus of points for which $R_d(C) = 0$. We will return to this point below.

The first goal of this thermodynamic analysis is the determination of the temperature-dependent equilibrium partial pressures for all the important vapor species, H_x,C_xH_y, corresponding to a given total pressure P_{tot}, and incoming fluxes of C and H atoms (for example, from a CH_4/H_2

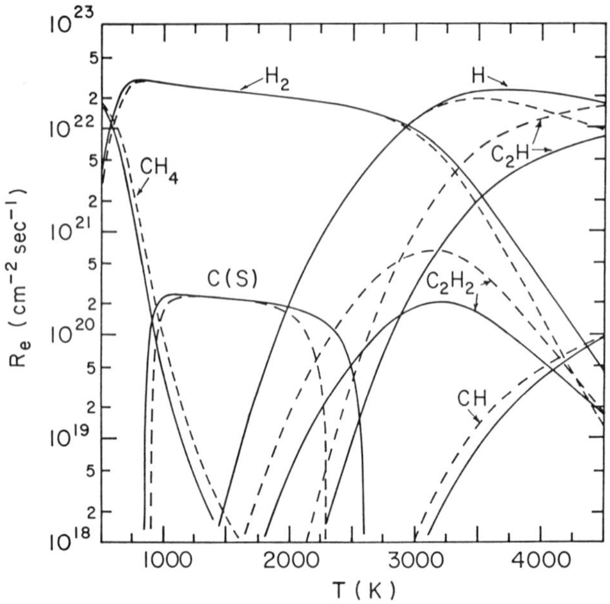

Re(cm^{-2}sec^{-1})
CH_4
C(s)
H_2†
H†
C_2H_2
C_2H ↑→
CH

Fig. 1. Predictions of the quasiequilibrium model for the evaporation rates R_e for H, H_2, CH_4, C_2H_2, C_2H, and CH and deposition rates R_d for solid carbon (graphite and diamond) as functions of temperature *T* for the interaction of a 1% CH_4 in H_2 mixture with solid carbon. The solid curves refer to graphite, whereas the dashed curves refer to diamond. The total pressure is 36 torr.

mixture). Since the free energies ΔG_o (C_xH_y) are known (*13*), all of the partial pressures P_{eq} (C_xH_y) can be expressed in terms of a single variable, $P(H_2)$ or P(H), via Eq. (5). From these values of P_{eq} (C_xH_y), the evaporation rates R_e (C_xH_y) can be determined using Eq. (2), whereas from P_{tot} and the CH_4/H_2 ratio, the incoming fluxes I can be determined from Eq. (1). When these rates R_e and fluxes I are substituted into Eq. (4) for the conservation of H, a polynomial equation of degree y_{max} in P (H) is obtained, where y_{max} is the maximum value of *y* of any C_xH_y species considered. For example, if CH_4 and C_2H_4, but not C_2H_6, are considered, then $y_{max} = 4$. The real, positive root of the resulting equation is the solution for P(H) for the conditions under consideration. The values of R_e (C_xH_y) corresponding to this solution for P(H) are then substituted into Eq. (7) for the carbon deposition rate R_d (C), from which it can be determined whether deposition ($R_d > 0$) or etching ($R_d < 0$) of solid carbon is occurring.

PREDICTIONS OF MODEL

We will now employ the QE model to predict for the C–H system the conditions under which solid carbon (diamond or graphite) can be deposited via CVD. In Fig. 1, the predictions of the QE model for the rates of evaporation R_e in units of molecules cm^{2}sec of H_2, H, CH_4, C_2H_2, C_2H, and CH from the surface of solid carbon are given as functions of

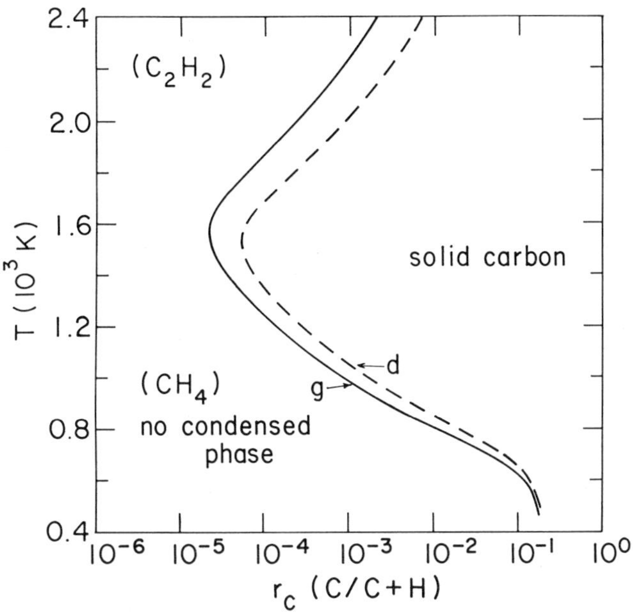

Fig. 2. Predicted phase diagram for the C–H system for a total pressure of 36 torr. Phase boundaries between the regions where solid carbon exists and where no condensed phases exist are shown for both graphite (g) and diamond (d). CH_4 dominates in the gas phase below about 1700 K, whereas C_2H_2 dominates at higher *T*.

temperature *T*. These values of R_e correspond to incoming fluxes of CH_4 (1%) and H_2 (99%) and a P_{tot} = 36 torr (4800 Pa). For both graphite and diamond, the dominant reaction product is CH_4 up to about 600 K. H_2 dominates from 600 K to about 2900 K, above which atomic H dominates. Above about 1700 K, C_2H_2 becomes the dominant hydrocarbon product, whereas above 2850 K C_2H is dominant. We note that, as expected, the partial pressures of the hydrocarbon species are higher for diamond than graphite. Other hydrocarbons (i.e., C_2H_4, CH_2, CH_3, and so on) have been included in the model, but are not shown in Fig. 1 since their evaporation rates are not significant when compared to those shown. For simplicity, at this stage of the model, all sticking and desorption coefficients have been set equal to one.

Since CH_4 is present in the incoming flux, solid carbon (diamond or graphite) will be deposited on the surface when the incoming C flux exceeds the flux of C atoms leaving the surface owing to etching. The rates of deposition R_d (C) of graphite and diamond for this mixture of 1% CH_4 in H_2 are also shown in Fig. 1. Under these conditions corresponding to an incident atom ratio $r_C = C/(C + H) = 4.9 \times 10^{-3}$, it is apparent that graphite will be deposited in the temperature range from 855 to 2560 K, whereas diamond can be deposited between 910 and 2295 K.

With the total pressure constant at 36 torr, the phase diagram for the C–H system, shown in Fig. 2, can be obtained. For this pressure, graphite and diamond can be deposited only for values of r_C greater than approximately 2×10^{-5} and 5×10^{-5}, respectively. Even when allowed,

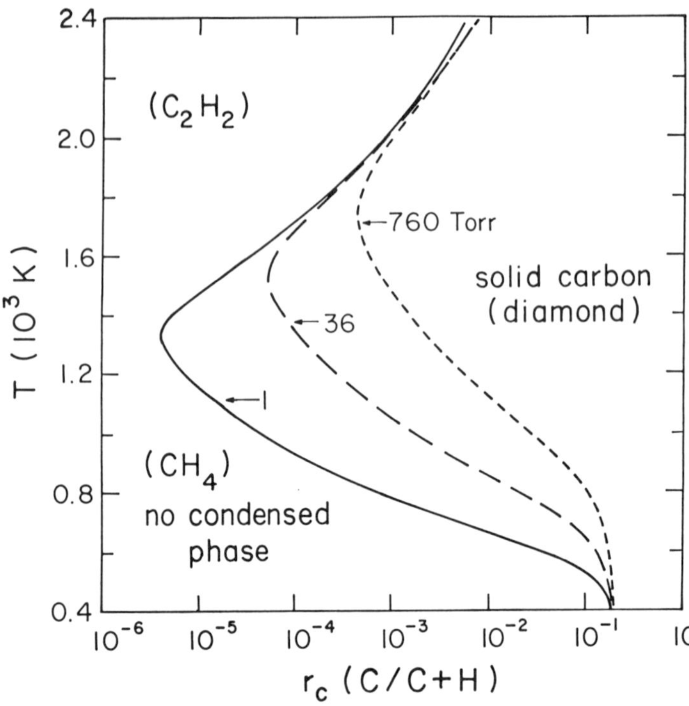

Fig. 3. Predicted phase boundaries for diamond at total pressures of 1, 36, and 760 torr.

solid carbon will only be deposited in the range of temperatures between T_L and T_H. Etching of solid carbon dominates below T_L via production of CH_4 and above T_H via production of C_2H_2. Note that the phase boundaries in Fig. 2 correspond to R_d (C) = 0, with etching occurring to the left and deposition occurring to the right of the boundary. Owing to the higher free energy of diamond in the temperature and pressure range of interest, the region of stability of diamond lies within the region of stability of graphite, as shown in Fig. 2.

The effect on the C–H phase diagram of varying the total pressure is illustrated in Fig. 3, where the phase boundary for diamond is presented for P_{tot} = 1, 36, and 760 torr. The boundary shifts dramatically with P_{tot}, with increasing P_{tot} corresponding to increased etching and a smaller stability region for diamond. The phase boundary for graphite shifts in a similar manner as P_{tot} is varied.

We note that results consistent with those of Fig. 3 have been obtained by Knippenberg et al. (*14*) for the case of the CVD of pyrolytic graphite. A useful analysis of the C–H system has also been carried out by Suris (*15*).

COMPARISONS WITH EXPERIMENT

We will now explore the extent to which this thermodynamic analysis can aid in our understanding of the observed CVD of diamond films for the conditions typically employed: CH_4/H_2 ratios of about 0.01, P_{tot} in the range 1–100 torr, and temperatures in the range 600–1000°C. Since graphite is the stable form of solid carbon for the pressures and temperatures now employed in the CVD of diamond, it is clear that some key kinetic factors will have to be introduced into the thermodynamic analysis in order to explain the observed relative stability of diamond. For example, the kinetic factors that lead to the enhanced etching by hydrogen of graphite relative to diamond will certainly have to be considered.

An analysis of the previous study of Balooch and Olander (*16*) of the reactions of atomic and molecular hydrogen with graphite yields the result that the etching of graphite in atomic H, via the production of CH_4, is actually enhanced above the rate predicted by the QE model between 450 and 800 K. This observation indicates that the actual region of stability of graphite must be smaller than shown in Fig. 2. Although it is not possible at this point to determine conclusively which kinetic factors are leading to the observed enhanced etching of graphite by hydrogen, possible candidates include: (1) a desorption coefficient $\eta'_{H_2}(g)$ for H_2 on graphite less than unity, and (2) H atoms tightly bound to the graphite surface. In the first case, production of CH_4 leading to etching would be enhanced, whereas production of H_2 would be diminished. In the second case, the recombination of H atoms to form H_2 would be inhibited. Further work is clearly necessary in order to identify the dominant kinetic factor(s).

In Fig. 4, we present a comparison between the experimental results of Matsumoto et al. (*3*) and the predictions of our thermodynamic analysis when we tentatively choose $\eta'_{H_2}(g)$ as the key kinetic parameter. Matsumoto et al. (*3*) studied the growth of diamond particles from CH_4/H_2 mixtures via hot filament CVD at total pressures of about 36 torr. They have presented SEM micrographs of the diamond particles deposited both at a fixed deposition temperature T of 800°C for varying CH_4 concentrations and also at a fixed CH_4 concentration of 1% ($r_c = 4.9 \times 10^{-3}$) for varying T. The data points presented in Fig. 4, shown as open squares, refer to the observation of well-defined diamond particles, whereas the open and filled circles correspond to small, poorly-defined diamond particles and large particles apparently covered with graphitic deposits, respectively.

In order to arrive at a possible explanation for the results of Matsumoto et al. (*3*), we have redetermined the phase boundary for graphite using a value of $\eta'_{H_2}(g) = 0.2$, which is within the range of values (0.03–1) needed to explain the results of Balooch and Olander (*16*). When this assumption concerning $\eta'_{H_2}(g)$ is incorporated into the model, the region of stability for graphite now can be seen in Fig. 4 to lie completely within that of diamond, which is unchanged from Fig. 2. The predictions of the modified QE model are consistent with the experimental results of Matsumoto et al. (*3*), shown in Fig. 4, as the conditions leading to the deposition of well-defined diamond particles fall within the region where only diamond is predicted to be stable. The deposition conditions for the small, poorly-defined particles lie near to the boundary for etching of diamond, whereas those for the graphite-covered particles lie in or near the region where graphite is stable.

An additional observation of Matsumoto et al. (*3*) concerned the fact that no diamond growth could be observed at a pressure of 760 torr under their deposition conditions. The QE model predicts that the etching-growth boundaries for both diamond and graphite shift to higher values of r_C as the pressure *P* is increased (*see* Fig. 3), and this is illustrated in Fig. 4, where the boundary for diamond corresponding to 760 torr is again presented. The deposition conditions used by Matsumoto et al. (*3*) now correspond to the etching region so that the observed absence of diamond growth at 760 torr is successfully explained by the QE model. Matsumoto et al. (*17*), however, have recently observed growth of diamond films using rf induction heating to generate a thermal plasma at P_{tot} = 760 torr. They observed diamond growth for up to 10% CH_4 in H_2, thus demonstrating that growth at 760 torr is possible for high enough CH_4 concentrations, in agreement with the predictions of the QE model shown in Fig. 4.

The recent careful study of Kobashi et al. (*9*) of the microwave plasma CVD of diamond films, using CH_4/H_2 mixtures with P_{tot} = 30 torr and T = 800°C, is also consistent with the predictions of the model as shown in Fig. 4. Kobashi et al. (*9*) observed a transition from isolated diamond particles for 0.2% CH_4 ($r_c = 1 \times 10^{-3}$) to continuous films for 0.3% CH_4 and higher. For 1.6% CH_4 ($r_c = 7.8 \times 10^{-3}$), they observed a microcrystalline film with a "cauliflower" surface structure, consistent with the presence of a strong graphitic component. It is clear from these results that the concentration of nondiamond components (graphitic, amorphous carbon) in their films increases as the CH_4 concentration increases, consistent with the approach to the graphite phase boundary as r_C increases (shown in Fig. 4).

It has also been pointed out recently (*10*) that there appears to be a lower pressure limit for diamond CVD of about 0.375 torr, below which only amorphous and graphitic carbon are deposited. This observation can be readily explained by the predicted shift of the graphite phase boundary to lower C/(C + H) ratios for decreasing values of P_{tot}, as also seen for diamond in Fig. 3. In fact, the QE model predicts that, at low enough pressures (below about 1 torr), CH_4 concentrations of 0.5–1% in H_2 will correspond to the region of stability of graphite.

The QE model may also provide some useful insights into the operation of the hot filaments typically employed for diamond CVD. Tungsten or tungsten carbide hot filaments, for example, may be thought of as acting as high temperature ($T \sim 2200°C$) substrates for the deposition of graphitic carbon from CH_4/H_2 mixtures, in addition to serving to dissociate the H_2 into atomic H and thermally convert CH_4 to other species, such as C_2H_2, that are more stable than CH_4 at high *T* (*see* Fig. 1).

In particular, as can be seen from Figs. 2 and 4, the phase boundary for solid carbon bends back around above 1600 K for the higher temperatures at which hot filaments typically operate. In this region, lower values of *T* and higher C/(C + H) ratios lead to the growth of solid carbon. Thus, depending on the specific operating conditions, hot filaments may either have bare, carbon-free surfaces or actually be covered

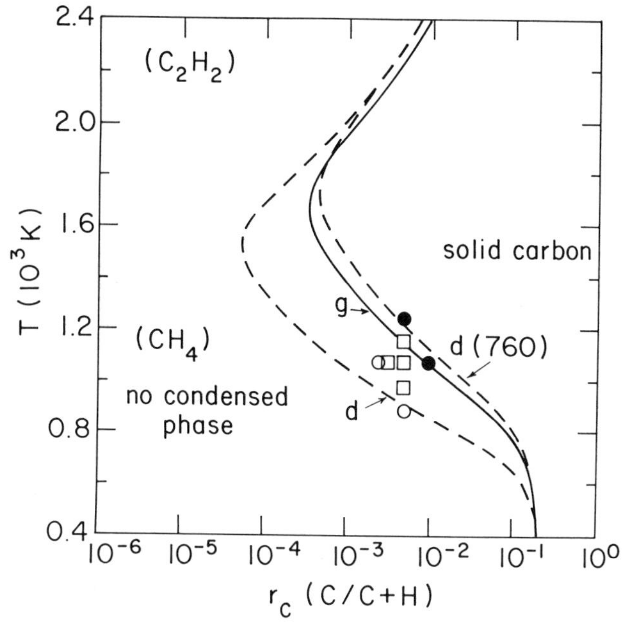

Fig. 4. Predicted phase diagram for the C–H system modified to include the enhanced etching of graphite by hydrogen. Also shown are the experimental results of Matsumoto et al. (*3*). Open squares refer to the observation of well-defined diamond particles, whereas the open and filled circles correspond to small, poorly-defined diamond particles and large particles apparently covered with graphitic deposits, respectively. The curve labelled "d" (760) corresponds to the phase boundary for diamond at 760 torr.

with carbon. The presence of carbon on the surface could affect the ability of the hot filament to dissociate H_2. Interesting filament-related effects have recently been observed by Celii et al. (*18*) when the atomic hydrogen concentration observed near a hot filament depended sensitively not only on the filament temperature, but also on the CH_4 concentration.

CONCLUSIONS

The QE model as modified to include the observed enhanced etching of graphite by hydrogen (*16*), thus, has been shown to be able to both clarify and even explain qualitatively the CVD of diamond films under the conditions presently being used. It is clear now why the observed ranges of parameters (600–1000°C, 1-100 torr, and 0.5–2% CH_4) have been successful for diamond deposition in terms of the diamond stability region present in the phase diagram of Fig. 4. The model can also provide insights into how the deposition of diamond can be expected to vary with the available experimental parameters. For example, the QE model shows that growth at lower temperatures or higher pressures will require higher carbon atom ratios r_C (at least 0.05) in order to remain in the diamond stability region.

In order for the QE model to make more quantitative predictions for growth or etching rates, realistic values for the sticking coefficients of the reactants (H, H_2, CH_4, and so on) and for the desorption coefficients of the products (H, H_2, CH_4, and C_2H_2) will be required. We are presently investigating the effects on diamond deposition of varying these parameters. Experiments are under way to test predictions of the QE model.

ACKNOWLEDGMENTS

This research has been supported by the US Department of Energy under grant DE-FG02-87ER45317. We wish to thank K. Mui for his contributions to this work and F. G. Celii (*18*) for sending us the results of his research prior to publication.

REFERENCES

1. Spitsyn, B. V., Bouilov, L. L., and Deryagin, B. V., *J. Crystal Growth* **52,** 219 (1981).
2. Varnin, V. P., Teremetskaya, I. G., Fedoseev, D. V., and Deryagin, B. V., *Sov. Phys. Dokl.* **29,** 419 (1985).
3. Matsumoto, S., Sato, Y., Tsutsumi, M., and Setaka, N., *J. Mater. Sci., ie.* **17,** 3106 (1982).
4. Fedoseev, D. V., Varnin, V. P., and Deryagin, B. V., *Russ. Chem. Reviews* **53,** 435 (1984).
5. Messier, R., Badzian, A. R., Badzian, T., Spear, K. E., Bachmann, P., and Roy, R., *Thin Solid Films* **153** 1 (1987).
6. Singh, B., Arie, Y., Levine, A. W., and Mesker, O., *Appl. Phys. Lett.* **52,** 451 (1988).
7. Chang, C.-P., Flamm, D. L., Ibbotson, D. E., and Mucha, J. A., *J. Appl. Phys.* **63,** 1744 (1988).
8. Celii, F. G., Pehrsson, P. E., Wang, H. T., and Butler, J. E., *Appl. Phys. Lett.* **52,** 451 (1988).
9. Kobashi, K., Nishimura K., Kawate, Y., and Horiuchi, T., *Phys. Rev. B* **38,** 4067 (1988).
10. Angus, J. C. and Hayman, C. C., Science **241,** 913 (1988).
11. Sommer, M., Mui, K., and Smith, F. W., *Solid State Comm.* **69,** 775 (1989).
12. Batty, J. C. and Stickney, R. E., *J. Chem. Phys.* **51,** 4475 (1969).
13. *JANAF Thermochemical Tables,* 2nd ed., Stull, D. R. and Prophet, H., eds., US Government Printing Office, Washington, DC, 1971, and subsequent supplements; for diamond: Bundy, F. P., Bovenkerk, H. P., Strong, H. M., and Wentorf, Jr., R. H., *J. Chem. Phys.* **35,** 383 (1961).
14. Knippenberg, W. F., Lersmacher, B., Lydtin, H., and Moore, A. W., *Philips Tech. Rev.* **28,** 231 (1967).
15. Suris, A. L., *Handbook of Thermodynamic High Temperature Process Data,* Begell, W., ed., Hemisphere, New York, 1987, p. 483.
16. Balooch, M. and Olander, D. R., *J. Chem. Phys.* **63,** 4772 (1975).
17. Matsumoto, S., Hino, M., and Kobayashi, T., *Appl. Phys. Lett.* **51,** 737 (1987).
18. Celii, F. G. and Butler, J. E., *Appl. Phys. Lett.* **54,** 1031 (1989).

Studies of Diamond Chemical Vapor Deposition

J. E. BUTLER,*[1] F. G. CELII,[1] D. B. OAKES,[1] L. M. HANSSEN,[2] W. A. CARRINGTON,[3] AND K. A. SNAIL[2]

[1]Chemistry Division, Code 6174; [2]Optical Sciences Division, Code 6522, Naval Research Laboratory, Washington, DC 20375-5000; and [3]University of South Florida, Tampa, FL

ABSTRACT

Diamond exhibits many desirable properties as a material, e.g., hardness, thermal conductivity, chemical and wear resistance, optical transmission, stiffness, high resistivity (pure), and high carrier velocities (doped). The application of these properties is often limited by the production and morphology of available material. The chemical vapor deposition of diamond offers hope that one might attain control over the quality and morphology of synthetic diamond. Diamond has been grown using chemical vapor deposition techniques at atmospheric pressure and in partial vacuum. The filament-assisted CVD and combustion flame growth techniques employed are described. Micro-Raman analysis of diamond deposited in flames and the results of *in situ* laser spectroscopic analysis of the filament-assisted diamond growth environment are presented. We have employed tunable IR diode laser absorption, resonance enhanced multiphoton ionization, and laser induced fluorescence spectroscopies to detect CH_3, C_2H_2, C_2H_4, H, and search for other species during the hot filament-assisted CVD of diamond from CH_4 in H_2. Simple models of the chemistry of diamond CVD are discussed.

Index Entries Diamond; chemical vapor deposition (CVD); flame deposition; *in situ* diagnostics; resonance enhanced multiphoton ionization spectroscopy (REMPI); IR diode laser spectroscopy; micro-Raman.

*Author to whom all correspondence and reprint requests should be addressed.

High Temperature Science, Vol. 27

INTRODUCTION

The cubic form of crystalline carbon, diamond, possesses many superlative physical properties, e.g., hardness, optical transparency, large electronic bandgap and high carrier mobilities, wear and chemical resistance, and thermal conductivity (*1*). It is the same group of properties that limits the exploitation of this material for technological applications because of the difficulty of cutting, machining, manipulating, and doping diamond into the desired devices. The chemical vapor deposition (CVD) of diamond is receiving considerable attention because of the desire to control the shape and impurity distributions for the production of devices that would be difficult to form or mold out of natural diamonds. This concept is now a well established phenomenon (*2*) and has been implemented by activating gas mixtures using hot filaments (*3–6*), microwave (*7–10*), radio frequency (rf) (*11,12*), and direct current (dc) discharges (*13,14*), and more recently, combustion flames (*15–19*). In spite of the success of the various CVD growth techniques, the mechanisms of growth and impurity incorporation of this thermodynamically metastable material are not well understood. The details of how a material, such as diamond, can be grown under conditions in which it is metastable are intriguing and can have far reaching consequences. One can hope that by understanding the gaseous and surface chemistry of diamond CVD, a deeper insight into CVD processes will be attained and this understanding will impact the development of various technical applications.

EXPERIMENTAL

The filament-assisted diamond deposition environment was chosen for study because of its successful implementation by various groups (*3–6*). The apparatus has been described previously (*20,21*). Typical growth conditions were 0.1 standard liters per min. (slm) total flow of 0.5% CH_4 in H_2 at 25 torr (1 torr = 133.32 Pa). All gases were research grade with purity in excess of 99.99%. The flow of reagents was controlled by mass flow controllers. The temperature of the filament (tungsten, GE218, 6–10 turns, 0.33 mm dia.) was monitored by a two-color optical pyrometer and that of the substrate was measured by a chromel-alumel thermocouple (and occasionally, an infrared (IR) two-color pyrometer through a fused silica window).

The IR diode laser diagnostics were performed by multiply passing the IR beam through the center of the cell and detecting the rotational-vibrational absorption spectra of the gaseous species in the optical path. The spectrometer and optical arrangement have been reported previously (*20–22*). The laser beam was aligned to sample the volume of gas between the filament and the substrate (2 mm high, 0.5 mm above the substrate). The filament to substrate distance was typically 19 mm for the IR measurements.

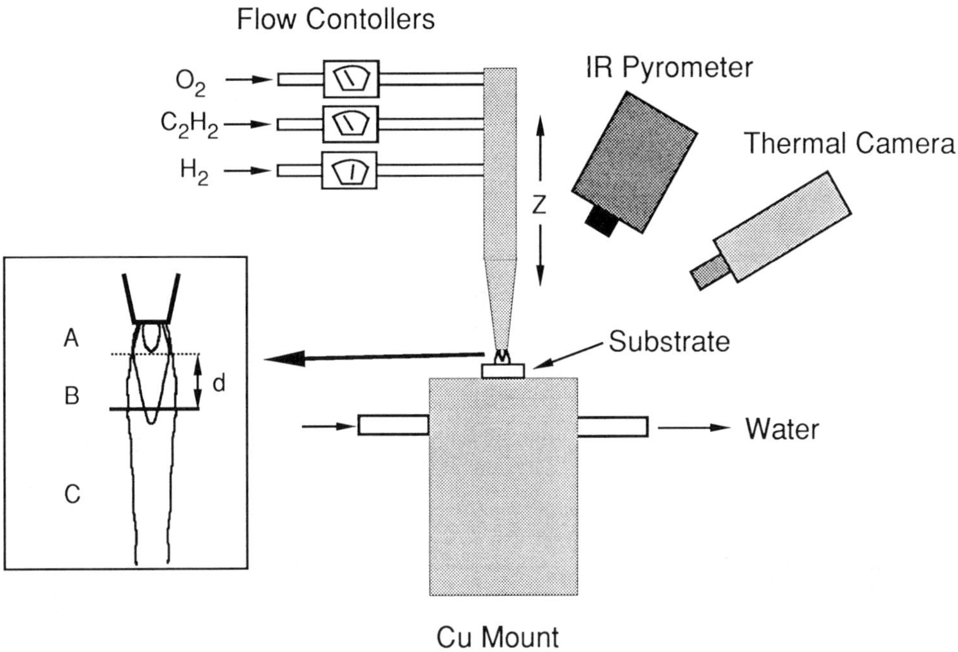

Fig. 1. Schematic for the oxy-acetylene torch deposition experiment. The inset is an expanded view of the oxygen-acetylene flame. Three distinct regions are: **A** the inner cone, **B** the acetylene feather, and **C** the outer flame. For diamond growth, the substrate is positioned inside the acetylene feather region.

The resonance enhanced multi-photon ionization spectroscopy (REMPI) measurements used a XeCl excimer laser pumping a tunable dye laser (0.25 cm^{-1} bandwidth, 15 ns pulse width, 20 Hz, <15 mJ/pulse) as the light source, with synchronous, gated detection of the photoelectrons collected on a Pt probe wire (*23–25*). A bias voltage of −150V on the surrounding surfaces gave a saturated photoelectron collection efficiency. The laser light was focused 1 mm from the collection probe with a 40 cm f.l. lens. Experiments were performed to suppress interference by the third harmonic field (*26*) by reflecting the laser light back upon itself to give a standing wave field in the probe volume. The collection probe was positioned 2–4 mm above the substrate and the filament position was variable between 6 to 20 mm above the substrate.

The combustion flames used in this work for diamond deposition were generated by an oxygen-acetylene welding torch (0.89 mm orifice) mounted above a water-cooled copper block on which the substrate was resting. The apparatus is shown schematically in Fig. 1. All torch experiments were conducted in the ambient atmosphere. The mass flowrates of both the oxygen and fuel gases were adjusted by mass flow controllers. The fuel gases used were acetylene, ethylene, ethane, and in some experiments, hydrogen was added to the hydrocarbon fuel. Typical growth conditions employed an oxygen to acetylene mass flow ratio, R, of 0.9 to 1.1, total mass flowrate of 2 to 3 slm, substrate temperature

Fig. 2. Scanning electron micrographs of diamond deposits generated in the filament assisted growth apparatus: **A** continuous polycrystalline film on a silicon substrate, **B** single cubo-octahedral crystal grown on a polycrystalline Ni substrate, **C** a multiply twinned crystal on the same substrate as **B**.

between 700 and 1100 C, and with the substrate immersed in the fuel rich intermediate cone ("acetylene feather") of the flame (*see* Fig. 1). The substrate temperature was monitored with a two color IR pyrometer, and the spatial temperature distribution was measured with an imaging IR camera.

Analysis of the deposited materials was performed principally by Raman and micro-Raman spectroscopy, optical and scanning electron microscopies, and, occasionally, by scanning Auger Microprobe spectroscopy. All Raman analyses employed 514.5 nm laser excitation. The spatial resolution of the micro-Raman analyses was less than one micron.

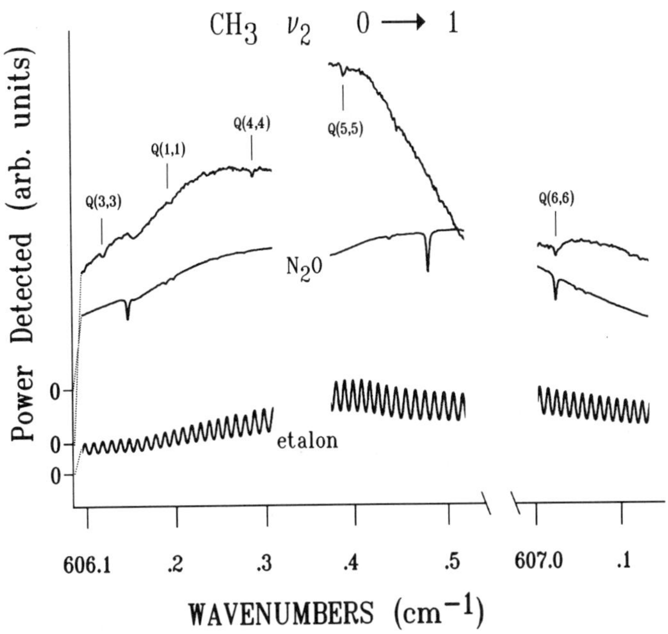

Fig. 3. The absorption spectra obtained in the diamond CVD chamber showing portions of the methyl radical v_2 fundamental Q(N = K) band head, N_2O reference gas for wavenumber calibration, and interpolation etalon. The pressure in the chamber was 25 torr, T_{fil} was 2350 C, and the reactant concentration was 2% CH_4 in H_2.

RESULTS

Diamond was deposited using the filament assisted technique on substrates of polycrystalline Ni, Ag-coated Ni, silicon (N type, 100 orientation, unpolished, or polished with diamond powder and cleaned), and highly ordered pyrolytic graphite (HOPG). The deposited material was characterized as diamond by Raman scattering (1332 cm^{-1} phonon), X-ray photo-electron spectroscopy (only carbon, 32 eV plasmon), Auger electron spectroscopy (only carbon, with diamond lineshape), and optical and scanning electron microscopies (single and twinned cubo-octahedra, polycrystalline films with well defined 100 and 111 facets, as shown in Fig. 2).

Portions of the IR spectrum between 600 and 1400 cm^{-1} were scanned for 0.5 to 2.0% CH_4 in H_2, T_{fil} = 2300 C and T_{sub} = 900 C. The IR absorption spectra of the methyl radical detected in this environment is shown in Fig. 3. Lines of the v_2 Q branch head are displayed along with absorption lines of the N_2O wavenumber calibration gas in a separate cell, and the interference pattern of an etalon used as a wavenumber marker. Many other individual rovibronic lines were detected, the majority of which were assigned to fundamental and hot band transitions of acetylene, C_2H_2. An example of an acetylene band head is given in reference 20. CH_4 and C_2H_4 were also detected. The following species

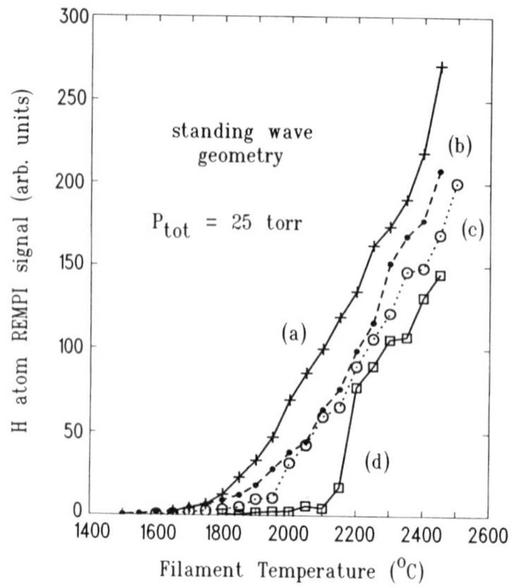

Fig. 4. Relative H atom REMPI signal intensities as a function of the filament temperature for a filament probe separation of 8 mm. The concentration of methane in hydrogen was: **A** 0%, **B** 0.5%, **C** 1.0%, and **D** 3.0%.

were searched for but not detected: CH_2 ($^3B_1$), and the stable molecules C_2H_6, C_3H_4, C_3H_6, C_3H_8, and C_4H_2. The methyl radical was also detected in experiments in which acetylene (1.0%) in H_2 was used as the initial carbon source.

The REMPI spectrum of the H atom (*24,25*) was obtained at 364.5 nm using a three photon resonance with the ls–2p transition, 121.57 nm, with ionization by an additional photon. The variation of the H atom REMPI signal with filament temperature for different concentrations of methane in hydrogen is shown in Fig. 4. The filament was 8 mm above the probe position and standing wave excitation geometry was employed for these measurements. The signal decreased when the filament to probe distance was increased.

The presence of the methyl radical was confirmed (*24*) by the detection of its 2 + 1 REMPI spectrum at 333.5 nm (3p $^2A''_2 \leftarrow\leftarrow X\ ^2A''_2$)(*27*).

Detection of the transient species C_2, CH, C_3, C_2^+, and $C_4H_2^+$ was attempted by laser induced fluorescence (LIF). Only C_3 may have been detected, with poor signal to noise ratio. Scattered light from the hot filament was a severe limitation with the present cell geometry.

The species examined using these diagnostics techniques are listed in Table1. Also listed in Table 1 are estimates of concentrations (or upper limits) of the species examined using IR absorption spectroscopy. The line strengths of stable species were determined by backfilling the cell with known pressures of the relevant specie. The estimate of the methyl radical concentration was made using the transition moment determination of Yamada (*28*). The major uncertainties in these estimates were the temperature, and the concentration and temperature inhomogeneities across the cell. A uniform temperature was assumed and set equal to the measured rotational temperature of acetylene, 200–300 C, estimated

Table 1
Summary of Molecules Examined, Technique Employed for Detection (No Entry Indicates the Technique Was Not Used), and Estimate of the Concentration for 20 torr of 1% CH_4 in H_2, 2200 C Filament Temperature, and ca. 900 C Substrate Temperature

	Detected by			Concentration estimate
Molecule	IR	REMPI	LIF	molec./cc
H		yes		
CH			no	
CH_2 ($^3B_1$)	no			
CH_3	yes	yes		$2 \pm 1.5 \times 10^{12}$
CH_4	yes			$8 \pm 3 \times 10^{14}$
C_2			no	
C_2^+			no	
C_2H_2	yes			$2 \pm 1 \times 10^{14}$
C_2H_4	yes			$6 \pm 2 \times 10^{12}$
C_2H_6	no			
C_3			yes	
C_3H_4 - allene	no			$< 8 \times 10^{13}$
C_3H_4 - methyl acetylene	no			$< 8 \times 10^{13}$
C_3H_6 - cyclopropane	no			$< 6 \times 10^{13}$
C_3H_6 - propylene	no			$< 2 \times 10^{13}$
C_3H_8	no			$< 3 \times 10^{13}$
C_3H_5		no		
C_4H_2	no			
$C_4H_2^+$			no	

from the relative intensities of the ($v_4 + v_5 - v_4$) Q-branch band head (*20*). This temperature is closer to that of the cold walls of the cell than either the filament or the substrate and is owing to the absorption of acetylene molecules distributed throughout the optical path.

Combustion flames from an oxy-acetylene torch have been used to successfully deposit diamond onto a variety of substrates, including: Si(100), Si(111), BN, Mo(100), Nb(100), TiC(100), Ta, and Cu. For oxygen-acetylene, the appropriate regime of flow ratio (of O_2/C_2H_2) was 0.9 to 1.1; for oxygen-ethylene it was 1.18–1.25; and for oxygen-ethane the flow ratio was 1.63. The stoichiometric ratios of oxygen to fuel gas for conversion to CO and H_2O at the inner flame cone are 1.5, 2.0, and 2.5 for acetylene, ethylene, and ethane, respectively. Thus, each combustion flame was fuel rich. Maximum growth rates for the oxygen-acetylene deposits were ~140 microns/h, and considerably less for the other two flames.

The variation in diamond growth with position in the acetylene flame was studied with an oxygen flow rate of 1.70 slm and an acetylene flow rate of 1.67 slm. When the distance (d) from the substrate to the oxygen-acetylene flame front (*see* Fig. 1) was 3.0–4.5 mm, the resulting nucleation density of diamond crystallites was low directly beneath the flame axis and significantly higher within an annulus centered on the flame axis. The outer edge of this annulus occurred where the edge of the

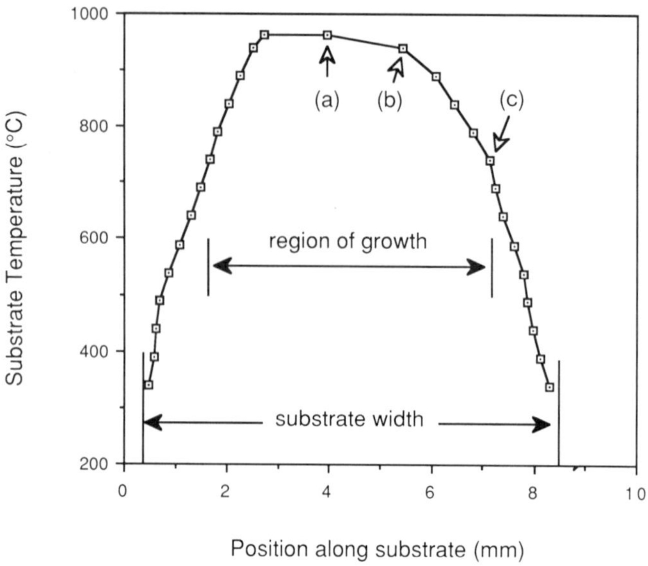

Fig. 5. Substrate temperature profile measured with a thermal imaging camera. The Si substrate was positioned 3 mm from the premixed flame front of an oxygen/acetylene flame ($O_2/C_2H_2 = 1.02$). Note the large (250 C) temperature difference across the region of diamond growth on the substrate. A typical annular growth profile has regions of low growth **A**, high growth **B**, and low growth **C**.

acetylene flame intersected the substrate. As d was increased to 7.5 mm, the nucleation density increased on the region of the substrate directly under the flame axis and the annular pattern of crystallites disappeared.

A thermal (8–12 micron) imaging camera having a spatial resolution on the substrate of ~⅓ mm (3 mrad) was used to measure substrate temperature profiles. The temperature profile observed during a typical deposition on a Si substrate in an oxygen-acetylene flame is shown in Fig. 5. For this data, the substrate to flame front distance (d) was 4.5 mm and the temperature difference across the region of diamond growth was approximately 250 C. The density of crystallites deposited in an oxygen-acetylene flame was frequently higher within an annular region [(b) in Fig. 5], and low in other regions [(a) and (c)] on the substrate.

An SEM and micro-Raman study of the spatial variation of diamond crystallite quality was performed on a sample grown in an oxygen-acetylene-hydrogen flame. For this sample, the (O_2/C_2H_2) flow ratio was equal to 1.02, the (H_2/C_2H_2) flow ratio was set at 0.10, and the peak substrate temperature ranged from 910 to 960 C. The substrate was positioned 4.5 mm below the premixed flame front. The quality of the diamond crystallites was documented by the ratio, R_{dc} of the peak intensity of the 1332 cm^{-1} (diamond) to the 1510 cm^{-1} (amorphous carbon) features. Individual crystallites show a significant decrease in R_{dc} with increasing distance from the growth center. One Raman spectrum and two SEM photos at each of three positions (corresponding to (a), (b) and (c) in Fig. 5 across the substrate are exhibited in Fig. 6. Crystallites

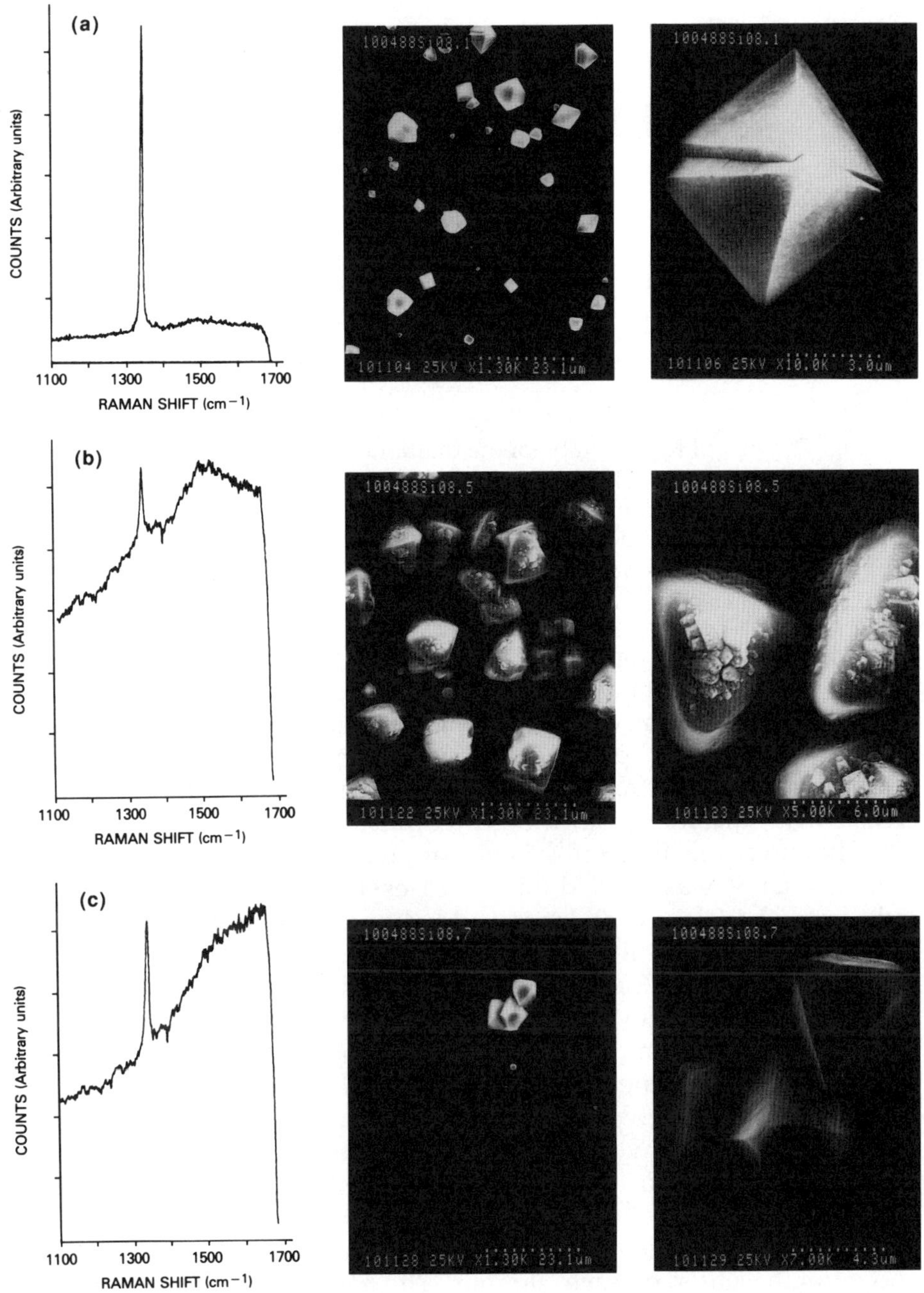

Fig. 6. Raman spectra showing the diamond Raman peak (1332 cm^{-1}) and the amorphous carbon peak (1510 cm^{-1}), corresponding to the SEM photos of growth for three radial positions on the substrate: **A** inside the annular growth region, **B** within the annulus, and **C** outside the annulus.

outside of the annular region (c), and at the edge of the growth, exhibit an increase in R_{dc} compared to those in the annulus (b). Electron and optical microscope analyses show a correlation between the decrease of R_{dc} across the annular growth region and the degree of secondary growth

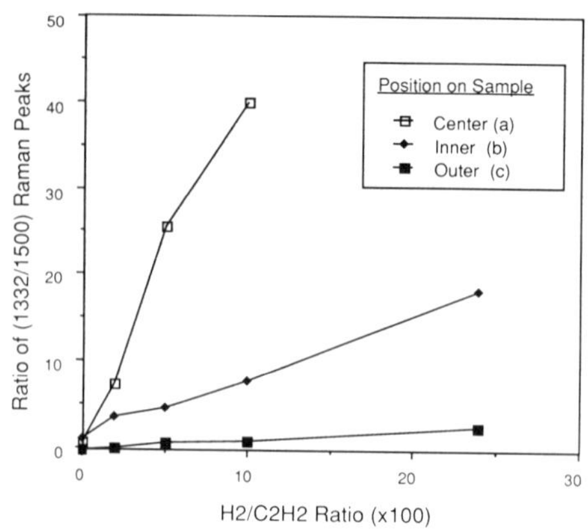

Fig. 7. Plot of R_{dc}, the ratio of the diamond Raman peak (1332 cm^{-1}) to the amorphous carbon peak (1510 cm^{-1}), for different hydrogen flowrates added to an oxygen-acetylene flame. Data are shown for growth at the center of the sample, and the inner and outer edges of the growth annulus. Note the monotonic increase in crystal quality, judged by R_{dc}, with increasing hydrogen flow rate.

observed on the diamond crystallite faces. Significant secondary growth can be seen on crystallites in the annular growth region (b), whereas "cleaner," better-formed crystallites are inside (a) and outside (c) of the annulus. The Raman spectra of all crystallites exhibit a fluorescent background signal level that increases with the wavelength shift and increases with distance from the growth center.

The effect on diamond growth of adding hydrogen to an oxygen-acetylene flame was studied for two cases: first, with the (O_2/C_2H_2) flow ratio fixed at 1.02, and hydrogen added to the flame, and second, with the ($O_2/(C_2H_2 + H_2)$) flow ratio fixed at 1.02, and hydrogen substituted for acetylene. In both cases the oxygen flowrate was fixed at 1.7 slm. The total flow rate for the hydrogen addition study ranged from 3.4 slm to 4.2 slm. Well-faceted diamond crystallites have been observed to grow in an oxygen-acetylene flame with an (O_2/C_2H_2) flow ratio of 1.05 and a substrate temperature between 700–1150 C. In the case where the (O_2/C_2H_2) flow ratio was fixed, hydrogen was added so that the (H_2/C_2H_2) flow ratio equalled 0.0–0.45. Most samples had an annular region of heavy crystallite nucleation.

Raman spectra were taken at the inner and outer edges of the annuli, and growth center of each sample. The ratio of the diamond to the amorphous carbon Raman peaks (R_{dc}) of individual crystallites is plotted in Fig. 7 vs the (H_2/C_2H_2) flow ratio for the three different positions on the samples. These results show a nearly linear increase in R_{dc} hence increasing diamond quality, with increasing hydrogen flowrate, although the diamond growth rate decreased significantly as the (H_2/C_2H_2) flow ratio exceeded 0.22. For a (H_2/C_2H_2) flow ratio equal to 0.40, no growth was observed. With a fixed ($O_2/(C_2H_2 + H_2)$) flow ratio, substitution of hydrogen for acetylene leads to an increase in R_{dc} for (H_2/C_2H_2) flow ratios up to 0.02. No diamond growth was observed for (H_2/C_2H_2)

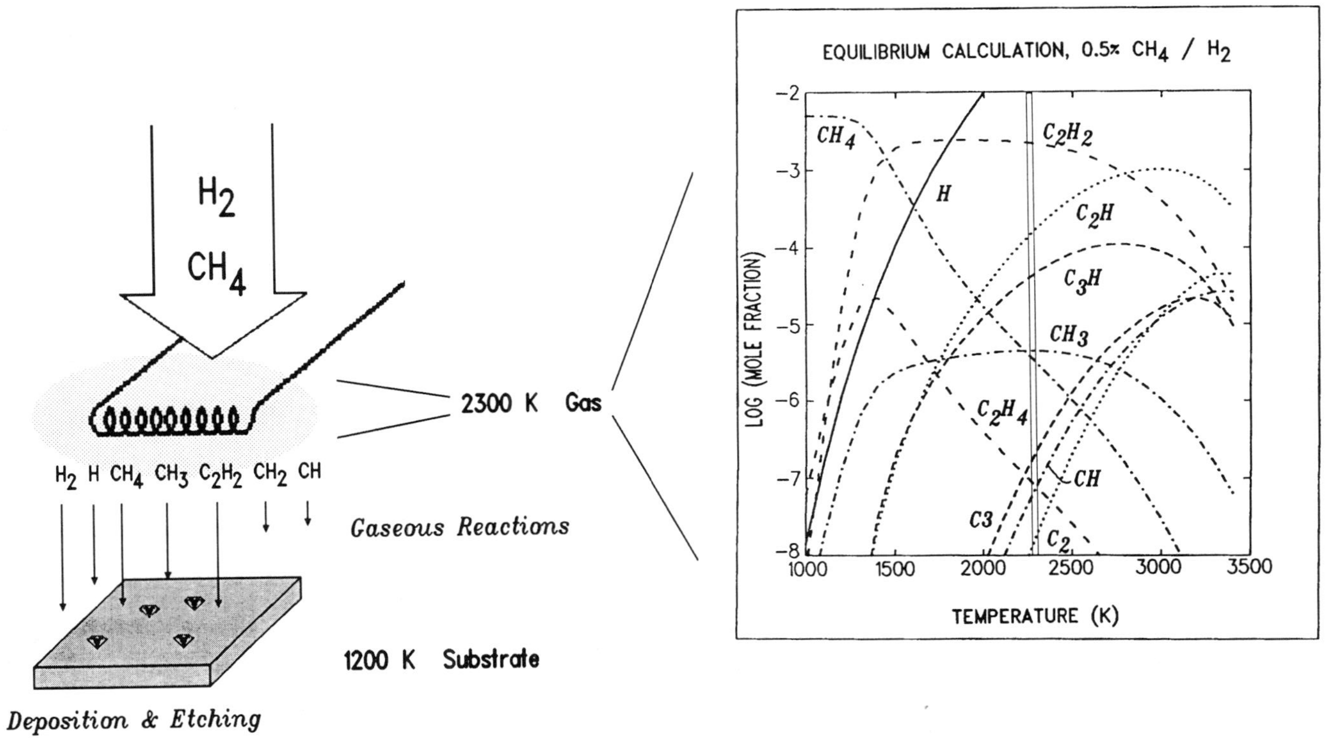

Fig. 8. A simple model of the gaseous processes occurring in filament assisted diamond CVD. On the right are the results of a thermodynamic equilibrium calculation of the mole fraction of various species present versus the equilibrium temperature. The schematic on the left summarizes the simple model described in the text.

flow ratios equal to or greater than 0.05. In oxygen-acetylene flames, very low diamond growth was observed for (O_2/C_2H_2) flow ratios greater than 1.08 when the substrate temperature was between 700 and 1000°C.

DISCUSSION

The observation of a large concentration of acetylene from reactants of H_2 and CH_4 in the filament assisted diamond CVD environment in this work and more recently by Harris et al. (*29*), demonstrates the complexity of the gaseous chemistry induced by the hot filament. It is noted that all of the observed gaseous hydrocarbon species are just those that are stable at high temperatures. This suggests that a role of the hot filament is to pyrolyze the gas inducing a complex series of reactions that result in the formation of chemical species at concentrations characteristic of a thermodynamic equilibrium at a temperature near that of the filament. On the right side of Fig. 8 is presented the results of a thermodynamic equilibrium calculation of the gaseous component concentrations vs temperature for an initial gas mix of 0.5% CH_4 in H_2 (*30,31*). Comparison of the concentration estimates reported in Table 1 with this equilibrium calculation shows qualitative, but not quantitative agreement. This would be consistent with a residence time of the gas near the hot filament that is insufficient for complete equilibrium to be established and/or for gaseous reactions which occur as the molecules are transported from the filament to the observation volume. It is likely that as the more reactive species generated at/near the filament (e.g., C, CH, CH_2) diffuse or flow to the colder deposition surface, reaction and depletion will occur. This simple model is shown schematically on the left side of Fig. 8. The utility of this thermodynamic equilibrium calculation as a "zeroth order" model for the role of the filament is further supported by the detection in this work of the methyl radical, CH_3, when acetylene, $C_2H_,$ was used as a reactant and the observations of several groups on the insensitivity of the diamond deposition to the precise nature of the hydrocarbon (*4,12,32*) (e.g., CH_4, C_2H_6, C_4H_{10}, C_2H_4, C_2H_2).

What is the role of atomic hydrogen in this deposition environment? This question will be very difficult to answer without detailed surface studies of the growing surface. The REMPI measurements (Fig. 4) indicate the presence of atomic hydrogen with increasing concentration as the filament temperature is raised, in agreement with the thermodynamic calculation presented above. The thermodynamic model does not predict the apparent decrease in atomic hydrogen with increasing methane concentration (Fig. 4). The data for 3% CH_4 in H_2 suggest a poisoning of the filament surface or suppression of the gaseous H atom concentrations at temperatures lower than 2250 C, as compared with reactant gas mixes with lower CH_4 concentrations. The depleted flux of atomic hydrogen to the surface correlates with the evidence for higher concentrations of graphitic/amorphous carbon deposits reported for the higher CH_4 concentration gas mixes (*33–35*). Thus one of the roles of the atomic hydrogen may be the preferential etching of the nondiamond forms of carbon which occurs simultaneously with the deposition of all forms of carbon (*36*). Other functions, such as surface stabilization, shifting of gas phase equilibria, and so on are also possible.

Hydrocarbon flames are known (*37*) to have similar species (C_nH_m,H, as well as oxygen species) present near the flame front when using a fuel rich stoichiometry. In particular, the oxygen-acetylene flame has a very high temperature (*38*) 3158 C, and thus might have significant concentrations of atomic hydrogen outside the inner flame cone. Hence, the observation of diamond deposition on an appropriately cooled substrate is not very surprising. As was shown in Fig. 7, the addition of H_2 to the flame, which might be expected to increase the atomic hydrogen concentration, also reduced the amount of amorphous carbon present in the diamond deposits. The measurements reported here are insufficient to attribute the large radial variations in the density, morphology, and quality of the diamond crystallites across the substrate to the measured substrate temperature profile because a similar distribution could be expected for the flux of reactive species in the flame.

CONCLUSIONS

This work has demonstrated the utility of *in situ* optical diagnostics in understanding the gaseous chemistry of the filament-assisted diamond CVD environment. The radical (H, CH_3) and stable (CH_4, C_2H_2, C_2H_4) species detected indicate that a simple thermodynamic equilibrium calculation is very instructive about the gaseous chemical composition above the diamond deposition surface, while the need for more refined modeling calculations is evident. A suppression of H atoms was observed with increasing CH_4 reactant concentration, with a marked depletion observed at 3.0% concentration for filament temperatures below 2200 C.

Diamond growth has been demonstrated in a variety of oxygen-hydrocarbon flames on a variety of substrates. For flames from a single orifice, the diamond crystallite growth density and quality can vary significantly depending on the position of the substrate in the flame, the substrate temperature and/or radial position in the flame. The addition of hydrogen to an oxygen-acetylene flame produces significant improvements in the diamond to amorphous carbon Raman peak ratio of the deposited crystallites.

ACKNOWLEDGMENTS

We gratefully acknowledge the assistance and discussions of many of our colleagues, and in particular, P. E. Pehrsson, A. Morrish, H. H. Nelson, and S. Dillich. This work was supported in part by the Office of Naval Research.

REFERENCES

1. Field, J. E., *The Properties of Diamond*, Academic, New York, 1979.
2. DeVries, R. C., *Ann. Rev. Mat. Sci.* **17,** 161 (1987).
3. Matsumoto, S., Sato, Y., Kamo, M., and Setaka, N., *Jpn. J. Appl. Phys.* **21,** L183 (1982).
4. Hirose, Y. and Terasawa, Y., *Jpn. J. Appl. Phy.* **25,** L519 (1986).
5. Singh, B., Arie, Y., Levine, A. W., and Mesker, O. R., *Appl. Phys. Lett.* **52,** 451 (1988).
6. Moustakas, T. D., Dismukes, J. P., Ye, L., Walton, K. R., and Tiedje, J. T., Proc. of the 10th International Conference on Chemical Vapor Deposition, The Electrochemical Society, Inc., 1164 (1987).
7. Kamo, M., Sato, Y., Matsumoto, S., and Setaka, N., *J. Cryst. Growth* **62,** 642 (1983).
8. Saito, Y., Matsuda, S., and Nigita, S., *J. Mat. Sci. Lett.* **5,** 565 (1986).
9. Badzian, A. R., Badzian, T., Roy, R., Messier, R., and Spear, K. E., *Mat. Res. Bull.* **23,** 531 (1988).
10. Saito, Y., Sato, K., Tanaka, H., Fujita, K., and Matuda, S., *J. Mat. Sci.* **23,** 842 (1987).
11. Matsumoto, S., *J. Mat. Sci. Lett.* **4,** 600 (1985).
12. Matsumoto, S., Hino, M., and Kobayashi, T., *Appl. Phys. Lett.* **51,** 737 (1987).
13. Suzuki, K., Sawabe, A., Yasuda, H., and Inuzuka, T., *Appl. Phys. Lett.* **50,** 728 (1987).
14. Matsumoto, S., Hino, M., and Kobayashi, T., *Appl. Phys. Lett.* **51,** 737 (1987).
15. Hirose, Y., and Kondo, N., Program and Book of Abstracts, Japan Appl. Phys. Spring 1988 Meeting, 434 (1988).
16. Hirose, Y. and Mitsuizumi, M., *New Diamond* **4,** 34 (1988).
17. Hanssen, L. M., Carrington, W. A., Butler, J. E., and Snail, K. A., *Materials Letts.* **7,** 289 (1988).
18. Hirose, Y., Proc. of First Int. Conf. on the New Diamond Sci. and Tech., October 1988, Tokyo, in press.
19. Snail, K. A., Hanssen, L. M., Carrington, W. A., Oakes, D. B., and Butler, J. E., Proc. of First Int. Conf. on the New Diamond Sci. and Tech., Oct. 1988, Tokyo, in press.
20. Celii, F. G., Pehrsson, P. E., Wang, H.-t, and Butler, J. E., *Appl. Phys. Lett.* **52,** 2043 (1988).
21. Celii, F. G., Pehrsson, P. E., Wang, H.-t., Nelson, H. H., and Butler, J. E., Proc. of First Int. Conf. on the New Diamond Sci. and Tech., Oct. 1988, Tokyo, in press.
22. Butler, J. E., Bottka, N., Sillmon, R. S., and Gaskill, D. K., *J. Cryst. Growth* **77,** 163 (1986).
23. Tjossem, P. J. H. and Cool, T. A., *Chem. Phys. Lett.* **100,** 479 (1983).
24. Celii, F. G., Pehrsson, P. E., Wang, H.-t., Nelson, H. H., and Butler., J. E., Proc. of 4th Interdisciplinary Laser Science Conf., Atlanta, GA, Sept. 1988.
25. Celii, F. G. and Butler, J. E., *Appl. Phys. Lett.* **54,** 1031 (1989).
26. Jackson, D. J. and Wynne, J. J., *Phys. Rev. Lett.* **49,** 543 (1982).
27. Hudgens, J. W., DiGuiseppe, T. G., and Lin, M. C., *J. Chem. Phys,* **79,** 571 (1983).
28. Yamada, C. and Hirota, E., *J. Chem. Phys.* **78,** 669 (1983).
29. Harris, S. J., Weiner, A. M., and Perry, T. A., *Appl. Phys. Lett.* **53,** 1605 (1988).

30. Lersmacher, B., Lydtin, H., Knippenberg, W. F., and Moore, A. W., *Carbon* **5,** 205 (1967).
31. Gordon, S. and McBride, B. J., "Computer Program for Chemical Equilibrium Calculations," NASA-Lewis Research Center, NASA SP-273, March, 1973.
32. Sato, Y., Kamo, M., and Setaka, N., Proc. 8th Int. Sym. on Plasma Chem., 2446 (1987).
33. Matsumoto, S., Sato, Y., Tsutsumi, M., and Setaka, N., *J. Mat. Sci.* **17,** 3106 (1982).
34. Sawabe, A. and Inuzuka, T., *Thin Solid Films* **137,** 89 (1986).
35. Mitsuda, Y., Kojima, Y., Yoshida, T., and Akashi, K., *J. Mat. Sci.* **22,** 1557 (1987).
36. Badzian, A. R. and DeVries, R. C., *Mat. Res. Bull.* **23,** 385 (1988).
37. Gaydon, A. G. and Wolfhard, H. G., *Flames,* Chapman and Hall Ltd., London, 4th ed., 1979.
38. Bernard, J. A. and Bradley J. N., *Flames and Combustion,* Chapman and Hall Ltd., London, 2nd. ed., 53 (1985).

Modeling of High Temperature Transport Reactions

H. GREINER* AND E. SCHNEDLER

Philips GmbH Forschungslaboratorium Aachen, Weisshausstrasse, 5100 Aachen, FRG

ABSTRACT

Chemical gas phase transport reactions very often determine the endurance of materials in contact with corrosive atmospheres at high temperatures. To control these processes, a good understanding of the underlying transport mechanisms is required.

In simple transport models, based on constant stoichiometry, transport rates are obtained from chemical equilibria. Algorithms allowing the rapid and interactive calculation of such equilibria with a virtually unlimited number of (ionic) compounds and (solution) phases are presented.

As an example transport of tungsten in a metal halide discharge containing the iodides of sodium, indium, and thallium is treated. It is shown that the presence of impurities like water and hydrogen has a decisive influence on prevailing transports.

Index Entries: Chemical transport; constant stoichiometry model; complex equilibria, calculation of; metal halide discharges; tungsten transport; impurities, influence of; molten metal halide salt mixtures.

INTRODUCTION

The design and optimization of many high temperature systems and processes requires a good understanding of mass transports occurring in corrosive gas phases. Examples treated in the literature include chemical reactors (*1*), nuclear reactors (*2*), chemical vapor deposition processes (*3*), and light sources (*4*).

*Author to whom all correspondence and reprint requests should be addressed.

In principle, the theory of transport processes in chemically-reacting flow systems provides the necessary methodology (*1*): the diffusional and convective transport of the various chemically-reacting species is modeled by a system of coupled partial differential equations with appropriate boundary conditions. But even in the era of super-computers the solution of such a system of partial differential equations remains a formidable computational task. Particularly in industrial research and development, it thus seems to be highly desirable to have available realistic models for transport processes simple enough to remain tractable on the computational and conceptual level.

In our contribution, we present such a model that is well-established in the literature dealing with the description of corrosive processes in light sources. Despite a number of obvious shortcomings, it has proven to be a valuable aid in understanding transport mechanisms in lamps. Here, we want to introduce the model and the chemical equilibrium routines underlying the calculation of transport currents.

The model has been widely applied to the description of tungsten and silicon transport in halogen incandescent and gas discharge lamps (*4–6*). In our contribution we treat tungsten transport in a metal halide discharge and show that the prevailing transports depend critically upon the level of impurities in the lamp.

A SIMPLE TRANSPORT MODEL

The simplified transport model is based on the following assumptions:

1. All chemical reaction rates are fast enough to maintain local thermal equilibrium in the gas phase and at the gas/condensate interfaces.
2. The model is geometry independent, i.e., the transport rates between two sites, A and B, depend only on the temperatures T_A and T_B and on the corresponding local gas phase compositions. As a consequence, spatial effects like diffusion and convection cannot be accounted for.
3. To describe the net transport of the elements (k) k = 1, . . ., m one introduces the so-called element solubilities

$$SP_{(k)}(T) = \Sigma R_{kj} p_j(T),$$

where the $p_j(T)$ are the partial pressures of the gaseous compounds (*j*) j = 1, . . . , n at temperature T and the R_{kj} the number of atoms of element (k) in a molecule of compound (*j*). The transport current for element (k) from site A to B is then proportional to the solubility difference

$$SP_{(k)}(T_A) - SP_{(k)}(T_B).$$

Thus, the solubility curves indicate the transport currents in the reaction volume.

The calculation of the solubilities for elements existing in condensed form in a whole temperature range is based on the following assumptions: the total pressure, P, in the reaction volume V_o is fixed by a background gas. The elemental composition (stoichiometry) of the gas phase is constant throughout the reaction volume and given by the equilibrium between the gas phase and condensed phases precipitated at the coldest spot T_0. The equilibrium is calculated for a reduced volume

$$V_{red} = \int_V \frac{T_0}{T(V)} \, dV$$

Once the total pressure and the stoichiometry have been determined, the equilibrium of the gas phase with the condensed phases existing at higher temperatures is computed. As a result, the partial pressures of the gaseous compounds and, hence, the solubilities of the condensed elements are obtained as a function of temperature.

These rather drastic assumptions severely limit the realism of the model. In particular, segregation and demixing effects owing to diffusion and convection cannot be explained, (*see 7–10*). Finally, we remark that ref. (*9*) shows how the effects of diffusion and convection can be incorporated in a geometry independent model.

CALCULATION OF COMPLEX CHEMICAL EQUILIBRIA

As we have shown in the previous section, our transport model reduces the determination of transport currents to the calculation of a series of complex chemical equilibria. From the computational point of view, this is a very attractive feature since we don't have to consider complicated partial differential equations whose solution demands a lot of computer storage and processing time. For simple reaction systems involving only a small number of species and phases, chemical equilibria can be calculated by a variety of free energy minimization methods well-documented in the literature (*11,12*). In our experience, however, these algorithms tend to become quite ineffective for complex systems, which involve many chemical elements forming a correspondingly large number of chemical compounds (including ions and electrons) and condensed phases (including solution phases). Computational problems result from the large number of variables (which in lamp applications may amount to several hundred), the determination of the stable phase set out of a multitude of possible phase combinations and the nonideal nature of the solution phases. The algorithms we have developed over the past years have proved to be very effective for the solution of complex equilibrium problems with respect to the following criteria:

1. Reliable convergence to the equilibrium, independent of an initial estimate.
2. Acceptable speed of convergence.

3. General applicability: the algorithm should not depend on specific thermodynamic models for the solution phases.
4. In case the system exhibits miscibility gaps, the algorithm should identify the absolute minimum (and not just a local one) of the free energy (i.e., calculate the equilibrium between subphases into which a phase can split).

In this section, we want to give a short introduction to the mathematical underpinnings of these algorithms: we consider a chemical reaction system involving m chemical elements. Their mole numbers are denoted by

$$\mathbf{b} = (b_1, \ldots, b_m).$$

The elements can combine to various chemical compounds that occur as components in the gaseous and condensed phases of the system. The j-th phase ($j = 1, \ldots, p$) is described by its total mole number N^j and its composition

$$\mathbf{x}^j = (x_1^j, \ldots, x_{s(j)}^j) \quad \text{with} \quad \sum_{i=1}^{s(j)} x_i^j = 1$$

defined in terms of mole fractions. The free energy of the j-th phase is given by

$$G^j = N^j G_m^j(\mathbf{x}^j),$$

where G_m^j is the molar free energy. The stoichiometry of the j-th phase is specified by a matrix

$$\mathbf{R}^j = (R_{ki}^j) \quad \text{with } k = 1, \ldots, m \text{ and } i = 1, \ldots, s(j).$$

According to the principles of equilibrium thermodynamics, the equilibrium of the system at constant temperature and pressure (the case of constant temperature and volume could also be considered) is characterized by the minimum of the total free energy)

$$\sum_{j=1}^{p} N^j G_m^j(\mathbf{x}^j) = \text{MIN} \tag{1a}$$

subject to the mass balance and nonnegativity constraints

$$\sum_{j=1}^{p} N^j \mathbf{R}^j \mathbf{x}^j = \mathbf{b} \qquad N^j \geq 0 \ \text{for} \ j = 1,\ldots,p \tag{1b,c}$$

The molar free energy functions G^j_m can always be assumed to be convex functions of their composition variables: for stable phases, this is a well-known fact of equilibrium thermodynamics. But even unstable phases, which possibly separate into subphases, can be described by convex free energy functions: rather than the original free energy functions G^j_m one considers their convex envelopes in the minimization problem described by Eq (1). This geometrical construction is equivalent to the application of the generalized lever rule describing the equilibrium between various subphases into which the original phase can split (*13*).

To calculate the heterogeneous chemical equilibrium between the gas phase and condensed phases (the most difficult part being the determination of the phases that actually precipitate), we have developed a method based on convex programming (*14*): the free energy minimization problem given by Eq. (1) represents a so-called generalized linear program for the N^j with variable cost coefficients $G^j_m(\mathbf{x}^j)$ and columns $\mathbf{R}^j\mathbf{x}^j$. A solution can be computed by linear programming techniques (revised simplex method). The problem is decomposed into subproblems (one for each phase)

$$\min\{G^j_m(\mathbf{x}^j) + \mathbf{u}(\mathbf{R}^j\mathbf{x}^j)] \quad j = 1,\ldots,p \tag{2}$$

which have to be solved repeatedly for "dual" variables $\mathbf{u} = (u_1, \ldots, u_m)$ given by a linear master program. The solution of the subproblems then defines a modified master program, new dual variables, and so on. It can be proven that this iterative procedure converges to the equilibrium solution. Equilibrium is characterized by the nonnegativity of expression (2) for all phases. For phases actually present at equilibrium, it must be zero. The equilibrium values of the dual variables u_k are equal to the chemical potentials of the elements.

Compared with other methods for the calculation of complex multiphase equilibria, the generalized linear programming algorithm has the following advantages

1. Convergence takes place independent of an initial estimate.
2. The problem is decomposed into subproblems that can be easily solved.
3. The scheme also applies for phases with nonconvex free energies (in that case, the absolute minimum of the subproblems (2) has to to be calculated at each iteration).

For the calculation of the equilibrium of the gas phase with condensed phases present at higher temperatures, a Newton-Raphson technique

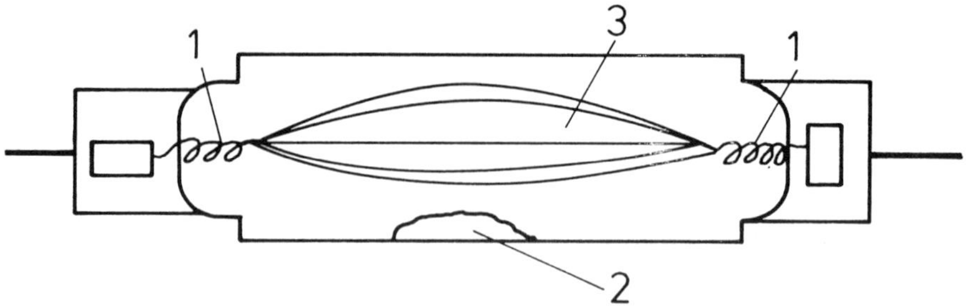

Fig. 1. Metal halide discharge lamp: (1) tungsten electrodes, (2) condensed phases at coldest spot, and (3) plasma discharge. The volume of the quartz vessel is 13.5 cm³.

has proved to be very efficient (*15*). In this case, the optimality conditions for the minimization problem (Eq.(1)) can be formulated as

$$\mu_j(T,P) = \mu_j^0(T) + kT \ln \frac{n_j}{N} + kT \ln P = \sum_{i=1}^{m} \lambda_i R_{ij} \tag{3a}$$

$$\sum_{j=1}^{n} n_j = N \quad \text{and} \quad \sum_{j=1}^{n} R_{ij} n_j = b_i \tag{3b}$$

Here, the n_j denote the mole numbers, the μ_j the chemical potentials of the gaseous species, P is the total gas pressure, and the λ_i are Lagrange multipliers that can be interpreted as the chemical potentials of the elements.

To guarantee a proper treatment of the species with exceedingly small mole numbers, the equations are transformed to logarithmic variables in n_j and N. Furthermore, the number of equations can be reduced by observing that the chemical potentials of the condensed species already determine a corresponding set of Lagrange multipliers. The calculation of a Newton-Raphson correction ($\Delta(\ln n_j)$, $\Delta(\ln N)$, $\Delta(\lambda_i)$) to a given iterate ($\ln n_j$, $\ln N$, λ_i) then amounts to the solution of a system of linear equations for the remaining $\Delta(\lambda_i)$. The notorious difficulties associated with the possible ill-conditioning of these equations (*see* (*11*)) can be avoided by expressing the stoichiometry matrix R_{ij} with respect to a basis consisting of the most abundant species in the gas phase. As shown in (*15*), this method for the calculation of gas–solid equilibria converges rapidly and reliably.

TUNGSTEN TRANSPORT IN METAL HALIDE DISCHARGES

Now we shall discuss the transport of tungsten in a metal halide discharge, as schematically shown in Fig. 1. The discharge consists of a quartz vessel and two tungsten electrodes and is filled with mercury and argon as a discharge medium. Furthermore, it contains the metal halides

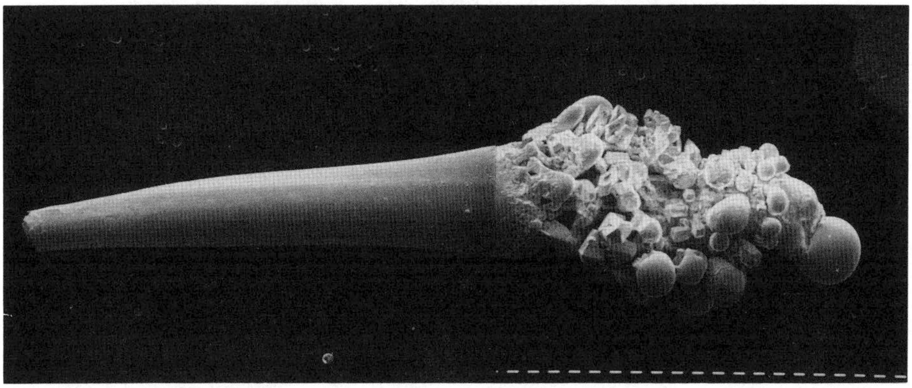

Fig. 2. Corrosion of a cylindrical electrode rod in a metal halide discharge (scale: one stroke equals 0.1 mm).

NaI, TlI, and InI, as additives. Their radiation in the discharge plasma determines the color-rendering properties of the lamp. The performance of such light sources can be severely impaired by the transport of tungsten along the electrodes and the deposition of tungsten on the quartz walls. A particularly drastic case of electrode corrosion is shown in Fig. 2.

To describe the chemical reactions in the lamp volume, oxygen and hydrogen impurities that arise from the decomposition of tungsten oxide and water have also to be taken into account. A standard filling of a 400W HPI (High Pressure Iodine) lamp consists of 14 mg NaI, 1.7 mg TlI, 0.19 mg InI, and 80 mg Hg. The chemical elements present in the lamp can combine to 128 gaseous and 78 condensed compounds. Liquid NaI, TlI, and InI form a molten salt mixture at the coldest spot of the lamp (about 1100 K), which can be described by a regular solution model (*16*). The equilibrium calculations display that the condensed phases present at the coldest spot are the molten salt mixture, solid WO_2, and liquid Na_2WO_4, which originates from the reaction

$$2NaI(g) + W(s) + 2O_2(g) = Na_2WO_4(l) + 2I(g)$$

Free iodine formed in this reaction has a decisive influence on tungsten transport since it combines with tungsten to form tungsten iodides and oxyiodides.

Figure 3 exhibits the major species contributing to the tungsten solubility as a function of temperature. The formation of these compounds and, hence, the tungsten solubility depends on the availability of free oxygen and iodine, which is controlled by the heterogeneous equilibrium at the coldest spot. Figure 4 shows the tungsten solubility for various oxygen contents in the lamp with no hydrogen present. It reveals that for an oxygen content of 0.65 μmol, the main tungsten transport is from the wall and the colder parts of the electrode (1100–1500 K) and electrode tip (2900 K) to the middle part of the electrode (around 2000 K). With decreasing oxygen content, the tungsten solubility at the lower temperatures (the main contribution coming from WO_2I_2) cannot be sustained and an appreciable transport of tungsten to the walls is observed. To avoid this highly undesirable wall blackening, the lamp must

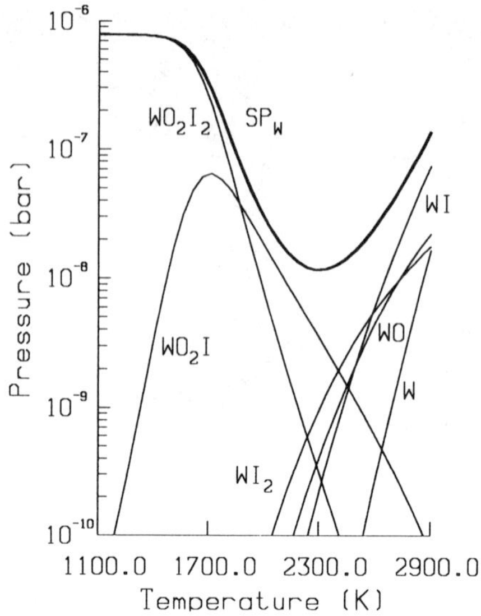

Fig. 3. Partial pressures of major species contributing to the tungsten solubility, SP_W, as a function of temperature (standard filling and an impurity content of 0.65 μmol O_2).

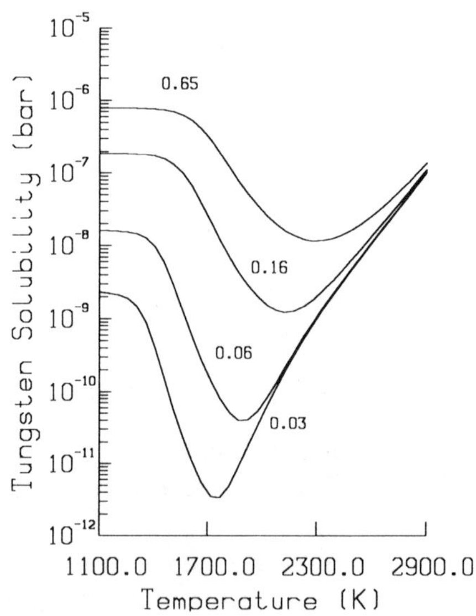

Fig. 4. Tungsten solubility as a function of temperatures of various O_2 contents given in μmol.

contain a certain level of oxygen (an excess of oxygen would lead to an undesired level of free iodine). Figure 5 shows that the addition of hydrogen enhances the tungsten solubility in the low temperature range 1100–1600 K (via the dissociation of water furnishing additional oxygen) and causes wall blackening. Thus, the hydrogen content of the lamp should be kept as low as possible.

The amounts of the metal halides in the discharge plasma are fixed

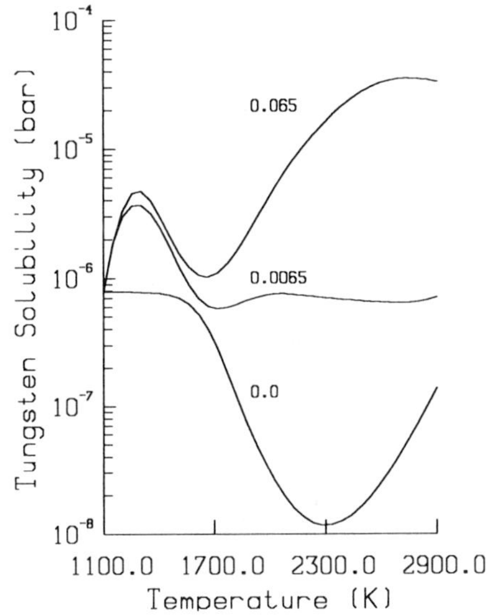

Fig. 5. Tungsten solubility as a function of temperatures for various H_2 content given in μmol and a O_2 content of 0.65 μmol.

by the respective vapor pressures above the molten metal halide salt mixture. Therefore, it is essential to perform the equilibrium calculations with a correct thermodynamic model for the activities of this solution phase (*16*). For a more complete discussion of tungsten transport in HPI-lamps, including experimental findings, we recommend ref. (*6*).

SUMMARY

We have presented a simple model for the description of mass transport in high temperature reaction systems. In combination with newly-developed algorithms for the calculation of complex multiphase chemical equilibria, the model allows the rapid and interactive calculation of prevailing transport currents.

The model has been applied to the description of tungsten transport in metal halide discharges. It has been demonstrated that tungsten transport critically depends on the presence of free oxygen and iodine whose partial pressures are fixed by the wall condensates.

REFERENCES

1. Rosner, D. E., *Transport Processes in Chemically Reacting Flow Systems* Butterworths, Stoneham, MA, 1986.
2. Kistemaker, J. and Boersma-Klein, W., *High Temp. Sci.* **24,** 37 (1987).
3. Jensen, K. F., *Chemical Engineering Science* **42,** 923 (1987).
4. Schnedler, E., *High Temperature Lamp Chemistry*, Zubler, E. G., ed., The Electrochemical Society, proceedings, vol. 85-2, 1985, p. 95.
5. Van Erk, W. and Rietveld, T., *High Temperature Lamp Chemistry*, Zubler, E. G., ed., The Electrochemical Society, proceedings, vol. 85-2, 1985, p. 57.

6. Fischer, E., *High Temperature Lamp Chemistry*, Zubler, E. G., ed., The Electrochemical Society, proceedings, vol. 85-2, 1985, p. 47.
7. Fischer, E., *J. Appl. Physics,* **47,** 2954 (1976).
8. Schnedler, E., *High Temp. Sci.* **19,** 237 (1985).
9. Rosner, E., *High Temperature Lamp Chemistry,* Zubler, E. G., ed., The Electrochemical Society, proceedings, vol. 88-4, 1988, p. 111.
10. Daikin, J. T. and Shyy, W., *High Temperature Lamp Chemistry,* Zubler, E.G., ed., The Electrochemical Society, proceedings, vol. 88-4, 1988, p. 139.
11. Smith, W. and Missen, R., *Chemical Reaction Equilibrium Analysis,* Wiley, New York, 1982.
12. Eriksson, G., *Chem. Scripta* **24,** 193 (1973).
13. Greiner, H., *CALPHAD* **12,** 133 (1988).
14. Greiner, H., *Mathl. Comput. Modelling* **10,** 529 (1988).
15. Schnedler, E., *CALPHAD* **8,** 265 (1984).
16. Greiner, H. and Schnedler, E., *High Temperature Lamp Chemistry,* Zubler, E. G., ed., The Electrochemical Society, proceedings, vol. 88-4, 1988, p. 162.

Principles of Area- and Phase-Selective CVD on Patterned Substrates

JAN-OTTO CARLSSON

University of Uppsala, Department of Chemistry, Thin Film and Surface Chemistry Group, Box 531, S-751 21 Uppsala, Sweden

ABSTRACT

Chemical vapor deposition (CVD) is known to be a large-area deposition technique. However, CVD can also be used for local deposition or so-called selective deposition. In this paper some principles of area-selective and phase-selective CVD on patterned substrates will be discussed. In area-selective CVD, only one phase is selectively deposited on desired substrate materials while the deposition is suppressed on other substrate materials. Phase-selective CVD means that several phases are deposited selectively and simultaneously. The systems discussed are area-selective growth of silicon, gallium arsenide, tungsten, and boron carbide, respectively. Phase-selective CVD is discussed for boron and boron carbides, respectively.

Index Entries: CVD; area-selective deposition; phase-selective deposition; silicon; gallium arsenide; tungsten; boron; and boron carbides.

INTRODUCTION

Chemical vapor deposition (CVD) is known to be a large-area deposition technique. However, CVD can also be used for local deposition or so called selective deposition, i.e., the deposition occurs only on some areas of the substrate surface. Selectivity may be attained by using different focused beams (photons, electrons, or ions). The beams induce local CVD reactions on those areas they hit. Another possibility is to stimulate the substrate surface through a mask with for instance a laser. The openings in the mask define the substrate areas where the deposition may take place.

Selective chemical vapor deposition may also be achieved on patterned substrates. The selectivity in this case is based on difference in the initial interfacial reactions between the different substrate materials and the vapor. The interfacial reactions on one substrate material should be inhibited completely to avoid nucleation, whereas the deposition reactions should be stimulated on those substrate areas where the deposition shall occur.

Principally there exist several categories of selective deposition systems. In the system described above, the deposition takes place on one substrate material, whereas no deposition is obtained on the other and so called area-selective deposition is achieved. However, different phases can also be deposited simultaneously and selectively on the different materials, resulting in phase-selective deposition. In analogy to the phase-selective deposition, films of different microstructures or different chemical compositions may be deposited on the different substrates materials and, hence, selectivity in microstructure or chemical composition is attained.

Selective deposition is an emerging field and there is a great demand of these processes in many application areas. With the steady reduction of IC feature sizes there is a need for self-aligned processes. Selective tungsten for metallization in VLSI and selective GaAs epitaxy for monolithic integration of optoelectronic devices are well known examples.

Other application areas may be in micronics, heterogeneous catalysis, engineering of film/substrate interfaces, and in growth of artificial 2-D and 3-D materials. Since selective deposition on patterned substrates is based on interfacial chemistry, there is practically no restrictions in the dimensions of the deposited materials islands. By using local kinetics and thermodynamic restrictions, a fascinating prospect of building up materials with artificial microstructures opens. In this paper the underlying principles of selective deposition are briefly discussed.

AREA-SELECTIVE GROWTH

Epitaxial Growth Conditions

There is a considerable technological interest today in area-selective epitaxy of both silicon and gallium arsenide. A brief discussion of area-selective growth with reference to silicon and gallium arsenide, respectively, is given below.

Epitaxial films can be grown at relatively high rates near equilibrium conditions, i.e., at at low driving force (low supersaturation) of the deposition process. For heterogeneous nucleation, a higher supersaturation is generally required. This means that conditions of selective growth are prevailing at a supersaturation lower than that for heterogeneous nucleation. This was used by Joyce and Baldrey for growth of silicon from $SiCl_4$ at 1200°C and atmospheric pressure in openings etched in the SiO_2 mask (*1*). A historical review of selective epitaxial growth, SEG, has recently been published by Borland (2). In SEG the growth is stopped when the surface of the growing film reaches the mask surface. A continued growth results in an overgrowth over the mask. The process is then called ELO (Epitaxial Lateral Overgrowth). For a review of the ELO process, the reader is referred to (3).

A key point in the SEG is the suppresion of the nucleation on the mask (usually silicon oxide or silicon nitride). The incubation time for nucleation varies with the substrate material and the deposition conditions. In an ideal case, this incubation time is longer than the deposition time required to prepare the desired structures. However, by using an alternating growth and etching process, SEG can be attained even for conditions of short incubation times for nucleation (4). The growth conditions are then prevailing for about the incubation time. After the growth cycle, the process is switched over to etching with for instance HCl. A minor etch is also obtained of the monocrystalline silicon in the SiO_2 openings.

With the successive increased integration of ICs and an increased topography, demands related to planarization arise. For instance for filling of contact holes polysilicon might be used. Furumura et al. (5) showed that SEG conditions, yielding perfect monocrystalline growth, also could be used for selective polysilicon growth. By adding a gas, which caused deterioration of the crystallinity, selective polysilicon growth was achieved. Trichloroethylene or methane was used as gas additives. A carbon contamination of the order of 10^{20} cm^{-3} was obtained in the films.

The GaAs SEG/ELO is nearly as old as the silicon SEG/ELO. Tausch and Lapierre (6) reported in 1965 on a GaAs ELO process based on a chloride vapor transport system. With the development of the purification techniques of metal–organic compounds like trimethyl gallium (TMG) and triethyl gallium (TEG), CVD as well as MBE, based on the use of these compounds together with AsH_3, are highly attractive for GaAs SEG. MBE and elemental sources yields monocrystalline growth in etched openings as well as polycrystalline GaAs on the mask (7) (microstructure-selective deposition). GaAs SEG has received much attention during the last few years as a technique for achieving monolithic integration of electronic and optoelectronic devices.

Growth of GaAs from AsH_3 and TMG by MBE or CVD is usually considered to be a nonequilibrium process. The perfection of the crystals grown, their morphology, and the correlation between the growth rate and thermodynamic parameters indicates that near-equilibrium conditions exist at the interface between the vapor and the solid. Hence, thermodynamics can be utilized to analyze selective growth as well as heterogeneous nucleation conditions in GaAs CVD.

According to nucleation theory, a certain supersaturation is needed for heterogeneous nucleation on the mask. From experimental selectivity data, the maximum supersaturation for maintaining selectivity can be calculated. The supersaturation is favorably expressed in terms of chemi-

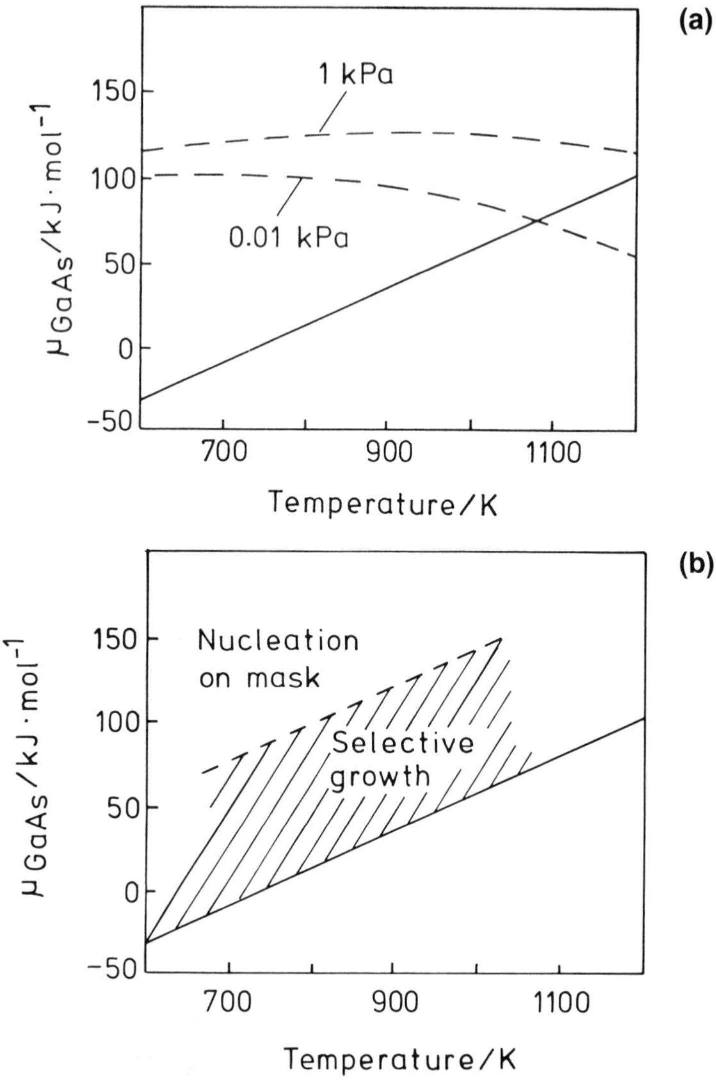

Fig. 1. a) Chemical potential of GaAs as a function of temperature for equilibrium with solid GaAs (full line) and equilibria without any solid present (dashed lines), respectively. $H_2/AsH_3/TMG = 500/10/1$.
b) Region of area-selective CVD or MBE of GaAs from AsH_3 and TMG.

cal potentials. The influence of temperature on the chemical potential of GaAs (expressed in elemental chemical potentials of Ga and As_2) at equilibrium with solid GaAs is shown in Fig. 1a. Growth will occur if the chemical potential of GaAs for the homogeneous equilibrium in the vapor is higher than that for the heterogeneous equilibrium.

The experimental technique used to determine, for instance, the temperature required to achieve SEG is to raise the temperature successively until no nucleation on the mask can be observed. Since the chemical potential of GaAs for the homogeneous equilibrium in the vapor has only a slight temperature dependence (*see* Fig. 1a), the driving force for the deposition (or supersaturation) will decrease upon a temperature increase and a driving force value, yielding no heterogeneous nucleation, will be reached.

Thermodynamics and MBE and CVD experimental SEG data were used in an effort to put experimental selectivity observations on a common basis. In MBE, a much lower pressure is used than in CVD. However, irrespective of the growth technique used, the experimental SEG data fall in the supersaturation region indicated in Fig. 1b. By using thermodynamics, the selectivity data from CVD can be converted to MBE and vice versa.

The trends in selectivity when changing the experimental conditions can also be predicted from thermodynamics. This is shown in Fig. 2a–c. The lower the driving force (the less negative $\Delta\mu$) the better selectivity can be expected. From the figures, it is deduced that a high temperature, a low total pressure, and a low AsH_3/TMG molar ratio will improve selectivity. This is in agreement with experimental observations (*see* e.g., *9–12*).

Substrate-Activated Selective Growth

When a substrate of different materials is exposed to the vapor in a CVD process, the materials represent areas of different activities or reactivities toward the vapor. One material may for instance act as an effective reducing agent or as a catalyst of, for instance, dissociative adsorption of gaseous reactants, which may favor the deposition. The other material may be relatively inert toward the vapor and growth may be inhibited. The inertness may be increased purposely by using gas additives that are preferentially adsorbed to one of the materials. Strongly adsorbed molecules may passivate a substrate surface considerably and supress the deposition process completely. A tendency to substrate-activated area-selective growth is frequently seen during the initial growth stage in CVD on polycrystalline, multiphase substrates. The different phases and the different crystallographic orientations of the grains exposed to the vapor, represent surface areas of different activities/reactivities and initial growth conditions. Coming to extremes this means that the deposition is inhibited on some substrate areas, whereas other areas are open for deposition.

Area-selective deposition of refractory metals is of highest interest for metallization in VLSI and ULSI. Selective deposition of refractory metals for metallization has been reviewed by several authors (*see*, e.g., *13–16*). The substrate-activated area-selective growth is well illustrated by the selective tungsten deposition from WF_6 and H_2 on Si/SiO_2 substrates. This process will be described principally below.

Tungsten can be deposited by CVD at low temperatures (300°C) from H_2 and WF_6 according to the reaction

$$3\ H_2(g) + WF_6(g) \rightarrow W(s) + 6HF(g)$$

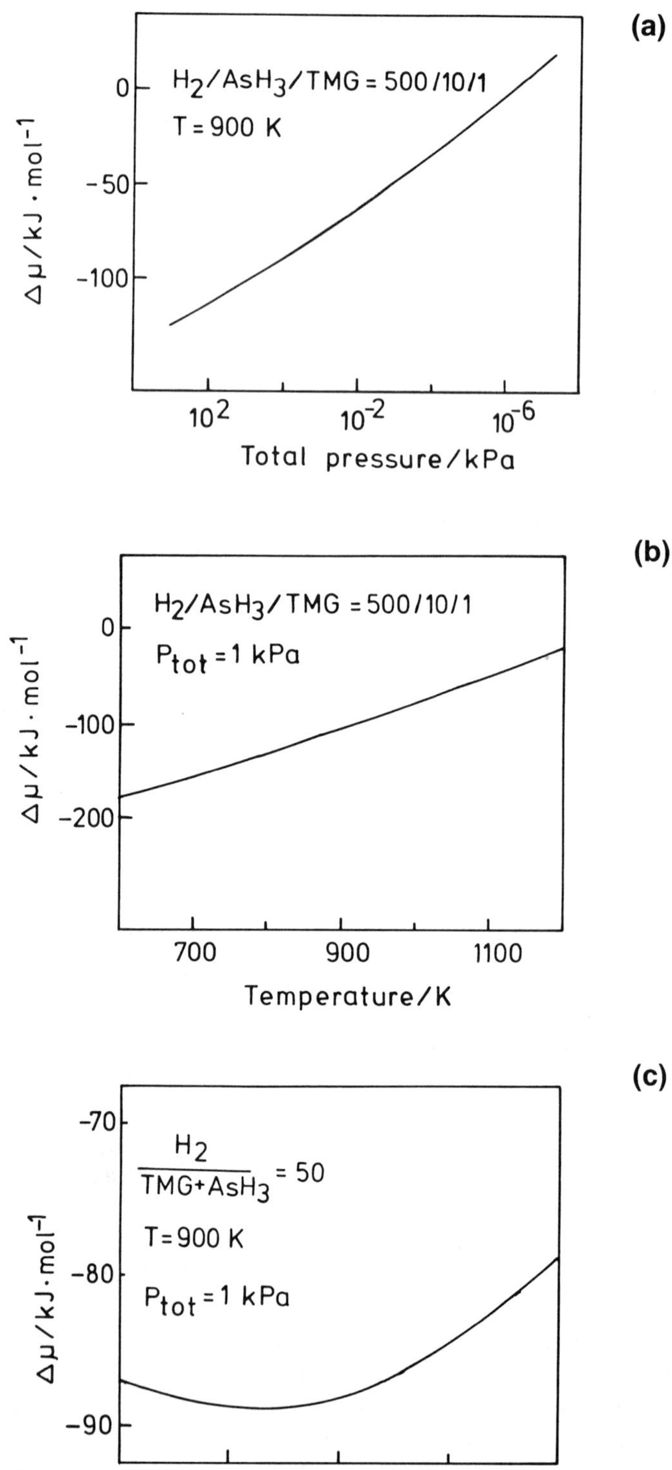

Fig. 2. Influence of total pressure (a), temperature (b), and gas composition (c), respectively on the driving force and the selectivity at GaAs growth from AsH_3 and TMG. A decrease in the driving force means an improved selectivity.

The deposition occurs on all substrate surfaces exposed to the vapor, since both the source material (WF_6) and the reducing agent (H_2) are gases. However, if the reducing agent was replaced by a solid reducing agent (like elemental silicon), the deposition should only occur on those substrate regions having a reducing agent. So, if a wafer exposes areas of elemental silicon and silicon dioxide to WF_6, tungsten deposition takes place only on the silicon areas and not on the adjacent silicon dioxide areas. Silicon in silicon dioxide can not act as a reducing agent, since this silicon has maximum in its oxidation number. This is the basis of the initial stage of selective tungsten CVD.

The selective tungsten deposition may proceed according to the scheme described above as long as elemental silicon is exposed to the vapor. After a while, however, the tungsten deposited onto elemental silicon will separate silicon from the vapor and hence a self-limiting growth process has been obtained. The mechanism of self-limitation is under discussion and might be also be owing to a polymerization reaction involving lower tungsten fluorides (*17*). The polymer formed may also separate silicon from the vapor and, hence, inhibiting the growth process. For growth of thicker tungsten layers, a reducing agent, H_2, has to be added to the reaction gas. If the WF_6 concentration is low and the H_2 concentration is high, i.e., conditions of low supersaturation is prevailing, deposition of tungsten will occur where tungsten already has been deposited (on elemental silicon) and not on the silicon dioxide. For deposition on the silicon dioxide, resulting in a loss of selectivity, tungsten nucleation must take place. The nucleation step requires a much higher supersaturation than only growth. Hence a deposition window, ranging from the supersaturation corresponding to equilibrium conditions up to the supersaturation value needed for heterogeneous tungsten nucleation on silicon dioxide, exists. Finally, the selective deposition of tungsten to substrate areas where tungsten already has been deposited, is favored by the dissociation of hydrogen molecules on these areas (*18*).

In summary, two main reaction steps can be distinguished in tungsten CVD (*see* Fig. 3).

1. In the first step, elemental silicon will act as the predominating reducing agent even if a large amount of hydrogen is used in the reaction gas. This results in tungsten deposition on those substrate regions where elemental silicon is exposed to the vapor. The reaction step includes an etching of elemental silicon, i.e., silicon is consumed.

$$2\ WF_6(g) + 3\ Si(s) \rightarrow 2\ W(s) + 3\ SiF_4(g)$$

Considering the stoichiometry of this reaction, about 200 Å silicon is consumed for 100 Å tungsten deposited. The topography of the Si/W interface is affected by this reaction. The etching and hence the topography can be reduced by for instance addition of SiF_4 to the reaction gas mixture (*19*).

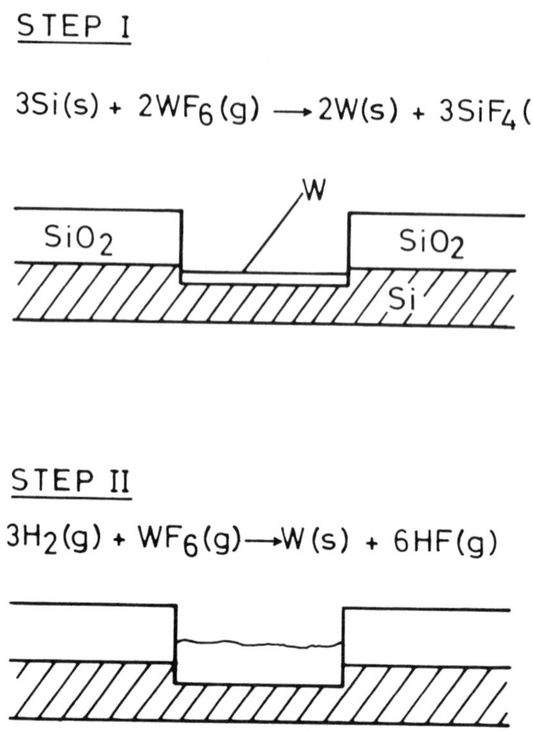

Fig. 3. The different process steps in selective tungsten CVD.

2. In the second step, another reducing agent, H_2, has to take over, since the tungsten film and probably the tungsten fluoro polymer obtained prevent the reactions between the vapor and the silicon.

The chemical reactions in the first step are usually extremely fast and a thermodynamically controlled CVD process is obtained. In the second process step, the deposition process was operated at a low supersaturation to avoid nucleation on the mask material (SiO_2). The growth conditions in the second step are close to those existing in the area selective epitaxy discussed above and can be analyzed from thermodynamics. Thermodynamics has been used as a guide for prediction of trends in selectivity and substrate etching when the deposition conditions are changed. It has also been used for identification of plausible (and often undesired) side reactions as well as of gaseous selectivity modifiers, improving selectivity (*20,21*).

For discussion of trends in selectivity, a selectivity number, SE, was defined.

$$SE = \Delta\mu(Si)/\Delta\mu(SiO_2)$$

$\Delta\mu(Si)$ is the driving force for the Si/vapor reaction, whereas $\Delta\mu(SiO_2)$ is the driving force for the SiO_2/vapor reaction. An SE number larger than unity favors selectivity to the silicon. For improvement of selectivity,

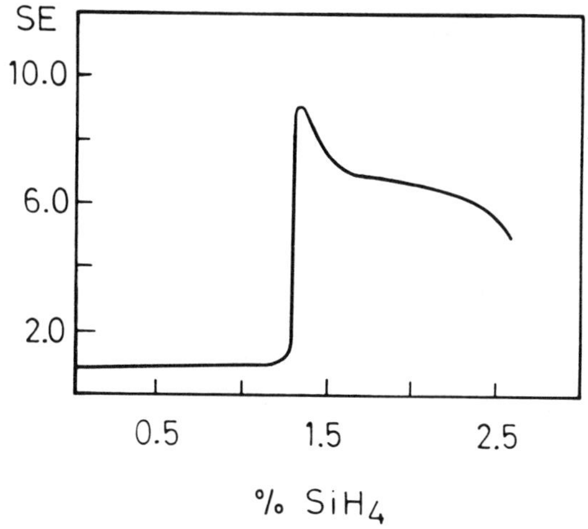

Fig. 4. Calculated selectivity number SE as a function of SiH_4 concentration. Temperature = 300°C, total pressure = 0.1 Torr (13.3 Pa), H_2/WF_6 = 39.

gaseous additives, increasing the chemical potential of tungsten in the vapor, can be used. SiH_4, which doesn't add more constituents to the vapor, is such an attractive selectivity modifier. Figure 4 shows the SE number as a function of SiH_4 concentration in the vapor. A drastic improvement in selectivity is obtained at an SiH_4 concentration of 1.2%. At this particular concentration a new reaction channel is opened (*21*).

Adsorption-Induced Selective Growth

As discussed above, the heterogeneous nucleation on one of the substrate materials must be suppressed during a relatively long time in an ideal selective growth system. The incubation time for nucleation is influenced by many factors; temperature, substrate reactivity, adsorption. Adsorbed molecules may reduce the rate of surface reactions and in extremes inhibit the nucleation completely. A concept of strongly adsorbed molecules to one of the substrate materials was used to achieve area-selective growth of boron carbide on a patterned substrate exposing areas of titanium and molybdenum to the vapor (*22*). Boron trichloride, ethylene, and hydrogen were used as reactants and the deposition temperature was 1400 K. The ethylene molecules (or fragments of them) were preferentially and strongly adsorbed on molybdenum and no nucleation of boron carbide was observed. On titanium, however, fast nucleation kinetics was obtained. The deposition was located only to those substrate areas having titanium. The boron carbide was amorphous and contained about 21 at-% carbon. This illustrates that adsorbed molecules may act as masks and can be used to inhibit the deposition on desired substrate areas.

PHASE-SELECTIVE DEPOSITION

A new dimension in the field of selective growth is created in phase-selective growth. Phase-selective growth means that several phases are selectively and simultaneously deposited on desired substrate materials/areas. This might result in growth of, for instance, a semiconductor together with an insulator, i.e., selectivity in properties is also obtained. Phase-selective deposition may be achieved in different ways. In this paper two principles of phase-selective growth will be discussed; phase-selective deposition attained by differential nucleation behavior and by secondary processes in or on the growing film, respectively.

Phase-Selective Deposition by Differential Nucleation Behavior

The initial substrate/vapor reactions and the nucleation kinetics is usually dependent on the substrate material. This may result in nucleation of different phases on different substrate materials. Provided that no secondary processes like phase-transformation in the solid state occurs in the film or that no new phase is nucleated on top of the growing film, the originally nucleated phases will continue to grow and a phase-selective deposition is obtained.

This principle was used for phase selective-growth of two boron carbides; Tl-BC_x and $B_{13}C_2$. The substrate used was that obtained after the area-selective growth of boron carbide described above, i.e., the substrate exposes molybdenum and amorphous boron carbide to the vapor. The vapor contained boron trichloride, methane and hydrogen and the growth temperature was 1400 K. Tl-BC_x was obtained on the amorphous boron carbide, whereas $B_{13}C_2$ was grown on molybdenum. This phase-selective growth was attributed to differential nucleation kinetics since no secondary processes were observed in or on the films (23).

Phase-Selective Deposition Achieved by Using Secondary Processes

Elemental boron has several crystalline polymorphs and in addition to that amorphous boron also exists. Phase-selective growth was studied in this system by using the Ti/Mo patterned substrates described above. Fast nucleation was observed on both Mo and Ti. α-rhombohedral boron was grown on Mo, whereas amorphous boron was obtained on Ti. The boron grown on Ti contained a small amount of Ti (about 500 ppm) throughout the layers, while no traces of Mo was detected in α-B (detection limit 1 ppm). The Ti stabilizes the originally nucleated amorphous boron (24). To obtain crystalline boron, a phase-tranformation in the solid state is needed and such a transformation can be assumed to have a high energy barrier to overcome. Amorphous boron may also be obtained initially on Mo. However, an immediate phase transformation is expected because of the deposition temperature used. Moreover, the film

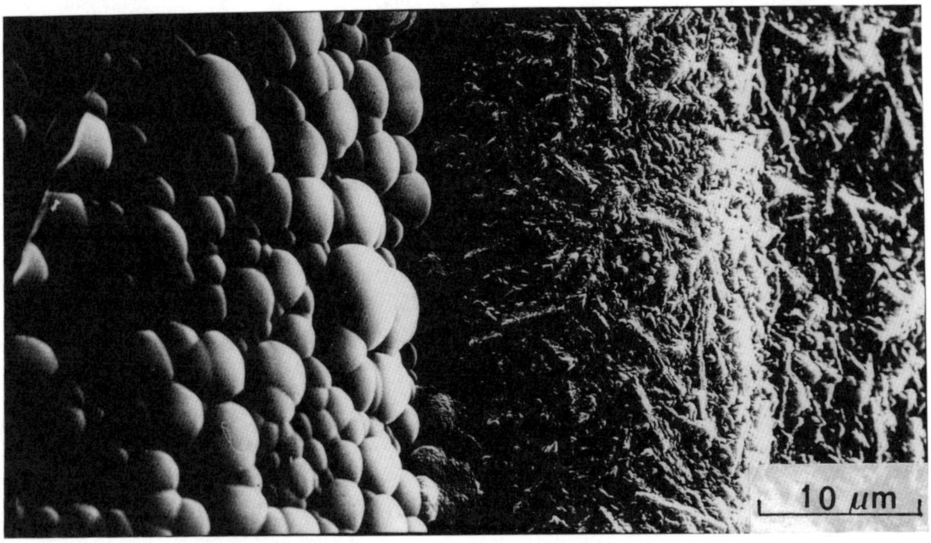

Fig. 5. Micrograph showing phase-selective CVD of amorphous and crystalline boron on a substrate exposing areas of titanium and molybdenum, respectively, to a hydrogen/boron trichloride gas mixture.

didn't contain any substrate contaminants contributing to a stabilization of the amorphous boron. The morphology of the phase-selectively deposited boron is shown in Fig. 5. The amorphous boron is characterized by the rounded nodules.

SUMMARY

The underlying principles of selective growth have been introduced. Area-selectivity can be achieved by using epitaxial growth conditions, variations in the substrate reactivity across a substrate surface or strongly adsorbed molecules as masks. Phase-selectivity can be obtained by differential nucleation kinetics or by differential secondary processes occurring in or on the films during growth. The materials systems discussed were silicon, gallium arsenide, tungsten, boron, and boron carbides.

Selective deposition by CVD is principally based on differences in interfacial kinetics between various substrate materials and the vapor. The interfacial reactions as well as the heterogeneous nucleation processes, which must be suppressed in area-selective deposition on at least one material, are very sensitive for contamination. However, with a proper design of the CVD system, use of high-purity gases and defect-free mask materials, the selective deposition will be a useful technique to prepare new material structures.

REFERENCES

1. Joyce, B. D. and Baldrey, *Nature* **195,** 485 (1962).
2. Borland, J. O., Proc. 10th Int. Conf. on CVD, Cullen, G. W. and Blocher, Jr., J. M. eds., The Electrochem. Soc., Pennington, NJ, 1987, p. 307.

3. Jastrzebski, L., *J. Cryst, Growth* **63,** 493 (1983).
4. Jastrzebski, L., Corboy, J. F., McGinn, J., and Pagliaro, Jr., R., *J. Electrochem. Soc.* **130,** 1571 (1983).
5. Furumura, Y., Mieno, F., Nishizawa, T., Maeda, M., *J. Electrochem. Soc.* **133,** 379 (1986).
6. Tausch, Jr. and Lapierre, III, A. G., *J. Electrochem. Soc.* **112,** 706 (1965).
7. Cho, A. Y. and Ballamy, W. C., *J. Appl. Phys.* **46,** 783 (1975).
8. J. O. Carlsson, to be published.
9. Heinecke, H., Brauers, A., Grafahrend, F., Glass, C., Putz, N., Werner, K., Wyers, M., Luth, H., and Balk, P., *J. Cryst. Growth* **77,** 303 (1986).
10. Ghosh, C. and Layman, R. L., *Appl. Phys. Lett.* **45,** 1229 (1984).
11. Kamon, K., Takagishi, S., and Mori, H., *J. Cryst. Growth* **73,** 73 (1985).
12. Tokumitsu, E., Kudou, Y., Konagai, M., and Takahashi, K., *Jap. J. Appl. Phys.* **24,** 1189 (1985).
13. Green, M. L., Proc. 10th Int. Conf. on CVD, Cullen, G. W. and Blocher, Jr., J. M., eds., The Electrochem. Soc., Pennington, N. J., 1987, p. 603.
14. Lami, Ph. and Pauleau, Y., *J. Electrochem. Soc.* **135,** 980 (1988).
15. Pauleau, Y., Lami, Ph., Tissier, A., Panyel, R., and Oberlin, J. C., *Thin Solid Films* **143,** 259 (1986).
16. Green, M. L., Ali, Y. S., Boone, T., Davidsson, B. A., Feldman, L. C., and Nakahara, S., *J. Electrochem. Soc.* **134,** 2285 (1987).
17. Lifshitz, N., *Appl. Phys. Lett.* **51,** 967 (1987).
18. McConica, C. M. and Krisnamani, K., *J. Electrochem. Soc.* **133,** 2542 (1986).
19. Levy, R. A., Green, M. L., Gallagher, P. K. and Ali, Y. S., *J. Electrochem. Soc.* **133,** 1905 (1986).
20. Carlsson, J. O., and Boman, M., *J. Vac. Sci. Technol.* **A3(6),** 2298 (1985).
21. Carlsson, J. O. and Hårsta, A., *Thin Solid Films,* **158,** 107 (1988).
22. Jansson, U. and Carlsson, J. O., *J. Vac. Sci. Technol.* **A6(3),** 1733 (1988).
23. Jansson, U., Carlsson, J. O., Markert, L. M., and Greene, J. E., submitted to J. Vac. Sci. and Technol.
24. Jansson, U., Carlsson, J. O., Markert, L. M., and Greene, J. E., to be published.

Fundamental and Practical Aspects of the Chemical Vapor Infiltration of Porous Substrates

R. NASLAIN* AND F. LANGLAIS

Laboratoire des Composites Thermostructuraux (UM 47 CNRS-SEP-UB1) Europarc, 3 Avenue Léonard de Vinci, 33600 Pessac, France

ABSTRACT

CVI is a new technique for depositing a solid from a gaseous precursor in the pore network of a porous substrate. It is particularly suited to the deposition of ceramics. The deposition rate depends on both the mass transfer of gaseous species in the pores and kinetics of the surface chemical reactions. In isothermal/isobaric CVI, mass transfer in the pores is only by diffusion (including Knudsen diffusion for the small pores). Thus, infiltration must be performed at low temperature and pressure to avoid an early sealing of pore entrances and to allow a significant filling of the pores. It is a rather slow process but it can be applied simultaneously to a large number of substrates even of complex shapes (e.g., to the fabrication of ceramic matrix fibrous composites). In forced-CVI, a pressure/temperature gradient is applied to the substrate with the result that (i) mass transfer in the pores is mainly by forced convection and (ii) the infiltration rate is increased by one order of magnitude. The use of forced-CVI raises a problem of tooling and seems to be only suited to the treatment of a single substrate of simple shape.

Index Entries: Chemical vapor deposition, CVD; chemical vapor infiltration, CVI; porous substrates; ceramic fiber preforms; densification process; ceramic matrix composites, CMC.

*Author to whom all correspondence and reprint orders should be addressed.

High Temperature Science, Vol. 27

INTRODUCTION

Chemical vapor infiltration (CVI) is a procesing technique, derived from chemical vapor deposition (CVD), according to which a solid (e.g., a ceramic material) is chemically deposited within the pore network of a porous substrate from a gaseous precursor. The main interest of CVI lies in the fact that it is performed at low temperatures. This feature is of particular interest when the solid is a refractory material. As an example, SiC, which melts peritectically at about 2700°C is deposited by CVI from chlorosilane/hydrogen gaseous mixtures at 800–1100°C.

The CVI process was initially designed and used for the fabrication of ceramic matrix composites (CMC), i.e., materials made of ceramic fibers embedded in a ceramic matrix. Starting from porous fiber preforms, near-net-shape CMC were obtained, according to a rather simple procedure, by filling their open porosity with the ceramic matrix deposited from suitable gaseous precursors (*1–4*). the most common CMC which have been obtained by CVI are C-C, C-SiC, and SiC-SiC composites (*1–11*). However, the feasibility of the CVI process has been also established for carbides (TiC, B_4C) (*8,12,14*), nitrides (BN, Si_3N_4) (*3,15–17*) and oxides (Al_2O_3, ZrO_2) (*18,19*). In this field, the fact that CVI must be performed at rather low temperatures is of particular interest since the reinforcing capability of most ceramic fibers is significantly lowered by exposures at high temperatures owing to a coarsening of their microstructure or chemical reactions with the matrix or/and the gaseous environment). As an example, such a strength lowering is known to occur above about 1100°C for the SiC-based Nicalon fiber (from Nippon Carbon) commonly used in CMC, i.e., well below the usual sintering temperatures of SiC, thus precluding the fabrication of SiC-SiC composites by hot pressing SiC (Nicalon) fibers in SiC powders.

CVI could also be used to improve the properties of those ceramics that exhibit a high open porosity due to the specific features of the process according to which they have been obtained. A typical example is given by reaction bonded silicon nitrides (RBSN) whose high residual open porosity can be CVI-filled either with Si_3N_4 or SiC improving thus both their mechanical properties and resistance to oxidation (*3*).

The aim of the present contribution is to give an overview on the CVI-processing of advanced materials focussing on (i) the specific conditions under which CVI must be performed and (ii) the practical aspects of its use in the field of fibrous CMC.

FUNDAMENTAL OF THE CVI OF A POROUS SUBSTRATE

The chemistry of CVI and CVD, for a given gaseous source species and substrate, has the same basis. However, the mechanisms leading to

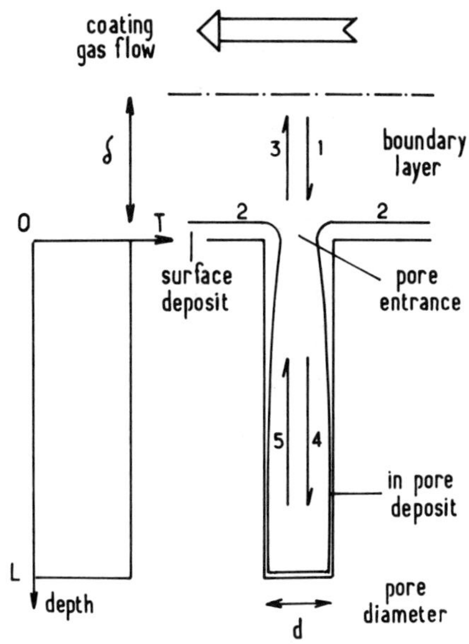

Fig. 1. Mechanisms involved in CVD and CVI.

the deposition of a solid on a pore wall (CVI) are more complex than those involved in CVD (where the solid is deposited on the external surface of the substrate).

As schematically shown in Fig. 1, the CVD of a solid on a substrate involves three main steps: (i) the gaseous source species reach the substrate surface usually by diffusing across a stagnant boundary layer whose thickness δ depends on the CVD-conditions (ii) surface phenomena take place on the substrate, including a chemical reaction between the adsorbed reactant species giving rise to the solid and to adsorbed gaseous products and (iii) these gaseous species leave the substrate surface by diffusion across the boundary layer. The overall deposition process can be rate-limited either by the kinetics of the surface phenomena (e.g., the chemical reaction occurring in step (ii)) or by the mass transport by diffusion through the boundary layer δ (steps (i) and (iii)) depending on the CVD conditions (temperature T, pressure P, and gas flowrate D). As an example, for given P, D values, the rate limiting steps are known to be the surface phenomena at low temperatures (the thermal variations of the deposition rate obeying an Arrhenius law) and the mass transport through the boundary layer at high temperatures (Fig. 2) (*13*). The driving force which allows the diffusion to and from the substrate surface is the concentration gradients in reactants and products that are present in the boundary layer owing to the chemical reaction occurring on the substrate. On the other hand, the chemical reaction taking place in step (ii) should be activated. Most CVD experiments used a thermal activation (the substrate being maintained at a high enough temperature). In order to obtain deposits homogeneous in thickness on substrates having a

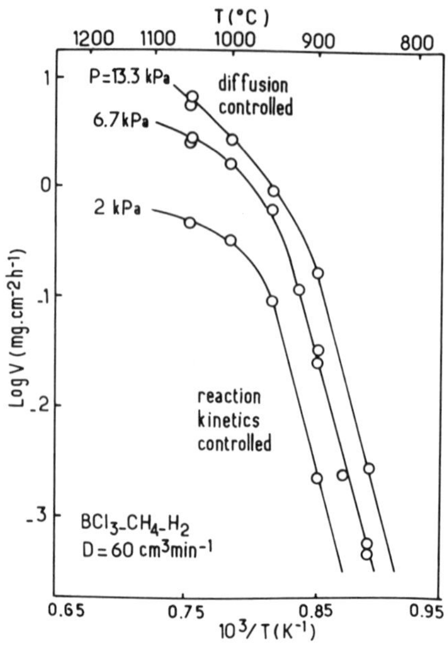

Fig. 2. Rate limiting steps in the CVD of a solid from a gaseous precursor from (*13*).

rough surface or a complex shape, CVD must be performed under conditions where deposition is rate-limited by surface reaction kinetics and not by diffusion. Otherwise, slow mass transport would involve low or high deposition rates depending on the length of the diffusion path to various points of the substrate.

In CVI, if the goal is to fill as completely as possible the open porosity of the substrate (which is actually the case for the fabrication of a CMC from a porous fiber preform), the deposition of the solid should occur preferentially: (i) as uniformly as possible on the pore wall whatever the pore morphology, and (ii) without an early sealing of the pore entrances (otherwise the densification process will stop). The first attempts made in this field have shown that the deposition of a solid in a pore network from a gaseous precursor requires specific conditions with respect to those commonly used in CVD (*1–3,20*).

As shown schematically in Fig. 1, for a single straight pore open at one end in a hot substrate set in a flow of gaseous precursor, there are two new steps in the deposition mechanism. On one hand, the reactant gaseous species, after having first diffused across the boundary layer δ, should reach any point on the pore wall (including those located at a depth L, L being the length of the pore) before being adsorbed and consumed by the chemical reaction (step 4). On the other hand, the gaseous products resulting from the deposition reaction occurring at any point on the pore wall should leave the pore and diffuse towards the boundary layer (step 5). It is worthy of note that these two new mass transport steps generally take place over long distances with respect to pore sections (the thickness of a fiber preform is typically of the order of

1–10 mm) and under almost isobaric and isothermal/conditions: gaseous diffusion in a stagnant medium can be considered as the only mass transport mode involved in the pores for most CVI experiments. Therefore, in order to allow the reactant molecules to diffuse into the pore, as far as possible from the pore entrance before reacting (thus avoiding the deposition of the solid to be limited to the vicinity of the pore entrance and an early sealing of the pore), CVI should be performed under conditions where the kinetics of the surface phenomena is slow with respect to the mass transport of the gaseous species by diffusion along the pore, i.e., at low temperature and low total pressure (*20,21*). Even under such conditions, one could not avoid the occurrence of a gradient in the deposit thickness along the pore axis which has been both observed experimentally and calculated (on the basis of different models) (*3–5,7,18,20,21*).

As an example, Figs. 3 and 4 show the effect of both temperature and pressure on the thickness profile of a SiC deposit (precursor: CH_3SiCl_3/H_2) within a pore (open at both ends) of 10 mm for two diameter values (i.e., 100 and 1 μm). The profiles were calculated on the basis of a model first proposed by Van den Breckel et al., for thin deposits in narrow tubes (*22*) and then adapted by Rossignol et al., to the densification of porous substrates with much smaller pore diameters (*21,23*). The main assumptions are the following:

1. There is no temperature and pressure gradient along the pore (i.e., the mass transport occur only by diffusion);
2. The concentration gradients along the pore are low; and
3. The chemical reaction giving rise to SiC is of first order.

Due to the pore diameter range considered here (i.e., $1<d<100$ μm), both ordinary and Knudsen diffusion were taken into account. The related diffusion coefficients were calculated according to (*21,24*). The data for the kinetics of SiC synthesis from CH_3SiCl_3 (MTS)/H_2 were those given by Schoch et al. (*25*). The profiles shown in Figs. 3 and 4 (which, as a matter of fact, correspond only to half the pore length) were calculated according to an iterative procedure (stopped when the deposit thickness at the pore entrance becomes equal to the pore radius, i.e., when the pore is sealed at both ends) (*23*).

Although such a model should be regarded as a first approximation, it leads to thickness profiles that (i) are in rather good agreement with those experimentally observed for porous fiber preforms (*see next section*), and (ii) clearly illustrate the effect of the three main parameters, i.e., T, P, and d. The thickness gradient along the pore decreases when the deposition temperature is lowered, the total pressure is reduced and the pore diameter is enlarged. As an example,a pore 100 μm in diameter will be almost fully filled at T = 900°C and P = 20 kPa (0.2 atm) at the pore sealing. On the contrary, a pore 1 μm in diameter will be only partly infiltrated under the same conditions. Therefore, it clearly appears that the infiltration of pores of very small diameters (e.g., 1 μm) require experimental conditions that involve very low deposition rates. As a

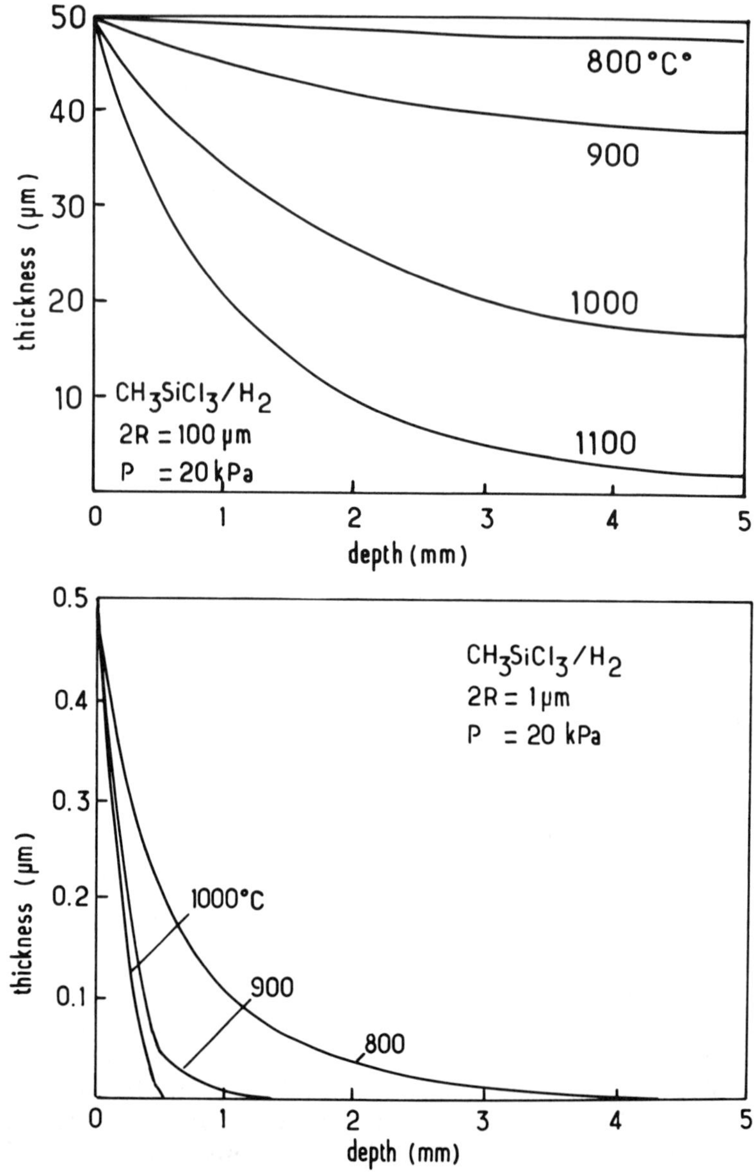

Fig. 3. Calculated thickness profiles of SiC deposited from MTS/H_2 in a cylindrical straight pore: influence of temperature and pore diameter.

result, the time necessary to achieve a significant degree of pore filling could be rather long in isothermal/isobaric CVI.

The infiltration rate of a pore (open at both ends) is highly increased when a pressure gradient is applied along the pore axis (Fig. 5). Under such conditions, the mass transport in the gas phase, which is now mainly by forced convection, is much more rapid (*26,28*). Moreover, a temperature gradient could be added along the pore. A higher temperature at the pore exit will result in faster deposition kinetics and thus compensate the effect of the depletion of the gas phase in source species

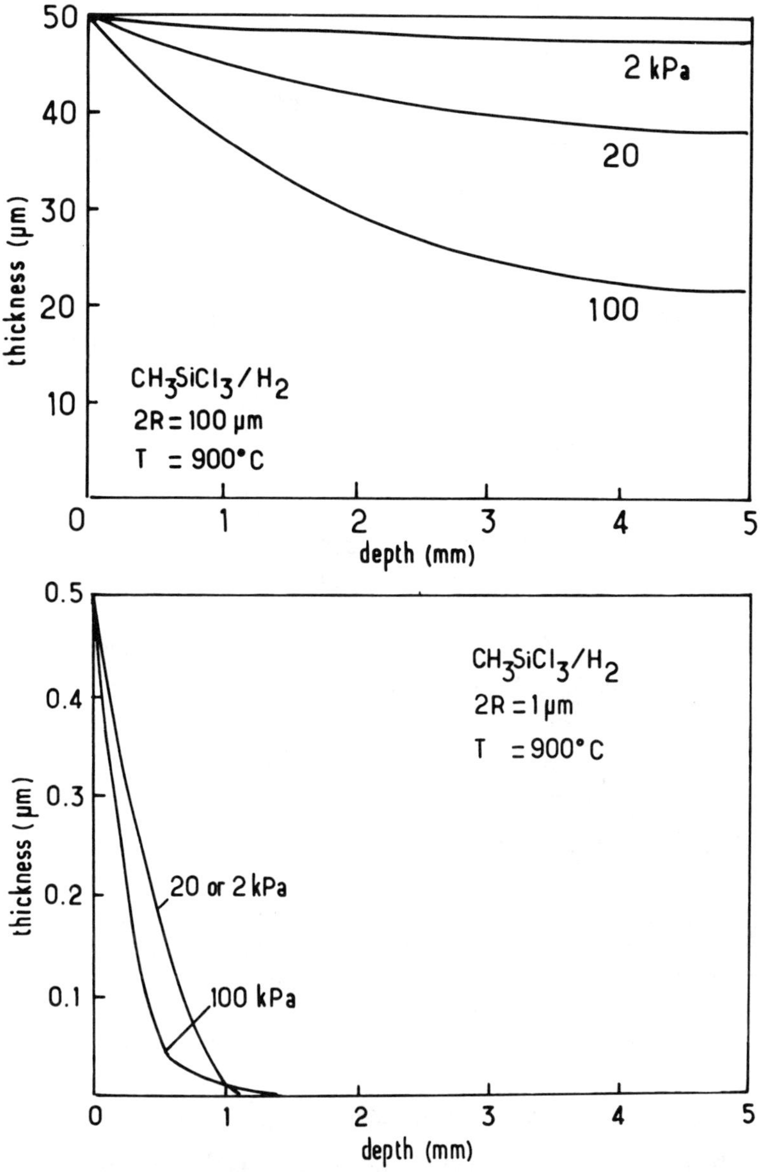

Fig. 4. Calculated thickness profiles of SiC deposited from MTS/H_2 in a cylindrical straight pore: influence of total pressure and pore diameter.

due to the chemical reaction along the pore. A model has been proposed, by Starr for the FCVI of a porous preform made of short fibers, by SiC deposited from MTS/H_2 mixtures. The calculations suggest that fine tuning of the infiltration process could be possible by controlling the cool face temperature of the substrate and the gas flowrate (*29*). Although this so-called Forced-CVI (FCVI) process is attractive, from a theoretical point of view, its use for the densification of substrates of large size and complex shape is not straightforward from an industrial standpoint, as discussed in the next section.

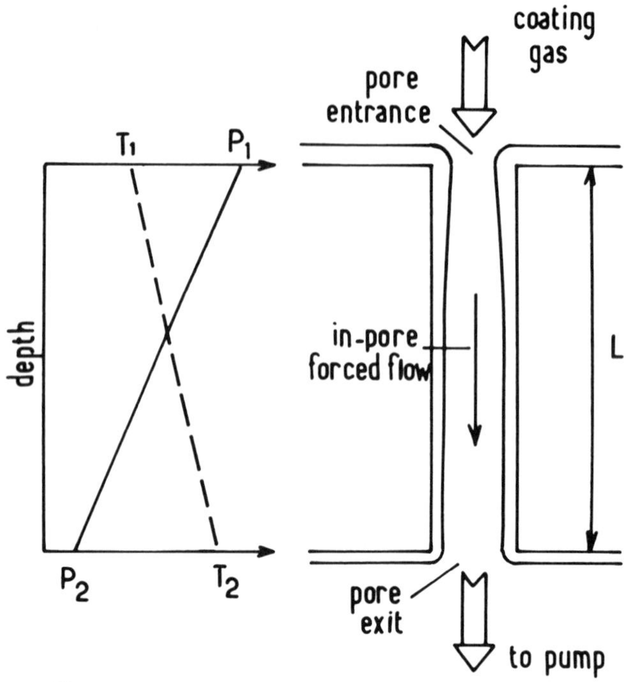

Fig. 5. Mechanisms involved in forced-CVI.

PRACTICAL ASPECTS OF THE CVI OF POROUS SUBSTRATES

Up to now, most CVI lab-scale experiments and plant-scale production have been performed according to the isothermal/isobaric process. The practical aspects of the process will be reviewed with reference to the elaboration of CMC for aerospace applications. Finally, the FCVI-process will be briefly described.

The Isothermal/Isobaric CVI-Process

As shown in Fig. 6, in isothermal/isobaric CVI, the porous substrates are stacked in a hot-wall deposition chamber (e.g., heated with a r.f. coil) fed with the gaseous precursor of the ceramic matrix. The pressure in the CVI-chamber is maintained at a constant value by pumping the exhaust gas. When the source species or the species resulting from the deposition reaction are corrosive (which is usually the case), they must be separated physically or chemically. Regarding the long times which are necessary to achieve high degrees of densification and the narrowness of the deposition conditions, all the CVI-parameters should be carefully controlled (otherwise, an eventual slight rise of either temperature or pressure will result, in a very short time, in a sealing of the pores).

The fiber preforms, which are used as starting materials, are characterized by a variety of fiber architectures, overall open porosities (V_{po})

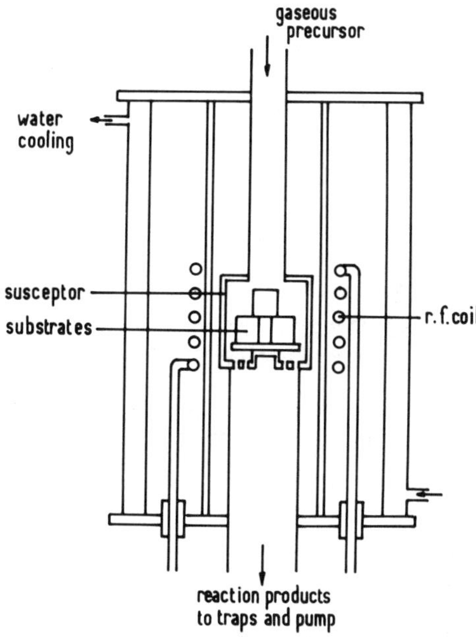

Fig. 6. The isothermal/isobaric CVI-process: the deposition chamber.

and pore size spectra. The most commonly used are made from a stack of ceramic fabrics (e.g., carbon, silicon carbide, or alumina-based fibers) consolidated with a small amount of a ceramic binder (according to a vapor or liquid routes). The overall-porosity of such preforms is of the order of 40% with pore diameters ranging from 10 to 500 μm (*6,8,20*).

As shown in Fig. 7a, when the CVI is performed at high T, P, the ceramic deposition, mainly occurs on the external surface of the substrate and the vicinity of the entrances of the pores which, as a result, become very rapidly sealed. On the contrary, when both T and P are simultaneously lowered, the deposition of the ceramic matrix takes place within the whole pore network, the entrances of the pores remaining open (Fig. 7b) (*21*). These features are qualitatively in agreement with the profiles calculated for a single straight pore, in section 2, for different T and P.

The densification rate of a given porous substrate, infiltrated under given CVI-conditions (i.e., fixed T, P, D), progressively decreases versus time as the residual porosity decreases (Fig. 8a). This feature is due to the fact that (i) the mass transport of the reactant and product species along the pores becomes more and more difficult as the pore size decreases, and (ii) the overall inner surface of the pore wall simultaneously decreases. Therefore, the infiltration rate is very low when coming close to the end of densification with the result that the full densification of a fiber preform (V_{po} = 40%) may require several hundreds of hours. Practically, the densification process is stopped when the residual porosity has fallen to about 10–15%. Furthermore, a surface machining may be applied to the substrates, if necessary, to reopen the pores that could have been

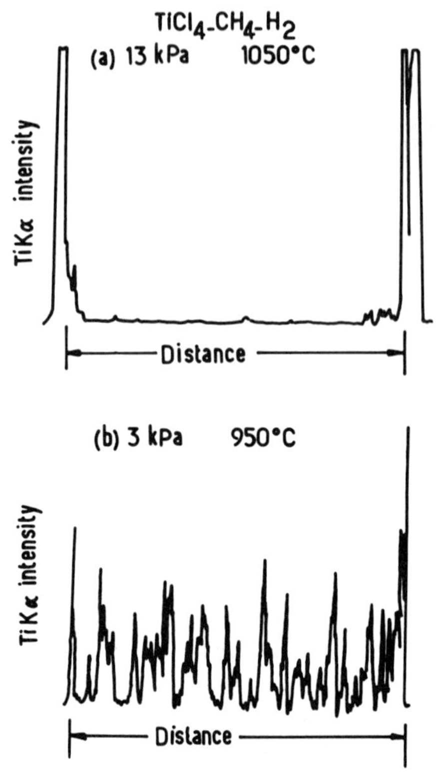

Fig. 7. Distribution profiles of a ceramic (i.e., TiC) deposited by CVD/CVI on/in a porous substrate: (a) at high T, P (CVD) and (b) at low TIP (CVI). Analyses performed with an X-ray electron microprobe (TiKα) on cross-sections, from (*21*).

sealed during the CVI-treatment (as discussed above, the thickness of the deposit is always higher near the pore entrance). It has been experimentally established that, under optimized CVI-conditions, the densification of a porous fiber preform obeys a linear Ln ((Mo-Mp)/Mo) = f(t) law (where Mp is the mass of ceramic deposited within the preform at time t and Mo the theoretical mass that could be deposited if all the initial open porosity of the preform is filled). A deviation from this law indicates that the pores have been sealed due to incorrect CVI-conditions, as shown in Fig. 8b. When properly controlled, the CVI process leads to a progressive filling of the pore network of the substrate and to a residual porosity of a few percents (the pores with a pore size higher than 10 μm that are the majority of the initial porosity being almost totally filled) (Fig. 9).

As mentioned above, the isothermal/isobaric CVI processing of CMC is a slow process and it leads to materials exhibiting some residual open porosity (V_p). Despite these drawbacks, its use in industry has been justified on the basis of the following advantages:

1. It requires rather simple technology and procedure;

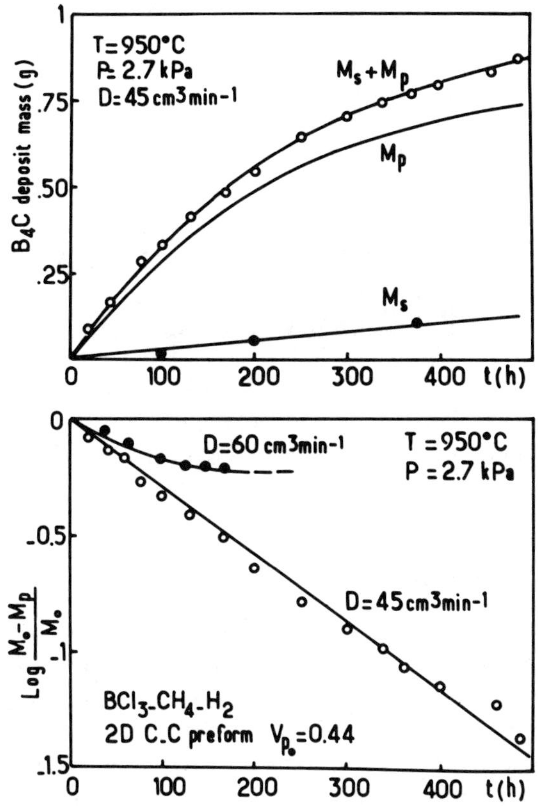

Fig. 8. Kinetics of densification by CVI (i.e., by B_4C) of a porous carbon fiber preform: (a) total (Ms + Mp), surface (Ms), and in-pore (Mp) depositions vs time, (b) identification of optimized CVI conditions (for D = 45 cm^3 min^{-1}) and pore sealing (for D = 60 cm^3 min^{-1}) from (*13*).

2. It does not lower the reinforcing capability of the fibers in the preform owing to the rather low temperatures at which CVI is performed;
3. It allows an *in situ* optimization of the fiber-matrix interphase (e.g., through the formation of a thin layer of say pyrocarbon or boron nitride) necessary to give the CMC a high toughness (*30*); and
4. A large number of parts of simple or complex shapes can be treated simultaneously (the number of parts being only limited by the size of the deposition chamber) (*11,31–34*).

The process is well suited to thin-wall parts as could be predicted from the theoretical considerations discussed earlier. Although the feasibility of the process has been established for C, SiC, B_4C, TiC, Si_3N_4, BN, Al_2O_3, ZrO_2 (and could be extended to other ceramic matrices), its use in industry is limited up to now and as far as we know to carbon and silicon carbide (*20*).

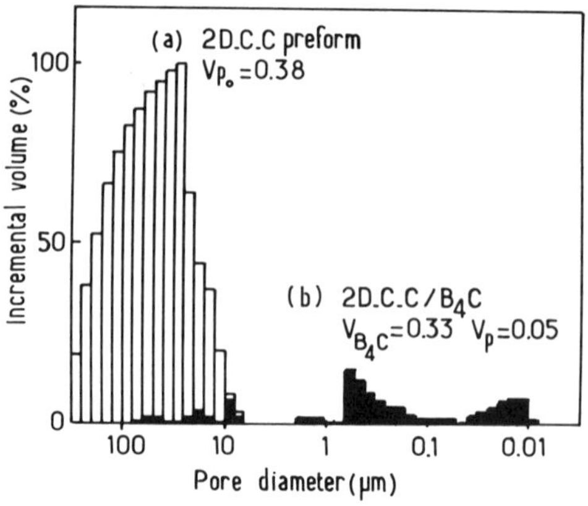

Fig. 9. Pore size spectra of a 2D-C-C fiber preform: (a) before CVI and (b) almost fully densified with B_4C deposited from BCl_3-CH_4-H_2 (in black), from (*13*).

The Forced CVI-Process

The FCVI-process has been designed to increase the rate of deposition (*26–28*). As mentioned earlier, the mass transport of the reactant and product gaseous species along the pore axis is accelerated by applying a pressure gradient in the porous substrate. As shown schematically in Figs. 5 and 10, the feed gas is injected in the pore network at a pressure P_1 through one side of the substrate, whereas the exhaust gas is pumped at a pressure P_2 (with $P_2 < P_1$) at the opposite side. Under such conditions, the mass transport through the pore is mainly by forced-convection, the role played by diffusion being minor.

In order to prevent an early sealing of the pores at the substrate surface through which the feed gas is injected, it has been suggested to superimpose a temperature gradient to the pressure gradient, that surface of the substrate being cooled at a low enough temperature (i.e., at a temperature where the rate of deposition is very slow) either directly (as shown in Fig. 10) or indirectly (*26–28*).

At the beginning of the infiltration process, the deposit occurs preferentially near the hot upper surface of the substrate. If the experiment is run at high temperatures (e.g. 1200°C for SiC deposited from MTS/H_2) to achieve a high deposition rate, the upper part of the pore network becomes rapidly sealed. Furthermore, when the deposit is a good thermal conductor (which is the case for many ceramics including SiC), the hot front moves downwards. Under such conditions, the exhaust gas no longer escapes from the cylindrical substrate through its upper surface but radially (since the pore entrances on the lateral cooled surface of the substrate remain open) (Fig. 10). Therefore, by combining axial and radial forced flows of the feed gas through the substrate, a high degree of densification can be achieved in a short time. Durations as low as 20–40 h (i.e., corresponding to an improvement of one order of magnitude) have been reported for the infiltration of SiC fiber preforms (V_{po} = 40–60%),

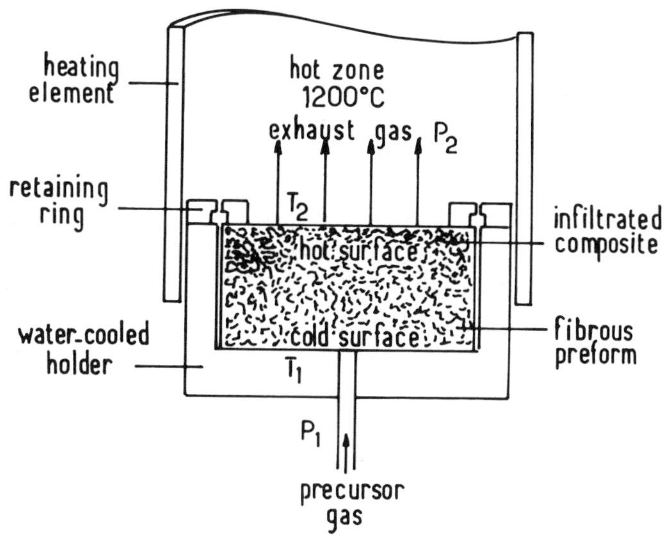

Fig. 10. Schematic of forced-CVI, adapted from (*28*).

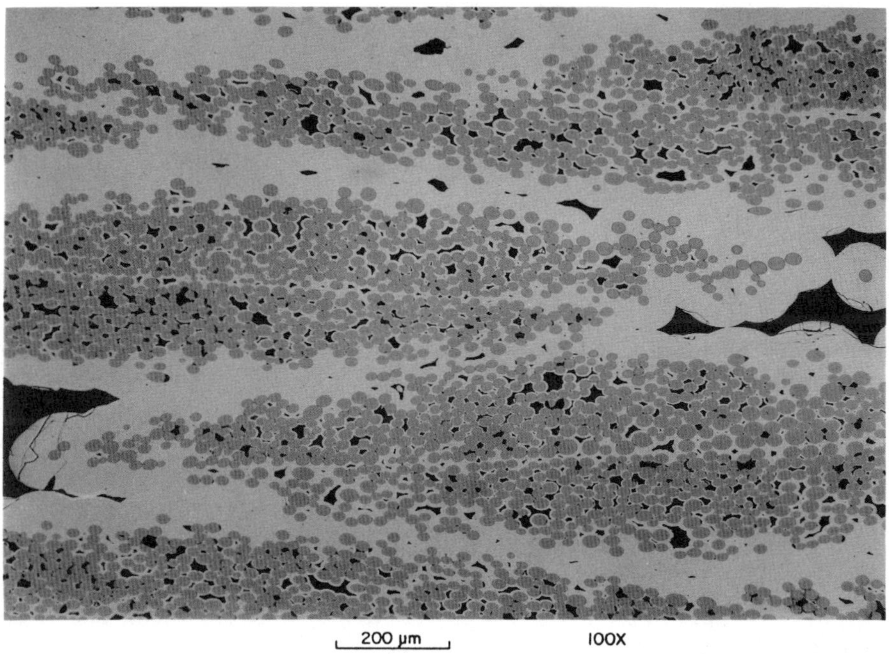

Fig. 11. Micrograph of a 2D-SiC-SiC composites obtained by FCVI of a porous 2D-SiC (Nicalon) preform, from (*26*).

by either SiC or Si_3N_4, resulting in CMC whose densities were 65–95% the theoretical densities (*28*).

As shown in Fig. 11 the ceramic is uniformly deposited on the pore wall and the residual porosity regularly distributed. As a matter of fact and as far as we know, the mechanical properties of the CMC measured on lab-scale samples prepared by FCVI are close to those reported for similar samples obtained by conventional CVI (*26,27,33,34*).

Although the FCVI process significantly improves the infiltration rate, its practical use may be limited to specific applications (e.g., CMC pipes) owing to tooling requirements. It is apparent from Fig. 10 that even in the simple case of one cylindrical substrate, a specific tooling is necessary to generate the pressure-temperature gradients. This problem will be still more difficult to solve for preforms with complex shapes. Finally, it seems doubtful that FCVI could be used to infiltrate simultaneously a large number of substrates of different shapes. It thus appears that the isothermal/isobaric CVI and FCVI processes have their own drawbacks and advantages and could be used in a complementary manner.

ACKNOWLEDGMENT

The authors acknowledge the assistance of R. Fedou in the adaptation of the infiltration model to the deposition of SiC from MTS/H_2, in a straight pore.

REFERENCES

1. Christin, F., Fr. Pat. 77.26979, September, 1977.
2. Christin, F., Naslain, R., and Bernard, C., Proc. 7th Int. Conf. CVD, Los Angeles, Sedwick, T. O. and Lydtin, H., eds., The Electrochem. Soc., Princeton, 1979, p. 499.
3. Fitzer, E. and Hegen, D., *Angew Chem. Int. Ed. Engl.* **18,** 295 (1979).
4. Fitzer, E., Hegen, D., and Strohmeier, H., *Rev. Int. Hautes Tempér. Réfract.* **17,** 23 (1980).
5. Christin, F., Heraud, L., Choury, J. J., Naslain, R., and Hagenmuller, P., *Proc. Euro-CVD-III,* Hintermann, H., ed., LSRH, Neuchâtel (CH), 1980, p. 154.
6. Naslain, R., Hagenmuller, P., Christin, F., Heraud, L., and Choury, J. J., *Proc. ICCM-3,* Paris, Advances in Composite Materials, Bunsell, A. R. et al., eds., Pergamon, 1980, vol. 2, p. 1087.
7. Heraud, L., Christin, F., Naslain, R., and Hagenmuller, P., *Proc. 8th. Int. Conf. CVD,* Paris, Blocher, J. M. et al., eds., The Electrochem. Soc., Pennington, NJ, 1981, p. 782.
8. Naslain, R., Rossignol, J. Y., Hagenmuller, P., Christin, F., Heraud, L., and Choury, J. J., *Rev. Chimie Minérale* **18,** 544 (1981).
9. Fitzer, E., Fritz, W., and Gadow, R., *Proc. Symp. Adv. Ceramic Mater.* Tokyo Inst. Technology, Yokohama (1983): *Advances in Ceramics,* Somiya, S., ed., KTK Scientific, Tokyo, 1985.
10. Fitzer, E. and Gadow, R., *Ceram. Bull.* **65/2,** 326 (1986).
11. Reagan, P., Ross, M. F., and Huffman, F. N., *Adv. Ceram. Mater.* **3/3,** 198 (1988).
12. Rossignol, J. Y., Naslain, R., Hagenmuller, P., and Heraud, L., *Proc. ICCM-IV,* Progress in Science and Engineering of Composites, Hayashi, T. et al., eds., Tokyo, 1982, p. 1227.

13. Hannache, H., Thesis #813, Univ. Bordeaux, Sept. 26, 1984.
14. Hannache, H., Langlais, F., and Naslain, R., *Proc. Euro CVD-5*, Carlsson, J. O. and Lindström, J., eds., Uppsala (5), 1985, p. 219.
15. Hannache, H., Naslain, R., and Bernard, C., *J. Less Common Met.* **95,** 221 (1983).
16. Hannache, H., Quenisset, J. M., Naslain, R., and Heraud, L., *J. Mater. Sci.* **19,** 202 (1984).
17. Pierson, H. O. *J. Composite Mater.* **9,** 228 (1975).
18. Colmet, R., Lhermitte-Sebire, I., and Naslain, R., *Adv. Ceram. Mater.* **1/2,** 185 (1986).
19. Minet, J., Langlais, F., and Naslain, R., *Composites Science and Technology*, accepted for publication, 1989.
20. Langlais, F. and Naslain, R., *Tailoring Multiphase and Composite Ceramics*, Tressler, T. E. et al., eds.; *Mater. Sci. Res.*, Plenum, **20,** 145 (1986).
21. Rossignol, J. Y., Langlais, F., and Naslain, R., *Proc. 9th. Int. Conf. CVD*, Cincinnati, Mc. Robinson, D. et al., eds., The Electrochem. Soc., Pennington, NJ, 1984, p. 596.
22. Van der Breckel, C. H. J., Fonville, R. M. M., Van der Straten, P. J. M., and Verspui, G., *Proc. 8th Int. Conf. CVD*, Paris, Blocher, J. M., et al., eds., The Electrochem. Soc., Pennington, NJ, 1981, p. 142.
23. Rossignol, J. Y., Thesis 833, Univ. Bordeaux, March 18, 1985.
24. Bird, R. B., Stewart, W. E., and Lightfoot, *E. N. Transport Phenomena*, Wiley, ed., New York (1960).
25. Schoch, G., Fritz, W., and Fitzer, E. N., private communication.
26. Caputo, A. J., Lowden, R. A., and Stinton, D. P., ORNL/TM-9651, Oak Ridge Nat. Lab., June 1985 available NTIS, Springfield, VA.
27. Caputo, A. J., Lackey, W. J., and Stinton, D. P., *Proc. 9th Annual Conf. Composites and Adv. Ceram. Mat.*, Cocoa Beach, FL, Ceramic Eng. and Sc., the Amer. Ceram. Soc., July–August 1985, p. 694.
28. Stinton, D. P., Caputo, A. J., and Lowden, R. A., *Ceram. Bull.* **65/2**, 347 (1986).
29. Starr, T. L., *Proc. 10th Int. Conference*, CVD Honolulu, G. W. Cullen, ed. 1987, p. 1147.
30. Naslain, R., *J. de Physique*, Colloque C1, Supl. 2, **47,** C1-703 (1986).
31. Fitzer, E., *J. Chimie Phys.* **81** (11/12) 717 (1984).
32. Lamicq, P. L., *J. Chimie Phys.* **81** (14/12) 735 (1984).
33. Bernhart, G., Lamicq, P. L., and Mace, J., *L'industrie Ceramique,* **790,** 1 (1985).
34. Dauchier, M., Bernhart, G., and Bonnet, C., *Proc. 30th Natl. SAMPE Symp.*, March 19–21, 1985, p. 1519.

Laser-Induced Chemical Vapor Deposition of High-Purity Metals

THOMAS H. BAUM* AND CARL E. LARSON

IBM Almaden Research Center, 650 Harry Road, San Jose, CA 95120-6099

ABSTRACT

The laser chemical vapor deposition (LCVD) of aluminum and gold has been achieved via pyrolytic decomposition of trimethylamine aluminum hydride and dimethylgold (β-diketonates), respectively. The rates of metal deposition, the purity of the deposited metals, and the electrical properties of the deposits are a direct result of the chemical and physical properties of the precursor and the metal.

Index Entries: Chemical vapor deposition; laser; metals; high-purity; precursors; thermal.

INTRODUCTION

Laser-induced chemical vapor deposition (LCVD) is an attractive technique for the selective, direct writing of micron-sized metal features. The ability of this method to repair clear defects in lithographic masks (*1*), repair or interconnect conducting circuitry (*2*), and 'personalize' integrated circuits (*3*) is of great practical significance. The testing of early engineering designs and the customization of limited-volume microelectronic components may provide a cost-effective alternative to conventional lithographic processes.

LCVD, also commonly known as direct writing, relies upon the localized pyrolytic decomposition of precursor molecules adsorbed at and/or colliding with the laser-heated surface. The reaction zone is de-

*Author to whom all correspondence and reprint requests should be addressed.

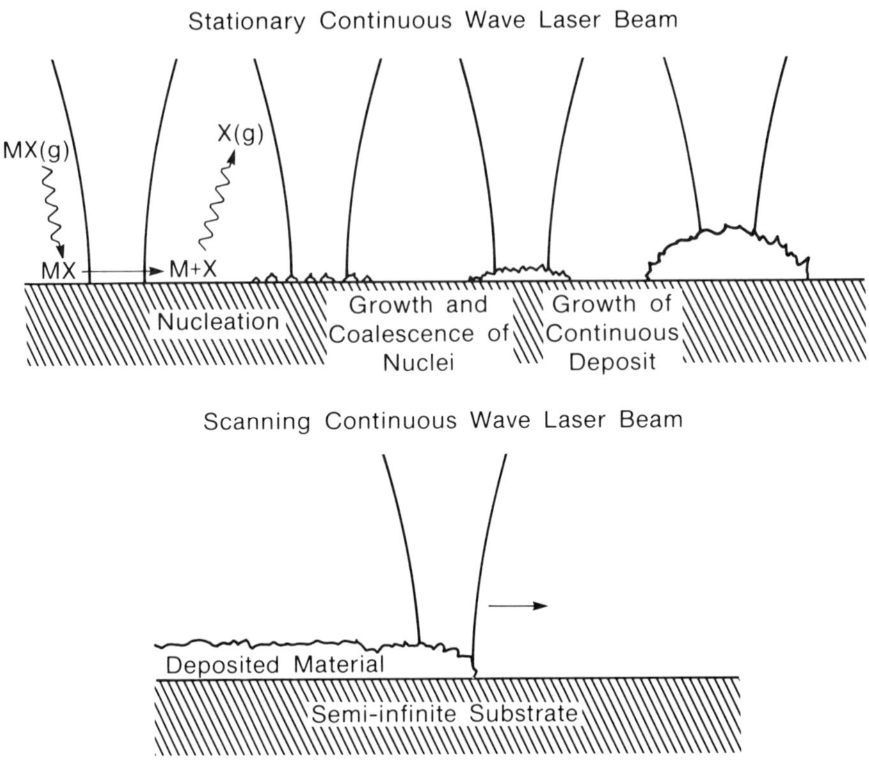

Fig. 1. Schematic diagram of the growth processes that occur during the laser-induced chemical vapor deposition of metals onto surfaces. The first case is for a stationary beam (or substrate) that incorporates surface nucleation, coalescence of nuclei, and steady-state growth of a continuous deposit. The second case illustrates the scanning of the laser beam (or substrate) to create conducting lines of metal. The same growth processes are occurring simultaneously for the scanning case.

fined by the diameter of the focused laser spot, the optical and thermal properties of the substrate, and the metal being deposited. Surface temperatures can be calculated for a given laser power after considering the aforementioned variables. In cases when the substrate consists of two or more layers, the surface temperature profile can be difficult to calculate and may vary by several hundred degrees once the deposition of metal is initiated (*4*). For a stationary beam deposition, the laser-induced deposition of metal can be divided into distinct growth regimes: surface nucleation, coalescence, and steady-state growth, as depicted in Fig. 1. The majority of the metal is deposited during the steady-state growth phase. During the direct writing of a metal line, all of these growth processes are occurring simultaneously.

An understanding of the physical processes and the chemical reactions occurring during heterogeneous decomposition of the precursor on the substrate and metal is necessary for a complete description of the LCVD process. For LCVD to be generally applicable, stable metal precursors that decompose to high-purity metals are a necessity. The precursor molecular properties play a critical role in the LCVD process, because they determine the vapor pressure, the adsorption/desorption isotherms, the decomposition temperature, the chemical reaction pathways, the purity of the deposit, and quite often, the rate of thin-film formation. Also, the stability and ease of precursor handling are important considerations for LCVD. All of these combined chemical and physical properties are a direct result of the precursor's chemical structure.

The best precursors for LCVD will perform well with regard to the criteria mentioned above. The clean decomposition of the precursor must occur over a wide range of temperatures, because the surface temperature is continuously changing as the surface is modified by LCVD. The formation of stable decomposition products that are removed by a low-energy pathway from the growing deposit is critical to achieving high purity. Thus, a basic understanding of decomposition mechanisms for organometallic precursors can stimulate the synthesis of precursors that are 'tailored' to produce high-purity metallic films. In some cases, new precursors must be synthesized to enable the deposition of a specific metal, increase the vapor pressure of the precursor, or just improve its ease of handling. Nevertheless, the chemical and physical properties of the precursor must be compatible with the LCVD process to achieve the desired goals of surface selectivity, high purity, and rapid rates of thin-film formation.

The importance of the precursor chemical structure is exemplified by the recent study of triisobutyl aluminum (TIBA) thermal decomposition (*5*). A well-established alkyl reaction mechanism in TIBA, namely β-hydride elimination, results in surface-adsorbed aluminum hydrides with the subsequent formation of stable reaction products (i.e., isobutene and hydrogen). Thus, a low energy pathway to ligand reaction facilitates the removal of the carbon-containing species as stable reaction products and high-purity aluminum is deposited. In contrast, other trialkyl aluminum complexes (i.e., trimethyl) decompose to carbon-contaminated films owing to the formation of reactive alkyl radicals (*6*) that cannot be easily removed from the growing deposit.

In this paper, dimethylgold β-diketonates (complexes 2a, b, and c) and trimethylamine aluminum hydride (complex 3), also known as trimethylamine alane, are used as precursors for the laser-induced pyrolytic deposition of high-purity gold and aluminum films, respectively. The gold complexes can be chemically substituted to increase the ambient vapor pressure and improve the observed rates of gold film formation. With the aluminum complex, the hydride structure results in a high ambient vapor pressure and affords high rates of high-purity aluminum thin-film formation. The chemical and physical properties of these precursors make them ideal candidates for the direct writing of high-purity metal films.

METHODS

The synthesis of the dimethylgold β-diketonate complexes was achieved by the literature procedure of Brain and Gibson (*7*). A second

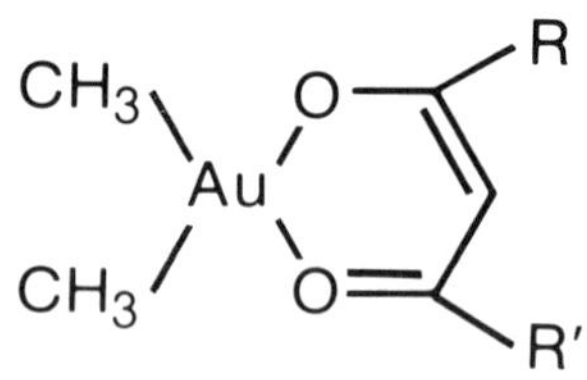

Complex	Structure	T_m(°C)	P(mT)
2a	R=R'=CH3	82	9
2b	R=CH3 R'=CF3	40	40
2c	R=R'=CF3	5	350

Fig. 2. The general molecular structure and chemical and physical properties of the dimethylgold β-diketonate complexes. CF_3 substitution into the β-diketonate ligand results in structures 2b and 2c, which exhibit increased ambient vapor pressures and undergo pyrolytic decomposition to high-purity gold.

simplified synthetic procedure was identified (*8*) and relies upon the reaction of trimethylgold (*9*) with the desired β-diketone at −78°C. Crude complexes 2a and b were purified by recrystallization and sublimation to obtain white, crystalline solids and complex 2c was distilled as a faintly yellow-colored liquid (*10*). The molecular structure, chemical, and physical properties of these materials have been reported (*11*) and are given in Fig. 2. An electron diffraction study has determined the molecular structure of 2a in the gas-phase (*12*). The synthesis of trimethylamine aluminum hydride complex (Figure 3) was achieved by reported procedures (*13*). The isolated crude product was purified by sublimation at 40°C (30 μm HG) to obtain a white crystalline solid. The ambient vapor pressure of the purified product is approximately 2 torr at 25°C. This measurement is consistent with that reported for the 2:1 adduct of trimethylamine to aluminum hydride (*14*). Although structure 3 is air and water sensitive, it is not pyrophoric, in significant contrast to the trialkyl aluminum precursors. The purified material has a melting point of 75°C and is thermally decomposed to aluminum metal below 300°C.

Laser-induced deposition was carried out using an argon ion laser operating at 5145 Å in the TEM_{00} mode. The beam was expanded (3×) and collimated with a telescope, then focused through a microscope objective (10X, 0.2 NA). The measured (*15*) beam diameter is approximately 9 ± 1.0 μm at the 1/e intensity points and the beam was nearly Gaussian. In the laser deposition experiments, a fresh sample of the precursor was placed in a stainless steel ampule. After several freeze-pump-thaw cycles, the sample was opened to a vacuum cell that previously had been evacuated to 10^{-5} torr. The system was allowed to

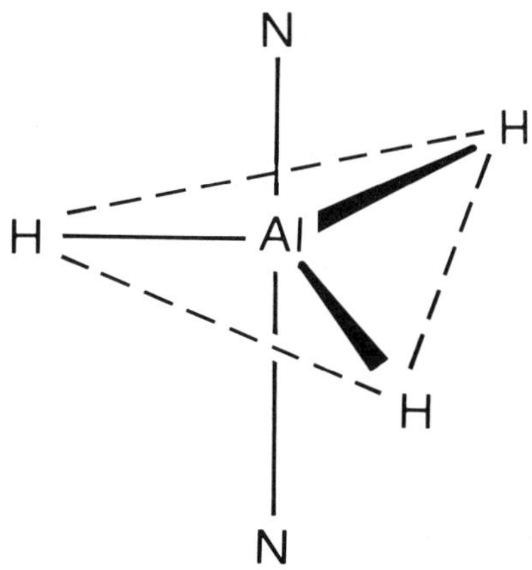

Fig. 3. Molecular structure of the trimethylamine aluminum hydride complex (2:1) used to laser deposit high-purity aluminum. The methyl groups on the trimethylamine ligands are not depicted. The 2:1 complex was spectroscopically shown to have D_{3h} symmetry. (ref. 25).

equilibrate for 10–15 min prior to any deposition experiments. The substrates used for deposition consisted of silicon wafers with a thermally grown oxide layer (0.2 μm thickness). In several cases, metal lines were deposited across Cr/Au pads of 0.2 μm thickness to measure the resistance of the laser-deposited lines. A four-point electrical probe was used to measure the resistance, and the resistivity was calculated after measuring the thickness and width by stylus profilometry. Several lines were examined by scanning Auger electron spectroscopy (SAES) (Surface Science Labs in Mt. View, CA). The lines were also examined by SEM using a Philips 505 microscope with computer enhanced imaging.

DISCUSSION

A precursor's vapor pressure may be increased by altering its chemical structure or by external heating. For practical applications, the chemical synthesis approach offers several advantages. Most importantly, the design and engineering of elaborate, thermally-controlled vacuum chambers can be completely avoided, and this greatly simplifies the LCVD equipment. It is well-known that fluorine-substituted organics have higher vapor pressures than their nonfluorinated analogs (*16*). With this in mind, CF_3 substitution of the β-diketone ligand in complex 2 (b and c) results in an increased vapor pressure of the precursor under ambient conditions (*11*). Thus, a slight chemical modification of the gold precursor increases the volatility and enables faster rates of gold film formation, as discussed below.

The nucleation of laser-induced metallization, a crucial first step in direct writing, is often slow and quite substrate-dependent. It was experimentally observed that the nucleation time was substantially reduced when the vapor pressure of the precursor was increased. Thus, in going from complex 2a to 2c, under identical laser and substrate conditions, the time of nucleation was shortened. It has also been observed that the nucleation time decreases with increasing laser power for a given precursor. The surface coverage and initial temperature rise appear to be critical parameters to the surface nucleation time. It may be concluded that the rate of surface nucleation is proportional to both surface concentration and laser power. The surface nucleation process during direct laser writing has recently been modeled (*17*).

Generally, laser-induced rates of deposition are governed by the rate of surface chemical reactions and/or by the mass-transport of material to the locally-heated surface. For slow surface chemical reactions during deposition, the slowest chemical reaction step becomes rate-limiting. If the surface chemical reaction is sufficiently fast, then the mass transport of material to the reaction zone is rate-limiting. The use of the gold complexes 2a–c for the laser-induced deposition of gold has been previously reported (*11,18*). A detailed study of the kinetics of gold deposition from 2c demonstrated that the surface chemical reactions were fast and the overall rates were mass-transport limited (*19*). Experimentally-measured rates of gold deposition were directly proportional to the precursor vapor pressure, and rates up to 5 μm/s were achieved. Thus, the greater the volatility of the precursor, the more rapid the rate of gold deposition.

Pyrolytic decomposition of the dimethylgold β-diketonates results in high-purity (≥95 at %) gold deposits. The thermal dissociation of these precursors yields stable reaction products via reductive elimination. A detailed study (*20*) of the decomposition pathways of 2a in solution demonstrated that concerted reductive elimination of ethane and the rate-determining liberation of 2,4-pentanedione occurred, as shown in Eq 1. An alternate chemical pathway for removal of reactive carbon-containing ligands (i.e., radical species) was the production of 2-methyl-2,4-pentanedione via radical-radical recombination (Eq. 2). Although one must be careful when comparing condensed-phase and gas-phase reactions, a separate study (*21*) indicates that similar decomposition pathways are observed during LCVD for 2c. Thus, specific chemical reactions during decomposition of the precursor allow the carbon-containing ligands to be removed via low-energy reaction pathways and result in high-purity gold deposits.

$$(CH_3)_2Au(C_5H_7O_2) \xrightarrow{\Delta} Au + C_2H_6 + C_5H_8O_2 \quad (1)$$

$$\bullet CH_3 + \bullet C_5H_7O_2 \rightarrow C_6H_{10}O_2 \quad (2)$$

Polycrystalline gold films have been formed via LCVD of dimethylgold β-diketonates in the absence of buffer gases (*17*). It was later shown that the specific geometry and structure of the gold deposits depends strongly upon the conditions during laser-induced growth. In the presence of buffer gas, thermal decomposition of 2c produced gold crystallites that varied from nodular to dendritic to large, oriented crys-

Fig. 4. Large crystal of gold metal formed via the laser-induced decomposition of 2c in air.

tals (*22*). The laser power, buffer gas, and buffer gas partial pressure are primarily responsible for the changes in deposit shape and size. Partial pressures of O_2 and air resulted in large, crystalline deposits of gold, as seen in Fig. 4. Inert buffer gases (i.e., argon) resulted in dendritic gold structures (Fig. 5). This strong gold crystal dependence with buffer gas type and pressure is indicative of mass-transport controlled growth, as reported by others (*23*). However, chemical reaction with the buffer gas during LCVD may also influence the shape of the deposited material.

Lastly, the dimethylgold β-diketonate complexes are stable, non-pyrophoric, easy to handle, and undergo thermal decomposition at relatively low temperatures (160°C). The low decomposition temperature enables gold LCVD to be performed on thermally sensitive substrates, such as polyimide (Fig. 6). The combined chemical and physical properties make 2a–c ideal precursors for the laser-induced chemical vapor deposition of gold. The rate of gold film formation can be controlled by the choice of the precursor to fit a specific LCVD application.

Another example of a precursor that leads to high-purity metal upon decomposition is 3, which was first prepared by Wiberg and coworkers in 1942 (*24*). The 1:1 and 2:1 adducts of trimethylamine to aluminum hydride have been synthesized and found to exist in a reversible equilibrium above the solid of the 2:1 complex (Eq. 3) (*14*). Early on, it was believed that 3 was a dimeric species (*13*), but it was later spectroscopically shown to be a monomer of C_{3v} symmetry for the 1:1 adduct and D_{3h} symmetry for the 2:1 adduct (*25*), as depicted in Fig. 3. The

Fig. 5. Dendritic structure of gold deposited by laser-induced decomposition of 2c in the presence of argon buffer gas.

dissociation of one trimethylamine ligand can occur rapidly in high vacuum or room temperature; an enthalphy for dissociation (ΔH_d) was determined to be -18.0 kcal/mol (*14*). Liberation of the first trimethylamine ligand results in formation of the 1:1 adduct of trimethylamine to aluminum hydride (Eq. 3). Further decomposition results in irreversible liberation of the second trimethylamine ligand, forming the coordinatively unsaturated and highly reactive aluminum trihydride. Thus, thermal decomposition of 3 to produce aluminum metal is believed to proceed, as depicted in the following chemical reactions

$$AlH_3(N(Me)_3)_2 \rightleftarrows AlH_3N(Me)_3 + N(Me)_3 \quad (3)$$

$$AlH_3N(Me)_3 \rightarrow AlH_3 + N(Me)_3 \rightarrow Al + \frac{3}{2} H_2 \quad (4)$$

As discussed above, rates of metal film formation are often proportional to the precursor vapor pressure. Precursor 3 has a measured ambient vapor pressure of 2 torr. Rapid rates of aluminum deposition and rapid scan velocities can be realized during LCVD from 3 when compared to LCVD from 2 and suggest that the growth rate is limited by gas-phase mass transport. Future experimental studies will examine the kinetics of aluminum deposition from the decomposition of 3 in greater detail.

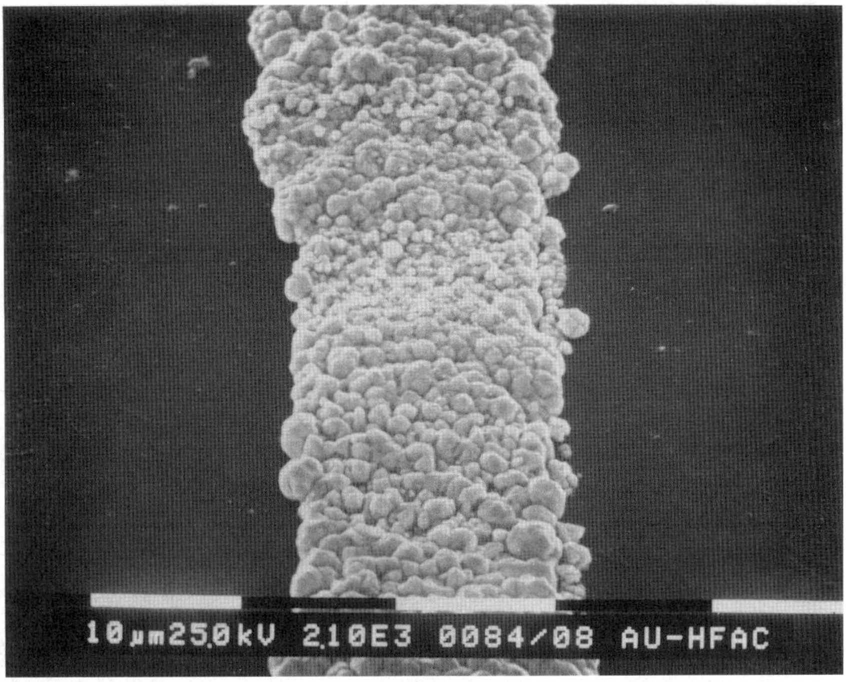

Fig. 6. A gold line was laser-deposited from 2c onto a polyimide substrate. The well-defined line, roughly 17 µm wide and 3 µm thick, is deposited onto the polymer film without any thermal damage.

The purity of the laser-deposited aluminum was examined by SAES. Three aluminum lines that had apparent resistivities from 3 to 5 times bulk aluminum were analyzed (*see below*). For all three samples, the surface was heavily oxidized (Al_2O_3), as would be expected after exposure to air prior to the analysis. Argon ion sputtering to a depth of 200 Å revealed relatively clean aluminum with traces of oxygen as the only impurity. Based upon the absence of carbon and nitrogen emissions in the SAES, no impurities from 3 are present in the deposited film. In the spectra, an oxidized silicon peak (1606 eV) is present from the SiO_2 substrate and accounts for some of the observed oxygen content. Oxygen was the only impurity observed and results strictly from environmental exposure. After accounting for the substrate contribution to the oxygen content, the purity of aluminum is 95–97 at. % with 3–5 at. % of oxygen being present at 200 Å depth. The Auger data are consistent for the three aluminum lines that were analyzed. The analyzed lines were deposited under varying conditions of laser power and scan velocity; thus, pyrolytic decomposition of 3 results in relatively high purity aluminum under a variety of deposition conditions.

Utilizing the described experimental setup, lines of gold and aluminum metal were deposited by translating the substrate/cell in the focal plane of the laser beam. Once high-purity metals can be laser deposited, the utility of LCVD and the electrical integrity of the laser-deposited lines

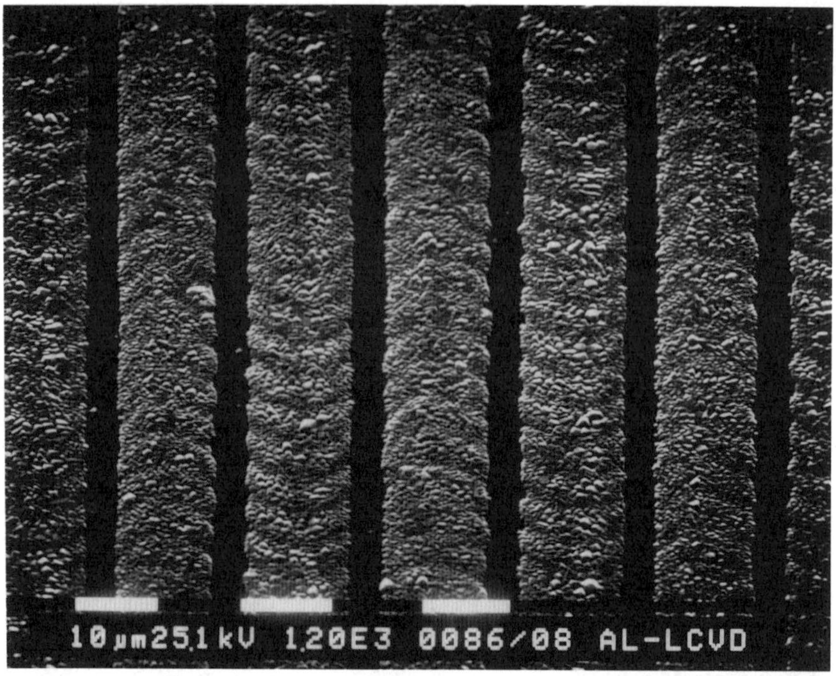

Fig. 7. Portion of a serpentine pattern of aluminum lines deposited onto SiO_2/Si. Each aluminum line is approximately 11 μm wide and 0.7 μm thick with a spacing of 3 μm and was written at 50 μm/s scan velocity. A clear periodic pattern is evident and coincides with the direction of the laser scan.

may be influenced by the ability to sharply define the metal features on the surface. As can be seen in Figs. 6 and 7, well-defined lines are produced from both the gold and aluminum precursors. The direct writing of an aluminum serpentine pattern (Fig. 7) exhibits the excellent spatial definition of the LCVD process. This exhibited resolution of laser-deposited surface structures, which results from pyrolytic decomposition of both 2 and 3, is a critical feature of the precursor decomposition during LCVD . The substrate properties and the precursor's activation energy for decomposition are intimately related to achieving high resolution of the laser-deposited features on a substrate. Thus, the activation energy of a precursor used for LCVD must be high enough to afford well-defined features on the surface, but low so as to enable deposition onto thermally-sensitive materials (i.e., polymers).

For practical applications, the electrical properties of the laser-deposited thin-films are of critical importance. The resistivity of laser-deposited metal features was determined by 4-point electrical probe measurements. The resistance of the metal line incorporates the resistance of the line, as well as the resistance associated with the contact between the laser-deposited line and the existing metallurgical pad. For the resistances reported here, the contact resistances are very low and the major contributor to resistance is the laser-deposited line. For gold deposition from 2a–c, the resistivities have been shown to range from 3–10 times that of bulk gold (2.4 μΩ-cm). Since the purity of the deposit is high, the morphological structure of the film appears to be responsible, in part, for the observed resistance. Thermal annealing was successfully used to reduce resistivities of laser-deposited gold lines (*18*). A decreased defect density results from densification of the deposit upon thermal annealing as evidenced by transmission electron microscopy.

The resistivity of aluminum lines was examined as a function of the 'writing' velocity for a constant laser power. Resistivities for several aluminum lines were consistently 2.5–4 times that of bulk aluminum (2.8 $\mu\Omega$-cm) and independent of the scan velocity from 5 up to 150 μm/s. The thickness of the deposited lines was found to decrease as the scan velocity increased from 5 to 150 μm/s. The resistivities were found not to be a function of the thickness of the deposited lines for thickness of 0.8 μm or greater. Also, using a smaller focal spot diameter (2 μm), velocities up to 600 μm/s have been utilized to produce thin films of aluminum (1000–2000 Å) on silicon with 2.0 μm minimum linewidths.

Highly crystalline deposits have been formed over a wide range of laser and scan conditions. The ratio of substrate temperature to melting temperature of the deposited metal have been shown to correlate with the structural growth of thin films (*26*) and often influence film properties, such as electrical resistivity, adhesion, and stress (*27*). The lower melting temperature of aluminum (T_m = 660°C) results in an increased T/T_m ratio during deposition, compared to gold LCVD. Thus, highly crystalline thin-films of aluminum with densely-packed structures are readily formed. As seen in Fig. 7, a periodic pattern is evident in the aluminum films and coincides with the scan direction relative to the laser beam. It is not clear whether this periodic structure results from a cycling of the surface temperature via the changing laser surface interaction during deposition or from the exothermic decomposition of 3. Further studies are necessary to better understand this phenomenon.

Lastly, it should be noted that the precursors utilized for pyrolytic laser-induced decompositions may also be decomposed by other focused energy sources. For example, the localized decomposition of dimethylgold β-diketonates by focused ion beams (*28*) and electron beams (*29*) has been reported. Scanning tunneling microscopes have also been used to induce localized metal depositions (*30*). All of these methods have great potential for submicron applications, but have produced highly-contaminated films thus far. The method of decomposition must be considered when choosing the precursor for these techniques as well. The ion beam method is further complicated by incorporation of the ion beam material into the growing deposited film. Synthetic approaches for metal precursors must account for the deposition technique and the resulting chemical reaction decomposition mechanisms.

SUMMARY

The laser-induced pyrolytic decomposition of dimethylgold β-diketonates and trimethylamine aluminium hydride precursors has enabled the direct writing of high-purity gold and aluminum, respectively. The chemical structure of the precursor is a critical feature of the LCVD process and specific chemical modifications may be used to increase the ambient vapor pressure and the rates of metal deposition. The formation of high-purity films relies upon the efficient removal of stable reaction products via low-energy chemical pathways. The utility of LCVD is only beginning to emerge and is critically dependent upon the availability of volatile precursors that can be selectively decomposed to high-purity metal films.

ACKNOWLEDGMENTS

A special thanks is extended to Carol Jones, Christopher Moylan, Toivo Kodas, and Paul Comita who were involved in some portion of the work reported here. A critical reading of the manuscript was performed by Christopher Moylan and Robert Jackson, resulting in many helpful discussions.

REFERENCES

1. Ehrlich, D. J., Osgood, R. M., Silversmith, D. J., and Deutsch, T. F., *IEEE Electron Device Letters,* **EDL-1,** 101 (1980).
2. Krchnavek, R. R., Cacouris, T., Gilgen, H. H., and Osgood R. M., *Mat. Res. Soc. Sym. Proc.,* Von Gutfield, R. J., Greene, J. E., and Schlossberg, H., eds., p. 179 (1985).
3. McWilliams, B. M., Hermann, I. P., Mitlisky, F., Hyde, R. A., and Wood, L. L., *Appl. Phys. Lett.* **43,** 946 (1983).
4. Kodas, T. T., Baum, T. H., and Comita, P. B., *J. Appl. Phys.* **61,** 2749 (1987).
5. Bent, B. E., Nuzzo, R. G., and Bubois, L. H., *Mat. Res. Soc. Symp. Proc.* **101,** 177 (1987); *J. Vac. Sci. Tech. A* **6,** 1920 (1988); *J. Am. Chem. Soc.,* **111,** 1634 (1989).
6. Squire, D. W., Dulcey, C. S., and Lin, M. C., *Chem Phys. Lett.* **116,** 525 (1985), *J. Vac. Sci. Tech. B* **3,** 1513 (1985); and Higashi, G. S., Rothbert, I. J., and Fleming, C. G., *Chem. Phys. Lett.* **115,** 2771 (1985).
7. Brain, F. H. and Gibson, C. S., *J. Chem. Soc* **762** (1939); and Komiya S., and Koch J. K., *J. Am. Chem. Soc.* **99,** 3695 (1978).
8. Baum, T. H., unpublished results (1985).
9. Gilman, H. and Woods, L. A., *J. Am. Chem. Soc.* **70,** 550 (1948).
10. The materials are commercially available from American Cyanamid under the Cypure CVD precursors label.
11. Baum, T. H., *J. Electrochem. Soc.* **134,** 2616 (1987).
12. Shibata, S., Iijima, K., and Baum, T. H., unpublished results (1989).
13. Ruff, J. K., and Hawthorne, M. F., *J. Am. Chem. Soc.* **82,** 2141 (1960).
14. Heitsch, C. W., *Nature* **195,** 995 (1962).
15. Skinner, D. R., and Whitcher, R. E., *J. Appl. Phys. E* **5,** 237 (1972).
16. Sheppard, W. A. and Sharts, C. M., *Organic Fluorine Chemistry,* W. A. Benjamin, New York (1969); and Reed, T. M., *Fluorine Chemistry,* vol. 5, Academic, New York (1964).
17. Hermann, I. P., Magnotta, F., and Kotecki, D. E., *J. Vac. Sci. Tech. A* **4,** 659 (1986).

18. Baum, T. H., and Jones, C. R., *Appl. Phys. Lett.* **47,** 538 (1985); and *J. Vac. Sci. Tech. B* **4,** 1187 (1986).
19. Kodas, T. T., Baum, T. H., and Comita, P. B., *J. Appl. Phys.* **62,** 281 (1987).
20. Klassen, R. B. and Baum, T. H., Organometallics, in press, (1989).
21. Comita, P. B., and Kodas, T. T., *Appl. Phys. Lett.* **51,** 2059 (1989); and Kodas, T. T. and Comita, P. B., *J. Appl. Phys.,* **65,** 2513 (1989).
22. Kodas, T. T., Baum, T. H., and Comita, P. B., *J. Cryst. Growth* **87,** 378 (1988).
23. Holman, W. R. and Huegal, F. J., *J. Vac. Sci. Tech,* **11,** 701 (1974).
24. Stecher, R. and Wiberg, E., *Ber.* **75,** 2003 (1942).
25. Fraser, G. W., Greenwood, N. N., and Straughan, B. P., *J. Chem. Soc.* **3742** (1963).
26. Thorton, J. A., *J. Vac. Sci. Tech.* **11,** 666 (1974); and *Ann. Rev. Mat. Sci.* **7,** 239 (1977).
27. Neugebauer, C. A., *Handbook of Thin-Film Technology,* Maissel, L. I. and Glang, R., eds., McGraw-Hill, New York, 1970, p. 8.
28. Shedd, G. M., Dubner, A. D., Lezec, H., and Melngalis, J., *Appl. Phys. Lett.* **49,** 1584 (1986); and Kubena, R. L., Stratton, F. P., and Mayer, T. M., *J. Vac. Sci. Tech. B* **6,** 1865 (1988).
29. Koops, H. W. P., Weile, R., Kern, D. P., and Baum, T. H., *J. Vac. Sci. Tech. B* **6,** 477 (1988).
30. McCord, M. A., Kern, D. P., and Chang, T. H. P., *J. Vac. Sci. Tech. B* **6,** 1877 (1988); and Silver, R. M., Ehrichs, E. E., and de Lozanne, A. L., *Appl. Phys. Lett.* **51,** 247 (1987).

Growth of Perovskite Films by Electrochemical Vapor Deposition

U. B. PAL* AND S. C. SINGHAL

Westinghouse R & D Center, 1310 Beulah Road, Pittsburgh, PA 15235

ABSTRACT

In the electrochemical vapor deposition (EVD) process, halides of the cations are volatilized and passed along with hydrogen over one side of a porous substrate, and oxygen is passed over the other side of the substrate. The reaction begins within the porous substrate (CVD stage). Subsequent electrochemical transport (EVD stage) through the reaction product continues the reaction, and the film grows over the porous substrate.

The electrochemical transport of ions during the EVD process is analyzed with respect to the defect model of the deposited perovskite. A complete illustration of the analysis is shown by considering the EVD growth of $LaCrO_3$ over a porous substrate. The analysis on the growth of $LaCrO_3$ revealed that the electrochemical transport consisted of oxygen ions diffusing from the oxygen side to the halide (chloride) side by vacancy mechanism and maintaining electroneutrality. Analysis of the $LaCrO_3$ film growth data also showed that the film growth is parabolic with time, and the reaction is diffusion (diffusion of oxygen ions) controlled. Previous conductivity measurements and the defect model of $LaCrO_3$ were combined with the EVD growth rate measurements to calculate the oxygen ion transport number, partial oxygen ion conductivity, and oxygen ion diffusivity as a function of the partial pressure of oxygen.

Index Entries: Perovskite; $LaCrO_3$; films; electrochemical; vapor; deposition; growth; defect; model; transport.

INTRODUCTION

Perovskites, ABO_3 (A is a lanthanide and B is a first row transition metal), have gained wide application because of their oxidation re-

*Author to whom all correspondence and reprint orders should be addressed.

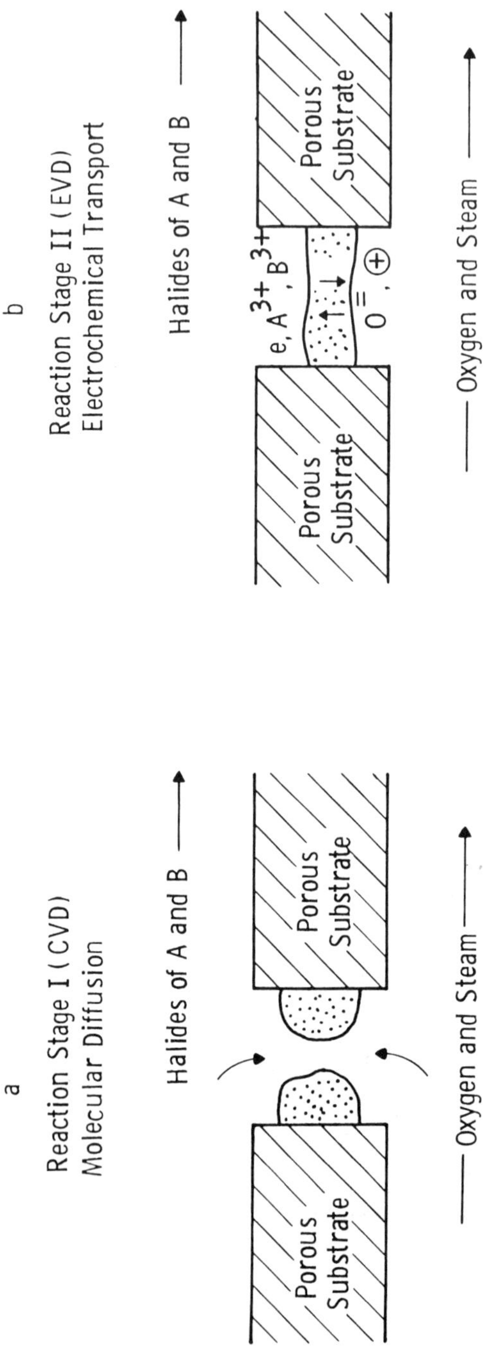

Fig. 1. Two stages of deposition of ABO_3.

sistance, high electronic conductivity, and capability of withstanding high temperatures (*1–4*). Their main areas of application include electrodes for magnetohydrodynamics power generation, electrodes for high temperature electrochemical devices, current leads for high temperature heaters, and superconducting ceramic materials. The electrical conductivity, thermal expansion, and chemical stability of these perovskites can be suitably tailored by appropriately doping the A ion sites and/or the B ion sites with a wide variety of elements (*5*).

Perovskites in the form of thin or thick films are desired in many of the aforementioned applications. Some of the present state of the art processes for producing such perovskite films are electrochemical vapor deposition (EVD), plasma spraying, sputtering, tapecasting, and so on. In this paper, the growth mechanism of the film by EVD process is described in terms of the defect chemistry of the perovskite structure; as a specific case, we have analyzed the growth mechanism of $LaCrO_3$ on a porous substrate during EVD. An understanding of such growth mechanism is necessary to determine the rate controlling step and to produce dense, pore-free thin or thick films.

ELECTROCHEMICAL VAPOR DEPOSITION (EVD) OF PEROVSKITE FILMS

Perovskite (ABO_3) film growth by EVD on a thin, porous substrate is achieved by volatilizing specific amounts of the halides of the cations (A and B) and passing them over one side of the porous substrate. Oxygen at a given activity is passed over the opposite side of the porous substrate (*6–8*). This is shown schematically, in Fig. 1.

In the first stage of the reaction (designated as the CVD stage) (Fig. 1a), molecular diffusion of oxygen and halides occurs through the porous substrate and they react to fill up the pores as follows

$$\text{Halide of A} + \text{Halide of B} + \text{Oxygen} \rightarrow ABO_3 + \text{Halogen gas} \ldots \quad (1)$$

The temperature, pressure, the type of halide, and the respective flows are so chosen that the above reaction is thermodynamically and kinetically favorable.

During the second stage of the reaction, after the pores are closed, electrochemical transport of one or more of the reacting species maintaining electroneutrality occurs through the ABO_3 deposited in the pores and continues the reaction (Fig. 1b). This second stage of the reaction is referred to as the EVD stage. Thus, the ABO_3 would grow over the porous substrate and form a film either on the halide side, the oxygen side, or both depending on the respective mobilities of the cations and the anions. If the anions are much more mobile than the cations, the ABO_3 film will grow over the porous substrate on the halide side and if the reverse is true, then the film will grow on the oxygen side. However, if the mobilities (μ) of the cations and the anions are comparable, then the film will grow over the porous substrate on both the sides. The effect

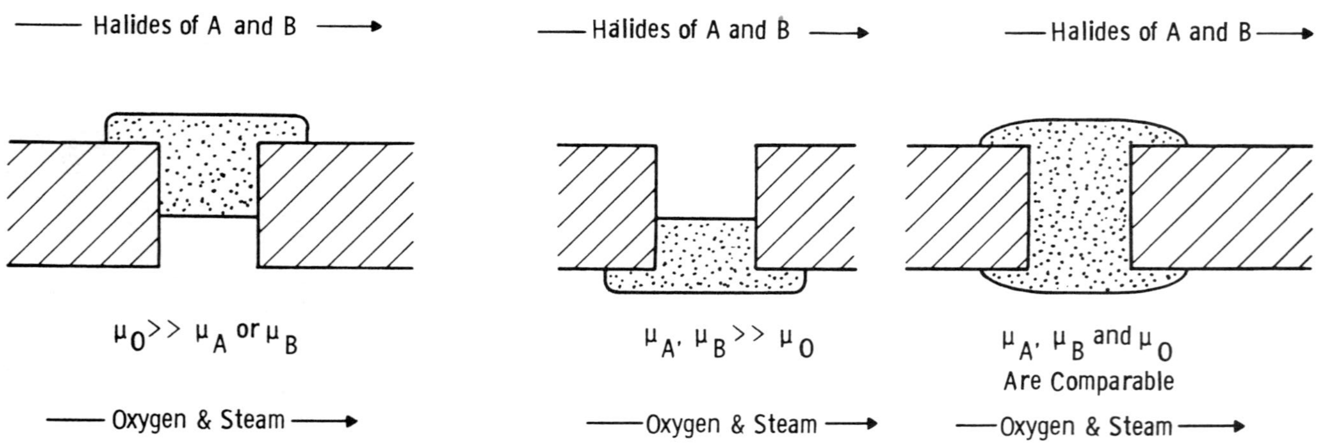

Fig. 2. Growth pattern during EVD of ABO_3 with respect to the mobilities (μ) of A, B, and O.

of the cation and the anion mobilities on the film growth is schematically, shown in Fig. 2.

Most of the perovskites ($LaCrO_3$, $YCrO_3$, $LaMnO_3$, and so on) are known to have Schottky-type defects (cationic and anionic vacancies), and the nonstoichiometry in these defects gives these materials their electronic characteristics (*9*). Therefore, for effective electrochemical transport of the more mobile ionic species during the second stage (EVD) of the reaction, one must establish a gradient in its vacancy concentration across the perovskite/halide interface and the perovskite/oxygen interface. This is achieved in the EVD process by maintaining the two interfaces at different partial pressures of oxygen. For instance, hydrogen could be introduced along with the halides to maintain the perovskite/halide interface at a much lower partial pressure of oxygen compared to the perovskite/oxygen interface.

The flows of the halide vapors, oxygen, and other carrier gases can be maintained above a critical level and the system pressure kept sufficiently low to eliminate any gas phase control of the EVD reaction. The growth of the perovskite film during the second stage of the EVD reaction will then be controlled by the diffusing cationic and/or the anionic species maintaining electroneutrality. Assuming that there are no interfacial or space charge effects, the film growth will be parabolic and the specific rate constant, K_γ, can be expressed as (*10*)

$$K_\gamma = \frac{RT}{2/Z_2/F^2} \int_{P'_{O_2}}^{P''_{O_2}} \frac{(t_A + t_B + t_O)\sigma_\Sigma t_e}{t_A + t_B + t_O + t_e} \, d \ln P_{O_2} \tag{2}$$

where R is the gas constant (1.987 calories/degree mol); T is the EVD reaction temperature (K); Z_2 is the valence of the anion (oxygen = 2); F is the Faraday constant (23,060 calories/volt mol); t_A, t_B, t_O, and t_e are transport numbers of cations A and B, anion O, and electrons (or holes), respectively; P'_{O_2} and P''_{O_2} are the partial pressures of oxygen at the perovskite/halide and perovskite/oxygen interfaces, respectively (atm) and σ_Σ is the overall electrical conductivity of the perovskite (Ω cm^{-1}).

Since, t_e for perovskites $\simeq 1$ and $t_A + t_B + t_O << 1$, Eq. (2) can be written as:

$$K_\gamma = \frac{RT}{2/Z_2/F^2} \int_{P'_{O_2}}^{P''_{O_2}} (t_A + t_B + t_O) \, \sigma_\Sigma \, d \ln P_{O_2} \tag{3}$$

Based on the analysis outlined above, the growth of Mg-doped $LaCrO_3$ on a porous substrate was investigated; respective mobilities of the cations and the anions were determined. The growth data were then combined with previously published conductivity measurements to calculate the transport number and the diffusivity of oxygen through Mg-doped $LaCrO_3$.

ANALYSIS OF THE ELECTROCHEMICAL VAPOR DEPOSITION OF Mg-DOPED $LaCrO_3$

EVD Process for Growing Mg-Doped $LaCrO_3$ Film Over a Porous Substrate

The porous substrate is loaded into the reactor and the reactor is evacuated to less than 200 μm Hg pressure. The system is then heated to over 1300°C to start the EVD process. Chlorides of lanthanum, chromium, and magnesium are volatilized in a predetermined ratio and passed along with hydrogen over one side of the porous substrate. Oxygen mixed with steam at a predetermined ratio is passed over the opposite side of the substrate. In the reactor design, the two sides of the substrate are isolated from each other. The EVD reaction proceeds, as described in Eq. (1), resulting in the deposition of the acceptor (Mg)-doped $LaCrO_3$ over the porous substrate.

Growth Mechanism of Mg-Doped $LaCrO_3$ During the EVD Process

In this section, the defect model of $LaCrO_3$ is described and used to interpret the transport mechanism involved in the EVD of Mg-doped $LaCrO_3$.

Defect Model for Mg-Doped $LaCrO_3$

It has been shown by earlier workers (*11,12*) that $LaCrO_3$ is a p-type oxide with acceptor dopant Mg^{2+} on the Cr^{3+} sites. The p-type nonstoichiometry in $LaCrO_3$ is owing to the formation of cation vacancies (*11,12*). The Schottky reaction for the $LaCrO_3$ can be described as:

$$Nil \rightarrow V(La)''' + V(Cr)''' + 3V(O)^{\cdot\cdot} \quad \ldots \tag{4}$$

$$K_s = [V(La)'''][V(Cr)'''][V(O)^{\cdot\cdot}]^3,\ [V(La)'''] = [V(Cr)''']\ \text{and}\ 3[V(Cr)'''] = [V(O)^{\cdot\cdot}]$$

where Nil represents electroneutrality; $V(La)'''$, $V(Cr)'''$, and $V(O)^{\cdot\cdot}$ are the vacancies of lanthanum, chromium, and oxygen, respectively; K_s is the schottky equilibrium constant; and $[V(La)''']$, $[V(Cr)''']$ and $[V(O)^{\cdot\cdot}]$ are the concentrations of the lanthanum, chromium, and oxygen vacancies, respectively. The *p*-type nonstoichiometry in $LaCrO_3$ is given by:

$$\frac{3}{2}\,O_2 \rightarrow 3O_o^x + V(Cr)''' + V(La)''' + 6\oplus \tag{5}$$

$$K_c = \frac{[V(Cr)'''][V(La)'''][\oplus]^6}{P_{O_2}^{\ 3/2}}\ \text{and}\ 6[V(Cr)'''] = 6[V(La)'''] = [\oplus]$$

where O_o^x is neutral oxygen sitting on oxygen site; $\oplus$ and $[\oplus]$ are the generated hole and the hole concentration, respectively; and K_c is the nonstoichiometric equilibrium constant. Considering the Mg on Cr site, the total neutrality of defect concentration can be written using Eqs. (4) and (5) as (*9,11*)

$$2\,[V(O)^{\cdot\cdot}] + [\oplus] = 6\,[V(Cr)'''] + [Mg(Cr)'] \tag{6}$$

Using Eqs. (4), (5), and (6), we can predict (*9,11*) the variation in the vacancy concentration of the defect species through $LaCrO_3$ as a function of P_{O_2} when the acceptor concentration dominates the defect reaction. They are as follows

$$[V(Cr)'''] \propto P_{O_2}^{3/4}$$

and

$$[V(O)] \propto P_{O_2}^{-1/2}$$

The above mentioned condition pertains to the Mg-doped $LaCrO_3$ ($LaCr_{0.98}Mg_{0.02}O_3$) deposited by the EVD process.

Transport Through Mg-Doped $LaCrO_3$ During EVD

From the aforementioned analysis, the variation trend in the concentration of vacancies of lanthanum, chromium, and oxygen through the growing $LaCrO_3$ during EVD is shown in Fig. 3. The concentration gradient of the vacancies shown in Fig. 3 suggests that, if the diffusion of ions is through vacancy mechanism, then the oxygen ions could move from the oxygen side to the chloride side, and the lanthanum and chromium ions could migrate in the reverse direction.

Experimentally, it is seen that, during the EVD of Mg-doped $LaCrO_3$, the film grows over the porous substrate on the chloride side, as shown in Fig. 4. As discussed earlier, this type of growth indicates that the mobility of oxygen ions through its vacancies is much greater than that of lanthanum or chromium. In order to further illustrate this point, $LaCrO_3$ markers in the form of particles were deposited on the porous substrate prior to EVD. After the EVD, it was seen that the $LaCrO_3$ EVD film grew around and over the markers (shown in Fig. 5) and the markers were still sitting over the porous substrate. The position of the markers indicates that the mobility of oxygen ions through the $LaCrO_3$ during EVD is much greater than the lanthanum or the chromium ions, and consequently, it also proves that the site of the $LaCrO_3$ EVD reaction is always at the chloride/$LaCrO_3$ interface.

Estimation of Transport Parameters Involved in the EVD of Mg-Doped $LaCrO_3$

In this section, the experimental data on the growth rate of Mg-doped $LaCrO_3$ during EVD are combined with the aforementioned analysis of the transport mechanism, and the oxygen transport number, along with oxygen diffusivity through the Mg-doped $LaCrO_3$ during EVD, are estimated.

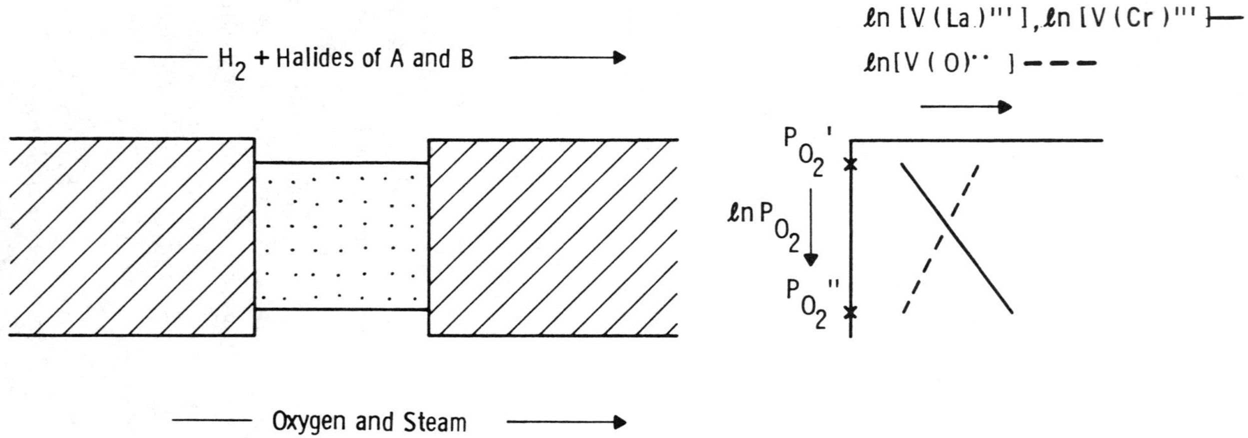

Fig. 3. Variation trend in the vacancy concentration of the defect species in $LaCrO_3$ during EVD.

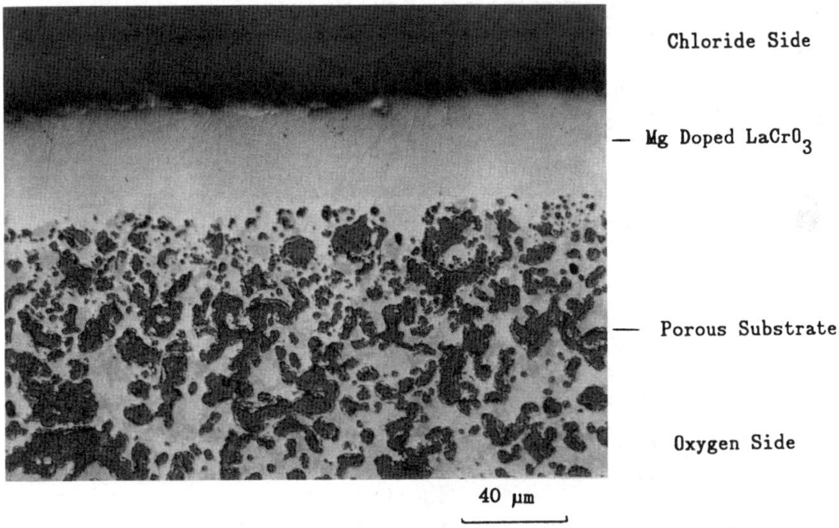

Fig. 4. EVD growth of $LaCrO_3$ over porous substrate showing that the film grows on the chloride side.

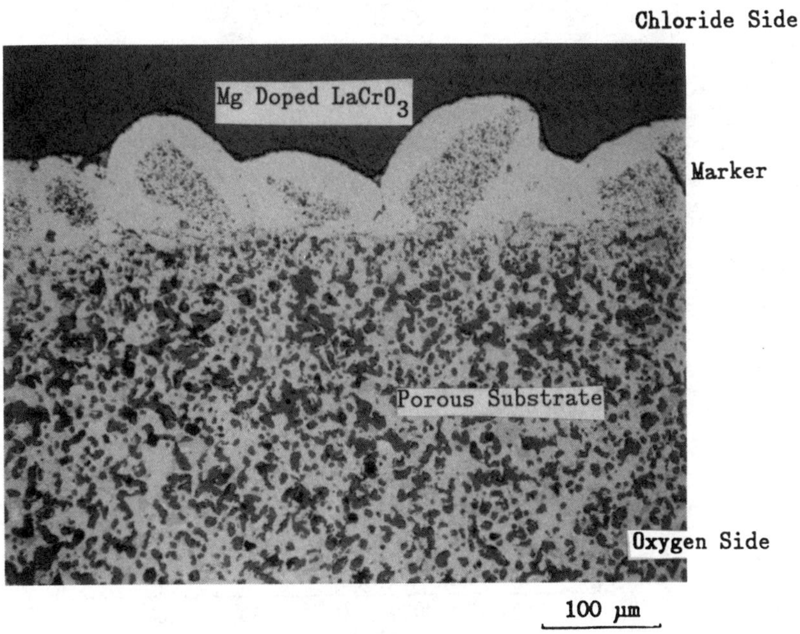

Fig. 5. Growth of $LaCrO_3$ film around and over the $LaCrO_3$ marker particles.

EVD Growth Rate Data of Mg-Doped $LaCrO_3$ Films

The portion of the porous substrate that first comes in contact with the fast-flowing halides and the carrier gases was used as the base for all Mg-doped $LaCrO_3$ EVD film thickness measurements; there is no gas phase control of the reaction in this region. The variation in thickness of the $LaCrO_3$ film, from over the aforementioned portion of the substrate, as a function of the EVD reaction time, is shown in Fig. 6. In Fig. 7, we

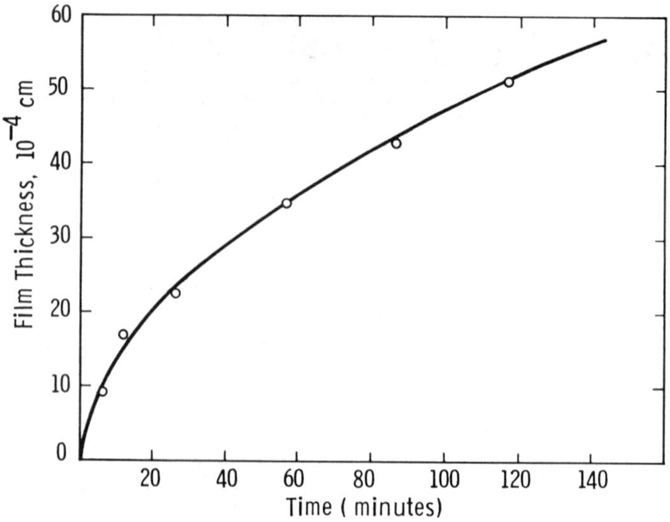

Fig. 6. EVD film growth profile of Mg-doped $LaCrO_3$.

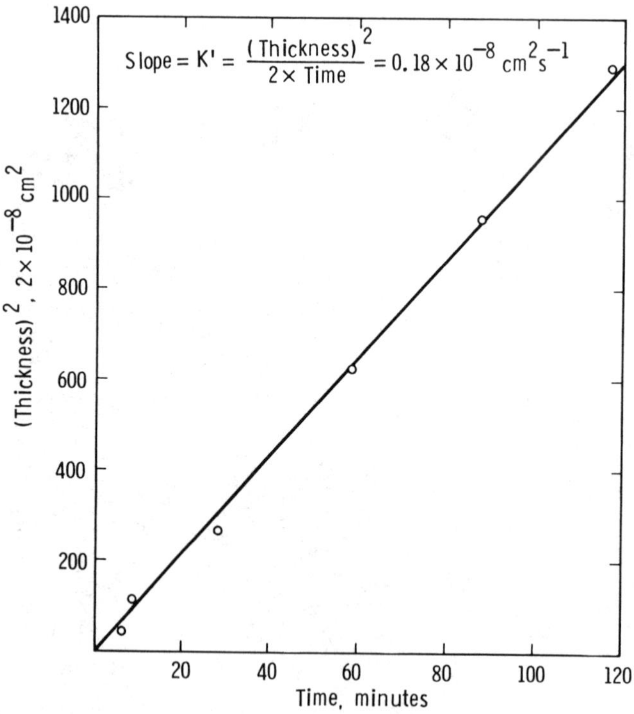

Fig. 7. A straight line plot of square of EVD $LaCrO_3$ film thickness vs time, indicating that the film growth is parabolic with time.

have plotted the square of the EVD film thickness as a function of the reaction time; the plot was found to be a straight line. This suggests that the EVD film growth is parabolic, and the EVD reaction is diffusion controlled. From the aforementioned analysis of the growth mechanism, this would mean that the EVD reaction is controlled by the diffusion of oxygen ions through oxygen vacancies. The rate of growth of the EVD film, (dx/dt), can therefore be expressed as

$$\frac{dx}{dt} = \frac{K'}{x} \text{ or } x^2 = 2K't \tag{7}$$

where x is film thickness in cm; t is the reaction time in seconds; and K' is the parabolic constant in cm^2/sec. From the slope of the straight line in Fig. 7, K' has been calculated as 0.18×10^{-8} cm^2/s. The parabolic rate constant K' can also be expressed in terms of the rate of equivalents of oxygen consumed in the reaction as follows:

$$\frac{dn}{dt} = \frac{K'}{V_{eq} \bullet x} = \frac{K_\gamma}{x} \tag{8}$$

where $\frac{dn}{dt}$ is the equivalent of oxygen used for reaction/cm^2/s; V_{eq} is equivalent volume (cm^3/equiv) of oxygen in $LaCrO_3$; and K_γ is the specific rate constant expressed in equiv/cm/s.

Oxygen Ion Transport Number (t_o)

From Eq. (8), we can calculate K_γ and then use Eq. (3) to calculate the oxygen transport number through $LaCrO_3$. This is done as follows:

$$K_\gamma = \frac{K'}{V_{eq}} = 3.2 \times 10^{-10} \text{ equiv/cm sec} \tag{9}$$

In the film growth of $LaCrO_3$ by EVD, it was proven earlier that t_{La} and t_{Cr} $<< t_O$. Therefore, using Eq. (3), we can write K_γ as

$$K_\gamma = \frac{RT}{2/Z_2/F^2} \int_{P'_{O_2}}^{P''_{O_2}} t_O \bullet \sigma_\Sigma \, d\ln P_{O_2} \tag{10}$$

The EVD growth rate of the Mg-doped $LaCrO_3$ reported here was measured at 1330°C (1603 K) with the oxygen side of the porous substrate at $P''_{O_2} \cong 10^{-4}$ atm and the chloride side at $P'_{O_2} \cong 10^{-15}$ atm. Assuming t_O to be independent of P_{O_2} in Mg-doped $LaCrO_3$ (*13*) and using the limits for P_{O_2} during EVD, the oxygen transport number (t_O) can be written from Eq. (10) as

$$t_O = \frac{2 \bullet /Z_2/ \bullet F^2 \bullet K_\gamma}{RT \bullet \left(\int_{10^{-15}}^{10^{-4}} \sigma_\Sigma \, d \log P_{O_2} \right) \bullet 2.303} \tag{11}$$

Using σ_Σ vs. Log P_{O_2} measurements of earlier workers (*11,13,14*) (shown in Fig. 8), and the value of K_γ from Eq. (9), t_O was calculated from Eq. (11) as 2×10^{-5}.

It may be noted that, using this value of t_O and the plot of σ_Σ vs P_{O_2}, one may also calculate the variation of partial oxygen ion conductivity, ($\sigma_O = t_O \cdot \sigma_\Sigma$), as a function of P_{O_2}.

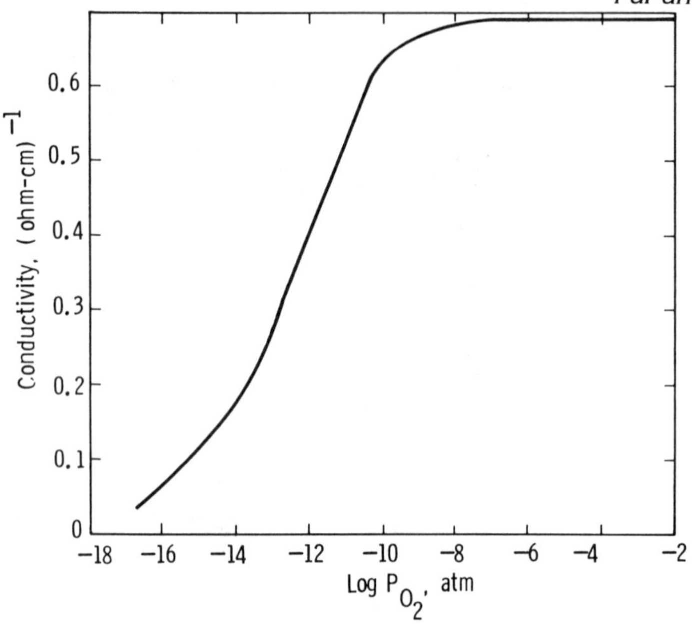

Fig. 8. Plot of conductivity of $LaMg_{0.02}Cr_{0.98}O_3$ vs Log P_{O_2} at 1330°C (1603 K).

Oxygen ion Diffusivity (D^*_O)

Film growth data were further used to obtain the oxygen ion diffusivity (D^*_O) in Mg-doped $LaCrO_3$ as a function of P_{O_2}. This was done as follows. First, the partial oxygen ion conductivity (σ_O) is written in terms of its diffusivity

$$\sigma_O = \sigma_\Sigma \bullet t_O = \frac{D^*_O \bullet Z_2^2 \bullet F^2 \bullet C_O}{RT} \quad (12)$$

where C_O is gram atoms of oxygen/cm^3 in $LaCrO_3$ = 0.08. Then, substituting Eq. (12) in Eq. (10) we have

$$K_\gamma = \frac{|Z_2| \bullet C_O}{2} \int_{10^{-15}}^{10^{-4}} D^*_O \, d\ln P_{O_2} \quad (13)$$

The $LaCrO_3$ film deposited by the EVD process has around 0.4 at % Mg in chromium sites. Therefore, the acceptor dopant concentration can be expected to dominate the defect reaction and, as discussed earlier, it implies $[V(O)^{\cdot\cdot}] \propto P_{O_2}^{-1/2}$. Also, since $D^*_O \propto [V(O)^{\cdot\cdot}] \propto P_{O_2}^{-1/2}$ we can write:

$$D^*_O = C_1 \bullet P_{O_2}^{-1/2} \quad (14)$$

where C_l is a constant. Substituting Eq. (14) in Eq. (13), we have

$$K_\gamma = \frac{|Z_2| \bullet C_O \bullet C_1}{2} \int_{10^{-15}}^{10^{-4}} P_{O_2}^{-1/2} \, d\ln P_{O_2} \quad (15)$$

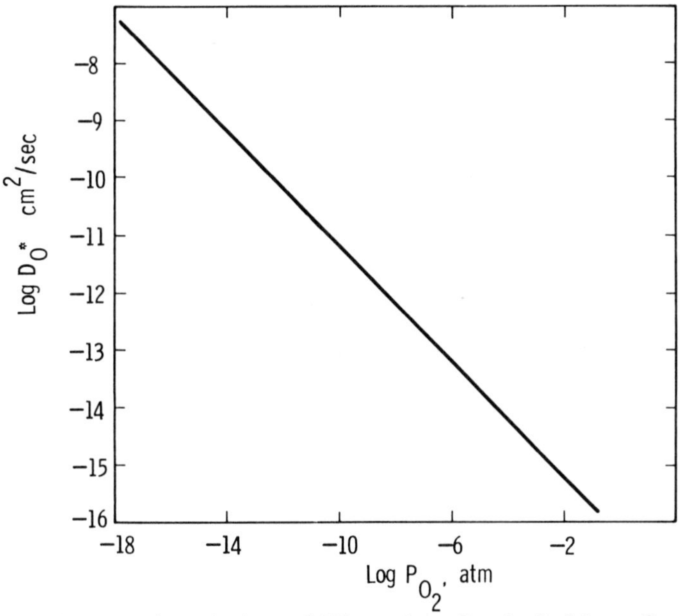

Fig. 9. Estimated variation of D_O^* vs Log P_{O_2} in $LaMg_{0.02}Cr_{0.98}O_3$.

Solving Eq. (15), we have $C_1 = 6.3 \times 10^{-17}$. Therefore, substituting for C_1 in Eq. (14), we can calculate D_O^* at the chloride/$LaCrO_3$ interface ($P'_{O_2} \cong 10^{-15}$ atm) and at the oxygen/$LaCrO_3$ interface ($P''_{O_2} \cong 10^{-4}$ atm). These values are: at the chloride/$LaCrO_3$ interface: $D_O^* = 2 \times 10^{-9}$ cm^2/sec and at the oxygen/$LaCrO_3$ interface: $D_O^* = 6.3 \times 10^{-15}$ cm^2/sec. The value of $\sqrt{D_O^* \cdot t}$ gives an estimate of the distance oxygen diffuses in Mg-doped $LaCrO_3$ film in a given time, t. Since the reaction involved in the EVD process is diffusion controlled, the value of $\sqrt{D_O^* \cdot t}$ also gives an estimate of the Mg-doped $LaCrO_3$ film thickness as a function of EVD reaction time, t. The diffusivity (D_O^*) calculated above at the chloride/$LaCrO_3$ interface was found to correspond to the experimentally observed growth rate of the film as shown in Fig. 6. This suggests that the oxygen potential through most of the film thickness is the same as the chloride/$LaCrO_3$ interface.

Finally, the value of C_1 and Eq. (14) is used to plot D_O^* as a function of P_{O_2} as shown in Fig. 9. From this figure, it is apparent that if the P_{O_2} at the chloride/$LaCrO_3$ interface increases by an order of magnitude, the diffusivity (D_O^*) would decrease by a factor 3.2. Hence, for uniformity of EVD film thickness over the entire length of the porous substrate, we must maintain uniformity in oxygen potential all along the length of the substrate.

SUMMARY

Electrochemical vapor deposition (EVD) of perovskites (ABO_3) is described—halides of A and B along with hydrogen are passed over one side of a porous substrate, and oxygen is passed over the other side of the substrate. The halides and oxygen react within the porous substrate, and the reaction product (ABO_3) seals the pores. Subsequent electrochemical transport of ions and electrons through the reaction product continues the reaction, and the film grows over the substrate.

The electrochemical transport during (EVD) of Mg-doped $LaCrO_3$ is analyzed using the measured growth rate, the defect structure of Mg-doped $LaCrO_3$, and the available conductivity data of Mg-doped $LaCrO_3$ as a function of P_{O_2}. Such an analysis revealed that the $LaCrO_3$ film growth is parabolic and, thus, diffusion controlled; diffusion of oxygen ions occuring by vacancy mechanism. The oxygen ion transport number is determined to be on the order of 10^{-5}. The oxygen diffusivity (D^*_O) in the Mg-doped $LaCrO_3$ is equal to $6.3 \times 10^{-17} \cdot P_{O_2}^{-1/2}$. At the $LaCrO_3$/chloride interface and through most of the film thickness, the oxygen diffusivity is determined to be on the order of 10^{-9} cm^2/s.

ACKNOWLEDGMENTS

The authors would like to acknowledge the dedicated efforts of G. R. Folser and J. A. Fraino in conducting the EVD experiments. This work was funded by the US Department of Energy under contract number DE-AC21-80ET-17089 and the Westinghouse Electric Corporation.

REFERENCES

1. Bowen, H. K., Halloran, J. W., and Petusky, W. T., *Corrosion Problems in Energy Conversion and Generation,* Tedmon, C. S., ed., Electrochemical Society, Princeton, NJ, 1974, p. 179.
2. Meadowcraft, D. B., *Brit. J. Appl. Phys.* **2,** 1225 (1969).
3. Anderson, H. U., Rossing, B., and Bates, T. L., *Rare Earths in Modern Science and Technology,* McCarthy, G. J. and Rhyne, J. J., eds., Plenum, New York, NY, 1978, p. 55.
4. Anthony, A. M., Benezech, G., Cabannes, F., Faucher, M., Foex, M., Loc, V., and Yerouchalmi, D., *Proceedings of the IUPAC 3rd International Symposium on High Temperature Technology,* Butterworths, London, 1969, p. 213.
5. Ruka, R. J., *US Patent 4,631,238,* December 23, 1986.
6. Isenberg, A. O., *US Patent 4,597,170,* July 1, 1986.
7. Isenberg, A. O., and Zymboly, G. E., *US Patent 4,609,562,* September 2, 1986.
8. Zymboly, G. E., *Proceedings of the Conference on High Temperature Solid Oxide Electrolytes,* vol. 1, Brookhaven National Laboratory, Long Island, New York, 1983, p. 37.
9. Anderson, H. U., Nasrallah, M. M., Flandermeyer, B. K., and Agarwal, A. K., *J. of Solid State Chemistry,* **56,** 325 (1985).
10. Wagner, C., *Z. Physikalische Chemie B,* **21,** 25 (1933).
11. Flandermeyer, B. K., Nasarallah, M. M., Agarwal, A. K., and Anderson, H. U., *J. of Am. Ceram. Soc.* **67,** 195 (1984).
12. Chikung, J. Y., Sparlin, D. M., and Anderson H. U., *J. of Am. Ceram. Soc.* **70,** 189 (1987).
13. Flandermeyer, B. K., Nasrallah, M. M., Sparlin, D. M., and Anderson, H. U., *High Temp. Sci.* **20,** 259 (1985).
14. Flandermeyer, B. K., Nasrallah, M. M., Sparlin, D. M., and Anderson, H. U., *Transport in Nonstoichiometric Compounds,* Simkovich, G. and Stubican, V., eds., Plenum, New York, 1985, p. 17.

Laser and Mass Spectrometric Studies of the Mechanism of Silicon (100) Single Crystal Etching Reactions

J. S. HORWITZ*,[1] AND M. C. LIN[2]

[1]Chemistry Division, Naval Research Laboratory, Code 6177, Washington, DC 20375-5000; and [2]Present address: Department of Chemistry, Emory University, Atlanta, GA 30322

ABSTRACT

The thermal etching of Si(100) single crystals by F_2 has been studied by electron impact (EI) and resonance enhanced multiphoton ionization (REMPI) mass spectroscopy. Laser ionization of the gas phase products observed under single collision gas-surface conditions indicates that SiF_2 radicals are formed with an apparent activation energy of 9.9 ± 1.65 kcal/mol. Similarly, EI measurements indicate that SiF_4 is desorbed from the surface with an apparent activation energy of 4.0 ± 1.0 kcal/mol. A comparison of the temperature dependence of the ion currents generated by EI for SiF_x^+ (x = (1–4)) with MPI of SiF_2 indicate that the SiF_x^+ signals are derived primarily from SiF_2 and SiF_4 (possibly vibrationally excited). Both n-type and p-type doping apparently have no significant effect in the measured rates and activation energies for SiF_2 and SiF_4 production.

Index Entries: Si; thermal etching; single crystals; electron impact; resonance enhanced multiphoton ionization (REMPI).

INTRODUCTION

The thermal etching of silicon by halogen containing compounds (F_2, XeF_2, and NF_3) remains an important model system for plasma etching. The ultimate goal in the study of these systems is the development of a detailed mechanism for the removal of silicon from the surface

*Author to whom all correspondence and reprint requests should be addressed.

and the formation of a volatile product. To that end, mechanistic studies of this reaction begin by determining what products are formed on both the surface and in the gas phase. In the study of the gas phase products, the effects of temperature and pressure are monitored to determine the energetics of the formation of these species. The morphology of the silicon substrate may also affect the etching mechanism by allowing the etchant to access different surface sites. And finally, the impurities (*n*-type and *p*-type) may also affect the mechanism. These impurity levels, though chemically quite small, have a large effect on the electrical properties of silicon and are believed to influence the chemical reactivity of silicon to halogens.

An analysis of the surface of a silicon single crystal following exposure to XeF_2 indicates that SiF_x (x = 1–3) are adsorbed on the silicon surface (*1*). If the exposure to XeF_2 is increased, SiF_4 is also observed to be adsorbed on the surface (2). No adsorbed XeF_2 or Xe is observed, indicating that this compound is dissociatively adsorbed, liberating F atoms that attach to available sites on the Si surface. Quartz crystal microbalance measurements indicate that the surface must be fairly saturated with fluorine before the surface adsorbed species desorb (*3*). Although surface analytic techniques seem relatively straightforward, there is considerable controversy concerning which of these species desorbs from the surface. One of the most commonly used diagnostic in the study of the gas phase products at low pressures involves electron impact mass spectrometry. Although this technique is the simplest to implement, the instability of the various SiF_x species to electron impact has made interpretation of the data difficult. SiF_4 is the thermodynamically-favored product however; SiF_3, SiF_2, SiF, and Si_2F_6 have also been proposed as gas phase products (*4–7*).

As a continuation of our ongoing investigation of silicon etching, we have examined the gas phase products in the thermally-activated reaction of silicon with F_2 by electron impact (EI) and resonance enhanced multiphoton ionization (REMPI) mass spectrometry. Gas phase products are detected under single collision gas-surface conditions for the reaction of F_2 with Si(100) undoped, *n*-type and *p*-type crystals. The combination of these two techniques allows us to distinguish neutral products desorbed from the surface from fragments formed by the electron impact. The temperature-dependent measurements yield an apparent activation energy for the production of SiF_4 of 3.9 ± 1.0 kcal/mol and an apparent activation energy for the production of SiF_2 of 9.9 ± 1.65 kcal/mol. A comparison of the ion currents generated by electron impact ionization and resonant laser ionization indicate that other gas phase products may be formed in this reaction. The origin of these products is discussed.

EXPERIMENTAL METHODS

The apparatus used in this study is shown in Fig. 1. Briefly, it consisted of a two chamber differentially-pumped system. The source chamber was separated from the detection chamber by a 2 mm skimmer.

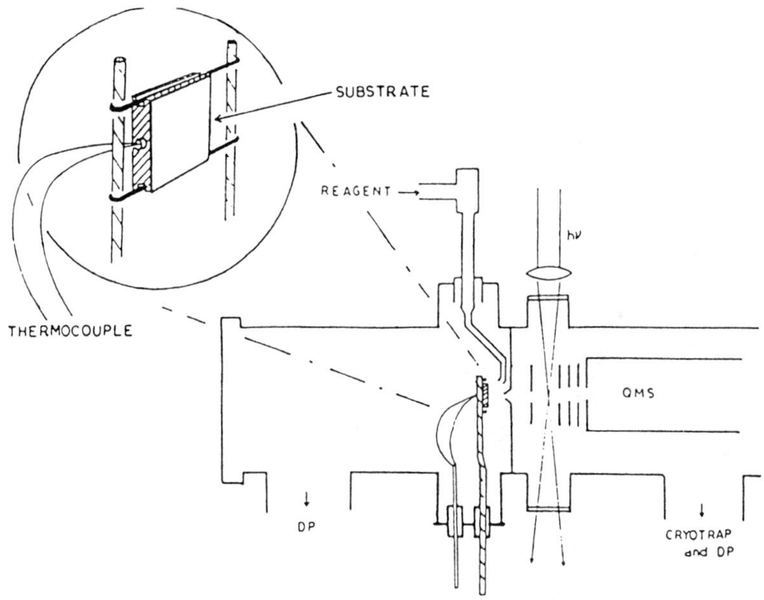

Fig. 1. Schematic of apparatus used in single crystal studies. *See* ref. (*8*) for details.

The quadrupole mass spectrometer (Extranuclear Laboratories, Pittsburgh, Pennsylvania) had been modified to allow for both electron impact and laser ionization. The system nominally had a base pressure of 10^{-6} and 10^{-7} torr in the source and detection chambers and was 10^{-5} and 10^{-6} torr, respectively, during the experiment. Si (100) crystals (undoped and n- and p-type, 0.1 ohm-cm) were purchased from Virginia Semiconductors. F_2 (10% in He) from Air Products was used without purification. Laser ionization was by either a Lambda Physik excimer pumped dye laser (EMG 102/FL2002) with output energies of 10 to 15 mJ/pulse or the frequency doubled output of a Quantel Nd:YAG pumped dye laser (YG481) with output energies of < 1mJ/pulse.

RESULTS

Shown in Fig. 2 are REMPI spectra SiF_2 radicals desorbed from a silicon single crystal at ca. 1000 K in the region from 360–400 and 320–321 nm. These spectra have been previously assigned to the (1 + 3) $\tilde{a}^3B_1 \longleftarrow \tilde{X}^1A_1$ (*9*) and the (2 + 1) $\tilde{B}^1B_2 \longleftarrow \tilde{X}^1A_1$ (*10*) transitions, respectively. These spectra provide unambiguous assignment of SiF_2 as a nascent product of the gas surface interaction between F_2 and Si.

Shown in Fig. 3 is the temperature dependence of the production of SiF_2 radicals in the F_2 + Si(100) reaction as measured by the SiF_2^+ ion current from photoionization at 320.5 nm. The solid curve represents a nonlinear least squares fit to the function $I(T) = A^*Exp[-E_a/RT] + C$. The apparent activation energy for the production of SiF_2 radicals from an Si(100) *n*-type, *p*-type, and undoped crystals is 9.9 ± 1.65 kcal/mol.

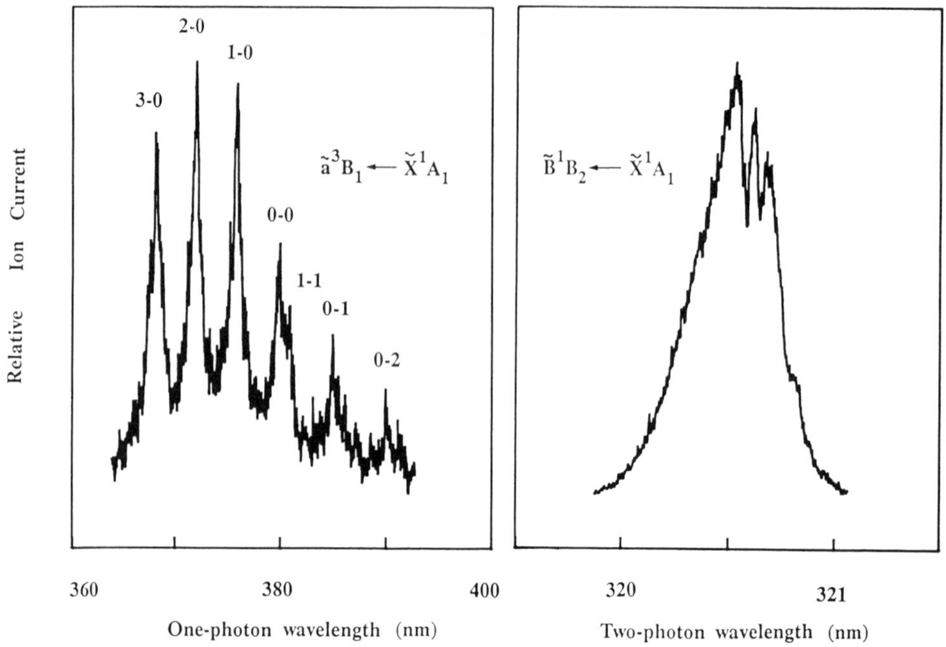

Fig. 2. REMPI spectra are shown for SiF_2 observed in the reaction of F_2 with a silicon single crystal. The crystal surface temperature was 1000 K. The $\tilde{a}^3B_1 \longleftarrow \tilde{X}^1A_1$ spectrum was obtained by monitoring the SiF^+ ion current. The $\tilde{B}^1B_2 \longleftarrow \tilde{X}^1A_1$ spectrum was obtained by monitoring the SiF_2^+ ion current.

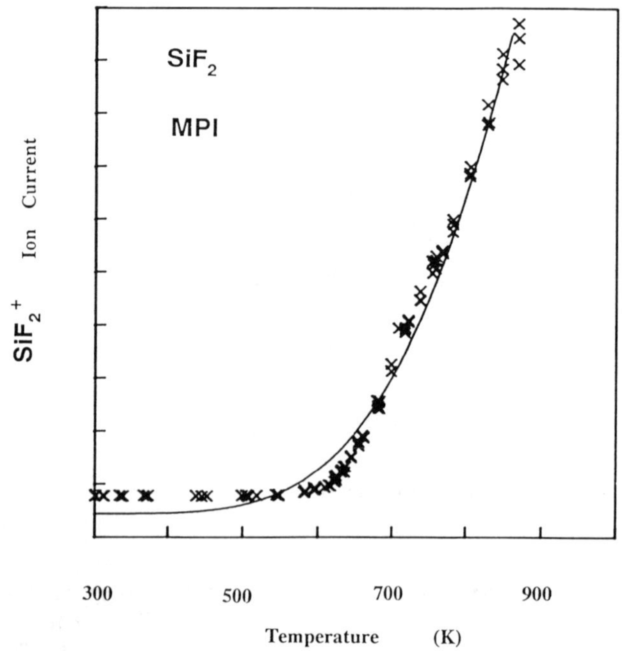

Fig. 3. The temperature dependence of the SiF_2^+ ion current from MPI of SiF_2 at 320.5 nm in the reaction of F_2 with a Si(100) single crystal is shown. The solid line represents a nonlinear least squares fit to the function $I(T) = A^*Exp(-E_a/RT) + C$. The activation energy for SiF_2 production is presented in Table 1.

Table 1
Apparent Activation Energies in kcal/mol for Products Observed in the Reaction of Si + F_2 Measured by Electron Impact and Multiphoton Ionization

Crystal Si(100)	EI Ion, SiF^+	EI Ion, SiF_2^+	EI Ion, SiF_3^+	EI Ion, SiF_4^+	MPI SiF_2
Undoped	8.04	7.91	4.91	4.90	9.37
	7.24	7.77	3.41	2.50	
	7.12	7.76	3.90	5.34	
	6.76	6.69	3.38	3.22	
n-type	5.18	8.18	3.74	4.38	8.05
	7.87	8.47	4.39	3.94	7.52
	7.82	7.31	4.09	4.34	12.43
	7.23		5.59	5.66	10.32
p-type	7.73	6.47	4.32	4.03	10.59
	6.53	5.50	3.87	2.89	12.07
	6.18	7.30	3.50	3.64	9.37
	6.31	6.80	3.06	2.90	9.32
Average	6.99 ± 0.85	7.29 ± 0.87	3.97 ± 0.70	3.98 ± 1.0	9.90 ± 1.64

The effect of doping is undiscernible in terms of measured apparent activation energies and the intensity of the SiF_2^+ signal. The temperature dependent measurements were made by monitoring the ion signal as a function of either an increase or a decrease in the temperature throughout a particular experiment. In general, the most reproducible data were obtained for decreasing temperature measurements. These results are summarized in Table 1 and are consistent with earlier measurements by MPI (*8*) and chemiluminescence (*11*). Etch depth and SiF_3^* chemiluminescence measurements (*11*) for the reaction of Si and F_2 at higher pressures of 1–15 torr yielded an activation energy for the removal of silicon of 8.9 kcal/mol. Based on this, it has been proposed that formation of SiF_2 is the rate-limiting step in the removal of silicon from the surface.

Shown in Fig. 4 is the EI/MS for the reaction of F_2 with an Si(100) single crystal held at 1000 K. Electron impact energies for all of the EI data reported here were obtained at 40 eV. Four features are easily assigned as resulting from gas phase products from this reaction. SiF_4^+ (m/z = 107), SiF_3^+ (m/z = 85), SiF_2^+ (m/z = 66), and SiF^+ (m/z = 47) with relative intensities of 0.004, 1.0, 0.5, and 0.16, respectively. SiF_4^+ and SiF_3^+ arise primarily from neutral SiF_4 generated in the reaction. This assignment is strongly supported by the cracking pattern of neutral SiF_4 at room temperature, also shown in Fig. 4. Under the conditions used here, SiF_4 yields very little parent ion. The majority of the ion current appears as SiF_3^+. None of the other SiF_x^+ (x = 1,2) fragments can be seen under these conditions (room temperature).

The temperature dependence of the ion currents, as measured by EI/MS, are shown in Fig. 5. These measurements were made under conditions identical to the MPI measurements for SiF_2 radicals. The apparent

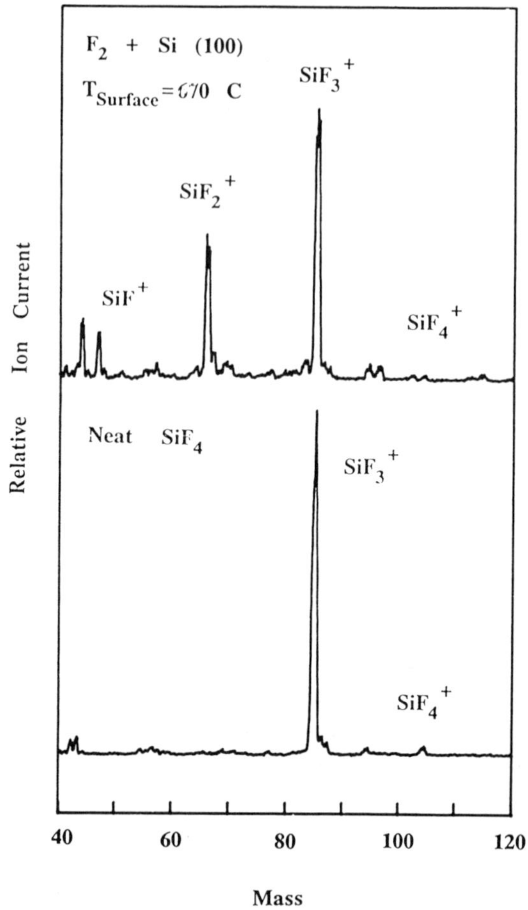

Fig. 4. Shown on the top is the EI/MS obtained at 40 eV for the reaction products of Si(100) and F_2. For comparison, the EI/MS of neat SiF_4 obtained under similar conditions is shown on the bottom.

activation energies fall into two categories. Both SiF_4^+ and SiF_3^+ appear with an apparent activation energy, ca. 4 kcal/mol. SiF_2^+ and SiF^+ appear with a slightly higher apparent activation energy of 7 kcal/mol. The activation energy results support the assignment that SiF_3^+ and SiF_4^+ ions originate from neutral SiF_4 desorbed from the surface. However, the dissimilarity between the apparent activation energy for production of SiF_2^+, measured by EI and MPI, suggests that these ions originate from different sources. That is to say that the neutral species that ionize in electron impact to give SiF_2^+ is different from the neutral species that is photoionized to give SiF_2^+. In addition, the similar nature of the activation energy for the production of SiF_2^+ and SiF^+ suggests that these ions are fragments of the same neutral species desorbed from the surface. These results are summarized in Table 1.

The pressure dependence of the ion currents detected by electron impact or resonant laser ionization are linear from 10^{-7} to 10^{-5} torr. This

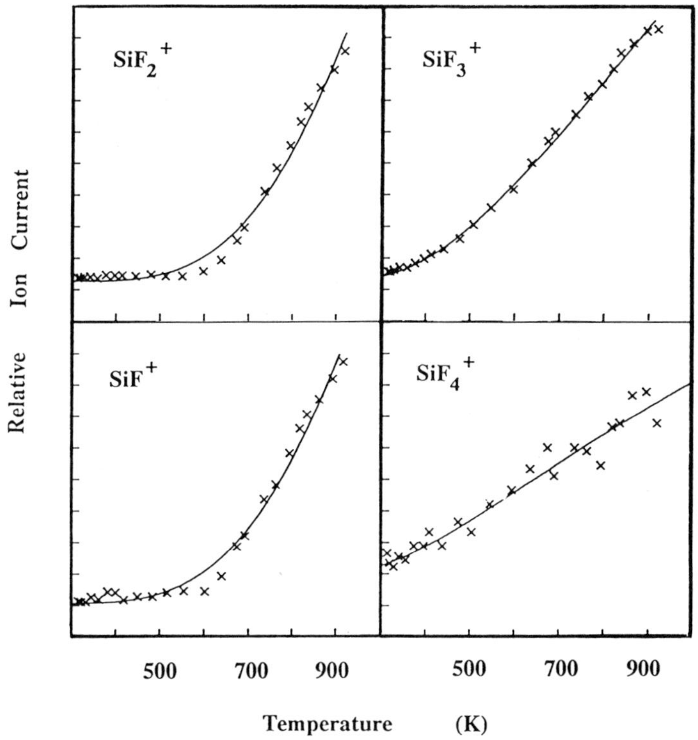

Fig. 5. The temperature dependence of the SiF_x^+ ions observed by electron impact in the reaction of Si(100) and F_2 is shown. The solid line represents the nonlinear least squares fit to the function $I(T) = A^*\text{Exp} -(E_a/RT) + C$.

indicates that the species detected are not formed as a result of gas phase chemical reactions. In general, the expression "single collision gas-surface conditions" means that the mean free path for collisions is long compared to the distance between the surface and the ionization region and that species formed on the surface do not undergo collisions between desorption and detection. The SiF_2^+ and SiF^+ ions are not observed in the room temperature reaction of Si and F_2.

We monitored the time evolution of the ion currents detected by EI and MPI. It is expected that coverage of the surface with F_2 should be relatively fast. For an unetched crystal, heated to 950 K and then exposed to F_2, ion currents detected by EI/MS appear very rapidly ($t < 1$ min). No difference can be seen in the appearance time of the SiF_x^+ ($x = 1$–4) species. However, the MPI-generated SiF_2^+ ion current does not appear with the same time dependence. The SiF_2^+ ion current, as measured by MPI of neutral SiF_2 radicals, is not present initially under the conditions mentioned above and grows to an equilibrium value relative slowly (15–20 min). A comparison of the time development of the ion currents detected by EI and MPI is shown in Fig. 6. This difference supports the proposal that the origin of the EI SiF_2^+ ion current is different from the origin of the MPI SiF_2^+ ion current.

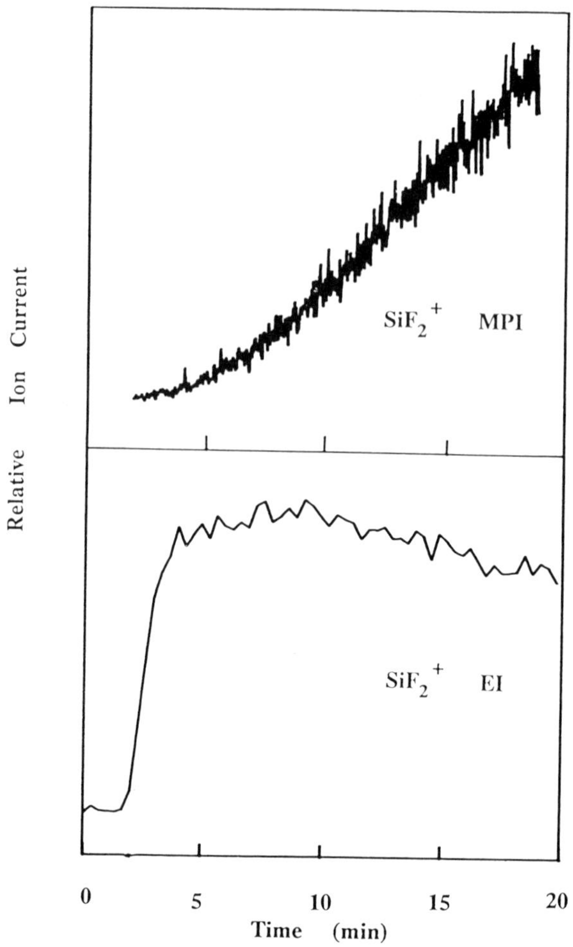

Fig. 6. The time dependence of the appearance of ion currents formed by electron impact or multiphoton ionization is shown. Clean, unetched single crystals were heated to ca. 1000 K and then exposed to F_2. SiF_2 radicals (top) were detected by MPI at 320.5 nm. SiF_2^+ ions (bottom) were formed by electron impact on the neutral products.

At pressures > 1 torr, chemiluminescence (*11*) has been reported in the reaction of silicon with fluorine-containing compounds. The chemiluminescence has been assigned to SiF_3^* formed by the reaction of SiF_2 with F or F_2. The chemiluminescence is then a measure of SiF_2 production. Flamm et al. (*12*) observed a time delay in the chemiluminescence for clean, unetched crystals following exposure to F_2 similar to our observed delay in the MPI of SiF_2. The delay was attributed to the removal of the surface oxide layer by fluorine, which was then followed by the production of the SiF_x species. This would seem reasonable were it not for parallel observation under our conditions of a delay in the production of SiF_2 at the same time we observed the prompt production of SiF_4. In fact, the production of SiF_4 is spontaneous at room temperature. The delay appears to be a part of the etching mechanism.

DISCUSSION

It has been well-established that the rate of thermal etching of silicon with reactive fluorine compounds, such as F_2 and XeF_2, below 700 K is controlled primarily by the dissociative adsorption of etchants. Flamm et al. (*12*) have demonstrated that the etching of silicon with atomic fluorine proceeds with an activation energy as low as 2 kcal/mol, based on their etch depth and $F + SiF_2 \longrightarrow SiF_3^*$ chemiluminescence measurements. On the other hand, when NF_3 was used as an etchant in our earlier experiments (*8*), we found that the activation energy for SiF_2 and SiF_4 production was 10 times higher, because of the strong N–F bonds and the much lower efficiency in providing F atoms.

The chemisorption of more reactive fluorine sources (F_2 and XeF_2) is known to generate readily multilayer of silylfluoride prior to the onset of etching (*3*). XPS analysis of F-etched silicon surfaces revealed that the average saturation of the SiF_x groups varied between $x = 2.6$ and 3.2 (*2*), suggesting that the top layers of Si atoms are highly fluorinated. How these silylfluoride layers break up and subsequently vaporize is basically an important and interesting question.

Numerous studies on the fragmentation mechanism of silylfluoride layers and the desorption of volatile products have been made by various laboratories using different complementary techniques. On the basis of the results obtained by chemiluminescence (*11–13*) REMPI (*8,10*), and EI/MS (*3,4,8*), the major gaseous products that have been conclusively identified are SiF_2 and SiF_4. Other thermal reaction products, such as SiF_3 and Si_2F_6, have also been reported (*4*). Our attempts to search for these products, specifically SiF_3 by REMPI/MS and EI/MS, have thus far been inconclusive.

Because of the relatively strong $\equiv$Si–SiF_3 bond, the SiF_3 surface species is expected to be quite stable. It is also the key precursor of SiF_4, the most abundant etch product. The fact that the results of the XPS analysis of postetched silicon surfaces indicated that SiF_3 was the most abundant species remaining on these surfaces (as alluded to above) (*2*) also supports its high stability. Otherwise, it would have desorbed prior to the postetching analysis. Thermochemically, the stability of the surface SiF_3 species, which is presumably attached to the bulk silicon by a strong $\equiv$Si–SiF_3 bond, may be estimated as strong as 80 kcal/mol (cf. D(SiF_3–SiF_3) $\simeq$ 80 kcal/mol (*14*)). Under steady state etching conditions, the SiF_3 group may undergo both decomposition (i.e., desorption) and fluorination reactions as schematically depicted below

$$\equiv Si\text{–}SiF_3(s) \longrightarrow \equiv Si\cdot(s) + SiF_3(g) \quad (1)$$

$$\equiv Si\text{–}SiF_3(s) \longrightarrow \equiv SiF(s) + SiF_2(g) \quad (2)$$

$$\equiv Si\text{–}SiF_3(s) \xrightarrow{+F} \equiv Si\cdot(s) + SiF_4(g) \quad (3)$$

$$\equiv Si\text{–}SiF_3(s) \xrightarrow{+F_2} \equiv SiF(s) + SiF_4(g) \quad (4)$$

On the basis of the known heats of formation of $SiF_x(g)$, x = 1–4 (*14*) and the approximation that D(≡Si–F) ≃ D(Si–F) = 129 kcal/mol, the following energy differences, measured with respect to the direct desorption reaction (1), can be reasonably estimated

$$\Delta H_2 - \Delta H_1 \simeq -31 \text{ kcal/mol}$$

$$\Delta H_3 - \Delta H_1 \simeq -128 \text{ kcal/mol}$$

$$\Delta H_4 - \Delta H_1 \simeq -238 \text{ kcal/mol}$$

Thus, the desorption of $SiF_3(g)$ via reaction (1) is thermochemically the least-favored process. In contrast, the production of $SiF_2(g)$ and $SiF_4(g)$ are thermochemically favored. In view of the large amounts of energy released from the final fluorination reaction (3) and (4), SiF_4 produced in the thermal etching reactions may be vibrationally excited. This internal excitation may lead to more fragmentation of SiF_4 by the electron impact resulting in more SiF_2^+ and SiF^+ than was observed in the room temperature mass spectrum of SiF_4. The large exothermicities for reactions (3) and (4) can also explain the abundance of SiF_4 in comparison with SiF_2 and the more likely absence of SiF_3. At lower temperatures, the production of SiF_4 is primarily limited by the availability of F atoms produced from the dissociative adsorption of the etchants.

Aside from reaction (2), $SiF_2(g)$ can also be generated by the direct breakup and fluorination of the $(SiF_2)_n$ polymeric chain, viz.

$$-(SiF_2)_n- \rightarrow -(SiF_2)_{\overline{n-1}} + SiF_2(g) \quad (5)$$

$$\xrightarrow{+F} -(SiF_2)_{n-m}-SiF_3 + SiF_2(g) + -(SiF_2)_{m-2} \quad (6)$$

Similar to the production of $SiF_4(g)$, the fluorination-induced fragmentation reaction of the $SiF_2(g)$ formation is favored by as much as 100 kcal/mol according to the estimate

$$\Delta H_6 - \Delta H_5 \simeq -[D(=SiF_2-F) - \Delta H_f(F)] \simeq -100 \text{ kcal/mol}$$

Since the generation of F atoms on the surface depends on the dissociative adsorption process, the measured activation energy for $SiF_2(g)$ production would naturally reflect the initial dissociation process. This is clearly manifested by Flamm's observation of the 2 kcal/mol activation energy for etching with F atoms (2) compared with 9–10 kcal/mol with F_2 and 22 kcal/mol with NF_3 (*8*).

The slow increase in the SiF_2 REMPI signal in comparison with the nearly instantaneous appearance of the SiF_x^+ ion signal shown in Fig. 6 suggests that a large fraction of desorbed SiF_2 radicals derive from decomposition of the partially fluorinated bulk silicon atoms in the subsurface layers, where the concentration of F atoms is lower because of the slow, rate-limiting diffusion process. Since the apparent desorption energy of SiF_2 is higher than that of SiF_4, most of the surface SiF_2 group must be consumed by further fluorination reactions forming SiF_4, owing to the continuous supply of F atoms at the surface. As the temperature of the surface is increased, however, the rate of SiF_2 desorption becomes competitive with SiF_4 production.

A close examination of the temperature dependence data presented in Table 1 and Fig. 3 and 5 indicates that the ions observed in electron impact can be described as originating from two neutral species, SiF_2 and SiF_4. Since the production of SiF_4 is highly exothermic, it may be desorbed from the surface with additional internal energy and under single collision conditions; this energy cannot be released before the electron impact. These "hot" SiF_4 molecules will have a different cracking pattern and could contribute to the SiF^+ and SiF_2^+ intensity. In Table 1, two apparent activation energies are unambiguously assigned for the production of SiF_2 (9.9 kcal/mol) and SiF_4 (4.0 kcal/mol). If we assume that only these two species produce the ions, SiF^+ and SiF_2^+, then the apparent activation energy for the appearance of these ions should be well described by a linear combination of the apparent activation energies for the production of SiF_2 and SiF_4 (assuming similar frequency factors for these two reactions). Shown in Table 2 is a summary of the relative contributions for these two neutral species in the observed temperature-dependent production of SiF^+ and SiF_2^+. On the average, the apparent activation energies that were previously calculated to be ca. 7 kcal/mol are also reasonably well described by a combination that is 98% 9.9 kcal/mol and 2% 3.97 kcal/mol. That is to say that the ions, SiF^+ and SiF_2^+, in this description come primarily from SiF_2 neutrals desorbed from the surface and contain only a small contribution from "hot" SiF_4. This is not an unlikely possibility. It should be mentioned that the temperature dependent data presented here were obtained under prolonged etching conditions under which the SiF_2 signal had reached a steady-state level.

It is also possible that the three apparent activation energies calculated from the temperature-dependent EI and MPI measurements may be the result of three different species (SiF_2, SiF_4, and Si_xF_y). Fragmentation of the third species by the electron impact would have to have significant yields of SiF_2^+ and Si^+. Since no species were detected at masses greater than SiF_4, we would propose as a third species SiF_3. Recently, electron impact cross-sections were reported for SiF_2 (*15*) and SiF_3 (*16*), with the latter fragmenting to produce SiF_2^+ and SiF^+ similar to our observation.

Energetically, we have shown that desorption of SiF_3 is not favored relative to SiF_2 and SiF_4; however, it is possible that this species is formed on the surface vibrationally excited and is desorbed prior to equilibration. Infrared luminescence has been observed on a silicon single crystal surface during exposure to XeF_2 by Chuang (*17*). Although the structured emission was assigned to vibrationally-excited SiF_4, not all the observed bands fit this assignment. The infrared emission is derived from a molecular species formed on the surface that does not emit following desorption.

Table 2
Preexponential Factors for the Expression
$I(T) = A^{*}Exp[-E_{a1}/RT] + B^{*}Exp[-E_{a2}/RT] + C$[a]

Si(100)	SiF^{+} A	SiF^{+} B	SiF^{+} $\left(\frac{B}{(B+A)}\right)^{*}100$	SiF_2^{+} A	SiF_2^{+} B	SiF_2^{+} $\left(\frac{B}{(B+A)}\right)^{*}100$
Undoped	10505	110	1.03	26848	316	1.16
	6603	120	1.78	20173	230	1.12
	14510	149	1.01	52371	494	0.93
	7253	180	2.42	39575	365	0.91
n-type	5849	66	1.11	21140	171	0.80
	16390	188	1.13	45072	284	0.63
	8436	158	1.83	21672	378	1.71
p-type	5452	89	1.60	6567	227	3.34
	2729	37	1.33	17515	307	1.72
	3928	173	4.21	11338	297	2.55
	3121	128	3.93			
Average			1.94 ± 1.13			1.48 ± 0.86

[a] E_{a1} = 9.9 kcal/mol; E_{a2} = 3.97 kcal/mol.

Recently, we reported a search for electronic states in SiF_3 by MPI. In that search, we detected a broad, unstructured photoion current in the ultraviolet at the SiF_3^+ mass for the reactions of SiF_3H + F and Si + F_2 (*9*). The absence of structure precludes assignment for this spectrum. We monitored the SiF_3^+ ion current in the reaction of a Si(111) single crystal with F_2 gas as a function of crystal surface temperature for laser excitation at 340 nm. An analysis of the temperature dependence yielded an apparent activation energy of ca. 7 kcal/mol. This is similar to the apparent activation energy calculated from the electron impact measurements of SiF_2^+ and SiF^+. The conclusion would be that the reaction product ionized by the laser to produce SiF_3^+ is the same species that fragments into $SiF_2{}^+$ and SiF^+ by electron impact, namely the neutral radical SiF_3. Further proof of this will require a direct diagnostic for SiF_3 radicals.

Finally, it is clear from the data given in Table 1 that n-type and p-type doping of the Si(100) crystal has no apparent effect on the measured activation energies for SiF_2 and SiF_4 production or the rate of the etching judging by the intensities of the SiF_2 REMPI and SiF_x^+ EI signals.

CONCLUSIONS

The gas phase products for the reaction of Si(100) n-type, p-type, and undoped crystals with F_2 have been monitored as a function of temperature under single collision gas surface conditions by EI/MS and REMPI/MS spectrometry. The apparent activation energy for SiF_2 desorption was found to be 9.9 kcal/mol. At high temperatures, desorption of this species is observed only after prolonged exposure. EI measurements indicate that SiF_4 is desorbed from the surface with an apparent activation energy of 4.0 kcal/mol and appears immediately upon exposure to F_2 at both room temperature and high temperatures. These values are consistent with the reported activation energy for etching of Si by F_2 of 8.9 kcal/mol (*12*) based on measurements under high pressure (1–10 torr) conditions since the rates of fluorination and fragmentation are expected to be strongly pressure dependent and should be quite different from what one would expect under single gas-surface collision conditions. We did not observe any effects of the silicon dopant on the rate of production of these two species. Energetically, production of these two species is favored over other SiF_x products. Production of SiF_4 is highly exothermic and, as a result, it is desorbed from the surface vibrationally excited. This additional energy may increase the production of SiF_2^+ and SiF^+ in the electron impact ionization of SiF_4.

ACKNOWLEDGMENT

The submitted manuscript has been offered by an employee(s) of the Midwest Research Institute (MRI), a contractor of the US Government under Contract No. DE-ACO2-83CH10093. Accordingly, the US Government and MRI retain a nonexclusive royalty-free license to publish or reproduce the published form of this contribution, or allow others to do so, for US Government purposes.

REFERENCES

1. Morar, J. F., McFeely, F. R., Shinn, N. D., Landgren, G., and Himpsel, F. J., *Appl. Phys. Lett.* **45,** 174 (1984).
2. McFeely, F. R., Morar, J. F., and Himpsel, F. J., *Surf. Sci.* **165,** 277 (1986).
3. Winters, H. F., Coburn J. W., and Chuang, T. J., *J. Vac. Sci. Technol.,* **B1,** 469 (1983).
4. Houle, F. A., *J. Chem. Phys.* **87,** 1866 (1987).
5. Vasile, M. J. and Stevie, F. A., *J. Appl. Phys.* **53,** 3799 (1982).
6. Mucha, J. A., Flamm, D. L., and Donnelly, V. M., *J. Appl. Phys.* **53,** 4553 (1982).
7. Houle, F. A., *J. Appl. Phys.* **60,** 3018 (1986).
8. Squire, D. W., Dagata, J. A., Hsu, D. S. Y., Dulcey, C. S., and Lin, M. C., *J. Phys. Chem.* **92,** 2827 (1988).
9. Horwitz, J. S., Dulcey, C. S., and Lin, M. C., *Chem. Phys. Lett.* **150,** 165 (1988).
10. Dagata, J. A., Squire, D. W., Dulcey, C. S., Hsu, D. S. Y., and Lin, M. C., *J. Vac. Sci. Technol.* **B5,** 1495 (1987).
11. Mucha, J. A., Donnelly, V. M., Flamm, D. L., and Webb, L. M., *J. Phys. Chem.* **85,** 3529 (1981).
12. Flamm, D. L., Donnelly, V. M., and Mucha, J. A., *J. Appl. Phys.* **52,** 3633 (1981).
13. Mitchell, M. J., Suto, M., Lee, L. C., and Chuang, T. J., *J. Vac. Sci.* **B5,** 1444 (1987).
14. Walsh, R., *J. Chem. Soc. Faraday Trans.* **I 79,** 2233 (1983).
15. Shule, R. J., Hays, T. R., Wetzel, R. C., Baiocchi, F. A., and Freund, R. S., *J. Chem. Phys.* **89,** 4042 (1988).
16. Hays, R. T., Shule, R. J., Baiocchi, F. A., Wetzel, R. C., and Freund, R. S., *J. Chem. Phys.* **89,** 4035 (1988).
17. Chuang, T. J., *Phys. Rev. Lett.* **42,** 815 (1979).

The Mechanism of Synthesis of High Temperature Materials by Combustion Processes

Z. A. MUNIR

Division of Materials Science and Engineering, University of California, Davis, CA 95616

ABSTRACT

Fundamental aspects of the self-propagating high-temperature synthesis (SHS) process are presented and discussed in terms of the application of this method for the preparation of high temperature materials. Recent investigations on the utilization of the SHS method for the synthesis of a variety of materials are discussed under the classifications of solid–solid and solid–gas reactions. The use of the SHS method to synthesize composite materials is discussed with particular reference to oxide/carbide composites.

Index Entries: Synthesis; combustion; high temperature materials.

INTRODUCTION

The possibility that highly exothermic reactions can become self-sustaining has been demonstrated through the mathematical analysis of Booth nearly four decades ago (*1*). More recently, the utilization of such reactions for the synthesis of high temperature materials has been the focus of attention of a considerable number of investigators in the Soviet Union, Japan, the US, and other countries. The synthesis of a large number of ceramic, metallic, and composite phases has been reported in a large number of investigations. Two recent reviews provide an overview of activities in this field of high temperature synthesis. (*2,3*).

Interest in this method, commonly referred to as the self-propagating high-temperature synthesis or SHS, focus on practical as

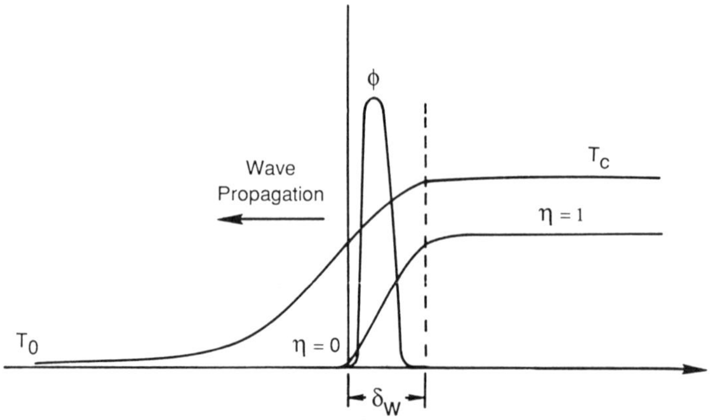

Fig. 1. Schematic representation of temperature profile of a combustion wave.

well as theoretical considerations. Attractive features of practical importance include the energy-efficient nature of the process (since once initiated the reactions are self-sustaining) and the relatively high purity of the materials produced by this method. This latter feature of the SHS process is a consequence of the high temperatures associated with the combustion wave and the expulsion of volatile impurities during its passage. Perhaps of greater practical potential for the SHS process is the feasibility of producing new nonequilibrium phases, both as single-phase and composite materials. Furthermore, with the simultaneous application of pressure during the passage of the combustion front, it has been shown that highly dense products can be produced (*4,5*). From a theoretical point of view, SHS reactions provide an area of investigation on the dynamics of highly nonisothermal synthesis reactions. Such reactions are associated with extremely high temperatures (adiabatic temperatures T_{AD}, of 5000 K are calculated for certain systems) and steep thermal gradients (as high as 10^5 $K.cm^{-1}$) and can propagate at rates as high as 25 $cm.s^{-1}$. Figure 1 shows a schematic representation of the temperature profile associated with a propagating combustion (reaction) wave. In this Fig., the wave is propagating from right to left with the temperature increasing from ambient, T_0, to the combustion temperature, T_C. Figure 1 also shows the spatial distributions of two additional parameters: The rate of heat generation, ϕ, and the degree of conversion of the reactants to the product, η. The latter increases from zero to one over the width of the combustion zone, δ_W. Over the same zone, the heat generation function, ϕ, increases from zero to a maximum, then decreases to zero as the reaction reaches completion. Because of thermal conduction, the temperature of the region ahead of the combustion zone is higher than the ambient temperature but decreases asymptotically to T_0. Thus, as the combustion front continues to move from right to left, the rate of heat evolution must include the contribution of heat transfer from the front before its arrival at any given point. Therefore, the heat flux to any given plane of combustion in the front is the sum of heat generation as a result of the chemical reaction and heat transfer from the front before its arrival.

This condition can be described by Fourier's one-dimensional heat conduction equation with a heat source that has the following simple form.

$$C_p \cdot \rho \cdot \partial T/\partial t = k\partial^2 T/\partial x^2 + q \cdot \rho\ \partial f(\eta)/\partial t \quad (1)$$

where

- C_p is the specific heat of the product
- ρ is the density of the product
- k is the thermal conductivity of the product
- q is the heat of the reaction
- T is the absolute temperature
- t is time
- x is the dimension along which the reaction is propagation
- $\partial f(\eta)/\partial t$ is the reaction rate

The rate of propagation of the combustion wave can be calculated from Eq. (1) if one assumes a temperature profile and a kinetic mode for the reaction (*6*). Assuming Arrhenius (chemical) kinetics (as opposed to diffusion kinetics), an expression for the rate of propagation is commonly reported in the following form.

$$v^2 = f(n) \cdot C_p \cdot k/q \cdot RT_c^2/E^* \cdot K_o \exp(-E^*/RT) \quad (2)$$

where

- v is the rate (velocity) of propagation of the front
- f(n) is a function of the order of the reaction
- T_c is the combustion temperature
- R is the gas constant
- K_o is a constant
- E^* is the activation energy of the reaction

The temperature dependence of the velocity, v, can, therefore, provide a means for the calculation of the activation energy, E^*, associated with the combustion process. In doing so, this approach can provide an insight on the mechanism of these high temperature reactions, at least in theory. In practice, however, the success of this approach depends on the validity of the assumed localization of the reaction. In Figure 1, the width of the reaction zone, δ_w, is depicted to be narrow in comparison to the heat-affected zone ahead of the wave. The physical significance of this is that the reaction starts and goes to completion within the boundaries of the narrow propagating wave. Obviously such an assumption is not valid in all systems. For example, in systems in which the typically powder reactants have relatively large particle sizes the reaction is not complete within the zone δ_w. In such cases and in others where the reactions are associated with high activation energies, the reaction continues to occur after the passage of the front, as has been observed in the case of the combustion synthesis of TiN (*7*) and NbB_2 (*8*). Figure 2 shows a schematic representation of combustion reactions of this general type.

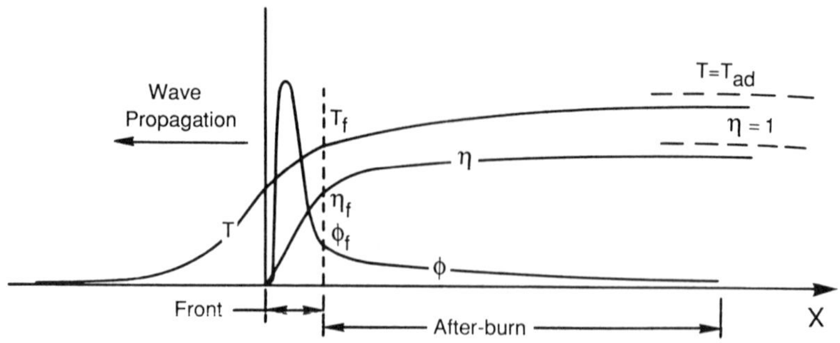

Fig. 2. Temperature profile of an incomplete combustion wave reaction.

The existence of cases in which η does not reach the value of one in the combustion front is indicative, of course, of incomplete reactions or the existence of an intermediate step in the conversion of the reactants to the product (*8*). The use of the temperature dependence of the wave velocity to obtain activation energies cannot be expected to provide the same results for the two general combustion reaction types discussed above (*31*). Other investigations must be conducted in order to determine the reaction dynamics of these combustion reactions. These include the coupling of temperature profile studies with synchrotron radiation experiments to determine the nature and sequence of reactions in the combustion front, as well as those taking place after its passage. The use of synchrotron radiation to obtain information on phase evolution with a time resolution of the order of one millisecond has the potential of elucidating the complex nature of combustion reactions (*9*).

Computer analysis of thermal profiles associated with the passage of combustion waves can provide functions for the parameters φ and η, and ultimately values for the activation energies of the processes. The degree of conversion and its spatial distribution, η(x), can be calculated from the first derivative of the temperature profile according to the following:

$$\eta(x) = [v \cdot C_p \cdot \rho(T - T_o) - k_1(dT/dx)]/[(k_2 - k_1)(dT/dx) + q \cdot v \cdot \rho] \tag{3}$$

where

- v velocity of the combustion wave
- C_p specific heat of the product
- ρ density of the product
- k_1, k_2 the thermal conductivities of the reactant and product phases, respectively
- q heat of reaction
- T absolute temperature
- T_o initial temperature of the reactants
- x dimension along which the wave is propagating

The rate of heat generation, φ(x), can be calculated from

$$\phi(x) = q \cdot v \cdot \rho(d\eta(x)/dx) \tag{4}$$

Finally, the activation energy of the process can be obtained from the temperature dependence of ϕ in accordance with

$$\phi(\eta,T) = q \cdot A \cdot f(n)\exp(-E^*/RT) \tag{5}$$

where A is a constant and f(n) is a function of the kinetic order, n, of the reaction. As indicated above, activation energies calculated by this method are not expected to agree with those determined from velocity measurements unless the reaction zone is narrow, i.e., the propagation of the wave constitutes the entire reaction. In many cases where this condition is not met, the activation energies calculated by the two methods differ by more than a factor of two as has been reported recently (*3*). An interesting aspect, with a great deal of practical significance, relates to the role of particle size and its influence on the temperature profile and hence the activation energy calculations. It has been reported that the reaction between fine particles of Ti and carbon is primarily confined to the combustion wave, with little or no "afterburn" (*8*). In contrast, afterburn, i.e., the continuation of the reaction after the passage of the wave was observed in samples with relatively large particle size of reactants. Since generally the goal of most synthesis reactions is the production of homogeneous phases, investigations on the effect of particle size on the dynamics of combustion reactions are of considerable importance.

SOLID–SOLID COMBUSTION REACTIONS

Reactions between solid particulates in self-propagating modes have been used to synthesize a large number of high temperature materials including carbides, borides, silicides, and intermetallic phases (*2,3*). As stated earlier the reactants are usually in compacted powder form and the process is initiated by an external heat source positioned near one end of the reactant mixture. Attempts to model the self-sustaining reactions between powders have not been notably successful primarily because of the necessity to invoke simplifying assumptions to a physically complicated system. In order to provide a configuration in which the initial boundary between reactants is well-defined, we have recently conducted experiments employing ensembles of laminated metallic films (*10*). Nickel (12.5 and 125 μm thick) and aluminum (12.5 μm thick) foils were laminated to give the required stoichiometry for one of the desired five "compounds" in the Ni-Al system. A schematic representation of the ensembles is shown in Fig. 3. To ensure the establishment of a self-sustaining combustion front for these ensembles, they were embedded in a 1:1 atomic ratio of a mixture of Ni and Al powders as shown in Fig. 4. Measurement of the ignition temperature, the temperature at which the combustion wave begins to propagate in the foil ensemble, showed it to correspond to the melting temperature of nickel, as seen in Fig. 5. This observation, which was independent of the nominal composition of the ensemble, provides contradictory results to observations on the propaga-

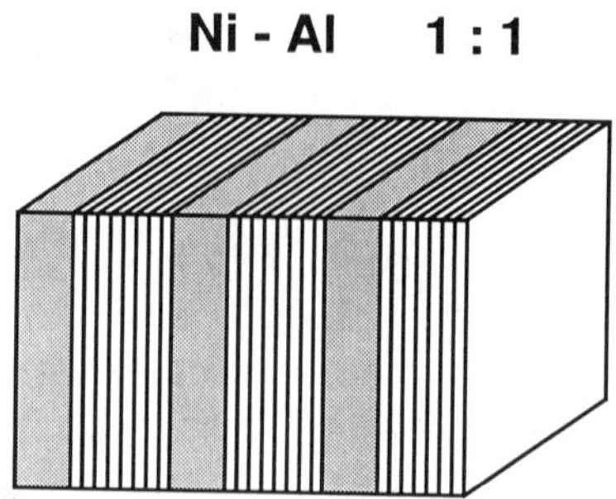

Fig. 3. Schematic representation of foil assembly.

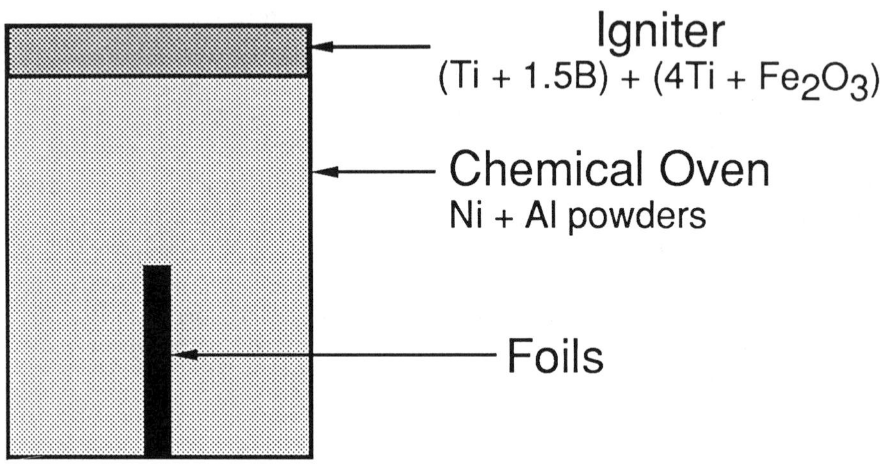

Fig. 4. Configuration of a chemical oven set-up for foil combustion

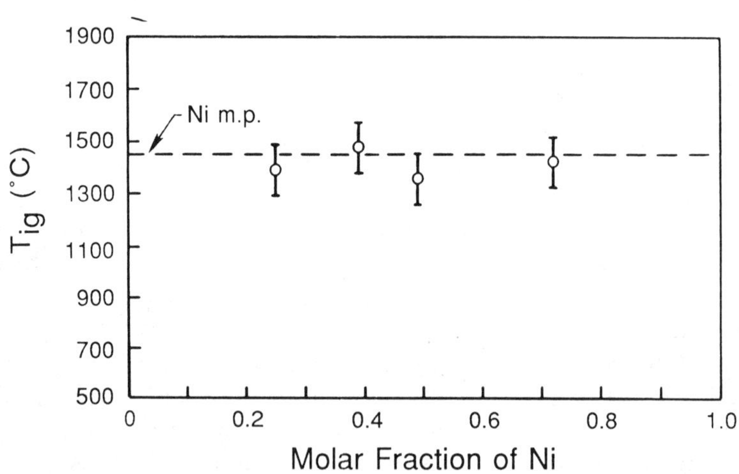

Fig. 5. The dependence of the ignition temperature on nickel content.

tion of a wave in Ni-Al powder compacts. In the latter case, it was suggested that the melting and spreading of Al is the event that triggers the advancement of the wave (*11*).

Experiments in which the advancement of the combustion wave was stopped by quenching provided insight into the mechanism of the self-propagating reaction in these foil ensembles. Figure 6 is a micrograph of 1:1 Ni-Al sample showing the combustion front and the region ahead of

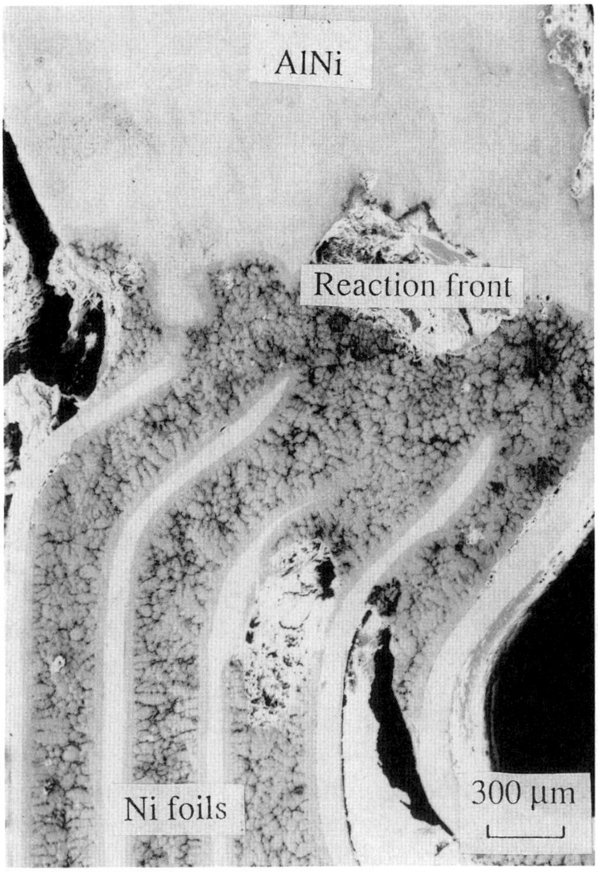

Fig. 6. Micrograph of quenched reaction front in a 1:1 Ni-Al foil ensemble.

it. Clearly seen ahead of the front are the remnants of the Ni foils (lighter regions), as well as evidence for extensive reaction between aluminum and nickel. In view of the observation that the propagation of the wave is initiated by the melting of nickel, it is assumed, therefore, that the reactions ahead of the front are those between molten aluminum and solid nickel. With one exception, the region behind the front, is found to be composed of the final product of combustion only. The only exception to this observation is in the case of the mixture 3 Al + Ni. Here, the region behind the front was found to contain Al_3Ni, as well as Al_3Ni_2, as shown in Fig. 7. Based on the assumption of a well-defined reaction interface, as in the case of a foil ensemble, recent investigations (*12*) have modeled the self-propagating process with the assumption that no interaction between the foils occurs in advance of the combustion front. The observations we report here show that such an assumption is not valid, at least in the case of Ni-Al foils combusted under the experimental configuration used in this study. The mechanism of the combustion reaction between nickel and aluminum ahead of the wave front includes the melting of aluminum and the subsequent dissolution of Ni to form intermediate compounds. The arrival of the wave results in reactions between molten nickel and these intermediates to form the final product.

Fig. 7. The products of combustion of a 3 Al + Ni foil ensemble.

SOLID–GAS COMBUSTION SYNTHESIS

Reactions between porous metals and gases are sufficiently exothermic to be self-sustaining in a number of cases. A notable example demonstrating this process is the combustion of refractory metals in nitrogen. For example, a self-sustaining combustion wave is easily obtained when titanium powder compacts are ignited in nitrogen, even at one atmosphere pressure (*7*). Since the local availability of nitrogen affects the reaction kinetics, the initial porosity of the sample and the pressure of nitrogen are expected to play dominant roles. From general principles, it is expected that higher porosity and pressure will enhance the conversion of the metal to the nitride (*12*). However, such expectations must be examined in light of possible changes in the porosity of the sample caused by the combustion reaction itself. Melting or sintering of the unreacted metal or the product of combustion can provide significant changes in the porosity of the sample and hence influence the degree of conversion.

Observations on the combustion synthesis of TiN and NbN provide contrasting behavior related to the influence of a molten phase (*13,14*). In the case of the former, the combustion temperature is higher than the melting point of Ti and thus the effects of the initial porosity and nitrogen pressure are more complex, as shown in Fig. 8. As the (initial) relative density, ρ, of the Ti sample increases, the conversion to the nitride increases, reaching a maximum of 65% at a density value of 57%. With further increases in ρ, the conversion decreases as seen in the figure. The lower conversion to the nitride at the lower densities (higher porosities) is a consequence of more extensive melting of Ti in the interior of the

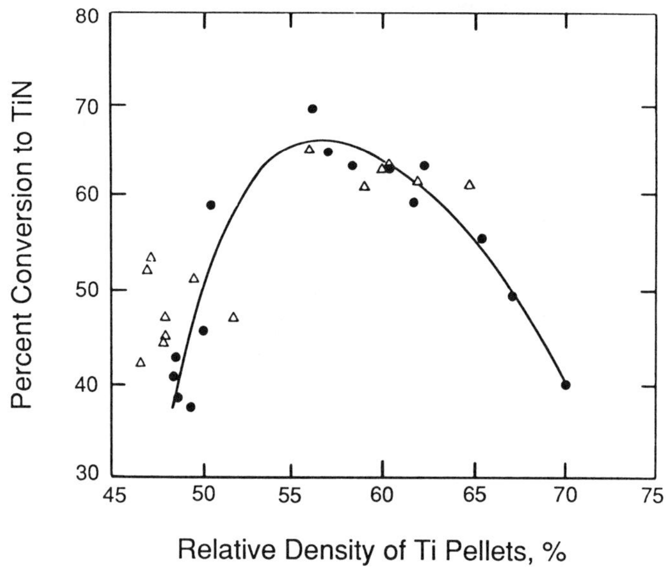

Fig. 8. The dependence of the percentage conversion to TiN on the initial relative density of the Ti compact.

Fig. 9. Micrograph of a cut combusted Ti sample showing evidence of melting.

sample as can be seen in a photograph of a sectioned sample, Fig. 9. As the density increases, the extent of melting decreases, and, hence, the degree of conversion to the product increases. However, as the density continues to increase to higher values, the conversion decreases, in accordance with the general expectation of the effect of reduced porosity. In contrast to this behavior, that of the combustion synthesis of NbN shows the expected dependence of conversion (and velocity) on porosity, as seen in Fig. 10. In this case, the combustion temperature is lower than the melting point of Nb and hence no significant change in porosity is expected. To demonstrate the effect of melting of Ti on the degree of conversion, experiments were carried out in which TiN is added to the Ti as a diluent. The addition of a diluent reduces the adiabatic combustion temperature (*4*) and thus can lower the extent of melting of the reactant. The addition of a sufficient amount of a diluent can reduce the combustion temperature to below the melting point of Ti.

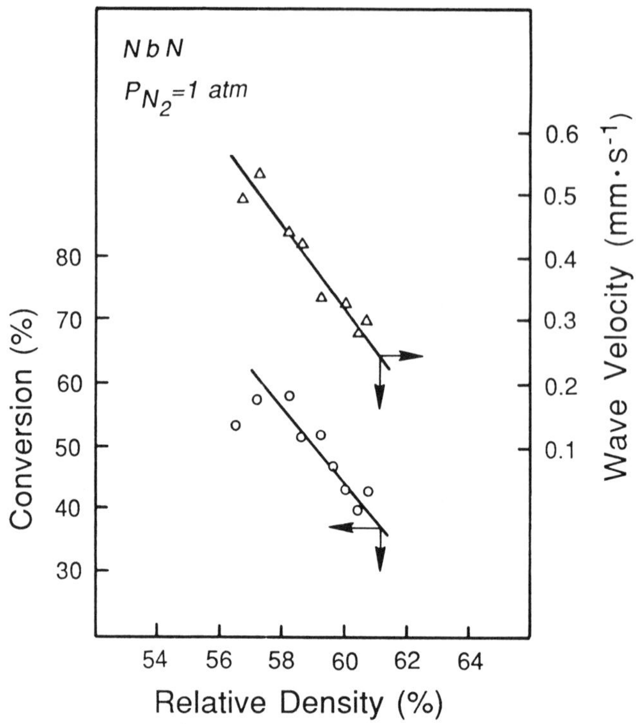

Fig. 10. The dependence on relative density of the degree of conversion to NbN and the velocity of combustion of Nb in 1 atm nitrogen.

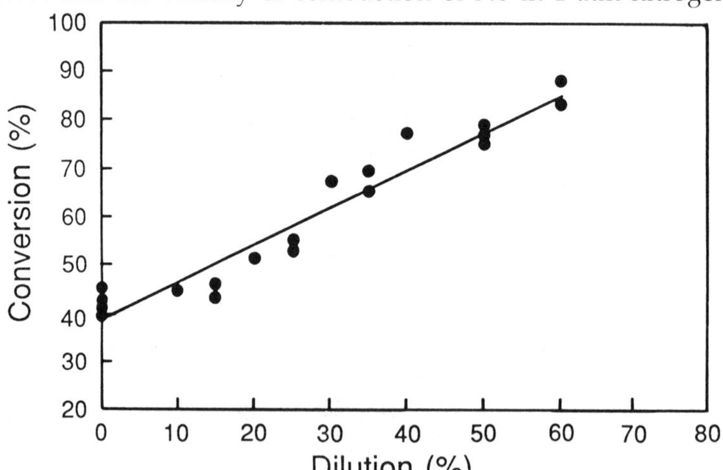

Fig. 11. The effect of dilution of Ti with TiN on the conversion to the nitride.

The results of such experiments, using Ti compacts with approximately 45% porosity are depicted in Fig. 11 for the combustion of TiN. A marked increased in the conversion to the product is seen as the percent of the TiN diluent is increased. In contrast to this behavior, a reaction system in which melting is not a consideration (e.g., that of the combustion of Nb in nitrogen) is affected in the opposite direction by the addition of a diluent. Figure 12 shows that the degree of conversion to NbN decreases

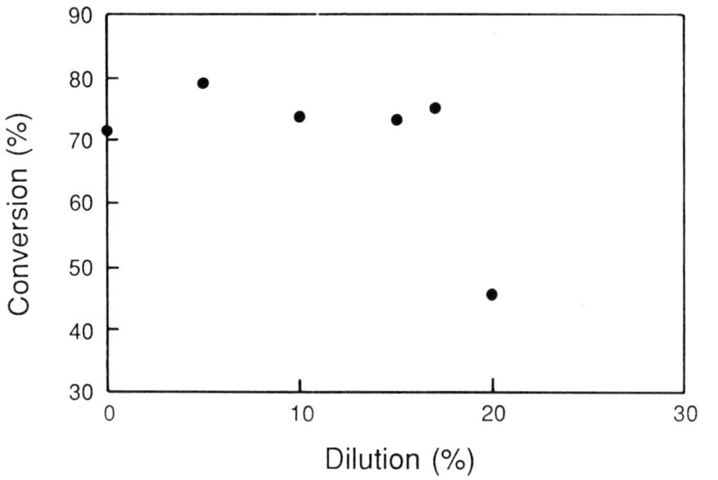

Fig. 12. The effect of dilution of Nb with NbN on the conversion to the nitride.

as the percent diluent is increased. Examination of combusted samples with a 20% diluent showed little evidence for reaction.

THE COMBUSTION SYNTHESIS OF COMPOSITES

As indicated earlier in this paper, one of the most attractive features of the SHS process is its potential for the preparation of composite materials. In the conventional methods of preparing these multiphase materials, the individual phases are typically prepared separately and are then mixed together and held at high temperatures for relatively long periods of time to ensure the formation of a desired interfacial bond. In contrast to this approach, the formation of composites by the SHS method accomplishes the two steps (phase formation and interfacial reactions) in one process. Whether this approach produces composites with significantly improved mechanical properties relative to those formed by conventional methods is not well established, however. There are reports that indicate that the microstructure of the SHS produced materials differs from that of conventionally prepared composites (*15*). With regards to microstructure, SHS-prepared composites are less likely to suffer from undesirable grain growth because of the relatively short duration the material is at high temperatures.

Our recent and current investigations have focused on the preparation of ceramic/ceramic and ceramic/metallic composites by the SHS process (*16,17*). The combustion synthesis of B_4C/Al_2O_3 and B_4C/MgO composites was investigated to provide an understanding of the mechanism of the process. Both of these composites are believed to form through a two-step process in accordance with the following:

$$x\,M + y/3B_2O_3 = M_xO_y + 2y/3\ B \qquad (6)$$

$$4\ B + C = B_4C \qquad (7)$$

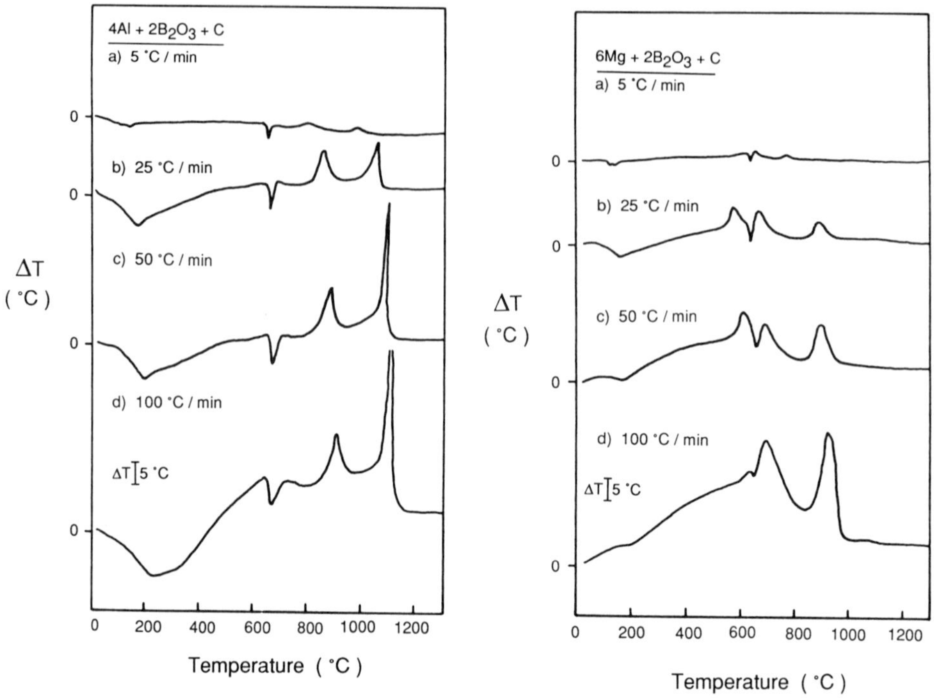

Fig. 13. The effect of heating rate on the DTA results of the reaction: (A) 4 Al + $2B_20_s$ + C; (B) 6 Mg + $2B_20_s$ + C.

where *M* refers to Al or Mg and *x* and *y* are the appropriate stoichiometric coefficients needed to produce Al_2O_3 or MgO through Eq. (6). The formation of B_4C from the elements (Eq. (7)) is not a highly exothermic reaction and does not proceed in a self-sustaining mode. However, in this case it is triggered by the first (thermite) reaction, Eq. (6). This is done by raising the temperature of the reactants, a circumstance that results in a higher adiabatic (or combustion) temperature (4), thus leading to the feasibility of a self-propagating reaction.

Differential thermal analysis (DTA) showed that the reaction of Eq. (6) is initiated prior to the melting of the metal reactant, in the case of Mg, and after the melting of the metal, in the case of Al, as can be seen from Figs. 13a, and b, respectively. In the case of Al, two exothermic peaks occur after the melting of this metal. X-ray analysis at different stages of the overall reaction has shown that the first exotherm is primarily that of the formation of aluminum borates, whereas the second exotherm is primarily that of the formation of Al_2O_3. Furthermore, as seen in Fig. 13(a), as the rate of heating in the DTA experiments is increased, the relative magnitude of the second peak increases. Ultimately, when the reaction is carried out in a combustion wave mode, the second exothermic reaction is apparently the only one taking place as judged by the presence of only minor amounts of the borate phases in the product.

An important observation in the combustion synthesis of the B_4C/MgO composite concerns the role of gaseous magnesium. Because of the

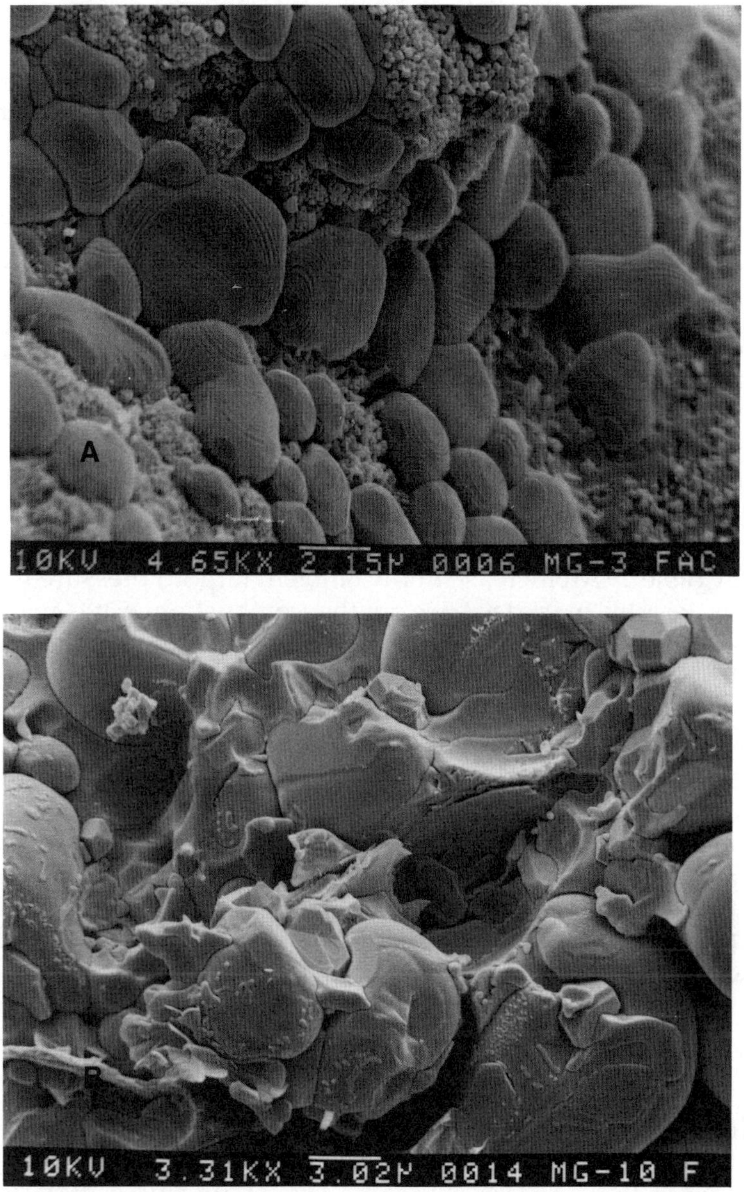

Fig. 14. Microstructure of the product of the reaction 6 Mg + $2B_2O_3$ + C: (A) Combustion at 1 atm argon; (B) Combustion at 1020 atm argon.

relatively high vapor pressure of Mg at the combustion temperature (1723 K), gas-solid reactions are likely to play a role in the formation of the B_4C/MgO composite. Evidence of gas-phase reactions are seen in the striations of the MgO particles formed under an ambient argon pressure of one atmosphere, Fig. 14(a). Increasing the pressure of the ambient inert gas to 1020 atm, results in a significantly different morphology (Fig. 14(b)), suggesting a decrease in the role of Mg(g) in the overall reaction.

Table 1
Effect of Argon Pressure on the Combustion Wave Characteristics in the Synthesis of MgO/B_4C Composites

Reactants	Argon pressure, atm	Wave velocity, $cm \cdot s^{-1}$	Combustion temperature, K
$6\ Mg + 2\ B_2O_3 + C$	1	—[a]	—[a]
	13	0.93	1723
	1020	0.56	2453

[a]Not obtained owing to excessive evaporation.

Analysis of condensates of volatile species from the combustion process, coupled with sample weight loss measurements, showed clearly that the application of a higher pressure had suppressed the volatilization of magnesium during combustion.

The transition from a combustion reaction in which the vapor phase plays a significant role to one in which the role of a gas phase is insignificant can be further appreciated by an examination of the results presented in Table 1. An increase in the ambient (argon) pressure from 13 to 1020 atm, results in a decrease in the combustion wave velocity, an increase in the combustion temperature, and a decrease in the amount of weight loss experienced by the samples. The trend in wave velocity with respect to that of the combustion temperature are inconsistent with general theoretical expectations and experimental observations (refer to Eq. (2)). We interpret these observations in terms of changes in the contribution of the vapor phase to the overall reaction. At low ambient pressure (13 atm) the rate of propagation of the reaction is enhanced by the permeation of the gas phase ahead of the wave. This contribution is diminished significantly (as judged by the low value of weight loss) when the pressure is increased to 1020 atm, and hence the velocity is decreased. In contrast, when the ambient pressure is high, the weight loss is low and thus more of the reactants are involved in the self-propagating reactions leading to a higher combustion temperature.

ACKNOWLEDGMENTS

This work was supported by a grant from the Office of Basic Energy Sciences of the US Department of Energy.

REFERENCES

1. Booth, F., *Trans. Farad. Soc.* **49,** 272 (1953).
2. Munir, Z. A., *Amer. Ceram. Soc. Bull.* **67,** 342 (1988).
3. Munir, Z. A. and Anselmi-Tamburini, U., *Mater. Sci. Repts.*, in press, 1989.
4. Holt, J. B. and Munir, Z. A., *J. Mater. Sci.* **21,** 251 (1986).
5. Yamada, O., Miyamoto, Y., and Koizumi, M., *Amer. Ceram. Soc. Bull.* **64,** 319 (1985).
6. Maksimov, E. I. and Shkadinskii, *Comb. Explos. Shock Waves* **7,** 392 (1971).

7. Munir, Z. A., Deevi, S., and Eslamloo-Grami, M., *High Temp. High Press.* **20,** 19 (1988).
8. Merzhanov, A. G., *Archiv. Combust.* **1,** 23 (1981).
9. Mesentsev, N. A., Sheronov, M. A., Smirnykh, V. V., Sokolov, A. S., Sukhorukov, A. V., Tolochko, B. P., Ushakov, U. A., and Vlasov, A. M., *Nucl. Instr. Meth. A246,* 604 (1986).
10. Anselmi-Tamburni, V. and Munir, Z. A., unpublished work.
11. Aleksandrov, V. V. and Korchgin, M. A., *Comb. Explos. Shock Waves* **23,** 557 (1987).
12. Munir, Z. A. and Holt, J. B., *J. Mater. Sci.* **22,** 710 (1987).
13. Eslamloo-Grami, M. and Munir, Z. A., unpublished work.
14. Zhang, S. and Munir, Z. A., unpublished work.
15. Miyamoto, Y. and Koizumi, *Proc. Int'l. Symp. on Sintering,* Tokyo (1987).
16. Wang, L. and Munir, Z. A., *Proc. Int'l. Symp. on Combustion and Plasma Synthesis of High Temperature Materials,* San Francisco, October, 1988.
17. Dunmead, S., Munir, Z. A., and Holt, J. B., *ibid.*

Application of a Thermochemical Data Bank System to the Calculation of Metastable Phase Formation During PVD of Carbide, Nitride, and Boride Coatings

P. J. SPENCER*[,1] AND H. HOLLECK[2]

[1]Lehrstuhl für Theoretische Hüttenkunde und Metallurgie der Kernbrennstoffe, RWTH Aachen, D-5100 Aachen; and [2]Institut für Material- und Festkörperforschung, Kernforschungszentrum, Karlsruhe, D-7500 Karlsruhe 1, FRG

ABSTRACT

In this work, which represents the preliminary stage of a larger research program, an attempt has been made to use thermodynamic calculation methods to predict not only stable, but also metastable phase formation during PVD of carbide, nitride, and boride coatings.

Theoretical investigations of phases formed from mixtures of TiN, ZrN, HfN, or TiC with AlN or SiC, and of phases formed in the system Ti-B-C have been made.

The necessary data for the thermodynamic calculations were in some cases obtained directly from a thermodynamic data bank, in other cases by critical evaluation of literature information using the data bank software, and in the case of metastable phases, by estimation methods based on the stored data for stable compounds.

Such phase diagram information as is already available has been used as a guide for the calculations, and from the resulting Gibbs energy vs composition curves for different temperatures, possible ranges of existence of metastable phases have been deduced.

Index Entries: PVD; metastable phases; thermodynamic calculations; carbides; nitrides; borides; data bank; phase diagrams.

*Author to whom all correspondence and reprint orders should be addressed.

INTRODUCTION

A major application of hard metal carbides and nitrides is as coatings for tool materials. Such coatings have frequently been produced by CVD methods, but there is an ever-increasing interest in PVD (Physical Vapor Deposition) techniques, in particular because PVD processes enable homogeneous, multicomponent phases with improved properties to be deposited easily. The stoichiometry of these phases can be controlled by adjusting the process parameters, whereas by controlling the substrate temperature, some materials can be deposited in different crystalline structures or as an amorphous phase.

In the present work, an attempt is being made to calculate the most likely phases to be formed during Physical Vapor Deposition of different coating systems. The basis for the calculations is the hypothesis made by Saunders and Miodownik (*1*) with regard to the deposition of metallic alloys on low-temperature substrates. They suggest that, due to the low substrate temperatures used, diffusion cannot take place to produce separation into a multiphase structure in the deposited materials. As a result a *single phase* coating is formed from the phase with the lowest Gibbs energy at the overall coating composition and substrate temperature concerned. Using this hypothesis, Saunders and Miodownik were able to predict accurately the ranges of stability (both stable and metastable) of different phases found in codeposited alloys at low substrate temperatures. In the case of mixed (Al,Ti)N coatings, it is known that, whereas the component nitrides AlN (hexagonal) and TiN (cubic NaCl) show virtually no solubility in each other at lower temperatures, PVD of a $^{50}/_{50}$ mixture of the two results in a homogeneous, single phase coating with the NaCl structure. The resulting material displays significantly improved properties over those of the component nitrides.

The present paper describes the results of a first series of investigations in which the feasibility and reliability of thermodynamic calculation techniques applied to the prediction of stability ranges of metastable phases in PVD-deposited mixed nitride, carbide, and boride coating materials has been studied.

The systems investigated are the binary combinations AlN-TiN, AlN-ZrN, AlN-HfN, AlN-TiC, SiC-TiN, SiC-TiC, AlN-SiC, in which AlN and SiC have a hexagonal structure and TiN, ZrN, HfN, and TiC a cubic NaCl structure, and the ternary system Ti-B-C, in which the TiB_2 (hexagonal), TiB (orthorhombic), TiC (cubic NaCl) and B_4C (rhombohedral) phases are of particular interest.

METHODS

To investigate the phases likely to be formed during PVD coating processes, Gibbs energy values must be obtained both for the stable and possible metastable crystalline phases formed by the components, as well as for amorphous mixtures in the systems concerned.

For the substances considered in the present work, Gibbs energy data are available from the SGTE (Scientific Group Thermodata Europe) thermodynamic data bank (2) for the pure substances in their stable crystalline form. No values are available, however, for the other main quantities needed, namely, the "lattice stabilities" of the pure components of the coatings in different metastable structures and the thermodynamic properties of mixing for the stable and metastable phases in the coating systems concerned.

These values must, therefore, be estimated. For present purposes, estimated values have been obtained assuming a particular stoichiometric composition for the pure nitride, carbide, or boride concerned. This represents a considerable simplification in some cases, because phases with a wide stoichiometry range may be involved. Nevertheless, since the present calculations are in the nature of a feasibility study, the results obtained should provide sufficient qualitative accuracy.

Lattice-Stabilities

In view of the dearth of phase diagram information for mixed nitride/carbide/boride systems, an attempt has been made to estimate lattice stability values for metastable phases of coating components by carrying out a simple comparison of thermodynamic data for different compounds with different structures across the Periodic System. The necessary data were again obtained from the SGTE thermodynamic data bank (2). The enthalpies of formation of different transition metal compounds in their stable structures were plotted as a function of the position of the transition metal in the Periodic System. The resulting plots for compounds with the cubic NaCl and hexagonal structures are shown in Figs. 1a and 1b, respectively. From these plots, it is possible to estimate a value of $\Delta_f H(298\ K)$ for a particular compound in a metastable NaCl or hexagonal structure by using extrapolation, interpolation, or comparison methods.

Entropy values may be derived in a similar way, and by approximating the heat capacity of a compound in a metastable structure by the heat capacity of the stable form of that compound, Gibbs energy values may be calculated for the metastable phase over a wide range of temperature.

This approach has been used to obtain lattice stability data for metastable forms of the nitrides, carbides, and borides of interest in the present study.

Mixing Properties

In the total absence of experimental information on the thermodynamic properties of mixing in the coating systems under investigation, the first assumption made might be that mixing between components in the same crystal structure is ideal. If this assumption is applied to mixtures of transition metal nitrides or carbides with AlN or SiC, the ranges of solid solution calculated from the resulting Gibbs energy curves are

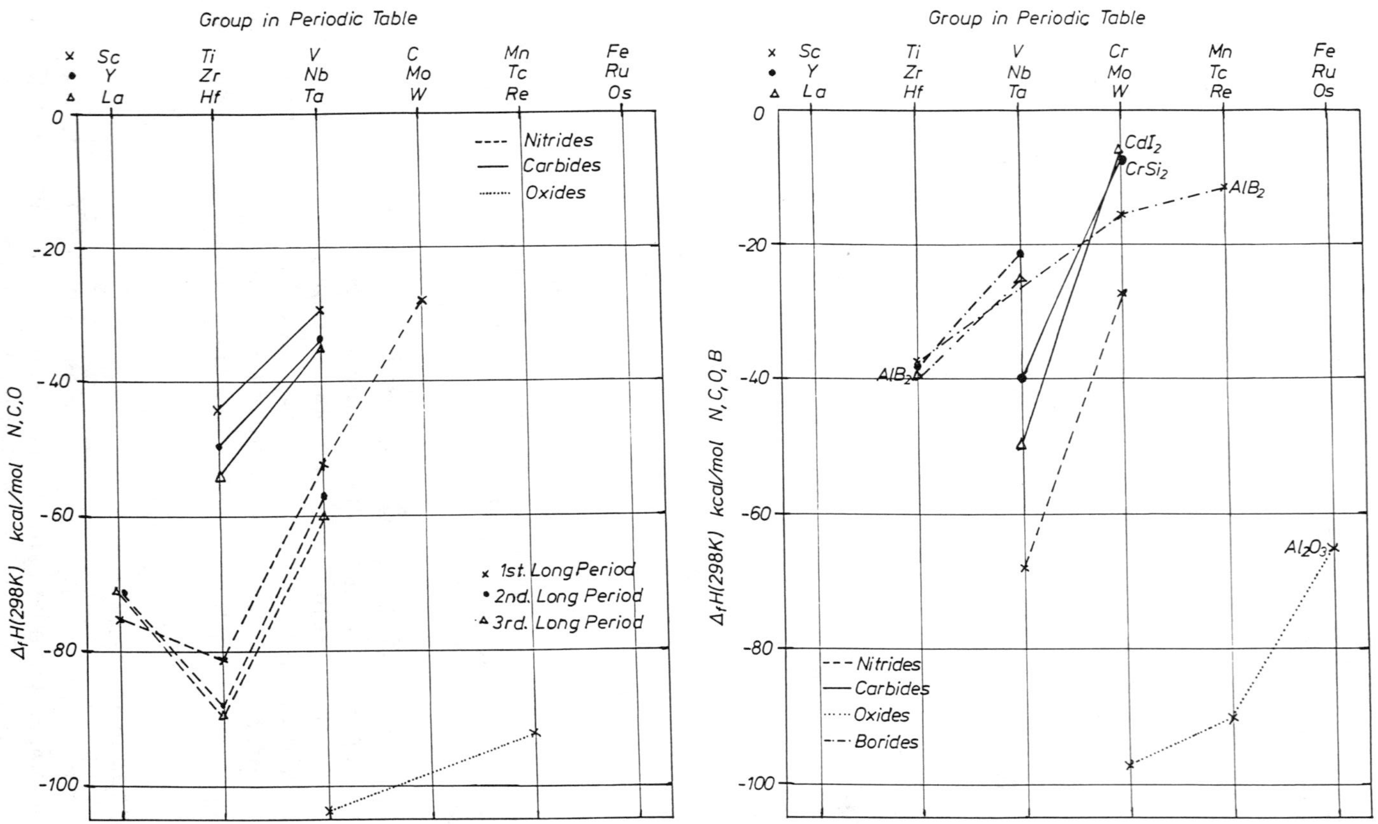

Fig. 1a. Values of $\Delta_f H(298\ K)$ as a function of Periodic sequence for compounds with the NaCl structure. **1b.** Values of $\Delta_f H(298\ K)$ as a function of Periodic sequence for compounds with the hexagonal structure. (The prototype hexagonal structure concerned is denoted by the appropriate compound name.)

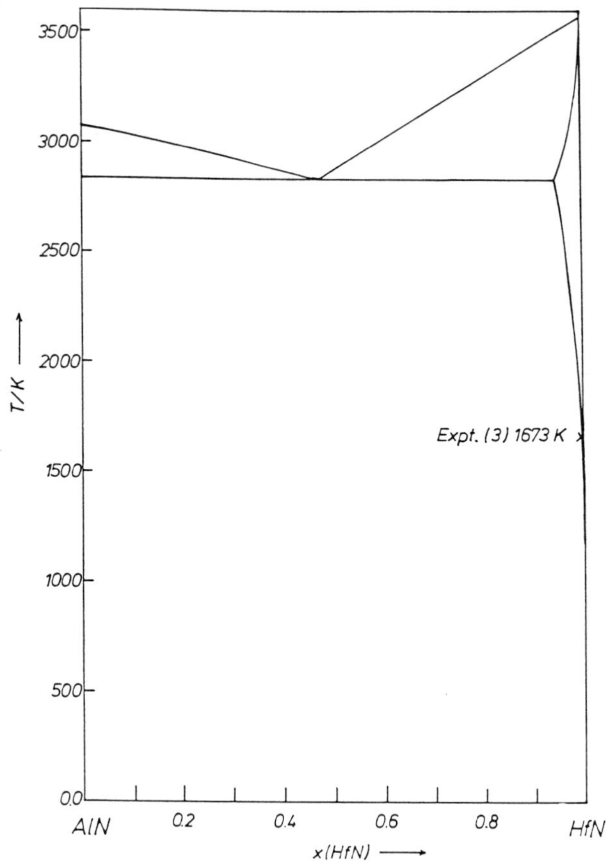

Fig. 2. The calculated AlN-HfN phase diagram using estimated lattice stabilities for metastable structures and regular mixing parameters of 0, 50 and 50 kJ mol^{-1} for the liquid, hexagonal and NaCl phases, respectively.

much too great and do not correspond to the very restricted solubilities observed in practice.

A more realistic description of the solubility range in the above-mentioned systems can be obtained by assuming regular mixing behavior in the hexagonal and NaCl phases. The regular parameter is selected so as to provide calculated solubilities that are consistent with the very limited amount of experimental phase diagram information available in the literature (e.g., 3–7). Thus, for example, the experimental solubility of AlN in HfN (3) can be approximated when a regular solution parameter of + 50 kJ mol^{-1} is used in conjunction with the lattice stability data for the stable and metastable forms of AlN and HfN to describe the enthalpies of mixing both in the "NaCl" and hexagonal phases. This consistency between the calculated and experimental equilibrium phase diagram information is a prerequisite for further calculations involving metastable phases. With the further assumption that mixing in the liquid phase is ideal, the phase diagram shown in Fig. 2 was calculated. This calculation was performed using the program SAGE (*see* ref. (*8*)). For purposes of the calculations, the gas phase that forms at high temperatures due to dissociation of AlN-rich mixtures, was suppressed.

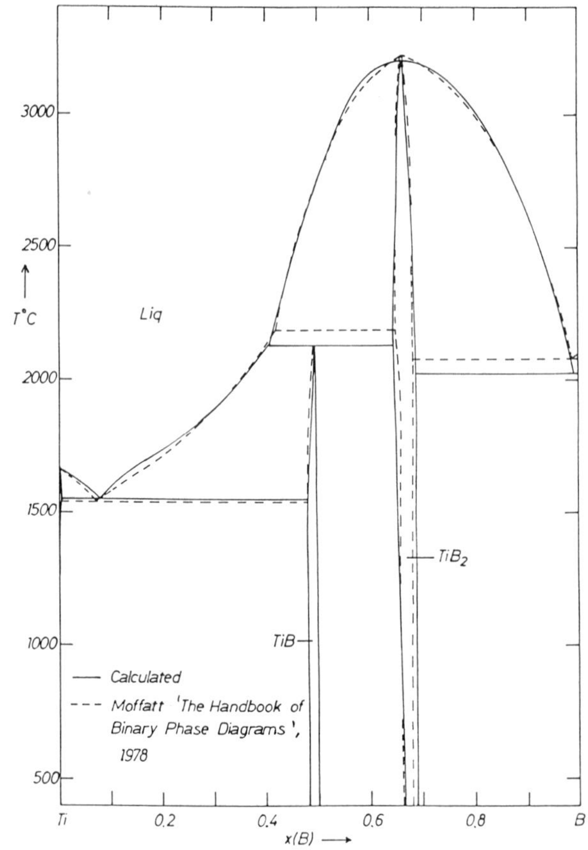

Fig. 3. The calculated Ti-B phase diagram.

Because experimental phase diagram information is very scarce for the systems being studied, the regular solution parameters used for AlN-HfN mixtures have been retained and used in conjunction with the appropriate lattice stability values to calculate thermodynamic properties and phase equilibria in other systems of interest here. The eutectic form of the phase diagram is very similar in all cases, except for AlN-SiC mixtures, for which the mixing of two compounds with a hexagonal structure results in a complete series of solid solutions at high temperatures. At lower temperatures, a wide miscibility gap is observed (*6*), and the mixing properties of the hexagonal phase in the AlN-SiC system require a regular solution parameter of $+44$ kJ mol^{-1} to give reasonable consistency with the experimental miscibility gap information.

The ΔG vs composition curves calculated using the estimated and derived thermodynamic data disclose possible ranges for the appearance of hexagonal and NaCl phases during PVD of the corresponding coating mixtures to be predicted.

In the case of the Ti-B-C system, thermodynamic calculations are assisted by the availability of experimental phase diagram information in the form of ternary isothermal sections and pseudobinary sections (*9–11*). Nevertheless, the basis for calculations in the ternary is provided by

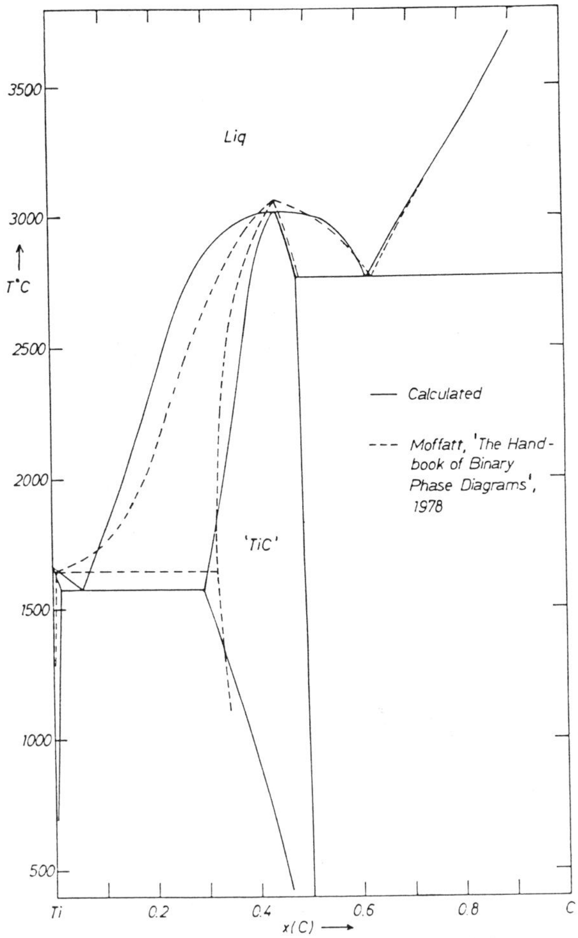

Fig. 4. The calculated Ti-C phase diagram.

evaluated thermodynamic data for the three bordering binary systems B-C, B-Ti, and C-Ti. A thermodynamic description of the B-C system is available from Kaufman et al. (*12*), but it was necessary to carry out evaluations of the B-Ti and C-Ti binary systems using experimental thermodynamic and phase diagram data. This was done using the optimization program written by Lukas et al (*13*). The calculated phase diagrams resulting from the evaluations are shown in Figs. 3 and 4. It should be noted that the ranges of stoichiometry of the carbide and boride phase are also described in these evaluations.

Thermodynamic properties of mixing for the different phases across the entire ternary system were now derived such that the experimental phase diagram information for the Ti-B-C system (*9–11*) could be satisfactorily reproduced from the resulting set of thermodynamic values. Figures 5a,b and 6a,b present comparisons of experimental and calculated isothermal sections of the Ti-B-C system at 2300 and 2000°C, respectively.

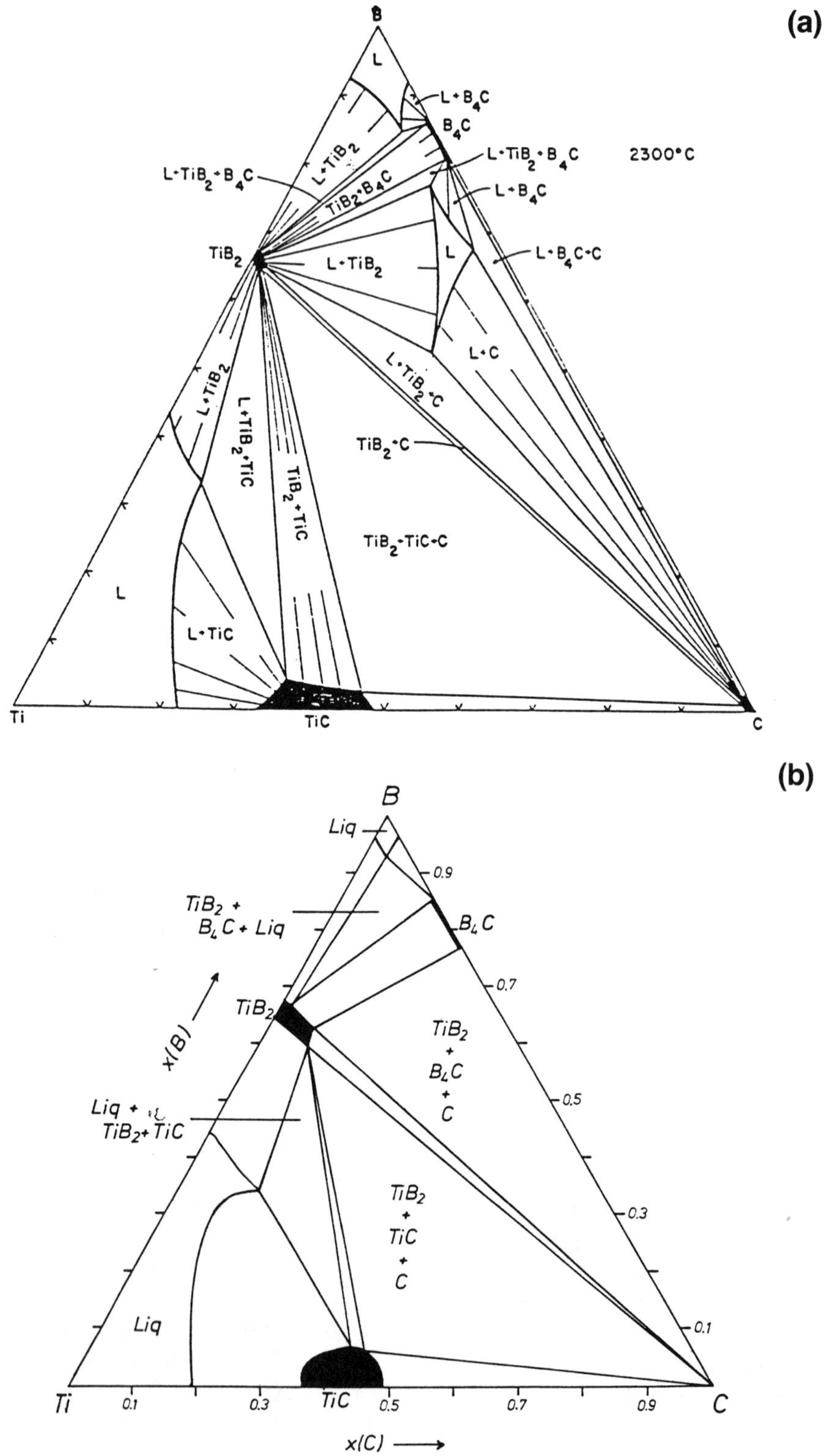

Fig. 5a. The experimental isothermal section at 2300°C (*10*) for the Ti-B-C-system. 5b. The calculated isothermal section at 2300°C for the Ti-B-C system.

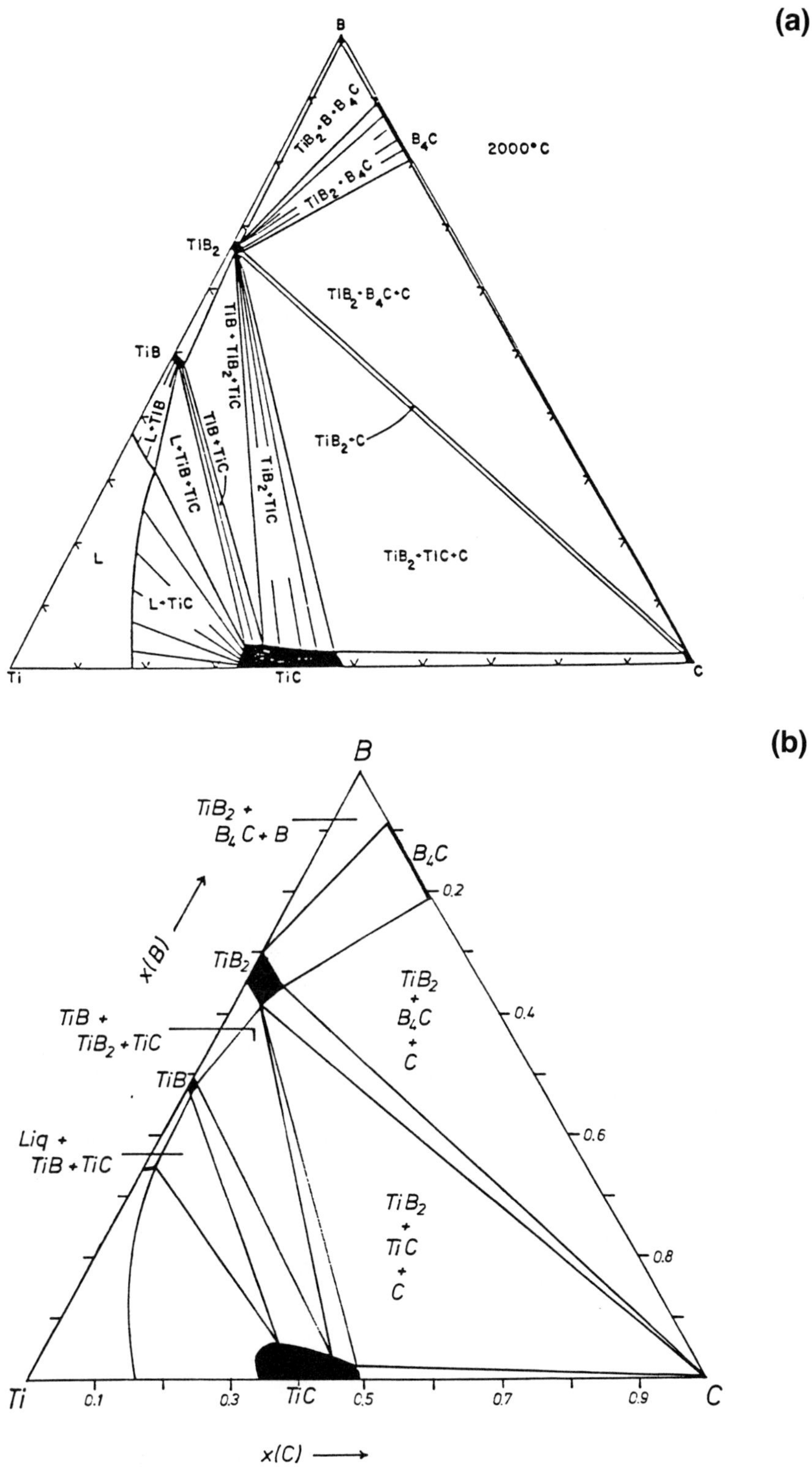

Fig. 6a. The experimental isothermal section at 2000°C (*10*) for the Ti-B-C system. 6b. The calculated isothermal section at 2000°C for the Ti-B-C system.

From the relative positions of the Gibbs energy surfaces for the different phases at a given temperature (calculated and plotted using the program SAGE (*8*) and other databank software), it is now possible to predict likely ranges of existence of particular phases in PVD deposited coatings of Ti-B-C mixtures.

RESULTS AND DISCUSSION

Mixtures of AlN or SiC with TiN, TiC, ZrN, or HfN

Very little experimental information is available concerning the composition and temperature ranges for deposition of metastable phases during PVD of these mixtures. However, an important "test system" is AlN-TiN, for which a number of investigators have produced coatings by sputtering techniques, but generally only at compositions close to the 50/50 composition. Knotek and Leyendecker (*14*) have recently carried out a more extensive study in the Al-Ti-N system. Their results show that the range of compositions in which the cubic NaCl structure is observed in PVD deposited coatings extends from TiN to about 60 mol% AlN in the TiN-AlN section (Fig. 7). They do not report on the nature of the phase at higher AlN concentrations.

From Fig. 8, it can be seen that the calculated ΔG curve at 500 K for the NaCl structure in AlN-TiN mixtures lies lower than the curve for the hexagonal structure at compositions from TiN to about 70 mol% AlN. At higher AlN concentrations the hexagonal structure has lower ΔG values. There is, thus, a very satisfactory correspondence between experiment and calculation with regard to the metastable range of the NaCl structure.

The form and relative positions of the Gibbs energy curves obtained for the mixtures of AlN with ZrN, HfN, and TiC are very similar, and the predicted range of the NaCl structure in these mixtures is also to about 70 mol% AlN.—in the case of AlN-TiC to somewhat higher AlN concentrations. For SiC-TiN and SiC-TiC mixtures, the NaCl structure is predicted to be stable to about 90 mol% SiC.

In all systems, the calculated equilibrium solubility of the component with the hexagonal structure (AlN or SiC) in the "NaCl" phase is greater than vice versa. This is in accord with experimental observation. It is, nevertheless, apparent that for reliable thermodynamic prediction of metastable phase formation in these systems, considerably more experimental information in the form of solubility data or thermodynamic properties of the metastable mixtures themselves, is required. This will greatly assist the estimation of unknown lattice stability values and mixing parameters.

At the present time, in the total absence of experimental data, no attempt has been made to introduce an amorphous phase into the calculations. This must be attempted as soon as possible, however, to provide consistency with unpublished observations for certain systems.

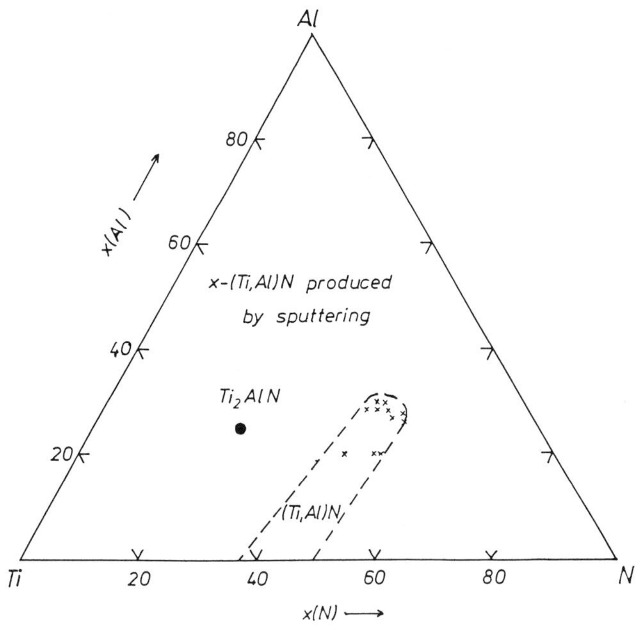

Fig. 7. Experimental range of the cubic NaCl phase in the Al-Ti-N system resulting from sputtering onto a low-temperature substrate (*14*).

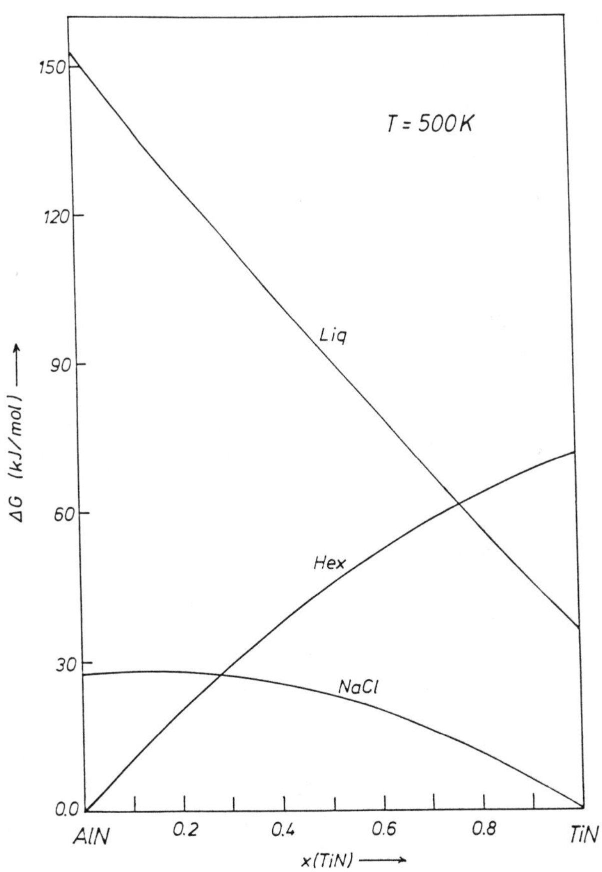

Fig. 8. Gibbs energy of formation curves at 500 K for the liquid, hexagonal and cubic NaCl phases in the AlN-TiN system.

The Ti-B-C System

The thermodynamic data for Ti-B-C mixtures, derived as described above, have been used to calculate Gibbs energy curves for the different phases of the system in selected sections across the ternary diagram. Thus, Fig. 9 shows the calculated Gibbs energy curves at 500 K for the hexagonal, NaCl, orthorhombic, and liquid phases in the section TiB_2-TiC.

Once again, assuming that the phase with the lowest Gibbs energy at any overall composition will be the phase formed during PVD of TiB_2-TiC mixtures, then the predicted phase formation is as follows:

From TiB_2 to approx. 23 mol% TiC—hexagonal
23 mol% TiC to 54 mol% TiC—orthorhombic
54 mol% TiC to TiC—cubic NaCl

These results may be compared with results from preliminary experimental investigations of coatings produced by sputtering of TiB_2-TiC mixtures. The experimentally observed phases are:

80 TiB_2—20 TiC—hexagonal
60 TiB_2—40 TiC—hexagonal
50 TiB_2—50 TiC—amorph./hexagonal?
40 TiB_2—60 TiC—amorph.
20 TiB_2—80 TiC—cubic NaCl

These first experimental observations are not definitive, but it is clear that there is a significant discrepancy between calculation and experiment in this case. The experimental results imply that the derived Gibbs energy curve for the orthorhombic phase lies too low and should perhaps be located above the curve for the hexagonal phase at most compositions. In addition, the necessity of describing the thermodynamic properties of an amorphous phase is clearly demonstrated. Such a description must again await more experimental information.

Figure 10 shows the corresponding calculated Gibbs energy curves at 500 K for the section TiB_2-C. In this section, the hexagonal phase is predicted to be stable to just over 50% C, whereas at higher C concentrations the liquid phase has lower Gibbs energy values than the hexagonal, NaCl or orthorhombic structures. This could mean that an amorphous phase is likely to be formed in this range, but no experimental information is as yet available to confirm or refute such a prediction.

These first theoretical investigations of the feasibility of applying thermodynamic calculation methods to the prediction of metastable phase formation in PVD coatings offer some grounds for optimism. More experimental information is required to give a firmer basis to the calculations, however. Some necessary experimental work is currently being undertaken. At the same time, better methods of estimating lattice stability values for metastable structures are being sought.

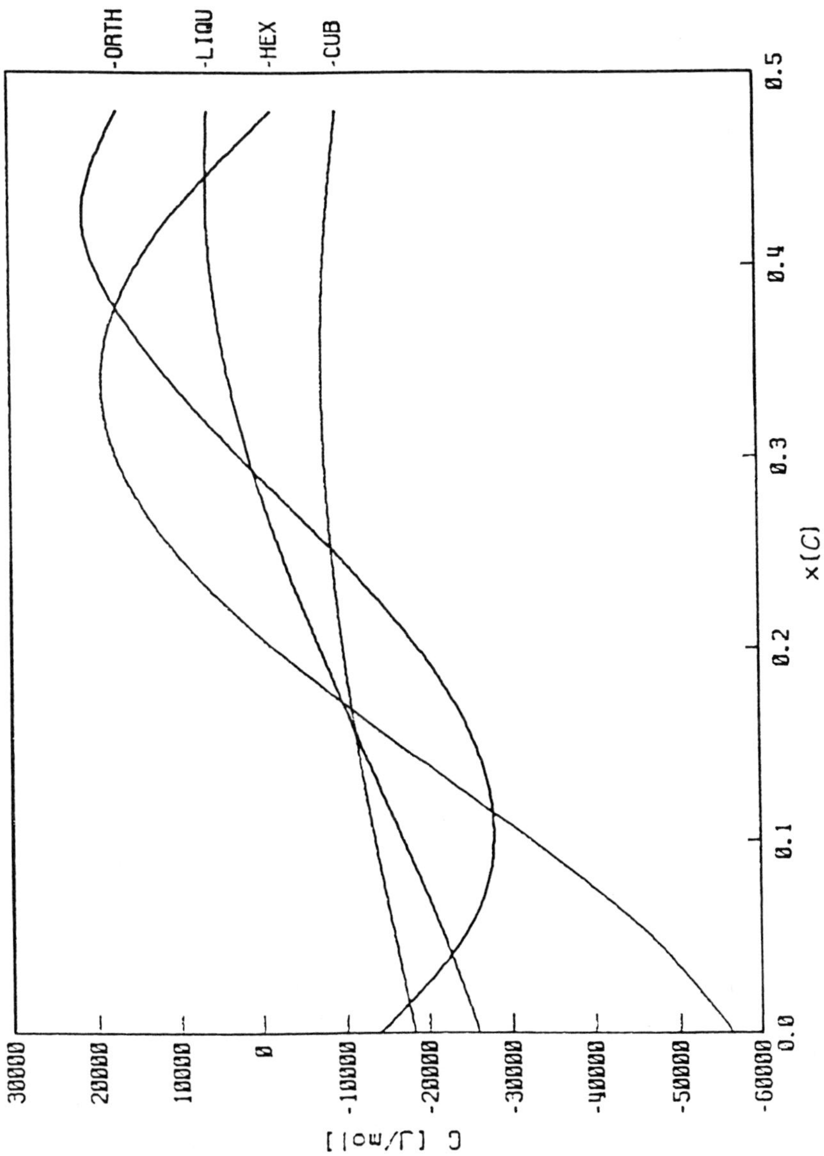

Fig. 9. Gibbs energy curves at 500 K in the TiB_2-TiC section of the Ti-B-C system.

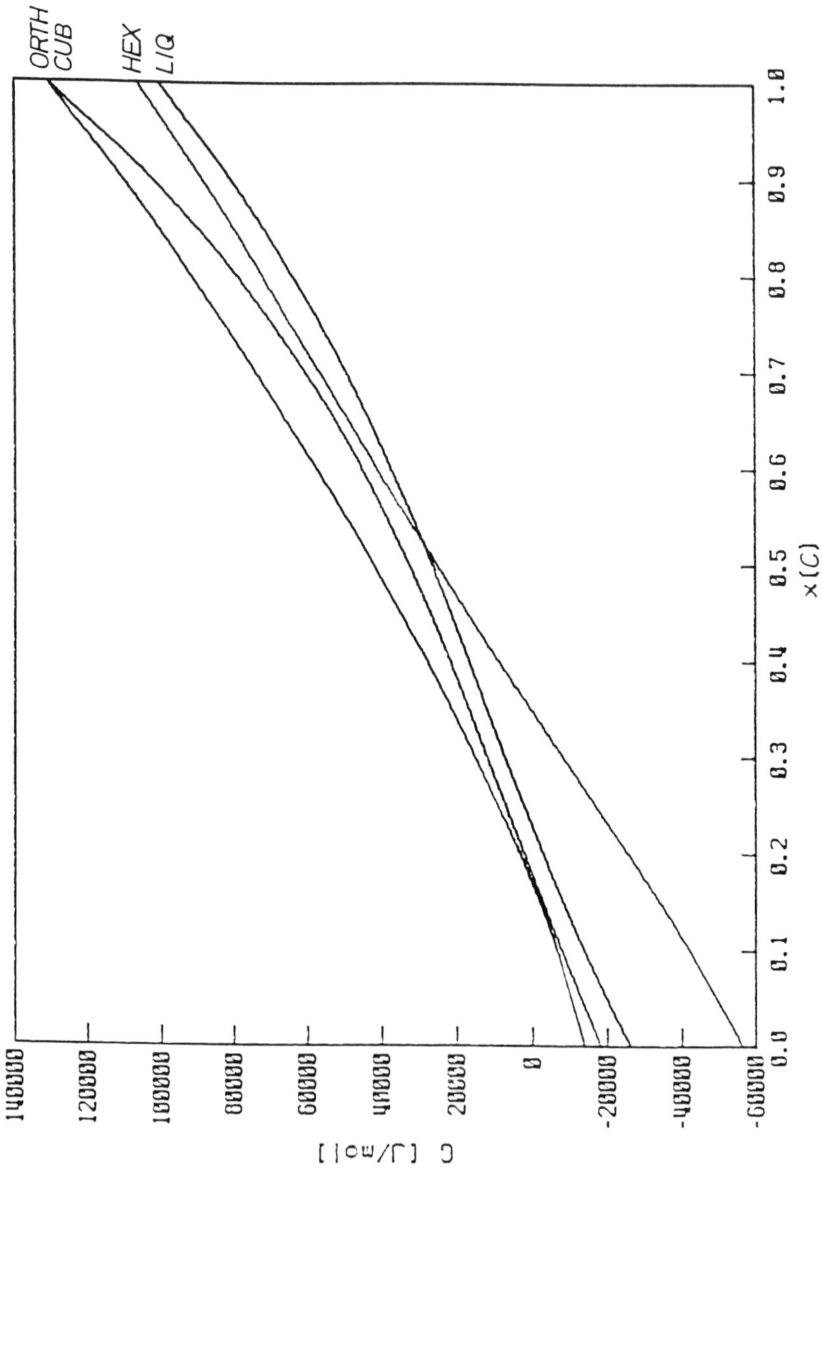

Fig. 10. Gibbs energy curves at 500 K in the TiB_2-C section of the Ti-B-C system.

ACKNOWLEDGMENTS

The authors would like to thank K. Hack for helpful discussions and Kaija Lehtimaki for carrying out calculations in the Ti-B-C system.

REFERENCES

1. Saunders, N. and Miodownik, A. P., *CALPHAD* **9,** 283 (1985).
2. Ansara, I., Sundmann, B., *Computer Handling and Dissemination of Data*, P. S. Glaeser, ed., Elsevier, CODATA, 1987.
3. Schuster, J. C. and Bauer, J., *J. Nucl. Materials* **120,** 133 (1984).
4. Schuster, J. C. and Bauer, J., *J. Solid State Chem.* **53,** 260 (1984).
5. Nickl, J., Schweitzer, K., and Luxemberg, P., *J. Less Common Met.* **26,** 335 (1972).
6. Rafaniello, W., Plichte, M. R., and Virkar, A. V., *J. Amer. Ceram. Soc.* **66,** 272 (1983).
7. Holleck, H., *Binäre und Ternäre Carbide und Nitride der Übergangsmetalle*, Bericht KfK 3087 B, 1981.
8. Eriksson, G. and Hack, K., *See CALPHAD* **8,** 15 (1984).
9. Nowotny, H., Benesovsky, F., Brukl, C., and Schob, O., *Mh. Chem.* **92,** 403 (1961).
10. Jeitschko, W., Nowotny, H., and Benesovsky, F., *Mh. Chem.* **94,** 565 (1963).
11. Ordanyan, S. S., Unrod, V. I., and Augustinik, A. I., *Porosh Met.* **9,** 40 (1975).
12. Kaufman, L., Uhrenius, B., Birnie, D., and Taylor, K., *CALPHAD* **8,** 25 (1984).
13. Lukas, H. L., Henig, E. Th., and Zimmermann, B., *CALPHAD* **1,** 225 (1977).
14. Knotek, O. and Leyendecker, T., *J. Solid State Chem.* **70,** 318 (1987).

Ultrafine Composite Synthesis by Laser-Induced Reactive Evaporation and Rapid Condensation

GAN-MOOG CHOW*[1,2] AND PETER R. STRUTT[1]

[1]*Box U-136, Institute of Materials Science, The University of Connecticut, Storrs, CT 06269; and* [2]*Current address: U.S. Naval Research Laboratory, Code 6371 Washington, D.C. 20375*

ABSTRACT

A new approach in nanocomposite synthesis involves rapid condensation of metallic and nonmetallic species produced by laser-induced reactions. These composite surface layers form by codeposition of fine (25–120 nm dia.) amorphous silica fibers and a metal matrix, at rates of $\geq$ 1 micron/s, and are strongly adherent to a nickel substrate material. The synthesis involves rapid condensation of amorphous silica from an intense plume, formed by the interaction of a continuous CO_2 laser beam with a ceramic target. This is either accompanied by codeposition of: tungsten from a heated filament via a chemical vapor transport mechanism, or aluminum formed by laser evaporation. These reactions occur in a 98.5% hydrogen, 1.5% methane atmosphere using a continuous flow reaction chamber, with access to an incident continuous CO_2 laser beam. In the composite layer thus produced, the silica fibers exist in the form of a random weave structure.

Careful analysis of free fiber assemblies, protruding above the composite layer surface, reveals no evidence of growth involving catalytic particles, as in the vapor–liquid–solid growth mechanism. An alternative theoretical model involving the fiber nucleation and growth has been developed. Basically, ultrafine silica particles (10–25 nm dia.) form by rapid condensation from the laser plume. These, by virtue of a strong dipole moment, coagulate to form clearly defined linear arrays, as in colloidal aggregation. Experimental observations provide strong evidence that these arrays transform into cylindrical

*Author to whom all correspondence and reprint orders should be addressed.

fibers. In the theoretical model, it is proposed that small linear particle clusters circulate within natural convection cells (above the substrate surface). Fiber growth continues by the progressive agglomeration of the ultrafine particles provided by laser rapid condensation. Eventually, the fibers adhere to the growing metal matrix either by chemical adsorption or electrostatic interaction. Effects of radiation–vapor interactions, diffusion-controlled coagulation are also discussed.

Index Entries: Laser plume; reactive evaporation; rapid condensation; chemical vapor transport; nanoscale fibers; nanometer-size composites synthesis.

INTRODUCTION

The intent in synthesizing composite materials is to produce unique properties by the selection of appropriate constituent phases, and then by achieving control of morphology and microstructural scale. With the diverse types of composite structures obtainable, using appropriate techniques, the possibility clearly exists of synthesizing desirable materials. Among many synthesis techniques, laser-assisted reactions are frequently employed. Some of the interesting features of the laser reactions include: (1) high intensity irradiation obtainable for rapid heating, (2) selective laser beam and material coupling, (3) small interaction zone resulting in localized reactions, thus acting effectively as cold-wall reactors with minimal contaminations, (4) high rates of chemical and physical transformations of reactants, and (5) generation of metastable phases.

To date, three laser-assisted mechanisms have been employed in surface modification and material synthesis, namely, laser–chemical vapor deposition (LCVD), laser–liquid reaction, and laser–evaporation. Little work has been done by the laser–evaporation technique, thus many potential applications remain (*1*). The nature of laser–evaporation depends upon the coupling of the beam with the material to be evaporated. Often, a material plume induced by absorption of the beam is observed in evaporation. The plume provides active vapor species that react in a reactive environment and form desirable products that can condense to produce surface layers.

From a practical standpoint of composite synthesis, control of the microstructural scale is critially important since, for example, methods leading to a drastic reduction in interfiber spacing will result in increased strength and fracture toughness. To date, methods for drastically reducing the microstructural scale have been on equiaxed structures, as in the pioneering work of Gleiter (*2*). Here, the synthesis of 1–100 nm particle materials resulted in the discovery of unique physical and chemical properties (*2–4*). Gleiter (*2*) synthesized nanoscale powder materials by evaporation, followed by rapid condensation. Conversion into bulk form was then achieved by removal from the processing chamber and powder compaction.

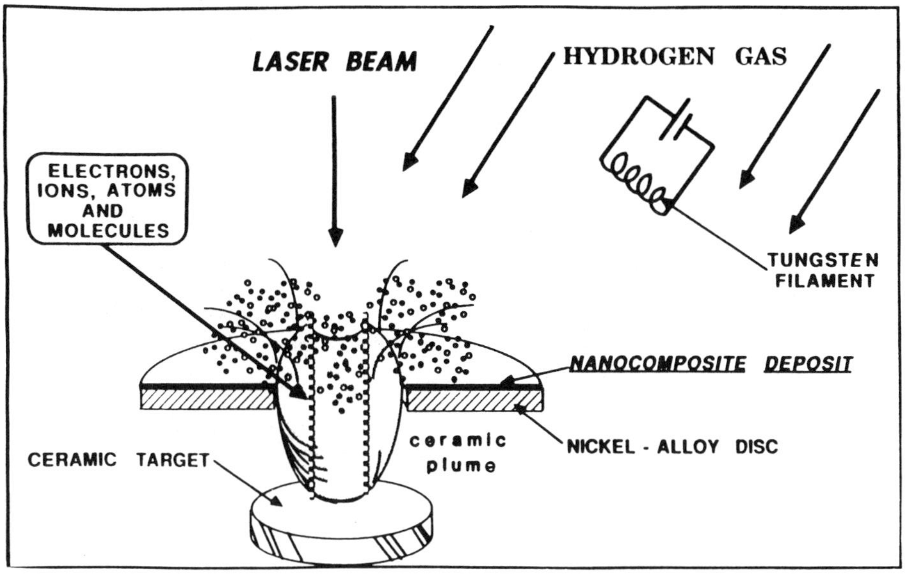

Fig. 1. A schematic showing the experimental processing setup.

This paper addresses a new approach of ultrafine composite synthesis by laser-induced coevaporation of a metallic target and a ceramic target, in the presence of a heated tungsten filament, within a reducing environment. Species produced by rapid condensation from the laser plume, along with the products derived from the chemical transport reactions involving the heated filament, form composite layers consisting of a metal matrix and randon-weave fiber network. Interestingly, the diameter of the fibers is in the range of 25–120 nm. It is shown here that the rapid condensation approach of Gleiter (2) can be extended to the synthesis of nanoscale fibers. Furthermore, it demonstrates that nanocomposites can be synthesized in a single process.

METHODS

The processing setup for codeposition of metal and nanoscale ceramic material is shown schematically in Fig. 1. In this arrangement, a nickel alloy disk (0.06 cm thick) of composition (wt%) 58% Ni, 4% Al, 9% Cr, and the balance of V, Ti, Mn, Co, and C is mounted above a ceramic target. The composition of the latter (wt%) is 55% SiO_2, 42% Al_2O_3, and a balance of TiO_2, MgO, and CaO. Prior to laser processing, the chamber is evacuated and flushed with nitrogen. Following this, and during laser irradiation, a continuous flow of 98.5% hydrogen/1.5% methane is maintained at a flow rate of 150 sccm. The 99.9% purity tungsten filament, in proximity to the laser beam-material interaction zone, is switched on 15 min before each experiment and is maintained at 1700°C; it remains on during actual laser irradiation.

The processing procedure involves irradiating the nickel-alloy disk with a cw carbon dioxide laser, using a power density of 2×10^4 W/cm^2.

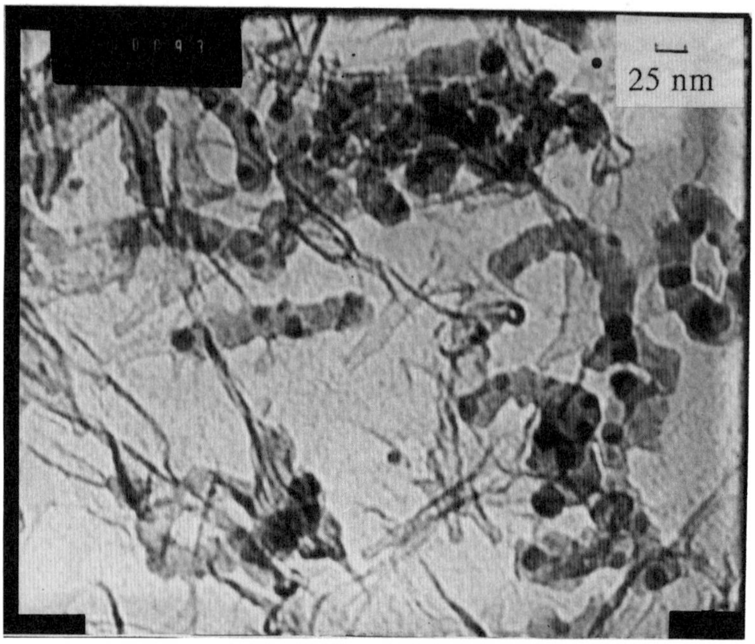

Fig. 2. TEM micrograph of silica fibers produced by a short 2-s laser–ceramic target interaction time.

This intensity is not only sufficient for rapid evaporation, but for complete beam penetration through the disk at the focal spot. This immediately results in interaction with the underlying ceramic target to produce a plume containing predominantly silica species. Subsequent condensation of these species, together with simultaneous:

(i) condensation of aluminum, from the aluminum containing nickel-alloy disk; and
(ii) deposition of tungsten, via chemical transport from the heated filament, results in deposition of a composite layer, on the cooler region of the nickel-alloy disk.

Experimentally it was found that (i) occurred for shorter (2–3 s), and (ii) occurred for longer (8–10 s) interaction times. Microstructural examination, using electron microscopy, revealed that silica incorporated in the deposited films frequently existed in the form of fibers with dimensions in the 25–120 nm range. These fibers were also analyzed by Auger spectroscopy, Fourier transform infrared spectroscopy, and X-ray and electron diffraction. In all cases, they were amorphous. Observation of fiber-extraction replicas in the electron microscope was particularly valuable in studying initial fiber growth and development of randomly distributed fiber networks. The effect of increasing the ceramic target–laser interaction time is clearly apparent by comparing fiber structures in Figs. 2 and 3. With the short interaction time of about 2 s (Fig. 2), 25 nm diameter fibers are seen to be forming by apparent coagulation of silica particles. However, a longer interaction time (about 7–8 s), results in well-developed 100 nm diameter fibers that form relatively dense net-

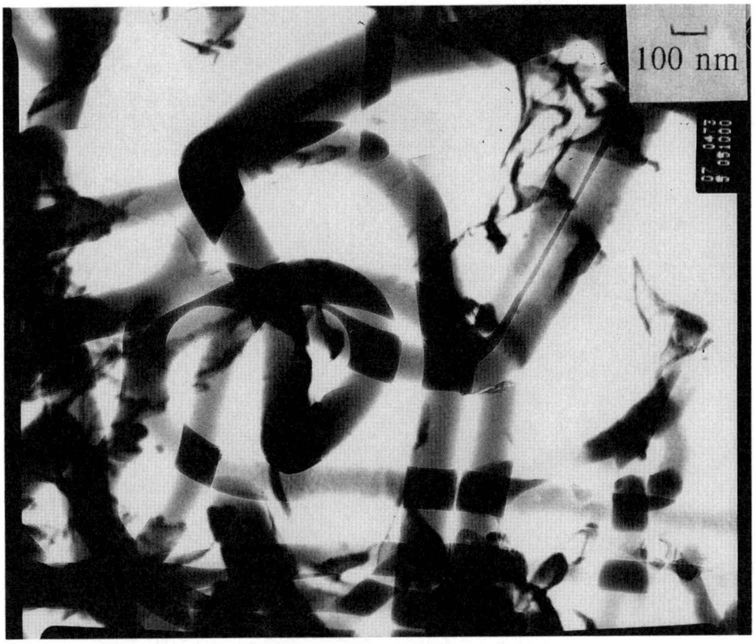

Fig. 3. TEM micrograph of silica fibers produced by a long 7-s laser–ceramic target interaction time.

works. The rapidity of fiber growth is particularly intriguing, and it has been established that composite films can grow at rates of ~ 1 μm/s.

In the experiments, a critical factor was found to be the position of the filament relative to the nickel-alloy disk (*see* Fig. 1). With the filament in close proximity to the disk, for instance, nanoscale silica particles no longer formed fibers, but aggregated into large crystalline particles. These followed chain-like structures, as shown in Fig. 4.

DISCUSSION

X-ray diffraction data indicated that the as deposited films contained either W or Al. Al was deposited from evaporation of the Ni alloy when short (e.g., 1 s) laser–ceramic target interaction times were used. The vapor pressure (10^{-1} torr) of Al in the molten metal allowed Al to escape and condense on the cooler surrounding region. Absence of noticeable Ni and Cr in the deposits is explained by their longer time of stay in the vapor cavity owing to their low vapor pressures at the temperature concerned, thus allowing these species to be superheated. Atoms of these superheated vapors have too much kinetic energy and would not be able to condense on the Ni alloy substrate. W was not deposited for short laser beam–ceramic target interaction times. This is because of the insufficient concentration of water vapor available in the interaction zone to facilitate W transport (*see following discussions*). When longer laser beam–ceramic target interaction times were used (about 8–9 s), W was deposited as a matrix. In this case, the underlying ceramic plume heated the Ni alloy substrate by blackbody radiation, making the substrate too

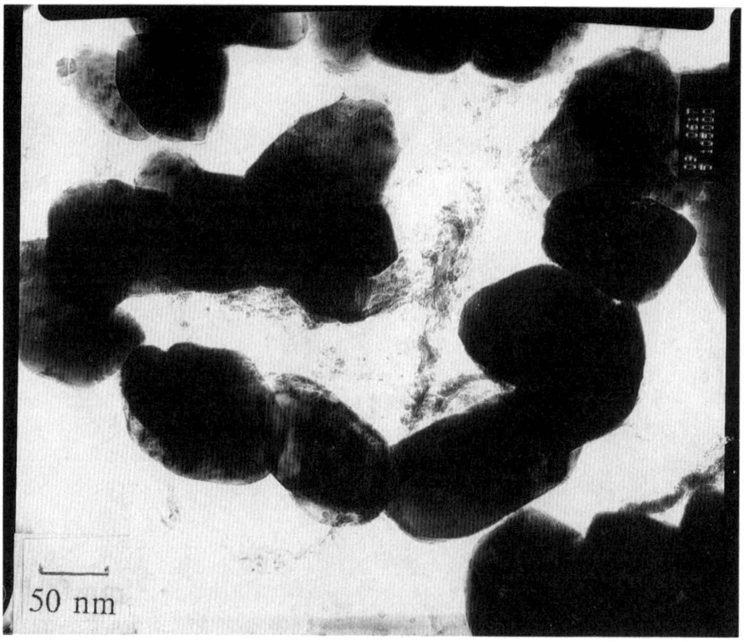

Fig. 4. TEM micrograph of linear silica aggregates produced by placing the heated W filament in closer proximity to the Ni alloy substrate disk.

hot for Al to condense on it. Simultaneous codeposition of a ceramic-dispersed phase was observed in both Al and W matrices. This ceramic phase was found to be amorphous silica fibers with diameters in the range of 25 to 120 nm, as revealed by the high resolution TEM extraction replicas (Figs. 2 and 3). Compared with a SiO_2 standard, the chemical composition of these fibers was found to be $SiO_{1.86}$. The slight discrepancy of Si–O stoichiometric ratio is explained by the reducing and high temperature environment that caused a small O-loss from these fibers. Crystalline linear chain-shape aggregates, rich in Si and O, were deposited when WO_n was adsorbed on the aggregates, in the course of their condensation (Fig. 4), and fibers were not produced in this case. Al_2O_3 was not produced from the evaporation of the ceramic target because of its negligible vapor pressure at the given temperature.

The source of W was the heated W filament. Though the filament temperature was well below the temperature range for any significant W evaporation to occur, W was deposited on the Ni alloy substrate via a chemical transport mechanism. It is well-known that a substance A (a solid or liquid), which reacts with a gas to form exclusively vapor-phase reaction products can, in turn, undergo the reverse reaction at a different place in the system, resulting in the reformation of A (*5*). W was reported to be transported in the presence of a temperature gradient and hydrogen-containing water vapor (*6*). In this process, the thermal evaporation of SiO_2 produced oxygen upon pyrolysis, and the oxygen atoms reacted with hydrogen to form water vapor. The water vapor molecules reacted with the heated W filament to form volatile WO_n species. These species were transported from the hot filament to the Ni alloy substrate and resulted in reformation of W on the substrate surface. Proposed chemical reactions are as follows:

$$SiO_2(s,l) = SiO(g) + \frac{1}{2}O_2(g) \tag{I}$$

$$H_2(g) + \frac{1}{2}O_2(g) = H_2O(g) \tag{II}$$

$$SiO_2(l) + H_2(g) = SiO(g) + H_2O(g) \tag{III}$$

$$W(s) + nH_2O(g) = WO_n(g) + nH_2(g) \tag{IV}$$

Reaction (I) is pyrolytic and is coupled with reaction (II). Reactions (III) and (IV) are coupled. Also reactions (III) and (IV) are the chemical transport processes. With continuous flow of H_2 into the reaction chamber, reactions (II) and (III) will continue to support the W transport. The rate-limiting step of the W transport, therefore, is the availability of SiO_2, which is the product of evaporation of the ceramic target.

In addition to serving as a W source, the W filament played other roles. It was experimentally observed that the Ni alloy was heated by thermal radiation from the heated filament. Calculations showed that the temperature of the Ni alloy would be at 975°C owing to radiative heating of the filament. The measured temperature was in the range of 200 to 300°C, indicating that the thermal loss owing to hydrogen conduction was significant. The dissociation of molecular hydrogen to form atomic hydrogen at the filament was calculated to be only 0.23%. Therefore, it is clear that significant dissociation of molecular hydrogen would only occur when the high temperature laser-assisted reactions took place. It was also observed that the melt-through time of the Ni alloy by the laser beam was less than 2 s when the heated tungsten filament was used. This melt-through time was three times faster than that obtained when the filament was not used. FTIR data indicated that a thin film of carbon species was deposited by the dissociation of methane at the heated filament. This film possessed an absorption band that matched with the wavelength of the laser beam. Since the Ni alloy has no absorption band in the infrared, the predeposition of this IR-active film enhanced the beam–metal coupling, resulting in a faster melt-through time. The W filament also had electron emission of 1×10^{-3} amp/cm^2 when heated to 1700°C. Without externally biasing these electrons, they were expected to form a cloud over the filament and were at thermal equilibrium with it. These electrons would experience thermal diffusion, collisions with gas molecules, and attachment to neutral molecules. These processes led to the possibility that, at any given time, free electrons could be found in the laser beam–material interaction zone. It has been shown (7) that a free electron loses only the fraction of $2m_e/M$ of its energy in an elastic impact with a molecule, where m_e is electronic mass and M is molecular mass. If the electron continues to drift, it will accumulate energy until this relatively slow loss process balances the rate at which it gains energy from the electric field. Therefore, the presence of few additional free electrons in the interaction zone (owing to the heated filament) could interact with the laser beam and promote inverse bremsstrahlung absorption. This would cause an increase in the rate of heating of the material plume and result in a greater degree of ionization. Under certain conditions, this collection of neutral and ionized particles may become a shield to the laser beam, acting as a reflecting body above the substrate surface (*8*).

The infrared absorption of the laser beam by the gas mixture and evaporants depends on: the ability of molecules to acquire permanent or induced dipole moments, and the matching of the molecular bond vibrational frequency with that of the electromagnetic radiation. In this study, the hydrogen/methane gas mixture does not couple with the laser radiation since molecular hydrogen in its vibrational state does not possess a dipole moment; neither can the laser radiation excite the C–H stretch and C–H bend vibrational modes. However, Si–O bonds are, to some extent, excited since the laser beam frequency lies within its absorption band. Another process that takes place in the interaction zone is the reaction of SiO vapor with H_2O vapor to form SiO_2, which condenses into droplets above the Ni alloy substrate. Coagulation of these droplets to form metastable linear chain shape aggregates is discussed below.

The apparent coagulation of 25 nm, and smaller diameter, particles to form fibers (*see* Fig. 2) is particularly intriguing. So too is the rapidity with which this process occurs. In transmission electron microscopy studies, there was no discernable evidence of catalyst particles at the fiber tips, neither were the fiber tips hemispherical. These studies, it should be noted, included chemical analysis using the Philips 420 STEM system. Thus, fiber nucleation and growth does not appear to occur via the vapor–liquid–solid (VLS) mechanism (*9*).

An alternative fiber growth mechanism is now proposed. This involves rapid condensation of particles immediately outside the laser plume, with subsequent particle coagulation to form linear arrays, leading to the eventual formation of fibers. First, the laser beam produces a plume upon interaction with the ceramic target to provide active species; furthermore, it also heats the nickel alloy substrate by blackbody radiation. This white ceramic plume obviously contains a variety of neutral and ionized species derived from the ceramic target material. Because of its relatively high vapor pressure, SiO vapor rapidly disperses from the plume so that a proportion of SiO species chemically reacts with oxygen atoms and H_2O to form SiO_2 at the region above the nickel alloy substrate. Within the laser optical path, stable silica nuclei cannot exist because of immediate revaporization owing to strong absorption of the laser radiation. This phenomenon, where absorptive nuclei are unable to survive in the laser optical path, has been observed in studies revealing an absence of Rayleigh scattering (*8*). Immediately outside the optical path, however, stable nuclei form by condensation in a cooling gas mixture environment, with subsequent growth to form speherical droplets. Using the model by Homer and Prothero (*10*), it is calculated that a droplet of 50 nm radius forms in 5×10^{-8} s. These silica droplets and the gas mixture form an aerodisperse (aerosols) system.

Coagulation of aerosol liquid or solid particles occur when they come into contact to coalesce or adhere to one another. The geometric form of particle aggregates depends upon the nature of various applied forces. Usually, aerosol droplets are spherical and, upon collision, fuse together to produce a single spherical particle (*11*), which of course is a natural consequence of favoring the lowest configurational energy of the system. However, aerosols such as Fe_2O_3 (*12*), MgO (*13*), and Al (*14*) can coagulate to form aggregates of thread- or chain-like shapes. Phenomenologically, these shapes arise from coagulation mechanisms involving: (1) induced or permanent electric dipole moments the particles acquire within a strong electric field (2) dipole moments arising from particle contact polarization, and (3) free charges on particle surfaces. In this work, it appears that silica droplets possess a permanent dipole moment arising from the fact that they are small, and the net fluctuation in charge on one side of a droplet can be relatively large. The probability of such occurrence is inversely proportional to the size of the droplets, and as a result, linear aggregates of silica droplets are formed when the droplets are small. These linear aggregates of silica have been observed (Fig. 4).

In the thermal coagulation of the aerosol droplets, the diffusion equation can be applied to brownian motion since the experimental time intervals are large compared with the relaxation time. Values of diffusion coefficients increase as the particle size decreases. For a particle of a radius of about 10^{-6} cm, the diffusion coefficient, D, is about 10^{-6} cm^2/s. In thermal coagulation of droplets, the droplet particles follow a Brownian trajectory. Let L be the average diffusion distance of the droplets contained in the interaction volume:

$$L = N^{-1/3} \tag{V}$$

where N is the droplet concentration in the interaction zone. Since coagulation only involves droplets that are close to each other, it is reasonable to subdivide the interaction volume into independent cells in which localized coagulation occurs. Consider an individual cell in which a number of droplets diffuse toward each other to coagulate and then transform into a fiber. For a fiber of 25 nm in diameter and 170 nm long, it will consist of 10 droplets of individual diameter of 25 nm. The total diffusion distance that these 10 droplets will travel in order to coagulate to form a linear chain is L′

$$L' = n^{1/3}L \tag{VI}$$

where n is the number of droplets in the fiber. Therefore, L′ is calculated to be 9.8×10^{-7} m. The total time of coagulation of these 10 droplets to form a linear aggregate is $t_{coagulation}$

$$t_{coagulation} = L'^2/2D \tag{VII}$$

Using $D = 10^{-6}$ cm^2/s, $t_{coagulation}$ is 0.005 s. In other words, it takes 0.005 s for 10 droplets of individual diameter of 25 nm to coagulate to form a

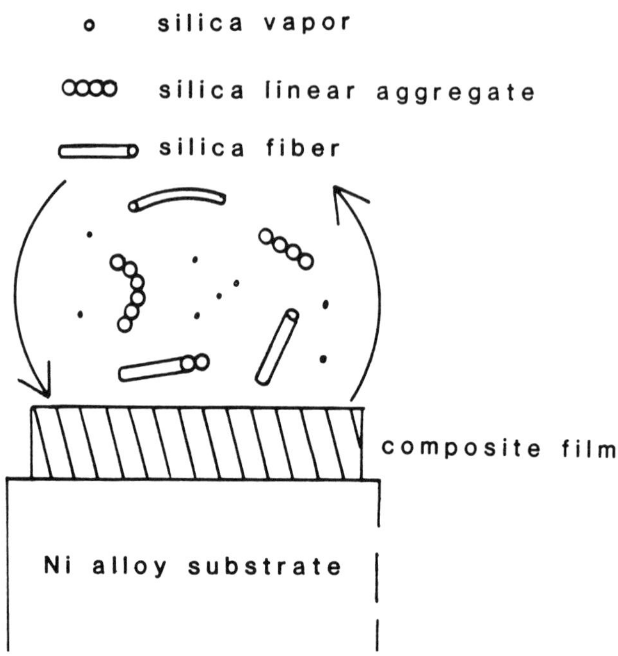

Fig. 5. A schematic showing a convective current above the Ni alloy substrate disk. The ceramic plume caused by the laser-ceramic target interaction provides blackbody radiative heating of the Ni alloy substrate disk. With the cooling effect of the hydrogen, a natural convection circuit is established above the substrate disk.

linear aggregate. The linear aggregate subsequently transforms into a fiber. The calculated $t_{coagulation}$ is an average time of aggregation. Actually, during the later stages of the thermal-diffusion process, a critically small separation distance of the droplets is reached. At this time, there is a sudden acceleration in coagulation and $t_{coagulation}$ will decrease quickly because of the effect of polarization.

Consider a single linear aggregate. This aggregate is metastable, because it will transform to a more energetically favorable cylindrical shape of equal mass, or even to the most favorable spherical shape, provided appropriate conditions are available. At the initial stage of aggregation, the linear chain can have its shape preserved if its surface energy is lowered. It has been known that the addition of a small concentration of W to form WO_3 lowers the surface energy of many ceramic material systems (*15*). In this work, the experimental evidence indicates that WO_3, chemically transported from the heated W filament, adsorbs on the linear chain. As a result, the surface energy of the metastable linear-chain aggregate is lowered and its shape is stabilized. The linear aggregate grows length-wise as each droplet is added to its end. The end of the aggregate is also hotter than the rest of its body. Similar to the VLS model, this liquid-like end has a larger accommodation coefficient for individual vapor molecules. Together with the polarization effect (and possibly surface charge effects), this surface energy effect causes length-wise growth of aggregate. If the degree of preservation of this chain is not high, owing to a low concentration of WO_3 species in the interaction zone, the linear chain can transform to a cylindrical fiber to lower the system energy, while aggregation simultaneously continues at its tip.

The fibers were deposited simultaneously with either Al or W. The fibers adhere to the growing composite film via three possible mechanisms: (1) settling owing to gravity, (2) Brownian motion closely above the growing film, and (3) natural convection circuit above the substrate. The settling velocity of a spherical particle of 50 nm radius in the experimental conditions is calculated to be 1×10^{-4} cm/s. The settling velocity of a nonsperhical particle is even smaller. Therefore, the settling owing to gravity is not important for fine particles. Brownian motions of the fibers closely above the growing film surface, however, can result in their adherence to the film. A convective current of the gas mixture is also established above the Ni alloy substrate owing to heating by the ceramic plume radiation. The silica fibers are carried by this convective current circulating above the substrate. Upon each pass of this gas parcel close to the surface of the growing film, there is a finite probability that a fiber can adhere to the surface via electrostatic interaction or chemical adsorption. It is calculated that the convective circulation accounts for 50% of the measured fiber density. Figure 5 is a schematic of this mechanism.

SUMMARY

In summary, the new technique involves the rapid condensation of nanoscale particles (in this case silica), from the plume formed by the interaction of a CO_2 laser beam with a ceramic target. As the process occurs in a hydrogen/methane gaseous mixture environment, the nanoscale particles are sustained in Brownian motion. Coagulation of these particles by electrostatic interaction finally results in the formation of nanoscale fibers. These become incorporated by codeposition of tungsten, where the tungsten has been transported from a heated filament by a chemical transport mechanism.

ACKNOWLEDGMENTS

The authors wish to thank the Office of Naval Research for its support: contract N00014-78-C-0580. In addition, they wish to thank Paul G. Klemens (Department of Physics, University of Connecticut) for helpful discussions and F. P. Massicotte (Department of Metallurgy, University of Connecticut) for computer graphics. They thank C. Burila, G. Lang, B. Laube, and D. Delong of the United Technologies Research Center for assistance with the microprobe and auger analyses. They also thank A. Strauss and M. Finn of Lincoln Lab for auger analyses.

REFERENCES

1. Boyd, I. W., *Laser Processing of Thin Films and Microstructures,* Springer-Verlag, New York, 1987, p. 78.
2. Birringer, R., Herr, V., and Gleiter, H., *Fall MRS meeting, symposium: Chemical Vapor Deposition of Ceramic Nanocomposites,* 1986.
3. Hayashi, C., *J. Vac. Sci. Technol. A.* **5,** (1987).
4. Pope, E. J. and Mackenzie, J. D., *MRS Bulletin* **231,** 20 (1988).
5. Schaffer, H., *Chemical Transport Reactions,* translated by H. Frankfort, Academic, New York, 1964, p. 1.
6. Alterthum, H., *Z. Physik. Chem.,* **110,** 1, 13 (1924).
7. Kennard, E. H., *Kinetic Theory of Gases,* McGraw Hill, New York, 1938, p. 474.
8. Peebles, H. C., Williamson, R. L., *Proceedings of International Conference on Laser Materials Processing,* Osaka, Japan, 1987.
9. Wagner, R. S. and Ellis, W. C., *Trans. Met. Soc. AIME* **233,** 1053 (1965).
10. Homer, J. B. and Prothero, A., *J. Chem. Soc. Faraday Trans. I.* **69,** 673 (1973).
11. Fuchs, N. A., *The Mechanics of Aerosols,* translated by R. E. Daisley and M. Fuchs, Pergamon, New York, 1964.
12. Schweckendiek, O., *Z. Naturforsch* **5a,** 397 (1950).
13. Beisher, D., *Z. Elektrochem.* **44,** 375 (1938).
14. Whytlaw-Gray, R., and Patterson, H. S., *Smoke,* London, 1932.
15. Editorial board, Principles of Glass Technology, (China publisher of building industry, 1984).

Synthesis of Ultrafine Si_3N_4 Particles by CO_2-Laser Induced Gas Phase Reactions

PETER R. BUERKI, THOMAS TROXLER, AND SAMUEL LEUTWYLER*

Institut für anorg., analyt. und physikal. Chemie, Universität Bern, Freiestrasse 3, CH-3000 Bern 9, Switzerland

ABSTRACT

The syntheses of ultrafine Si_3N_4 powders by cw CO_2-laser induced gas–phase reactions of SiH_4, SiC, and $SiCl_4$ with NH_3 at pressures ranging from 500 to 1500 mbar were investigated (10.591 μm 10P(20), 10.741 μm 10P(34), and 10.304 μm 10R(12) radiation). The reaction of SiH_4 and NH_3 produced Si_3N_4 powders at a rate of up to 18 g/h (2 kWh/kg Si_3N_4 powder) with diameters determined by TEM ranging from 180 to 250 Å and BET surface areas ranging from 55 to 121 m^2/g. The powders are amorphous or partially crystallized and have very high purity. Correlation of stoichiometry (excess Si) and gas flow rate was found under a wide range of conditions. Heating of solid SiC in the presence of NH_3, using the 10R(12) CO_2-laser line, produced traces of Si_3N_4. Direct synthesis of Si_3N_4 from $SiCl_4$ and NH_3 was not successful, owing to the efficient spontaneous reaction forming silicon diimide, $Si(NH)_2$.

Index Entries: Si_3N_4 powders; laser-induced reactions; gas-phase reactions; IR-laser photochemistry; ultrafine powders; ceramic materials.

INTRODUCTION

The reliability of modern ceramics in terms of hardness, rupture toughness, brittleness, and reproducibility of specific properties is still far from being satisfactory owing to disadvantageous properties of the

*Author to whom all correspondence and reprint orders should be addressed.

powders used as starting materials, such as high contamination levels, large particle diameters, broad size distributions, and nonspherical shapes. Therefore, many attempts have been made during the last 10 yr to produce ceramic powders with improved properties, e.g., by reactions in solutions (sol-gel processes, precipitation), thermal decomposition of reactants, and gas-phase reactions induced by plasmas or lasers (*1*).

The demands on ceramic powders with superior properties are high purity, small particle diameters below about 1 μm, narrow size distributions (monosized powders), spherical particle shapes, and agglomerate-free powders (2). Two years ago, an investigation of laser-induced, gas-phase reactions to produce ceramic powders was started at our department (3). The goal of our research is to develop widely applicable techniques for the production of ultrafine powders of ceramic and related materials. Up to now, three types of laser-induced gas-phase reactions to produce Si_3N_4 powders are investigated, for which the proposed overall reactions balances are

a) $3\ SiH_4 + 4\ NH_3 \xrightarrow{h\nu} Si_3N_4 + 12\ H_2$ (3,4,10–14)

b) $3\ SiC + 7\ NH_3 \xrightarrow{h\nu} Si_3N_4 + 3\ HCN$

c) $3\ SiCl_4 + 16\ NH_3 \xrightarrow{h\nu} Si_3N_4 + 12\ NH_4Cl$

EXPERIMENTAL METHODS

Powder Synthesis

For reaction a) NH_3 and SiH_4 were mixed and vertically introduced into a stainless steel reaction chamber through capillary or slit nozzles. The gas flow was horizontally intersected by the beam of a line tunable cw CO_2-laser (Apollo mod. 570) 2–3 mm above the nozzle orifice (Fig. 1a). The maximum output power at the 10P(20) line was 50 W. The beam is focused by a ZnSe lens (focal length 508 mm). The beam power is measured by a power meter from Coherent Radiation (mod. 201, Palo Alto, CA).

Two nozzle geometries were used. For lower production rates, a capillary nozzle with an inner diameter of 1.40 mm was used, whereas for higher production rates, a slit nozzle with an aperture of 0.40 × 8.45 mm was employed. The long axis of the slit was parallel to the laser beam.

The reaction products were entrained in a continuous flow of argon. Total pressures in the reactor ranged from 500 to 1500 mbar (1 mbar = 100 Pa). The Si_3N_4 powder transported by the argon flow is trapped in a borosilicate-glass fiber filter (Schleicher & Schuell, Feldbach ZH, Switzerland, Art. no. 603g) mounted before a two-stage rotary vacuum pump (Alcatel, type 2012AC, ALCATEL GmbH, Effretikon, Switzerland). The gas flows were determined by several flow meters, to an accuracy of

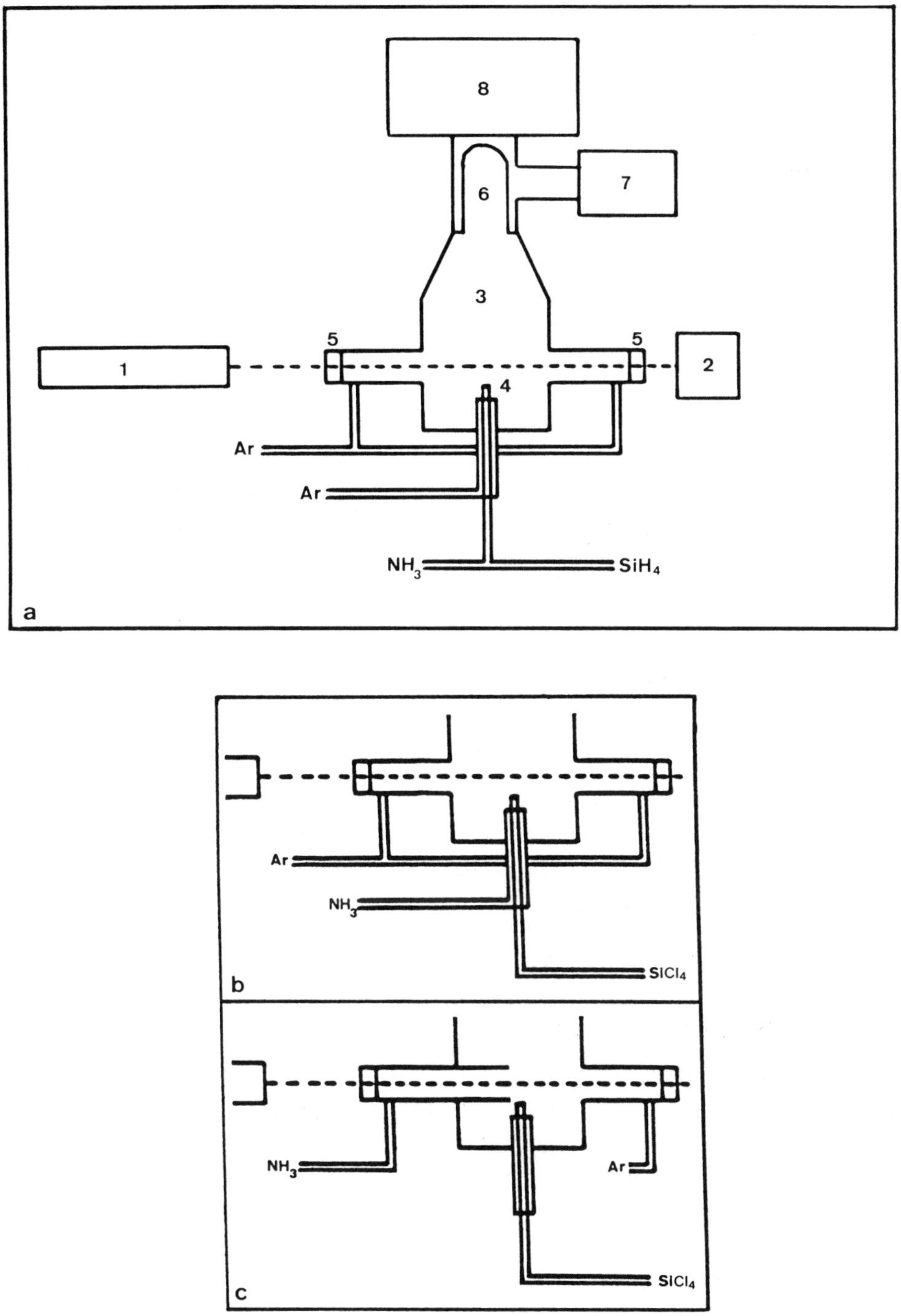

Fig. 1. Reactor and gas lines used for the synthesis of Si_3N_4: (a) reaction of SiH_4 and NH_3, (b) and (c) gas lines used for the reaction of $SiCl_4$ and NH_3; (b) collinear flow; (c) cross-flow configuration. (1) line tunable CO_2-laser; (2) power-meter; (3) reactor chamber; (4) reactant gas inlet (capillary or slit nozzle); (5) NaCl-window; (6) borosilicate-glass fiber filter; (7) rotary vacuum pump; (8) inert gas glove box.

±10% (Fischer & Porter GmbH, Goettingen, FRG, series 10A3239S). The total pressure during the reaction was measured using a spring manometer (Haenni & Cie., Jegenstorf, Switzerland, DRF 160/131.133, −1 to 0.6 bar).

The reactant gases had the following purities: NH_3 (CARBAGAS, Switzerland, quality 36), contamination by $H_2O < 200$ ppm; SiH_4 (CARBAGAS, Switzerland, quality epitaxie) contamination by $O_2 < 10$ ppm, $H_2O < 2$ ppm, CO and $CO_2 < 10$ ppm, and $CH_4 < 40$ ppm; and Ar (CARBAGAS, Liebefeld, Switzerland, quality 46) contamination by O_2 and $H_2O < 15$ ppm. Typical gas flows were: capillary nozzle, NH_3, $3 \cdot 10^{-5}$–$10 \cdot 10^{-5}$mol/s; SiH_4, $0.4 \cdot 10^{-5}$–$2 \cdot 10^{-5}$ mol/s; and Ar, $1 \cdot 10^{-3}$–$2 \cdot 10^{-3}$ mol/s; and slit nozzle NH_3, $10 \cdot 10^{-5}$–$30 \cdot 10^{-5}$ mol/s; SiH_4, $3 \cdot 10^{-5}$–$16 \cdot 10^{-5}$ mol/s; and Ar, $0.5 \cdot 10^{-3}$–$12 \cdot 10^{-3}$ mol/s.

For reaction b), the same reaction chamber was used. SiC was used as cylindrical green bodies with 9 to 11 mm diameter and 5 cm length. The SiC cylinders were introduced into the reaction chamber and heated by the laser beam. Ammonia was directed through a small nozzle to the intersection of SiC cylinder and laser beam.

For reaction c) the experimental set-up was partially changed. $SiCl_4$ (Merck AG, Buchs, Switzerland, quality puriss.) was introduced from a heated Pyrex glass vessel through the inner tube (stainless steel) of the nozzle into the reaction chamber (Fig. 1b). A flow of ammonia passing through the outer tube sheathes the inner $SiCl_4$ flow. The gases mix and pass through the laser beam. A cross-flow configuration (Fig. 1c) was studied too.

The reactor chamber is connected directly to a glove box (MBraun GmbH, Garching, FRG, type MB-150 B/G-II) with an argon atmosphere. During synthesis, the flow apparatus is separated from the glove box by an intervening bulkhead, which also acts as a support for the glass-fiber filter. The O_2 contamination level in the glove box is continously registered (O_2 analyzer, digital, MBraun GmbH, Garching, FRG).

Powder Handling and Storage

Following synthesis, the reactor chamber is shut off in front of the glass-fiber filter by a gate valve and the filter transferred into the glove box through the connecting bulkhead. Then, the Si_3N_4 powders are transferred into Pyrex glass bottles (Schott, Mainz, FRG). They were never exposed to O_2 levels exceeding 1.5 ppm.

Powder Characterization

The powders were characterized by powder diffractometry (Guinier type camera; $CuK\alpha_1$ radiation) and IR spectroscopy (Perkin-Elmer IR spectrometer type 599B). The specific surface areas were determined using the multipoint BET method (*5,6*) (Carlo Erba, Sorptomatic series 1800). The morphology and size distributions of the powders were characterized by transmission electron microscopy (TEM) (Hitachi H-600). The absolute magnification of the microscope was calibrated to ±7%

using bovine liver catalase specimens (*7*) (Balzers Union, Liechtenstein). The concentration of nitrogen in the powders and the contamination level by oxygen were determined by LONZA AG, Waldshut, FRG, using a combustion analysis method (LECO, system TC-136). The silicon content was calculated as the difference between total sample weight and the weights of nitrogen and oxygen.

RESULTS AND DISCUSSION

Reaction of Silane with Ammonia

Laser-heated SiH_4 reacts with NH_3 in a flame of variable colors varying from red to white-yellow at temperatures above 1000°C (*4,8*). The flame shape is paraboidal for the capillary nozzle and wedge-shaped for the slit nozzle. The reaction takes place at several wavelengths of the CO_2-laser, e.g., 10P(20), 10P(34), and 10R(12) (10.591 μm, 10.741 μm, and 10.304 μm, respectively) (*4*). The laser radiation at the 10P(20) and 10P(34) line is mainly absorbed by SiH_4 and at the 10R(12) line by NH_3 (*4*). The reaction could be initiated with each of these wavelengths. Most of the experiments were done using the 10P(20) line (10.591 μm). The resulting silicon nitride primary particles have a spherical shape and quite a narrow size distribution, as shown in Fig. 2a.

Depending on synthesis conditions, the powders are amorphous to X-rays or partially crystalline. Sometimes, there exist two particle populations, a main population of spherical particles and a minor population of small whisker-type particles attached to the other population. This second phase could not be characterized owing to its extremely low concentration (Fig. 2b), but may be β-Si_3N_4 recrystallized from the amorphous spherical particles.

The size distributions and the specific surface areas of the powders are tabulated in Table 1. The absolute uncertainty in size is 10% for synthesis runs SN 5–SN 17 and 7% for synthesis runs SN 18–SN 25.

The purity of the synthesized silicon nitride powders is high. Contamination with oxygen is in the range of 0.5 to 1.5% (by weight). Since the reaction takes place without wall contact and the energy is transferred by laser radiation, the contamination level by metallic impurities should be extremely low and influenced mainly by the contamination levels of the reactant gases. However, the absolute concentrations of metallic impurities as yet have not been determined analytically.

The Si_3N_4 production rate using the capillary nozzle was about 1 g/h (absorbed CO_2–laser power 4–7 W). It could be raised continuously to 18 g/h using the slit nozzle, which allows optimum interaction of the gas flow with the laser radiation, and higher flow rates. With this configuration, the absorbed laser power was 16–27 W. The energy input was about 2 kWh/kg Si_3N_4 powder. We project that the production rate can be raised by another factor of two to about 40 g/h with an energy input of 1 kWh/kg Si_3N_4 powder.

The influence of the following parameters on the powder charac-

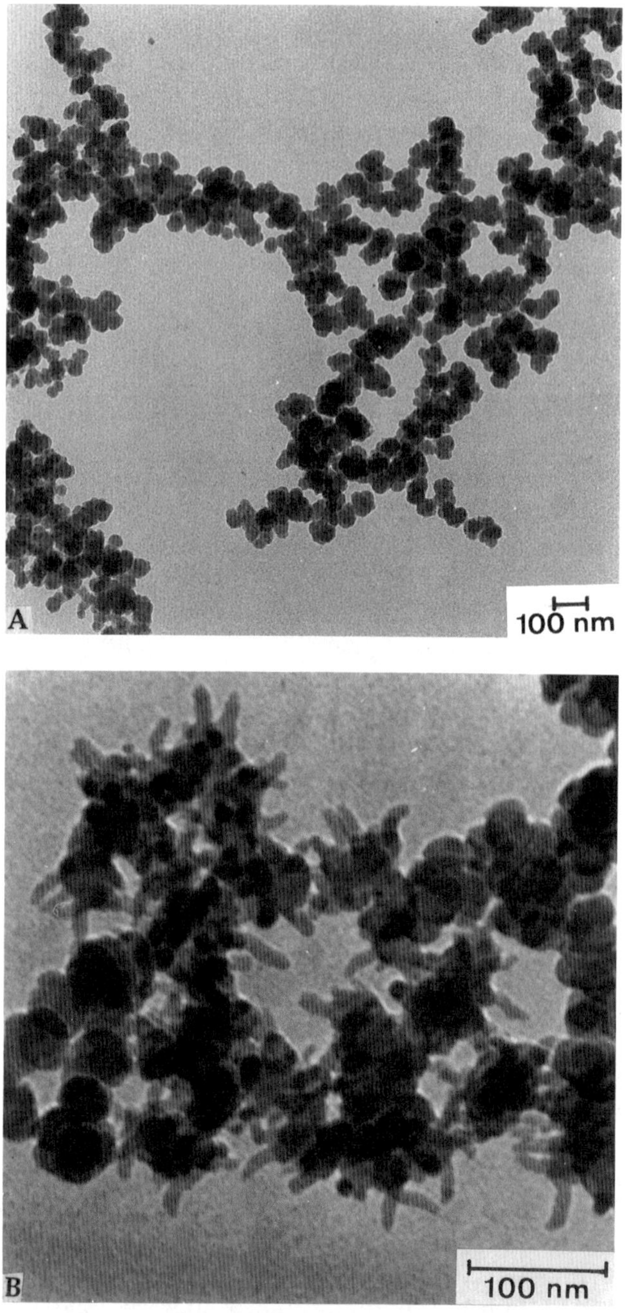

Fig. 2. TEM pictures of Si_3N_4 powders; (a) synthesis run no. SN 7; (b) synthesis run no. SN 13. A second population of small, whisker-type particles occurs in the powder shown in Fig. 2b.

Table 1

Run Label	Pressure, mbar	Laser intensity, W/cm^2	Wavelength, μm	Gas flow velocity, m/sec	Excess silicon, weight %	Size distribution, $x \pm 2\sigma$ (%) Å	Specific BET-surface, m^2/g
SN 4	500	6500	10.591	4.1	16.9	221 ± 48 (22)	121.7 ± 6.1
SN 5	500	700	10.591	1.6	6.6	260 ± 49 (19)	75.3 ± 3.8
SN 6	500	7500	10.591	0.8	5.1	281 ± 57 (20)	69.5 ± 3.5
SN 8	500	700	10.591	1.4	8.4	230 ± 47 (20)	89.0 ± 4.5
SN 9	500	700	10.591	3.4	16.8	246 ± 47 (19)	77.8 ± 3.9
SN 13	500	7400	10.591	1.0	1.4	241 ± 60 (25)	71.3 ± 3.6
SN 14	500	5600	10.591	1.0	0.3	264 ± 53 (20)	61.1 ± 3.0
SN 15	500	4700	10.741	1.0	1.1	264 ± 54 (20)	65.6 ± 3.3
SN 17	500	6500	10.304	0.4	5.3	276 ± 65 (23)	56.0 ± 2.8
SN 21	500	9100	10.591	3.9	8.2 ± 2.2		107.8 ± 5.1
SN 7	900	6900	10.591	0.6	4.4	298 ± 54 (18)	55.4 ± 2.8
SN 10	900	650	10.591	0.6	1.8	250 ± 48 (19)	66.9 ± 3.3
SN 11	900	680	10.591	1.0	1.2	205 ± 46 (22)	61.1 ± 3.0
SN 12	900	6400	10.591	1.0	~0	213 ± 43 (20)	72.5 ± 3.6
SN 16	900	7400	10.591	0.4	2.2	272 ± 59 (22)	54.8 ± 2.7
SN 18	900	7600	10.591	0.3	12.1 ± 4.4	234 ± 47 (20)	57.5 ± 1.3
SN 19	900	7750	10.591	1.0	15.9 ± 5.6	246 ± 50 (20)	57.1 ± 2.8
SN 20	900	8100	10.591	1.6	4.0 ± 1.4	233 ± 38 (16)	69.4 ± 2.6
SN 22	900	8600	10.591	2.8	8.1 ± 1.8		86.6 ± 3.2
SN 25	1150	8260	10.591	2.8			76.5 ± 2.3
SN 24	1300	8100	10.591	2.3			77.1 ± 1.3
SN 23	1500	5300	10.591	2.0	9.43 ± 0.39		87.2 ± 5.2

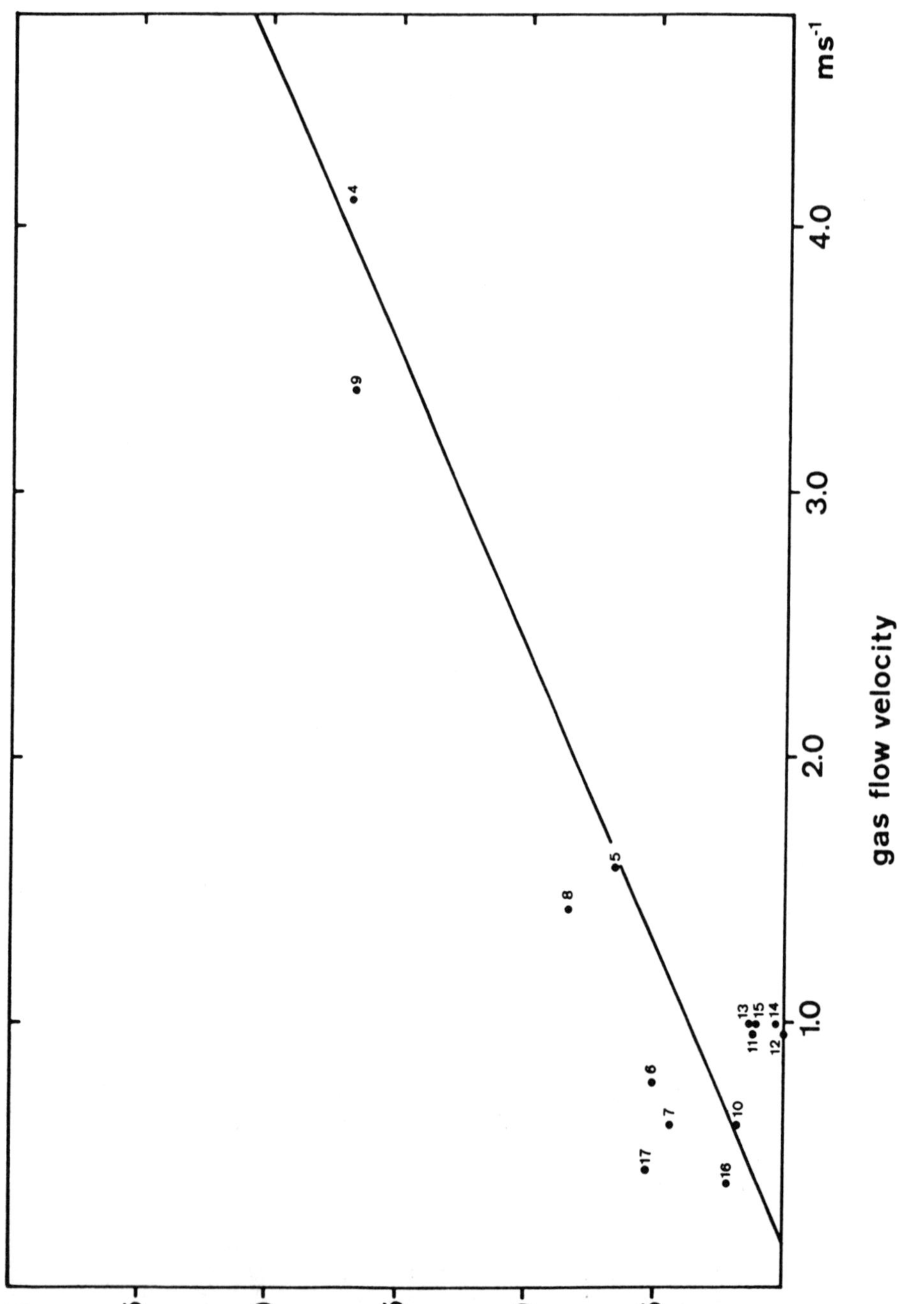
excess silicon [weight %]
%
25
20
15
10
5
gas flow velocity
ms-1
1.0
2.0
3.0
4.0
4
5
6
7
8
9
10
11
12
13
14
15
16
17

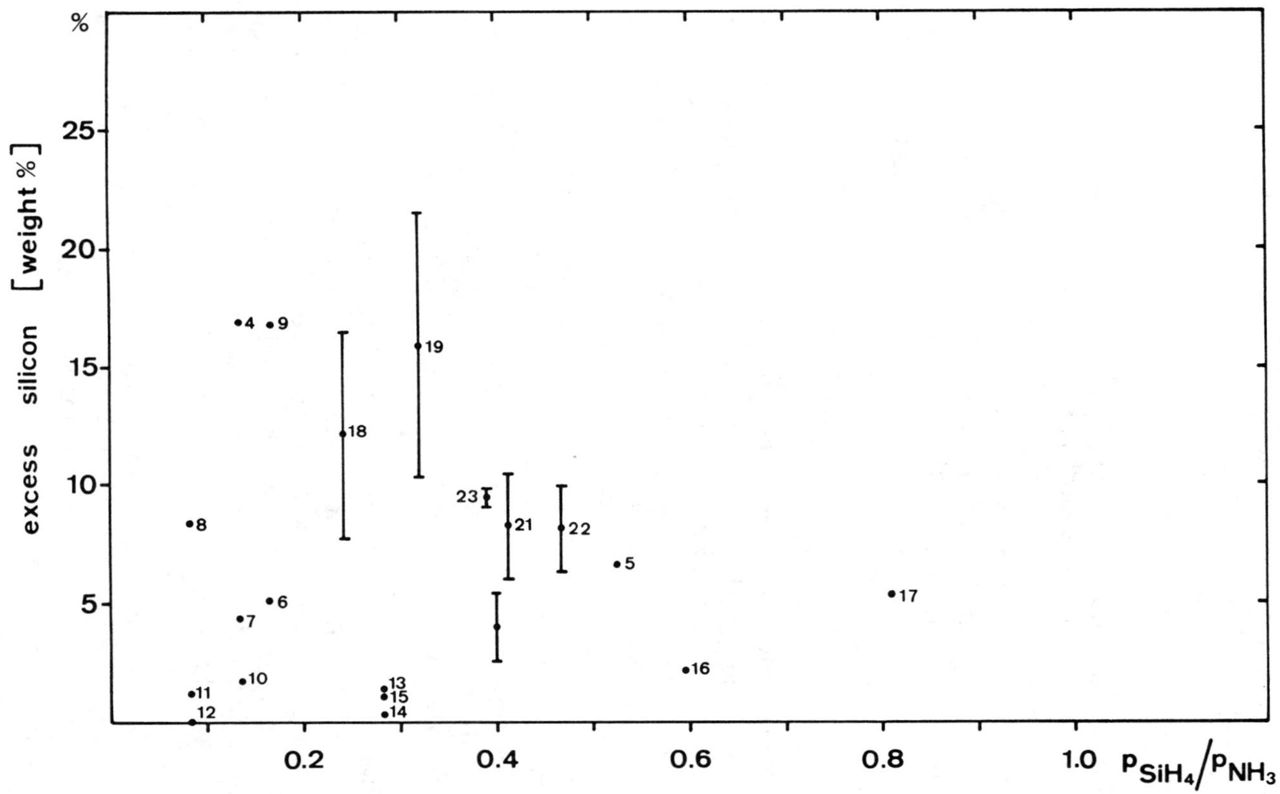

Fig. 3. Dependence of excess silicon content of the Si_3N_4 powders synthesized with the capillary nozzle (SN4–SN 17) on gas flow velocity (above); and reactant gas composition (opposite page) (ratio of partial pressures of SiH_4 and NH_3.

teristics were investigated: laser wavelength, laser intensity, pressure, flow velocity of reacting gases, and gas composition of the reacting mixture.

The CO_2-laser wavelength was varied in a series of experiments (10P(20), 10P(34), and 10R(12) line), keeping pressure and gas flow rates constant. Powders with excellent properties were obtained using all three lines (Table 1). No influence of laser wavelength on any powder characteristic was observed. Color and shape of the reaction flame did not change either.

For synthesis runs nos. SN 4–SN 17 (capillary nozzle), an approximately linear correlation was found between the stoichiometry of the Si_3N_4 powders and the flow velocity, v, of the reactant gases over a wide range of pressures and reactant gas ratios, as shown in Fig. 3a. Lowering the mean gas flow velocity lengthens the residence time of the reactants in the reaction flame and improves the stoichiometry of the resulting powders. Linear regression analysis of excess silicon (percent by weight) against v yields a slope of 4.48 ± 0.69 [(% by weight) s/m] and an intercept on the y-axis of -0.75 ± 1.16 (% by weight) with a correlation coefficient $r = 0.88$. The correlation is independent of laser power intensity and residence time of the reactants in the laser beam. The composition of the reactant gas mixture does not have a significant influence on the stoichiometry of the resulting powders, as shown in Fig. 3b.

The occurrence of a crystalline phase in the powders is also dependent on the gas flow velocity, as shown in Fig. 4. Lowering the gas flow velocities increases the amount of a crystallinic phase in the powders. Again, the occurrence of a crystalline phase is a function of the residence time in the flame and not the residence time in the laser beam. During preparation and powder diffractometry, the samples were exposed to air. Therefore, small amounts of α-SiO_2 were detected. The intensity of the peaks related to silicon varies with the excess silicon content of the powders.

We do not observe a significant influence of pressure on the powder stoichiometry, as shown in Table 1. The previously reported pressure dependence (*4,11–13*) seems to be weak. However, there exists a minimal pressure which is necessary to initiate the reaction (*3*). Below 200 mbar, it is not possible to ignite the reactant gases, because the gas flow velocities are too high. A stable reaction flame cannot be obtained at pressures below 300 mbar.

Laser radiation intensity does not influence the powder stoichiometry either (Table 1). The radiation intensity was varied over one order of magnitude (650–7500 W/cm^2) by varying the laser output power level and focusing or defocusing the laser beam.

Reaction of Solid Silicon Carbide with Ammonia

Using the 10P(20) and the 10R(12) line of the CO_2-laser, the silicon carbide green body cylinders were heated to white heat. At the 10R(12)

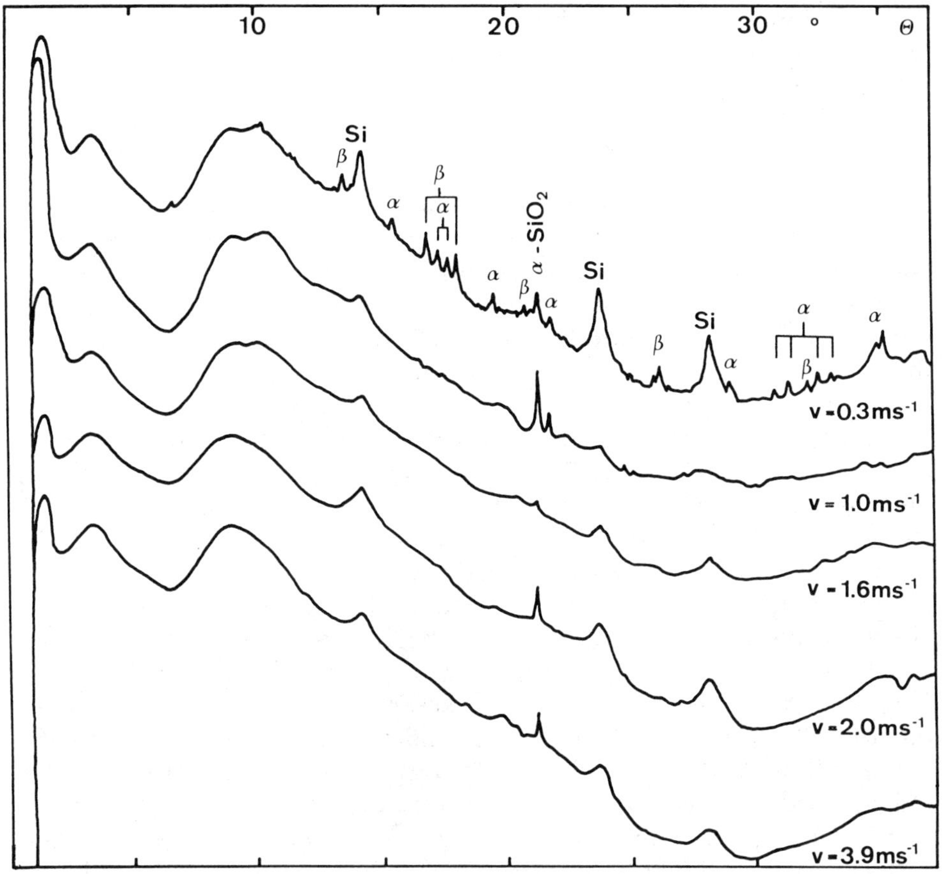

Fig. 4. Powder diffractometry diagrams of Si_3N_4 powders (α: α–Si_3N_4; β: β–Si_3N_4). The occurrence of a crystalline phase (narrow peaks) is a function of the gas flow velocity.

line, ammonia has a strong absorption band (*4*). A reaction between solid SiC and NH_3 was not observed. However, at the burn point, small amounts of a white crystallinic material could be seen. Using diffractometry, this material was identified as Si_3N_4. Because of the high thermal conductivity of SiC (86.46 J/msK), it was not possible to attain temperatures that initiated an appreciable reaction between silicon carbide and ammonia. However, there are some indications that it should be possible to initiate a reaction of a fine SiC powder dispersed in an ammonia gas flow, because the radiation energy deposited in the SiC particles is not dissipated and therefore much higher temperatures can be attained.

Reaction of Silicon Tetrachloride with Ammonia

The spontaneous reaction of silicon tetrachloride with ammonia to form silicon diimide, $Si(NH)_2$, and ammonium chloride (a) was reported by Persoz 160 years ago (*9*).

$$\text{a)}\ SiCl_4 + 6\ NH_3 \longrightarrow Si(NH)_2 + 4\ NH_4Cl$$

This reaction is strongly exothermic. The topic of the following study was to try to bypass this reaction pathway by a laser-induced process and open a new one, (b), forming Si_3N_4.

$$\text{b) } 3\ SiCl_4 + 16\ NH_3 \xrightarrow{h\nu} Si_3N_4 + 12\ NH_4Cl$$

Reaction (b) should be exoergic ($\Delta G = -964.08$ kJ/mol at T = 298 K); however, the exoergicity decreases as temperature increases. Since the IR spectrum of gaseous $SiCl_4$ shows an absorption peak of intermediate intensity at 9.6 μm, which matches the 9P band of the CO_2 laser, there are two conceivable methods to induce a reaction: (1) heating NH_3 with the 10R(12) line (10.304 μm), and (2) heating $SiCl_4$ with one of the 9P band lines at 9.6 μm.

Using the first method, the highest temperature attained was 550°C. No Si_3N_4 was detected. The only products that could be detected were NH_4Cl and $Si(NH)_2$. As reported in (*15*), it is to be expected that the system $SiCl_4/NH_3/Si(NH)_2$ forms Si_3N_4 at temperatures around 1250°C; therefore, the temperatures attained are much too low.

Applying the second method was a failure too. At 300 mbar, the absorption of all the 9P band lines by $SiCl_4$ was investigated. Maximum absorption was measured at the 9P(26) and the 9P(28) lines (9.604 and 9.621 μm, respectively), but did not exceed 6.4%. Several different methods to bring the two reactants into contact (different nozzle geometries, colinear flow, and rectangular reactant gas flow intersection) were studied, but with no success. Again, the only reaction products detected were NH_4Cl and $Si(NH)_2$. It has to be concluded that the spontaneous reaction between $SiCl_4$ and NH_3 to form $Si(NH)_2$ is thermodynamically and kinetically favored compared with reaction (b) to form Si_3N_4. A practicable alternative would be to use a mixture of $SiCl_4$ and SiH_4 instead of pure $SiCl_4$ vapor. If SiH_4 were heated with the 10P(20) line, it is probable that $SiCl_4$ would react with ammonia. However, the SiH_4 fraction in the mixture has to be about 10 times the $SiCl_4$ fraction to produce a flame. Hence, this method is not an economically viable alternative to produce Si_3N_4, compared with the reaction of pure SiH_4 and NH_3.

ACKNOWLEDGMENTS

This study was financially supported by the Kommission zur Foerderung der wissenschaftlichen Forschung (KWF project no. 1554). Technical support by ALUSUISSE AG, Neuhausen, Switzerland and LONZA AG, Waldshut, FRG is gratefully acknowledged. We like to thank R. Giovanoli, Markus Faller, Rudolf Steiner, Beatrice Wild-Frey, and Emma Ettinger (laboratory for electron microscopy, University of Berne, Switzerland) for their support during the electron microscopy and powder diffractometry measurements and P. W. Schindler and Bernhard Trusch (department of analytical chemistry, University of Berne, Switzerland) for their support with the BET measurements.

REFERENCES

1. Francis, T. L., *Keram. Zeitschrift* **37,** 341 (1985).
2. Barringer, E., Jubb, N., Fegley, B., Pober, R. L., and Bowen, H. K., *Ultrastructure Processing of Ceramics, Glasses and Composites,* Henck, L. L. and Ulrich, D. R., eds., Wiley, New York, 1984, p. 315.
3. Troxler, T., *M. sc. thesis,* University of Berne, Switzerland, 1988.
4. Haggerty, J. S. and Cannon, W. R., in *Laser-induced Chemical Processes,* Steinfeld, J. I., ed., Plenum, New York, 1981, p. 165.
5. Langmuir, I., *J. Am. Chem. Soc.* **40,** 1361 (1918).
6. Brunauer, S., Emmett, P. H., and Teller, E., *J. Am. Chem. Soc.* **60,** 309 (1938).
7. Wrigley, N. G., *J. Ultrastructure Research* **24,** 454 (1968).
8. Flint, J. H., Marra, R. A., and Haggerty, J. S., *Aerosol Sci. Technol.* **5,** 249 (1986).
9. Persoz, G., *Ann. Chim. Phys.* **2/44,** 315 (1830).
10. Symons, W. and Danforth, S. C., *Adv. Ceram.* **21,** 249 (1987).
11. Cannon, W. R., Danforth, S. C., Flint, J. H., Haggerty, J. S., and Marra, R. A., *J. Am. Ceram. Soc.* **65,** 324 (1982).
12. Cannon, W. R., Danforth, S. C., Haggerty, J. S., and Marra, R. A., *J. Am. Ceram. Soc.* **65,** 330 (1982).
13. Kizaki, Y., Kandori, T., and Fujitani, Y., *Jap. J. Appl. Phys.* **24,** 800 (1985).
14. Cauchetier, M., Croix, O., Luce, M., Michon, M., Paris, J., and Tistchenko, S., *Ceramics International* **13,** 13 (1987).
15. Glemser, O. and Naumann, P., *Z. Anorg. Allg. Chem.* **298,** 134 (1959).

Chemical Clusters from Solid State Systems at High Temperatures

Interstitials as a Means to Stability and Versatility

JOHN D. CORBETT,* EDUARDO GARCIA, YOUNG-UK KWON, AND ARNOLD GULOY

Ames Laboratory-DOE and Department of Chemistry, Iowa State University, Ames, IA 50011

ABSTRACT

The reactivity of many "naked" clusters and the means by which these may be isolated as stable "chemical" clusters are briefly considered for the orbitally-rich transition metals, where clusters must be sequestered by coordination with ligands, and for the heavier main group elements that utilize only three p orbitals in bonding. Numerous "naked" clusters with favorable closed-shell configurations are known for the latter group both as solids at room temperature and in cluster beams. A few binary transition-metal clusters stable at elevated temperatures are obtained with chloride, sulfide, and so on as ligands. Greater versatility and variety have recently been found for electron-precise clusters that require an interstitial atom Z within M_6X_{12}-type clusters for stability, as with the $A^I_X[Zr_6(Z)X_{12}]X_n$ family or rare-earth-element analogs where 22 different examples of Z and many structure types are known for $X = Cl, Br, I$. A newer and even broader area of cluster chemistry involves interstitials bonded within chains of confacial octahedra in a variety of polar intermetallic phases that occur in the Mn_5Si_3 structure, where the large differences in valence state energies of the component atoms generally assure that valence bonding levels on the main-group element are filled first. The electron-rich Zr_5Sb_3 ($\simeq^1_\infty[Zr_3Sb_3{\cdot}Zr_2]$) host is found to form Zr_5Sb_3

*Author to whom all correspondence and reprint requests should be addressed.

Z phases with at least 16 examples of Z. Band calculational results for the binary host and for the sulfide are considered. Similar investigations are described for the hosts Zr_5Sn_3, Zr_5Pb_3, La_5Ge_3, La_5Pb_3, and M_5B_3, M = Ca, Sr or Ba and B = Sb or Bi. Valence-precise (Zintl) phases are achieved in some M_5B_3Z compounds from the last two groups.

INTRODUCTION

Clusters of metal atoms are of great interest in materials chemistry as well as in impinging areas of chemistry and physics. At one end of the spectrum are the "naked" homoatomic clusters studied in the gas phase and at the other, condensed nanoscale to micron-sized particles with considerable potential for catalysis, microdevices, etc. (*1*). However, metal clusters seen in beam studies usually exhibit little stability discrimination with respect to size in the range of roughly 5–25 or more atoms, many appearing to have more of the character of van der Waals or "jellium" molecules than of "chemical" clusters with well defined, closed shell electronic configurations and high stabilities such as are known in more molecular systems, P_4 and C_{60} for example (*2*). The equilibrium electronic structures of small "chemical" clusters of metals should naturally be better described by molecular orbital rather than by band methods suitable to the bulk. However, it appears likely that few if any metal clusters with these closed shell configurations and relatively high stabilities are achieved in most single component beams because of natural limitations on the electron counts that can be achieved. Examples of these same clusters are, of course, unknown in condensed systems because of their intrinsic reactivity. This article will briefly explore the bases for the above behaviors and some exceptions thereto and then will outline the newer aspects of the more stable "chemical" clusters that may be achieved in binary and ternary systems. A new feature of this chemistry is the added versatility and stability achieved for transition metal cluster halides through manipulation of the cluster electron configurations and thereby the phases formed via the encapsulation of interstitial atoms within the clusters. We will also relate the beginnings of a new cluster chemistry that can be achieved in infinite chain structures of some electron-rich polar intermetallic phases with the Mn_5Si_3 structure and the broad interstitial chemistry that is possible therein.

NAKED CLUSTERS

Transition metals owe much of their prolific chemistry to the profusion of good bonding orbitals, namely (n-l)d, ns, and np, but it is this same orbital abundance that precludes the isolation of small naked transition-metal clusters that are stable in bulk. As an example, a six-metal cluster with octahedral symmetry will exhibit 7 or 8 good cluster-bonding orbitals, and an equal number that are strongly antibonding, but this will still leave about half of the d orbitals unused, roughly nonbonding, and, for most metals, partially occupied by valence electrons. Many of these orbitals will be outward pointing and clearly the source of further reactions, condensation, and so on. As a result, transition metal clusters will be stable in dense systems only when these orbitals are fully occupied, as is commonly encountered when the clusters are effectively sheathed by neutral ligands (CO, phosphines, and so on) or, sometimes, anions. The usefulness of the 18-electron rule for each metal atom in so many organometallic examples attests well to these orbital requirements and to the existence of a sharply limited number of electronic configurations of special stability in what we chose to call "chemical" clusters. Naturally, the presence of fewer ligands or anions on a cluster surface can be thought of as furthering the condensation of these clusters through additional metal-metal bonding into large clusters, oligomers, and so forth. Tantalizingly large clusters may be so achieved, e.g., with Rh_{13}, Pt_{26}, or Au_{55} units suitably encased by ligands (*3*). However, most of the useful ligands in these cases fall into the molecular/organic category, and stability for the upper limits of these clusters are low, seldom exceeding a couple of hundred degrees.

Before moving to the more limited variety of cluster systems that are stable at elevated temperatures, it is worth noting that there is an area of exception to this naked cluster reactivity, and this occurs largely because the constituent elements have a very limited number of good bonding orbitals available. This circumstance applies best to post-transition or main-group elements from the silicon, phosphorus and sulfur families in low oxidation states where p orbitals are the principal means for bonding, the increasing s-p separation for elements in the later groups or heavier periods leading to s orbitals that exhibit essentially only nonbonding functions. As in boranes, the three valence orbitals generate strongly bonding, closed shell configurations in several categories; those with $2n + 2$ electrons in closo (regular deltahedral) geometries containing n atoms, $2n + 4$ electrons in the related nido polyhedron where one vertex is missing, and so on. Another problem in the isolation of these, as illustrated by the Sb_4 and Bi_4 analogs of P_4, is that the neutral species inevitably rearrange to the common elemental structures in the condensed state. This tendency can be altered substantially by working with the isoelectronic ions so that disproportionation is the limiting reaction, viz, $Na_2Bi_4 \rightarrow 2NaBi + 2Bi$, and by reducing the lattice energy change that favors the decomposition product through strong complexing of the sodium cations in both binary phases with 2,2,2 crypt, a polyether amine. Examples of such naked clusters, the so-called Zintl anions, that have been isolated in this way include the closo Sn_5^{2-}, Pb_5^{2-}, and $TlSn_9^{3-}$, the nido $Pb_2Sb_2^{2-}$ and Sn_9^{4-}, and square planar Bi_4^{2-} (*4*). Of course, high temperature stability is lost by this last trick, but some of these anions, or their isoelectronic analogs, may still be found in a limited number of high temperature systems. Thus, Bi_8^{2+} and Bi_5^{3+} can be isolated from molten $AlCl_4^-$ systems (*5*), and some of the smaller or more stable polyanions are found as high melting alkali-metal compounds, viz, as As_7^{3-} and Te_3^{2-} as well as the P_4 analogs Ge_4^{4-}, Sn_4^{4-}, and so on (*6*).

Naked cluster species with Zintl ions above like the boranes electronic configurations exhibit stabilities that far exceed those with other electronic counts. However, it appears that none of the known, or predictable, closed shell examples cited above can be achieved as neutral or low charged clusters in a beam generated from a single type of atom, except, of course, for analogs of the long known P_4. On the other hand, these special "chemical" configurations with their very large stability enhancements can be readily observed in mixed element beams as what have been called "compound clusters", for example, as Pb_3Sb_2, $Pb_2Sb_3^+$ ($\triangleq Sn_5^{2-}$), Pb_5Sb_4 ($\triangleq Pb_9^{4-}$), $Cs_5Sn_9^+$ ($\triangleq Sn_9^{4-}$), and Bi_4In^+ ($\triangleq Bi_5^{3+}$) (*7–10*).

SOLID STATE CLUSTERS BY HIGH TEMPERATURE ROUTES

Inorganic groups that sheath transition metal clusters and render them stable at higher temperatures have traditionally involved the chalcogenides and halides, although examples of the pnictides (As, Sb, and so on) will also be considered later. The classic examples among the chalcogenides are the famous Chevrel phases built from units like Mo_6S_8, in which a nominal metal octahedron has sulfur strongly bound over each triangular face. Strongly bonding orbitals that remain at each metal vertex cause these to link up further, six of the face-capping sulfurs in one cluster also bonding to vertices in six other clusters and vice versa, an arrangement that also provides a Mo-Mo intercluster conduction path and metallicity. Many derivatives such as $M^I_xMo_6S_8$ ($x \leq 4$), oligomers $S(Mo_3S_3)_nS^{m-}$, and infinite polymer chains $^1_\infty[Mo_3S_3^-]$ as well as molybdenum and niobium oxide examples have been discovered and are well described elsewhere (*11–13*). Cluster halides of the early transition metals in which interstitials provide an additional versatility represent a still newer chemistry and provide broader insights on cluster stability.

Traditional cluster halides are almost always based on transition metal octahedra that are edge-bridged by 12 halide atoms. The number of binary cluster halides possible is relatively low, reflecting a highly restricted range of stable electronic configurations, particularly at elevated temperatures where equilibrium considerations apply. Accordingly, the known examples involve only $(Nb,Ta)_6X_{12}^{2+,3+}$ units for X = F, Cl, Br, and some I, corresponding to clusters with 15 or 16 bonding electrons after the lower lying halogen valence levels are filled. The binary compounds that contain these clusters utilize the additional halide anions in intercluster bridging roles, e.g., in Nb_6Cl_{14}, Ta_6Cl_{15}, whereas ternary versions with the same electron counts may be formed with additional alkali-metal halide, $CsNb_6Cl_{15}$ and $Na_4Nb_6Cl_{18}$, for example. The anion in the last is written more clearly as $(Nb_6Cl_{12})Cl_6^{4-}$ to show that the last six chlorides are individually bonded at the cluster vertices. Only six structure types and three X:M ratios are known for the lot. An even smaller group of face-capped octahedral cluster halides are known in $(Nb_6I_8)I_3$ and $(Mo_6X_8)X_4$ ($X \neq F$), but the list can be expanded by including mixed halide-chalcogenide clusters of molybdenum and rhenium (*11,12*).

Several new types of cluster units, as well as a much greater versatility in stoichiometry and structures, have recently been found for the electron-poorer transition metals in groups 3 (rare-earth elements R) and 4 (Zr, Hf). Since the average oxidation state of the metal in an $M_6X_{12}^{m+}$ cluster with a positive charge of $O \leqslant m \leqslant 3$ (like the group 5 examples above) is +2 to +2.5, analogous clusters of these earlier elements would contain only 3 − 6 (R) or 9 − 12 (Zr) metal-metal bonding electrons, significantly below those in any known examples stable at high temperature. The generation of the zirconium clusters with an appreciable negative charge might help, but in fact only those in the rare-earth-metal group seem to form cluster anions. Nature instead provides a new and remarkable way around this shortage of cluster bonding electrons, the encapsulation of an interstitial atom (Z) within the clusters to furnish not only additional electrons but also strong host - Z bonding. The example shown in Fig. 1A emphasizes the strong Zr-Z bonding and the halide bonded at each vertex. A considerable range of valence characteristics among the workable Z examples now allows access to a large variety of cluster charge types (compositions), structures, and new chemistry, in marked contrast to the small variability known with niobium and tantalum.

Three relatively distinct families of centered octahedral (M_6X_{12}-type) clusters are encountered: zirconium chlorides, zirconium iodides, and rare-earth-metal iodides. Some of these provide remarkable parallels in host-Z combinations with intermetallic systems to be described shortly. The cluster halide work is quite recent, and only limited amounts have been published, especially for the rare-earth-metal iodides.

Centered Zirconium Chlorides

The breadth of zirconium chloride examples that can be achieved arises from three aspects of bonding orbital conservation. The first comes from internal Zr-Zr and Zr-Z bonding and the variability in the number of valence electrons provided by the interstitials Z that may be accommodated, namely H, Be, B, C, or N in a large range of structures or the transition metals Mn, Fe, Co, or Ni in just a few. The structural variety arises from two other remarkable features of these (and other) clusters, the dominance of six-metal octahedra, always with twelve edge-bridging halides (M_6X_{12}), and the evident necessity of bonding some sort of halide at the six metal vertices of each cluster (Fig. 1A). As the Cl:Zr ratio increases from 12 to 18, the chloride used for the last type of bonding changes from that which is also edge-bridging halide in other clusters through chloride that bridges between clusters (Zr_6-Cl-Zr_6) to chloride that is bound at only one vertex. Finally, alkali metal cations A^+ may also be accommodated within the cluster network. All three variables allow for a large family with compositions

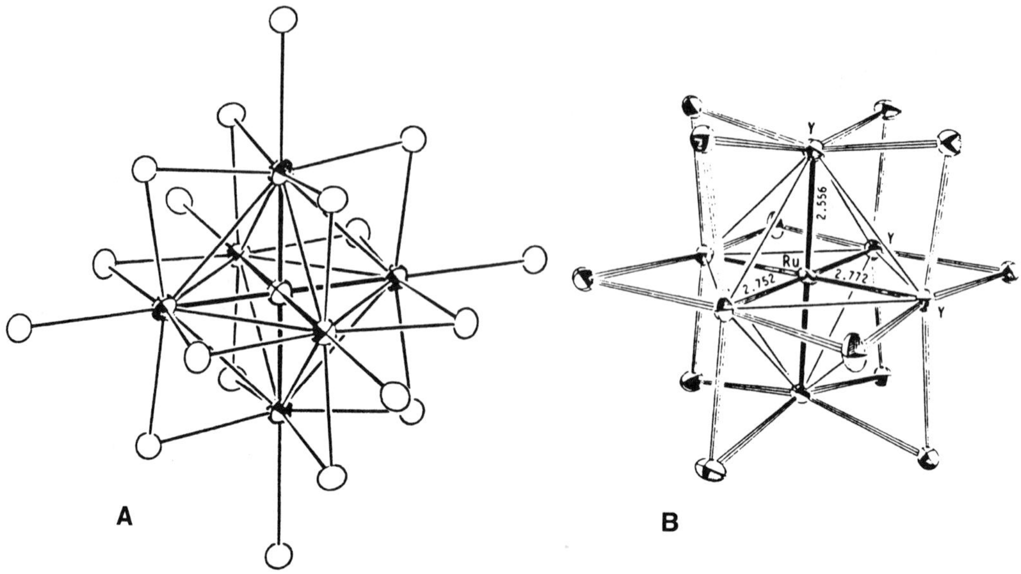

Fig. 1. **A** An example of a centered zirconium cluster $(Zr_6(Z)X_{12})X_6$ with 12 edge bridging and 6 terminal halide atoms. Zr—crossed ellipsoids; Z—solid, with Zr-Z bonding emphasized; —open ellipsoids. **B** A heavy metal interstitial example in the $Y_6I_{12}Ru$ cluster.

$$A^{I}_{x}(Zr_6Cl_{12}Z)Cl_n$$

where examples are found with x and n each ranging through all values between zero and six.

Much of the driving force that affords this versatility appears to originate with the need to achieve an optimum electron count within the Zr_6Z cluster. M.O. calculations show, and experimentation confirms, that this number is about 14 with a main-group Z atom and 18 with a centered transition metal. Two isoelectronic series that exemplify the former class are $Zr_6Cl_{12}Be$, $Zr_6Cl_{13}B$, $Zr_6Cl_{14}C$, $Zr_6Cl_{15}N$, or $Zr_6Cl_{13}B$, $RbZr_6Cl_{14}B$, $Rb_2Zr_6Cl_{15}B$, $Rb_3Zr_6Cl_{16}B$, and so on. (The 14-electron count in each case comes from $6 \cdot 4 + x - (n + 12)$ plus those contributed by Z.) From another viewpoint, the bonding can be thought of in terms of the bonding orbitals and valence electrons that are contributed by Z to the formation of four strong Zr-Z bonds, together with three filled orbitals (t_{2g} or t_{1u}) in each octahedron that are only Zr-Zr bonding. The transition metal-interstitial examples hold four more electrons because of a non-bonding e_g^4 orbital centered on Z (*14–18*). A general description will appear elsewhere (*19*).

Centered Zirconium Iodides

The zirconium cluster iodides seem to be somewhat more flexible than the chlorides in both the size of Z that can be accommodated and deviations from the optimum electron count. Both effects probably originate with the larger size of iodide (matrix effect) (*17*). The list of known interstitials in a periodic table format is as follows.

				B	C	N
				Al	Si	P
Cr	Mn	Fe	Co		Ge	

These all occur with either Zr_6I_{12} or Zr_6I_{14} compositions and structures, in contrast to the chlorides, and without incorporation of the heavier transition metals, in contrast to the following rare-earth-metal systems. The different interstitial sizes may be accommodated in this group is surprising, corresponding to a range of 0.27 Å (11%) in $\bar{d}$(Zr−Z), which produces a difference of 0.38 Å in $\bar{d}$(Zr−Zr) (*20,22*).

Centered Rare-Earth-Metal Iodides

Examples of centered clusters for group 3 metals (R) exhibit a somewhat different collection of structure types appropriate to their electron-poorer status, forming $R_6X_{12}Z^{3-}$ anions, for example, and with an even more surprising variety of interstitials than before. Iodides are again the more prolific. Only the small scandium is known to exhibit some variety of Z within the chloride clusters (B, N), whereas yttrium chlorides seem to be quite sterile as far as centered clusters are concerned. On the other hand, $R_6I_{12}Z$ type clusters, typically sampled for but likely not limited to R = Y, Pr, or Gd, occur in $R(R_6I_{12}Z)$ or $R_6I_{10}Z$ structures for

					B	C (C_2)
Mn	Fe	Co	Ni	Cu		
	Ru	Rh	Pd			
Re	Os	Ir	Pt			

(*23–26*). The main-group Z possibilities have not been thoroughly researched beyond the gadolinium halide carbides (*12*). The tetragonally compressed $Y_6I_{12}Ru$ example is shown in Fig. 1B without the interconnecting iodides that are bonded at all vertices.

Condensed Clusters

Condensed examples involve clusters that share opposite metal edges (and halogens) and are found in a number of binary systems, for example, as infinite chains in $(Gd,Y)_2Cl_3$, double chains in Sc_7Cl_{10}, and infinite double metal sheets in Zr(Cl,Br), as discussed before (*13,27*). More examples occur only as interstitial examples, including in some hitherto supposed binaries that are really ternaries like $Sc_5Cl_8(C,N)$ and $Sc_4Cl_6(B,N)$. Recent examples include Y_4I_5C (single chains), a portion of which is shown in Fig. 2, and $Y_6I_7C_2$ (double chains) (*28*). The binary zirconium monohalides in which cluster chains have now been con-

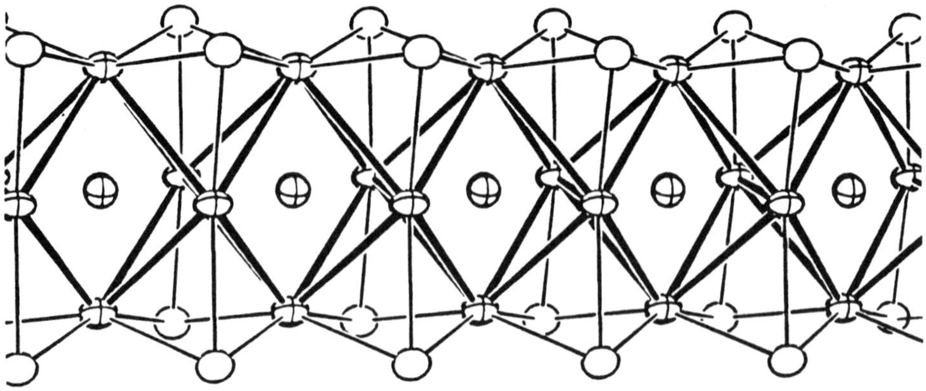

Fig. 2. A portion of the infinite cluster chain in Y_4I_5C. Note that the chain is constructed of carbon-centered $Y_6I_{12}C$ clusters that have been condensed through sharing of opposite metal edges and adjoining halides. The Y-C bonding is not shown (*28*).

densed into double metal sheets have an extensive interstitial chemistry as well, involving H, B, C, N, or O but not heavier examples (yet) (*13*). Rare-earth-metal monohalide analogs with H, B, C, and so on are also found but now only as the ternary phases (*12*). Tetrahedral interstices are available in these layered monohalide structures as well and are utilized by H, N, and O.

A striking feature in all of this chemistry is the predominance of only six-metal-atom cluster halides. Surprisingly, no larger cluster units have been found except in the chain and sheet condensation products just noted; even isolated or condensed tetrahedra are very rare. The larger halide-sheathed Zr_{13} cluster (close-packed 3-7-3) must be unstable with respect to the Zr_6X_{12} unit plus zirconium metal. The smaller six-atom clusters may be favored by strong, polar Zr-X bonding in an effective space-filling result together with energetically better bonding in the bulk metal, as will be discussed elsewhere (*29*).

Substantially all of the compounds discussed above are obtained in >90% yield from reactions of the components in welded niobium or tantalum containers at 750–950°C for 1–4 wk. Their high thermodynamic stabilities relative to alternate products make virtually quantitative yields common, which in turn provide useful assessments that the stoichiometries have been correctly deduced. An attempt to insert an unsuitable Z accordingly gives only a binary, and often rather ordinary phase, ZrCl, ZrI_3, YI_3, and PrI_2, for example.

POLAR INTERMETALLICS

Our inquiry regarding both clusters and their interstitial chemistry now extends beyond halides and chalcogenides to transition-metal-rich compounds of the pnictides, As, Sb, and Bi particularly but also even further to compounds of Si, Ge, Sn, and Pb. These compounds do not involve any major changes in concepts, although the last two types are apt to be a good deal less familiar. All compounds formed between one of the earlier transition metals (M) and one of these main-group (B) nonmetals, metalloids, or neighboring metals possess some common characteristics regarding electronic and bonding features, namely, those that arise because of sizable differences in valence state ionization energies (~electronegativities) of the two types of atoms. As a result, the main-group elements in all of these will be the major components of low lying (and, therefore, filled) valence p bands, and strong heteroatomic M-B bonding therewith will contribute greatly to the stability of the phases. Although it would be simplistic and misleading to identify all of the antimonides for example as "salts," their polar nature does make the foregoing electronic relationship very probable. The metal-rich phases of interest generally contain isolated main group elements and therefore exhibit no B-B bonding, so particularly simple (8-N) valence expectations apply to these anions. (Use of the term "anion" naturally refers to the polarity, not something approaching a "free" ion.)

Phases that are rich in the transition-metal component will likely also exhibit significant M-M bonding. Valence electrons that remain after the predominantly main group element (B) valence states (bands) are filled will occur in a conduction (or metal valence) band with major contributions from electropositive metal (M). Just how far to the left in the periodic table the main-group element can lie and still have the compound exhibit no overlap between the filled, nominally B, valence band, and the metal conduction band remains to be determined. However, we can anticipate that the more electropositive metals and many of the main-group metals or metalloids will form what we have termed "polar intermetallics," compounds in which strong heteroatomic bonding and the above valence generalities apply (*30*).

The change from halide to chalcogenide to pnictide clusters or other arrays requires, all other things being equal, an increased number of electrons to meet the valence requirements of the anions, or a decrease in the relative number of anions. For the same structural unit and approximate electron count, the chalcogenide analog of a halide cluster should in the first case be found for a later transition metal, as is observed (Mo vs Nb), although a change from a M_6Y_{12} to a M_6Y_8 cluster type, and, thus, in the optimum electron count, also pertains. Continuation of this trend to give isolated cluster pnictides of the platinum metals evidently does not occur; instead, another variable comes into play. Alternate ways to compensate for an electron count per cluster that is too low are through the inclusion of an interstitial, as we have seen, or via cluster condensation in the presence of fewer anions. The latter seems to apply regularly to metal-rich pnictides of the first four transition-metal groups with their paucity of valence electrons. Increases in anion polarizability (~formal charge, covalency) are probably also important in these results. Of course, these "explanations" are not sufficient to predict phase stability. Nonetheless, what is found may often be viewed as small, often six-metal, clusters that have been condensed through sharing of vertices, edges, or faces, most often into infinite arrays. Sometimes the condensation is so extensive that the sense of a simple cluster is virtually lost. The variety of possibilities has been thoroughly discussed by Simon (*11*). An interstitial chemistry comparable to that in the halides described above does not seem to have been investigated (or stumbled upon) for most of the pnictides (or chalcogenides) other than for a few hydrides. One major exception is compounds with the Mn_5Si_3 structure, and a broad one it is.

The Mn_5Si_3 Structure

The compounds in Mn_5Si_3 structure have long been known or suspected to take up interstitials, carbon and oxygen most often, or even to require the same for stability (*31–34*). Very few quantitative studies of structure or composition have appeared in support of these ideas, however. The breadth of the binary M_5B_3 phases that form with this structure and appear to be candidates is remarkable as well, some 155 by one count (*33*), in phases that encompass transition (M) metals from groups 2 through 6 together with main-group elements (B) involving all members of the boron, carbon, and nitrogen families. Pearson's handbook has 264 entries for this structure type when mixed systems are included (*35*). It would be surprising indeed if other metal-rich structures involving many of these same or neighboring elements, chalcogenides in particular, did not also exhibit a similar interstitial chemistry.

The important structural features of the Mn_5Si_3 arrangement are illustrated in Fig. 3 for the particular case of Zr_5Sb_3 (*30*). The hexagonal structure contains two types of infinite metal chains. On the left is depicted a portion of the chain of confacial (face-sharing) metal octahedra $Zr_{6/2}Sb_{6/2}$ that lies along 0,0,z in the unit cell, and on the right, the linear chain of metal (Zr_2) along ⅓,⅔,z and ⅔,⅓,$\bar{z}$ that is reminiscent of those in the A15 structure. The two chain types are not as isolated and the bonding is not as anisotropic as implied by these views; the antimony (or B) elements that bridge edges of the shared faces of the octahedra (left) are the same as those that are six-coordinate about zirconium (M) atoms on the linear chain (right). In fact, the Zr-Zr distances between the chains are comparable to those within the shared octahedra, though the overlap population (~ bond strength) is not as large. The confacial chain is very similar to those in Chevrel phase derivatives involving $^{1}_{\infty}[Mo_3(Ch)_3^-]$ units (Ch = chalcogenide); in fact, the structure of KMo_3S_3 structure can be derived from that of Mn_5Si_3 by replacing pairs of atoms in the linear (Zr_2) chains by single potassium ions.

Interstitial Derivatives

Many single crystal studies have now settled the questions of the amounts of interstitials incorporated and where they go—in the rather obvious octahedral cavities in the chain on the left (small circle, Fig. 3) and often at full occupancy (*36*). This confirms what had been estimated earlier in other systems, largely on the basis of powder diffraction and lattice constants.

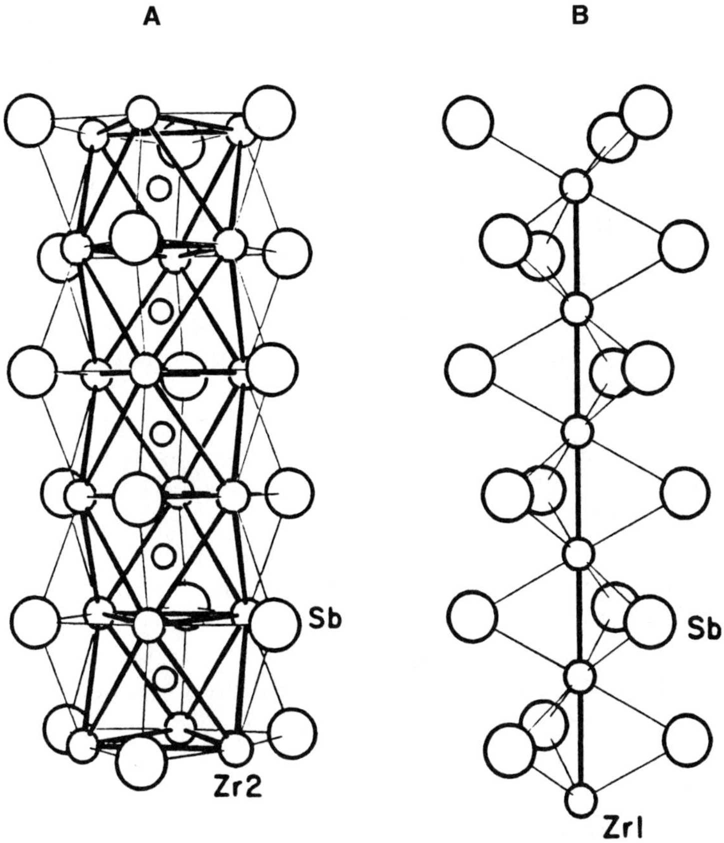

Fig. 3. The separate chains in the Mn_5Si_3 type structure, as determined for Zr_5Sb_3 (*30*). Antimony—large circles, zirconium—medium circles, interstitial Z—small circles. The confacial chain of zirconium octahedra in **A** contains the interstitial site for Z. The antimony atoms about the linear zirconium chain in **B** are the same as those sheathing the octahedral chain in **A**.

Syntheses of both the binary M_5B_3 hosts and ternary M_5B_3Z compounds containing interstitial Z have generally been aimed at achieving quantitative yields so that the stoichiometries are clear, the high sensitivity of Guinier powder diffraction being routinely employed for this determination. Traditional methods of arc melting (≈3000°C) followed by annealing as well as powder sintering have been employed most frequently, although metal fluxed and vapor phase transport reactions have also been useful. Temperatures of 900–1200°C are typically necessary to get equilibrium. Self interstitials like Zr_5Sb_{3+X}, $0 \leq X \leq 0.4$, or Zr_5Sn_4 are sometimes found (the fully occupied example is called the Ti_5Ga_4 structure). Lattice constant changes produced by self-(B) interstitials may sometimes mimic those produced by foreign (Z) elements, so more than a qualitative change in lattice dimensions may be necessary to establish the formation of a given ternary phase. The wide range of main-group elements that form the binary phases also means that some B-Z interchange may occur in some systems at arc temperatures, but ordering seems to be achieved fairly regularly on annealing.

The chemistry and valence properties of these extended cluster systems are significantly more variable than for the fairly electron-precise cluster and extended cluster halides noted above. On the other hand, the local Zr-Z bonding interactions should be rather similar. As might be anticipated from the earlier remarks regarding the valence characteristics of "polar intermetallics," many of these host M_5B_3 phases are metallic, and we shall keep book not on the cluster-centered bonding population, but on the total number of electrons in the conduction band in the hosts and in the ternary products with Z. Thus Zr_5Sb_3 has $5 \cdot 4 - 3 \cdot 3 = 11$ electrons per formula unit in the conduction band, a quantity that we designate as n. The ability to tune these systems so as to either approach or achieve the $n = 0$ state through choice of either the host or Z, or both, has also been investigated in several systems. The $n = 0$ limit naturally represents the formation of a valence compound (Zintl phase), conventionally a semiconductor. Resistivity studies relating to this end point are underway. Distortions in low-n systems that might be of a Peierls or other character and lead to band splitting have not been seen; on the other hand, fractional occupancy by Z in an ordered manner does lead to some interesting superstructures (*36,37*).

Zr_5Sb_3 Host: $n = 11$

The amazing ability of the Zr_5Sb_3 host to bond interstitial atoms with a wide range of chemical properties is summarized by the following list of those that have been incorporated (*37*).

					C		O
				Al	Si	P	S
Co	Ni	Cu	Zn	Ga	Ge	As	Se
		Ag				Sb	

The majority of the Zr_5Sb_3Z products have been identified on the basis of the formation of single phase samples with appropriate and reproducible shifts in Guinier-based lattice parameters following stoichiometric reactions by powder sintering or vapor phase transport. Several single crystal studies have naturally been much more informative about both dimensional changes within the host and the stoichiometry. On the other hand, the occurrence of self interstitials in the solution Zr_5Sb_{3+x}, $0.0 \leq x \leq 0.4$ $(x \neq 1)$, in this particular system correlates with an apparent ease of mixed Sb-Z interstitial formation in single crystals grown under arc-melting conditions. For example, an early report of the formation of $Zr_5Sb_3O_x$ (*31*) probably pertained to a Zr_5Sb_{3+x} sample instead; an authentic oxide sample exhibits significantly smaller dimensions than the binary solution. The system with Z = Fl is particularly complicated by both self interstitials with antimony and an antimony-poorer ternary phase (*36*).

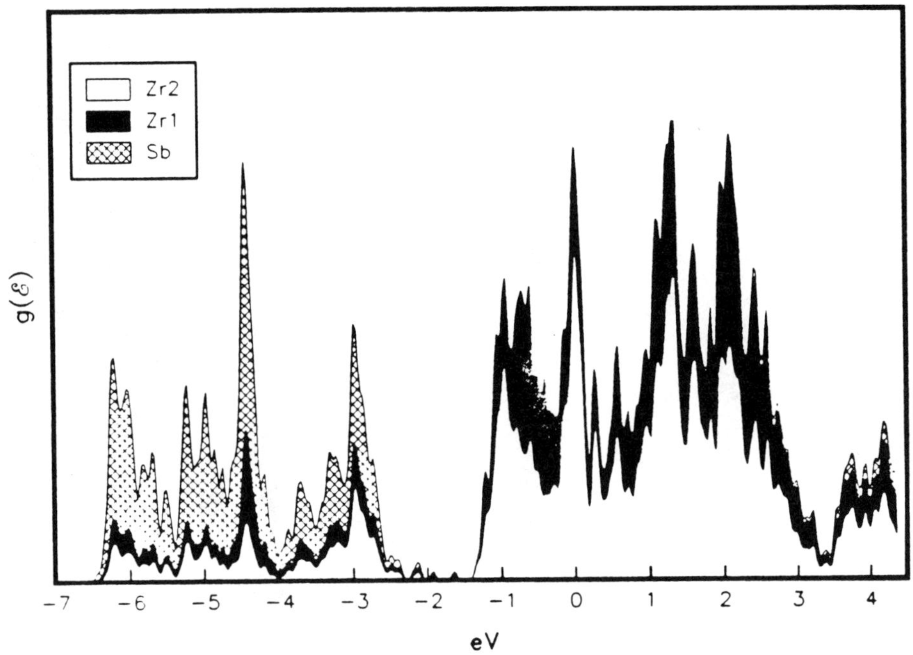

Fig. 4. The densities-of-states results for Zr_5Sb_3 (extended-Hückel calculation, 24 k points, $E_F = 0$) with the Zr 4d and Sb 5p orbital contributions projected out. The Zr 2 and Zr 1 atoms are in the confacial and linear chains, **A** and **B**, respectively, in Fig. 3. Other atomic orbitals are unimportant in this region.

The densities-of-states results from an extended-Hückel band calculation for the Zr_5Sb_3 host are shown in Fig. 4. This emphasizes the valence characteristics already anticipated in a general way. Projections of the Sb 5p and Zr 4d orbital contributions show that all of the antimony 5p (and 5s) bonding and nonbonding states occur in a low lying valence band(s), whereas only the Zr 1 (linear chain) and Zr 2 (chain of octahedra) atoms (Fig. 3) make significant contributions to the conduction band. The formation of an interstitial derivative from this will in a general sense utilize some of the electrons and Zr orbitals in the (approximately nonbonding) conduction band to form Zr-Z bonds while retaining with little alteration the dimensions and strong valence bonding of the host.

Figure 5 shows for comparison the comparable calculational result for Zr_5Sb_3S. As generalized above, the principal result is the loss of two electrons (and some Zr 4d states) from the conduction band and the appearance of S 3p (and some more Zr 4d) bonding states that overlap the lower energy part of the antimony valence band. (The Sb 5s and S 3s levels have greater binding energies (~ -20 eV) and are relatively unimportant.) Calculated overlap populations emphasize the gain of Zr-Z bonding and the loss of what was only modest Zr-Zr bonding in the confacial chains in forming the sulfide. On the other hand, the overlap populations for Zr-Sb, Zr 1-Zr 1, and Zr 1-Zr 2 interactions and all Zr-Sb

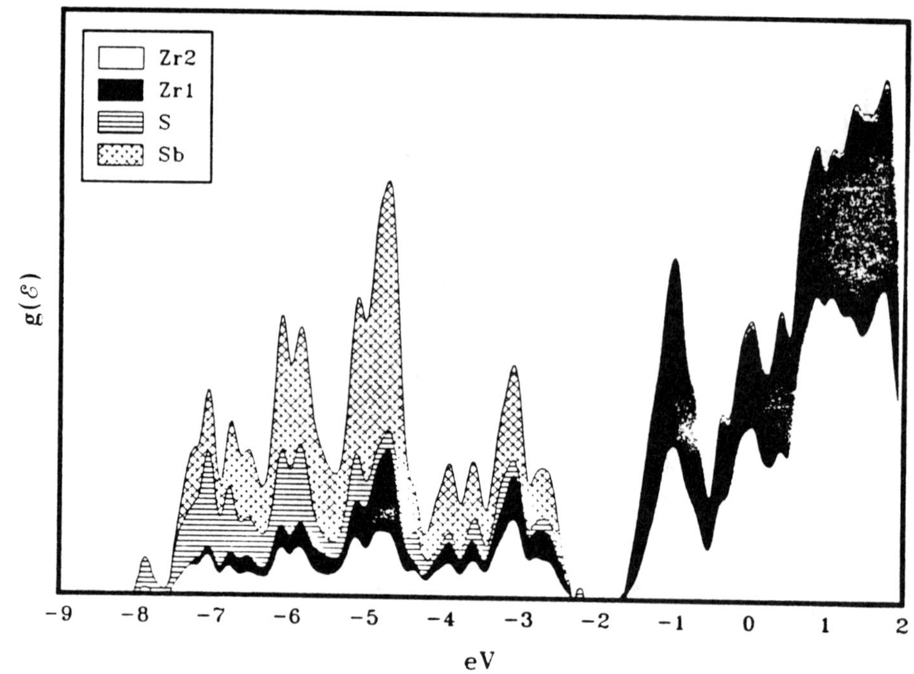

Fig. 5. The densities-of-states results for Zr_5Sb_3S with Zr 4d, Sb 5p, and S 3p orbital contributions projected out. The sulfur bonding and nonbonding states overlap the valence band of the host; *compare* Fig. 4.

distances are virtually unchanged in the process. Dimensionally, the greatest changes on interstitial inclusion in general show up as a lateral expansion of the shared faces of the host chain, while the Zr 2-Zr 1 distances along the linear chain and, therewith, the c axis exhibit much less increase, perhaps because this dimension is more determined by strong bonding in this chain. The driving force for these reactions seems to derive, as with the earlier isolated cluster examples, from the formation of strong Zr-Z bonds, but with the necessary valence electrons required coming from the conduction band supply. The Zr-Z bonding levels for all of the Z examples listed above will naturally not lie within the Sb 5p (−Zr) valence band, as described above for sulfur.

Zr_5Sn_3, Zr_5Pb_3 Hosts: n = 8

The tin system at equilibrium near 1100°C contains only the line phases Zr_5Sn_3 and Zr_5Sn_4, although homogeneous samples near $Zr_5Sn_{3.3}$ that contain interstitial tin can be obtained by arc melting. Segregation of samples into ~Zr_5Sn_3 and ~$Zr_5Sn_{3.3}$ compositions induced by temperature gradients during arc melting was apparently responsible for an earlier report of the formation of two distinct phases with this structure (*38*). The list of Z elements accommodated by Zr_5Sn_3 is similar to that given for antimony, viz

					B	C	N	O
					Al	Si	P	S
Fe	Co	Ni	Cu	Zn	Ga	Ge	As	Se
			Ag			Sn		

Further examples will doubtlessly be found, but the majority of the omissions in this table represent negative results. The Zr_5Pb_3 host systems have been investigated less thoroughly, but the behavior seems similar, with the following positive results for Z.

					Al	Si		S
Mn	Fe	Co	Cu	Zn	Ga	Ge		Se
			Ag		In			Te
						Pb	Sb	

La_5Ge_3, La_5Pb_3 Hosts: n = *3*

The lanthanium systems are still under exploration, but several aspects are already evident. Expanded Mn_5Si_3 structures readily result for La_5Ge_3Z with Z = Fe, Ru, Co, Ni, Cu, B, C, P, or As, and new supercells occur with some metals for ordered, fractional Z. Likewise, the lead analog forms isostructural compounds with B, C, Mn, Fe, Ru, Co, Ni, Cu, Sb, or Bi. It is interesting that La-Fe binary compounds are unknown.

Reactions of these $La_5(Ge,Pb)_3$ (n = 3) hosts with equimolar amounts of Z = Si, Ge, Sn (and Z = Pb for M = Ge), all of which require four more valence electrons, uniformily give a quantitative yield of $La_5(Ge,Pb)_3Z$ in the Sm_5Ge_4 structure. (Whether there is ordering of Ge or Pb and Z is not known.) This result is in complete accord with valence expectations already outlined since occupation of the valence levels of the main group element (B) and Z should produce $n = -1$, i.e., with one hole in the valence band. The usual consequence of this possibility is the same as with many other Zintl phases (*6*), the formation of B-B (B-Z or Z-Z) dimers equal in number to one-half the number of potential holes, and this is exactly what the Sm_5Ge_4 structure represents. A phase with an interstitial occupancy of 0.75 or less that would still allow retention of the Mn_5Si_3 structure ($n \geqslant 0$) does not appear to be stable for Z=Si, Ge. The occupancy with Z=B, C is not clear at this time.

M_5B_3 Hosts: n = *1*

Finally, the same Mn_5Si_3 structure is known to result for most members of the family $(Ca, Sr, Ba)_5(As,Sb,Bi)_3$. A few also exhibit the relevant β-Yb_5Sb_3 type structure—below. All of the $M_5(Sb,Bi)_3$ examples have recently been found to also form isostructural chlorides $M_5(Sb,Bi)_3Cl$ (n = 0) and, for the three examples studied, bromides as well. The volume increases observed on chloride formation are only one-fourth (with calcium) to one-half (barium) of standard value for a chloride ion, indicating how much of the cavity is already available in the binary phase. Again, the principal dimensional change appears to be in the shared face of the octahedral chain. The $Ce_5(Sb,Bi)_3$ compounds ($n = 3$) also form analogous chlorides and bromides ($n = 2$). The 5:3 alkaline-earth-metal compounds with Si, Ge, and so on ($n = -2$) occur in the Cr_5B_3 structure with B-B dimers, in complete analogy with the lanthanium examples cited above.

It was noted above that some of these alkaline-earth-metal binaries form a relatively complex β-Yb_5Sb_3 structure as well. A suitable preformed but small interstitial cavity and one valence electron are evidently also available in the known or hypothetical examples of this structure type for the binary calcium phases, and the fluorides $Ca_5(Sb,Bi)_3F$ with this structure can also be synthesized (but not Ba_5Sb_3F). Appropriate amounts of iodide, oxide or sulfide all form Zintl phases with other structures (*39*).

Both fluoride and chloride are plausible impurities in the alkaline-earth-metal samples that were available in past decades, and so earlier reports on the A_5B_3 binary compounds may have actually pertained to the ternary halides. Unfortunately, lattice dimensions are the only practical means of comparison, and some of the older values were not very accurate. However, in a larger context, there are probably many reported phases that unknowingly contained, or were stabilized by, adventitious impurities in similar roles. Whether there are such chemical and structural roles for essential interstitials in important alloy systems is generally unknown but a tantalizing prospect.

ACKNOWLEDGMENT

This research was supported by the Office of Basic Energy Sciences, Materials Sciences Division, US Department of Energy. The Ames Laboratory is operated for DOE by Iowa State University under Contract No. W-7405-Eng-82.

REFERENCES

1. Andres, R. P., Averback, R. S., Brus, L. E., Goddard, W. A., Kaldor, A., Louie, S. G., Moscovits, M., Peercy, P. S., Riley, S. J., Siegel, R. W., Spaepen, F., and Wang, Y., *J. Mat. Res.* **4,** (May/June), in press (1989).
2. Jena, P., Rao, B. K., and Khanna, S. N., *Physics and Chemistry of Small Clusters,* Plenum, New York (1987).
3. Schmid, G., *Structure and Bonding* **62,** 51–85 (1985).
4. Corbett, J. D., *Chem. Rev.* **85,** 383–397 (1985).
5. Corbett, J. D., *Inorg. Chem.* **7,** 198–208 (1968).
6. Schäfer, H., *Ann. Rev. Mater. Sci.* **15,** 1–41 (1985).
7. Martin, T. P., *J. Chem. Phys.* **83,** 78–84 (1985).
8. Wheeler, R. G., LaiHing, K., Wilson, W. L., Allen, J. D., King, R. B., and Duncan, M. A., *J. Am. Chem. Soc.* **108,** 8101–8102 (1986).
9. Schild, D., Pflaum, R., Sattler, K., and Recknagel, E., *J. Phys. Chem.* **91,** 2649–2653 (1987).

10. Wheeler, R. G., LaiHing, K., Wilson, W. L., and Duncan, M. A., *J. Chem. Phys.* **88,** 2831–2839 (1988).
11. Simon, A., *Angew. Chem. Int. Ed. Engl.* **20,** 1–22 (1981).
12. Simon, A., *Angew. Chem. Int. Ed. Engl.* **27,** 159–183 (1988).
13. Corbett, J. D. and McCarley, R. E., *Crystal Chemistry and Properties of Materials with Quasi-One-Dimensional Structures,* Rouxel, J., ed., pp. 179–204, D. Reidel, Dordrecht (1986).
14. Ziebarth, R. P. and Corbett, J. D., *J. Am. Chem. Soc.* **107,** 4571–4573 (1985).
15. Ziebarth, R. P. and Corbett, J. D., *J. Am. Chem. Soc.* **109,** 4844–4850 (1985).
16. Ziebarth, R. P. and Corbett, J. D., *J. Am. Chem. Soc.* **110,** 1132–1139 (1988).
17. Ziebarth, R. P. and Corbett, J. D., *J. Am. Chem. Soc.* **111,** in press (1989).
18. Zhang, J. and Corbett, J. D., unpublished research.
19. Ziebarth, R. P. and Corbett, J. D., *Acct. Chem. Res.,* accepted (1989).
20. Smith, J. D. and Corbett, J. D., *J. Am. Chem. Soc.* **107,** 5704–5711 (1985).
21. Hughbanks, T., Rosenthal, G., and Corbett, J. D., *J. Am. Chem. Soc.* **110,** 1511–1516 (1988).
22. Rosenthal G. and Corbett, J. D., *Inorg. Chem.* **27,** 53–56 (1988).
23. Dudis, D. S., Corbett, J. D., and Hwu, S. -J., *Inorg. Chem.* **25,** 3434–3438 (1986).
24. Hughbanks, T. and Corbett, J. D., *Inorg. Chem.* **27,** 2022–2026 (1988).
25. Hughbanks, T. and Corbett, J. D., *Inorg. Chem.* **28,** 631–635 (1989).
26. Payne, M. and Corbett, J. D., unpublished research.
27. Corbett, J. D., *Pure Appl. Chem.* **56,** 1527–1543 (1984).
28. Kauzlarich, S. M., Hughbanks, T., Corbett, J. D., Klavins, P., and Shelton, R. N., *Inorg. Chem.* **27,** 1791–1797 (1988).
29. Rogel, F., Zhang, J., Payne, M. W., and Corbett, J. D., *Adv. Chem. Ser.,* submitted.
30. Garcia, E., and Corbett, J. D., *Inorg. Chem.* **27,** 2353–2359 (1988).
31. Rossteutscher, W. and Schubert, K., *Z. Metallkde.* **56,** 813–816 (1965).
32. Rieger, W., Nowotny, H., and Benesovsky, F., *Mh. Chem.* **95,** 1413–1423 (1964).
33. Parthé, E. and Rieger, W., *J. Dentl. Res.* **47,** 829–835 (1968).
34. Nowotny, H. and Benesovsky, F., *Phase Stability in Metals and Alloys,* Rudman, P. S., Stringer, J., and Jaffee, R. I., eds., pp. 319–336, McGraw-Hill, New York (1966).
35. Villars, P. and Calvert, L. D., *Pearson's Handbook of Crystallographic Data for Intermetallic Phases,* vol. 1, pp. 580–581, Amer. Soc. Met., Metals Park, OH (1985).
36. Garcia, E., Kwon, Y. -U., Guloy, A., and Corbett, J. D., to be submitted for publication.
37. Garcia, E. and Corbett, J. D., *Inorg. Chem.* **27,** 2907 (1988).
38. Gran, G. and Andersson, S., *Acta Chem. Scand.* **14,** 956 (1960).
39. Hurng, W. -M. and Corbett, J. D., *Chem. Mat.,* **1,** in press (1989).

Performance Under Extreme Environments

Hot Corrosion of Materials

ROBERT A. RAPP

Department of Materials Science and Engineering, Ohio State University, Columbus, OH 43210

ABSTRACT

Alloys and ceramics may experience accelerated corrosion at high temperatures when their surfaces are coated by a thin film of fused salt in an oxidizing environment. Because of their high thermodynamic stabilities, fused alkali sulfates are frequently deposited or condensed onto hot substrates from the combustion product gases or the oxidation of heavy metal contaminants in the fuel, e.g., vanadium.

Solid-state electrochemical probes have been used in extensive determinations of the solubilities, and their dependencies on oxygen and Na_2O activities, for many important oxides in fused Na_2SO_4 at 1200 K. The expectation for accelerated synergistic dissolution kinetics for two oxides of differing basicity has been confirmed experimentally.

Solid-state electrochemical probes have been attached at the surface of coupons of preoxidized nickel, which were subjected to hot corrosion by a fused Na_2SO_4 film in SO_2/O_2 gas environments. The probes were able to identify the onset of hot corrosion, and the local values for oxygen activity and the basicity in the salt film during the attack. The resulting data supported the Goebel-Pettit model for basic fluxing of nickel oxide and the negative solubility gradient criterion of Rapp-Goto.

The acidic solubilities of thermal barrier candidate oxides Y_2O_3, CeO_2 and HfO_2 in a fused 0.7 Na_2SO_4–0.3 $NaVO_3$ solution greatly exceed their solubilities in pure Na_2SO_4, a problem that should exist for all oxides. A modeling of the constituent vanadates in the Na–V–S–O system permits a prediction of the dependence of oxide solubility on vanadium content and melt basicity.

Index Entries: Thin film; Na_2SO_4; Na_2O; Corrosion; oxidizing environment; melt basicity.

INTRODUCTION

Many very important engineering systems operating at high temperatures (650–1100°C) involve contact of metallic or ceramic materials with combustion product gases or other oxidizing gases containing inorganic impurities, e.g., gas turbines, steam generators, incinerators, and numerous petrochemical process vessels. As the gases are cooled, fused salt films may condense on the hardware to generate a highly corrosive condition analogous in some aspects to aqueous atmospheric corrosion. Some other engineering systems, such as the molten carbonate fuel cell, heat treatment baths, and so on, use high temperature salts to increase the kinetics of certain electrochemical and chemical reactions. Again, materials come into contact with corrosive fused salt films that affect an accelerated degradation known as hot corrosion. The engineering application of high temperature systems, with their associated problems, is understood to be inherent in advanced technologies. As the availability of high quality fossil fuels becomes limited, and as the world seeks to incinerate solid wastes and create more efficient energy conversion processes and so on, the occurrence and damage of hot corrosion is likely to increase.

A wealth of literature exists on the kinetics of hot corrosion, and on the resulting microstructures. Also, the chemistry and electrochemistry of hot corrosion have been reviewed rather recently (*1*). In this paper, more recent developments in the chemistry and electrochemistry of hot corrosion are presented, and earlier papers presenting well-known and accepted theories and experimental methods are referenced. Perhaps the three most important facts about the problem are: (1) generally fused salts are electrolytic conductors so that the attack must be electrochemical in nature; (2) oxyanion-fused salts (sulfates, carbonates, nitrates, and so on) exhibit an acid-base behavior that is crucial to most mechanistic aspects of hot corrosion; and (3) generally, metallic engineering alloys have virtually no regimes for mutual thermodynamic stability with a fused salt, such as Na_2SO_4, so that a corrosion product must form upon contact of the salt with the metallic substrate. Clearly, the avoidance of hot corrosion is achieved by forming a slow growing, compact, and only slightly soluble oxide scale.

PHASE STABILITY

This subject has been treated in a manner exactly analogous to aqueous solutions by the formalism of Pourbaix, i.e. plots of electrode potential (oxidizing potential) vs basicity, which for the pure Na_2SO_4 treated here, is defined as $-\log a_{Na_2O}$. The phase stability diagram for the Na–S–O system at 1200 K shown in Fig. 1 was derived from the standard Gibbs energies of formation for the compounds involved (*2*). The broad stability field of Na_2SO_4 has been divided into regimes denoting the dominant minority ionic species, upon the assumption of ideal solutions.

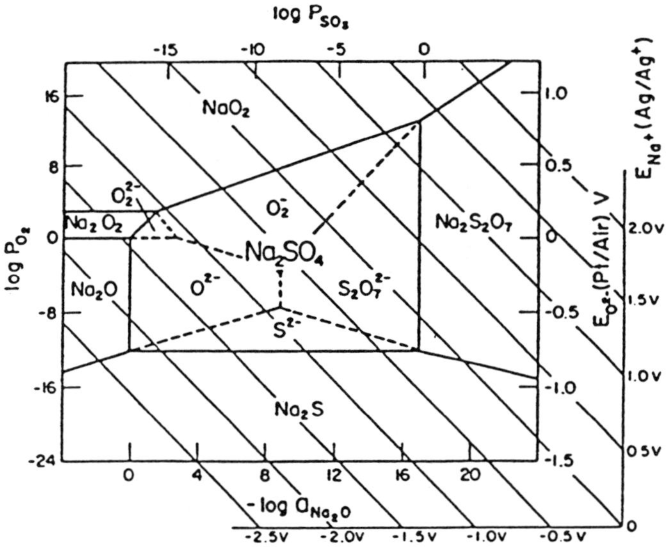

Fig. 1. Phase stability diagram for Na–S–O System at 900°C.

This procedure is especially advantageous for the development of realistic interpretations for chemical and electrochemical reactions using the criterion that spontaneous reactions involve the consumption of unstable species and the formation of stable products for the local environmental conditions (coordinates of Fig. 1). To measure these conditions accurately, a pair of solid-state tubular electrodes has been developed, forming a sodium-ion conducting electrolyte cell and an oxygen-ion conducting electrolyte cell, whose open-circuit potentials reveal the local oxidizing potential and the basicity according to well-accepted expressions (2). The emf's for these two reference electrodes in the regime of Na_2SO_4 stability have been indicated on Fig. 1.

With this understanding for the regime of stability for the electrolyte (or solvent) itself, one inquires about the phase stability for metals, oxides, or sulfides in contact with the fused salt. Figure 2 presents a phase stability diagram for the Ni–S–O system superimposed onto Fig. 1. Over a wide range of environmental conditions, the oxide NiO is stable in contact with fused Na_2SO_4, although it exhibits a variable solubility as a basic solute $NaNiO_2$ and the acidic solute $NiSO_4$, as indicated by the dashed lines (-1, -2, and -3) for the logarithm of the activities for these (fully ionized) species in fused Na_2SO_4. Notably, there is only a negligible regime of conditions where the metal nickel could coexist with fused Na_2SO_4 without reaction, usually to form a liquid nickel sulfide if an acid melt should contact the metal directly.

OXIDE SOLUBILITIES

The solubility of NiO in fused Ni_2SO_4 at 1200 K and 1 atm O_2 pressure has been measured by atomic absorption analysis of NiO-saturated melts for well-characterized basicities, as shown in Fig. 3 (3).

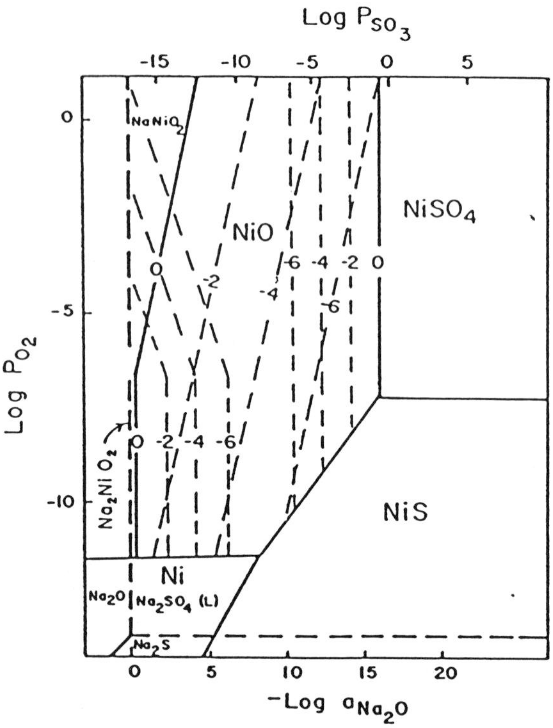

Fig. 2. Na–Ni–S–O phase stability diagram for 1200 K.

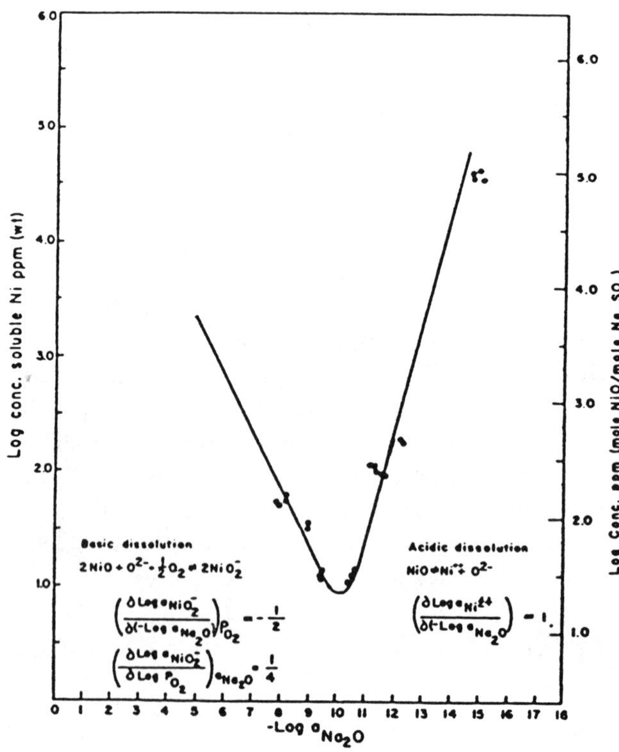

Fig. 3. Solubility of NiO in fused Na_2SO_4 at 1200 K.

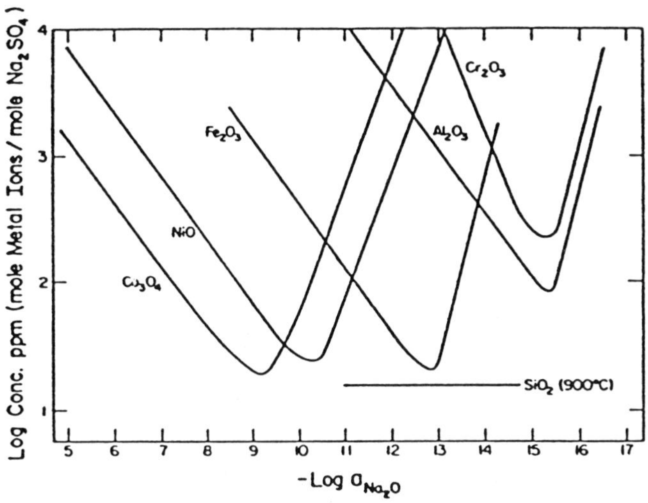

Fig. 4. Experimentally established oxide solubilities in fused Na_2SO_4 at 1200 K and 1 atm O_2.

The dependencies of the acidic and basic solubilities on salt basicity (slopes in Fig. 3) agree exactly with those expected for constant activity coefficients for the dilute solutes. The calculated Henrian activity coefficients are 2.54 and 1.54 for the acidic and basic solutes, respectively. This illustrative example demonstrates the predictable solubility behavior for a simple oxide in a pure oxyanion solvent. According to the same methods, the solubilities of the oxides Al_2O_3 (*4*), Co_3O_4 (*3*), Cr_2O_3 (*5*), iron oxides (*6*), Y_2O_3 (*7*), SiO_2 (*8*), and CeO_2 (*9*) have been established in pure-fused Na_2SO_4 at 1200 K. Figure 4 presents a plot for the solubilities for most of these oxides at 1 atm oxygen. In every case, the dependencies for the acidic and basic solubilities on melt basicity are very well described by simple dissolution expressions analogous to those written in Fig. 3. Because of the valence changes for iron and chromium with changing oxygen activity, their solubilities have been shown to depend also on oxygen activity, again in a very predictable way. Because the oxide SiO_2 does not form any ionized acidic solute for the basicity range indicated in Fig. 4, the solubility of SiO_2 does not depend upon melt basicity. (However, in more basic melts, SiO_2 is subject to basic dissolution.)

HOT CORROSION MECHANISMS

This quantitative knowledge of these oxide solubilities represents an important background data bank for the interpretation of the mechanisms for Type 1 hot corrosion based on a scale fluxing/reprecipitation argument as first proposed by DeCrescente and Bornstein (*10,11*) and Goebel and Pettit (*12*). Figure 5 presents a schematic diagram illustrating the fluxing of a protective scale on a pure metal and the precipitation of the same oxide as nonprotective particles in the salt film. Rapp and Goto (*13*) explained that the electrochemical nature of the hot corrosion pro-

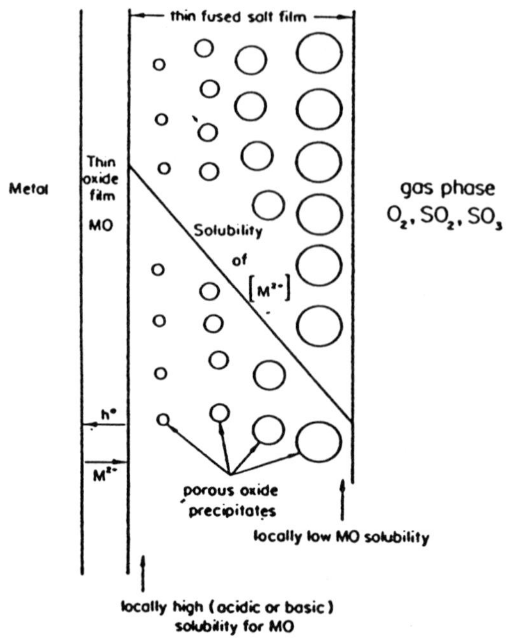

Fig. 5. Dissolution–reprecipitation of MO oxide driven by a negative solubility gradient in a fused salt film.

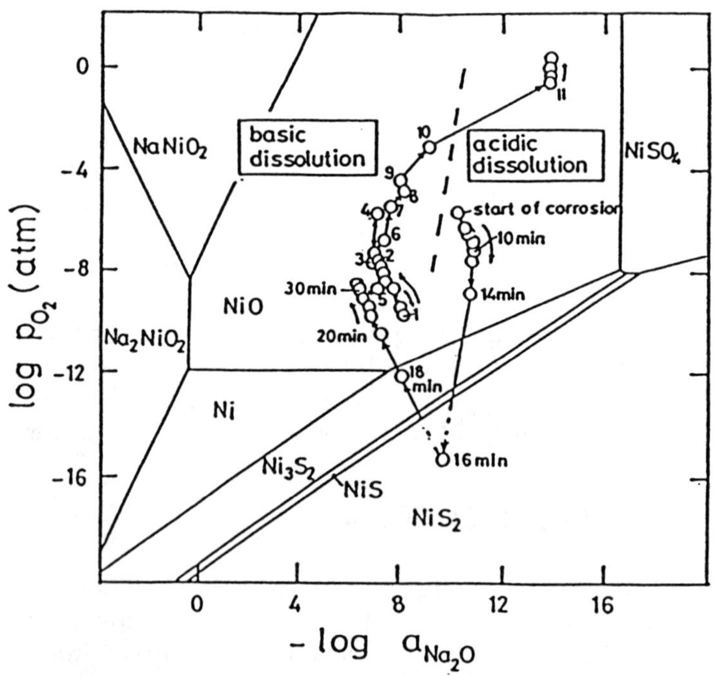

Fig. 6. Trace of basicity and oxygen activity measured for preoxidized 99% Ni covered with a Na_2SO_4 film at 900°C in 0.1% SO_2–O_2 gas atmosphere (preoxidized at 900°C for 4 h in O_2). Numbers designate reaction time in hours except as indicated; severe corrosion.

cess would lead to increased basicity at the site of the reduction reaction, so that basicity gradients, and associated solubility gradients, would be expected across the thin fused salt films. Accordingly, they suggested a negative solubility gradient for the protective oxide in the salt film as a criterion for a sustained hot corrosion attack. Goebel and Pettit (*12*) had pointed out that contact between a nickel substrate and sodium sulfate would lead to sulfidation of the metal with an associated local increase in the melt basicity. Indeed, for the hot corrosion of nickel at high temperatures (Type 1), sulfides are found in the substrate, and oxide precipitates are formed within the salt film, as schematized in Fig. 5.

Shores (*14*) explained that the acidic dissolution of NiO cannot satisfy the "negative solubility gradient" criterion, and suggested that the salt film chemistry would be dominated by the acidic gaseous environment of gas turbine combustion products. Recently, Otsuka and Rapp (*15*) used solid electrolyte probes attached to preoxidized nickel coupons that were then exposed to a thin film of Na_2SO_4 in an acid gas at 1173 K. The trace of the average local chemistry (coordinates for Fig. 1) for the salt film could then be followed as a function of time. As shown in Fig. 6, which plots the "reaction trace" as a function of time, a nickel coupon that suffered rapid, sustained hot corrosion actually was attacked by a basic salt to form a basic solute for NiO, to the left of the solubility minimum shown in Fig. 3. Indeed, the sulfidation of the Ni substrate greatly increased the basicity of the salt film, which was dominated by the interaction with the substrate and not the gas phase. This observation of basic hot corrosion of nickel was consistent with the views of several authors (*12–14*). A different experiment whereby a preoxidized nickel coupon was exposed to a quite basic salt also resulted in rapid hot corrosion, although substrate sulfidation did not occur. In other experiments, Otsuka and Rapp (*15*) were able to avoid hot corrosion by the formation of more protective initial NiO scales, and in that case, the local melt chemistry always remained to the right (acidic) side of the solubility minimum of Fig. 3. These studies established the validity for the "negative solubility gradient" criterion, which might, or might not, result from substrate sulfidation.

Another prediction from the paper by Rapp and Goto (*13*) has been recently tested (*16*). From an inspection of the widely spaced solubility curves of Fig. 4, and the model dissolution reactions written in Fig. 3, one would suppose that the simultaneous contact with fused sodium sulfate of two oxides with minima at differing basicity values could result in a synergistic accelerated dissolution of both the oxides. Namely, for an environment of basicity lying between the minima for two oxides, the oxide ions released upon the acidic dissolution of the more basic oxide would supply the reactant anions needed for the basic dissolution of the more acidic oxide. Hwang and Rapp (*16*) measured the kinetics for the individual and the coupled dissolution for powders of Fe_2O_3 and Cr_2O_3. In accord with prediction, as illustrated for Cr_2O_3 in Fig. 7, each of the oxides dissolved more rapidly in the presence of the second oxide for identical conditions. Synergistic dissolution could represent an impend-

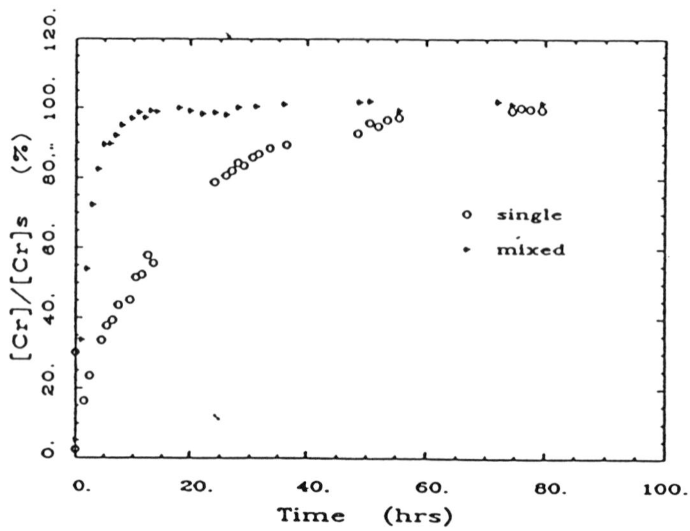

Fig. 7. Time dependence of the normalized concentration of dissolved Cr_2O_3 (pure and mixed with Fe_2O_3) in fused Na_2SO_4 in 1% SO_2–O_2 gas at 1200 K.

ing danger for the usual sort of high-temperature alloys comprised of the basic base metals Fe, Ni, or Co in combination with the acidic protective components Cr or Al. If the initial stages of protective oxidation to develop Cr_2O_3 or Al_2O_3 scales would fail to exclude the presence of the transient oxides Fe_2O_3, NiO, or CoO, then the two-phase scale should lead to the synergistic accelerated attack of both the acidic and the basic oxide, with the probable consequence of scale penetration by the salt and sulfidation of the substrate alloy.

The absence of any acidic ionic solute of silica, as indicated in Fig. 4, has led to a clear understanding for the excellent resistance to hot corrosion by fused Na_2SO_4 by the ceramics SiO_2 and Si_3N_4 in acidic salts (*17*). However, Jacobson found that the ceramics were rapidly attacked by the salt in the presence of (low acidity) air since the silica formed the basic solute $NaSiO_3$, which was then reprecipitated in the salt film in accordance with a negative solubility gradient for SiO_2. Basic sintering aids, such as Y_2O_3 and MgO, proved to assist in the attack of the ceramics by acidic-fused Na_2SO_4. Interestingly, both calculations and experiments showed that the combustion of lower quality fuel containing higher sulfur content was much less aggressive to the silica protective layer since more acidic gaseous atmospheres were formed by combustion of high-sulfur fuel.

VANADATE CHEMISTRY AND OXIDE SOLUBILITIES

As high quality petroleum becomes more scarce and expensive, the use of fuel contaminated by heavy metals, especially vanadium, intro-

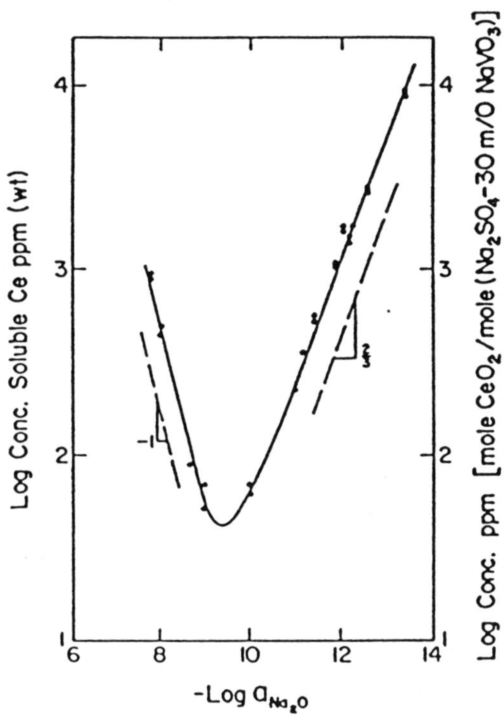

Fig. 8. Solubility of CeO_2 in fused 0.7 Na_2SO_4–0.3 $NaVO_3$ solution at 900°C in 900°C in 1 atm O_2.

duces a serious variation and complication to the hot corrosion of materials. Goebel et al. (*18*) have shown that alloys containing Mo, W, or V, which oxidize to form the strongly acidic oxides MoO_3, WO_3, and V_2O_5, were subject to an acidic dissolution attack of the protective oxides. Jones et al. (*19,20*) have studied the acid-base reactions between vanadate melts and various oxides to show qualitatively that the reactivity of Na–V–O melts is understood according to the following sequence of decreasing melt activity: $V_2O_5 > NaVO_3 > Na_3VO_4$.

Zhang and Rapp (*9*) measured the solubilities of the oxides CeO_2, HfO_2, and Y_2O_3 as a function of melt basicity at 1200 K in a solvent of fused 0.7 Na_2SO_4–0.3 $NaVO_3$ and the solubility of CeO_2 in pure Na_2SO_4. The solubility plot for CeO_2 in the sulfate–vanadate solvent, shown in Fig. 8, obeys basicity dependencies for the basic and acidic solubilities that correspond exactly to the dissolution reactions

$$2CeO_2 + Na_2O \rightarrow 2NaCeO_3 \quad \text{(basic dissolution)} \tag{1}$$

$$4NaVO_3 + 3CeO_2 \rightarrow Ce_3(VO_4)_4 + 2Na_2O \quad \text{(acidic dissolution)} \tag{2}$$

For the latter reaction, the final state may be envisioned as a dilute solution of Ce^{4+} ions dissolved in the mixed solution, whereby the complexing of O^{2-} ions by the reactant VO_3^- ions to form product VO_4^{3-} ions, is the important driving force for the acidic dissolution. In fact, the

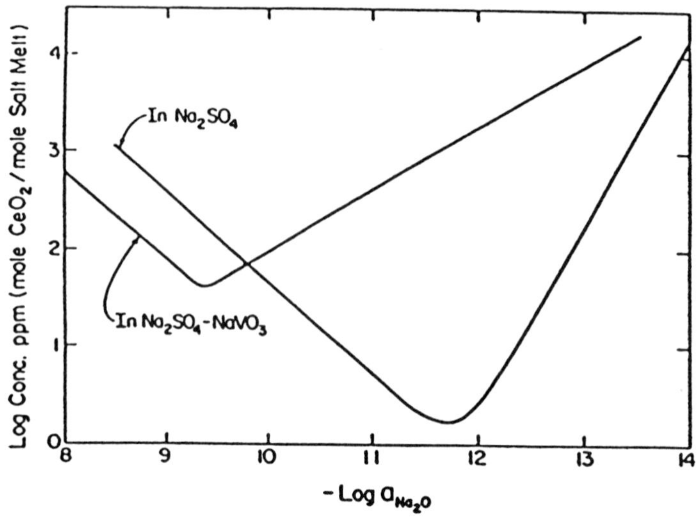

Fig. 9. Comparison of solubilities for CeO_2 in pure fused Na_2SO_4 and in 0.7 Na_2SO_4–0.3 $NaVO_3$ at 900°C in 1 atm O_2.

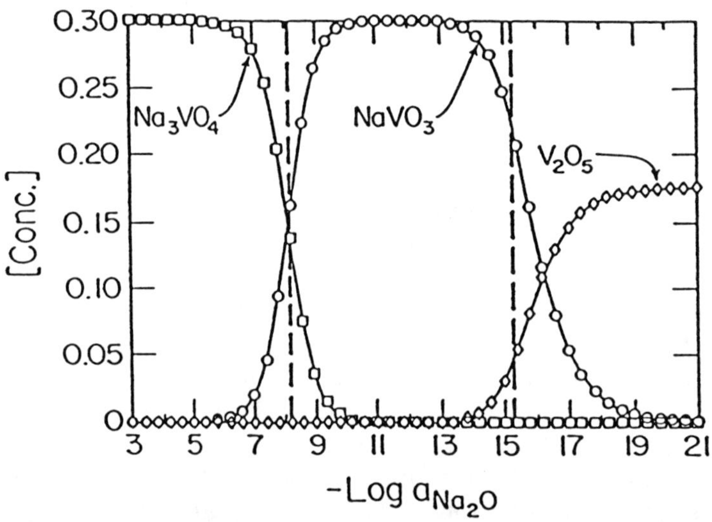

Fig. 10. Equilibrium concentrations of Na_3VO_4, $NaVO_3$, and V_2O_5 in 0.7 Na_2SO_4–0.3 $NaVO_3$ in 1 atm O_2 at 900°C.

solubility minimum for CeO_2 in the mixed solution occurred at surprisingly high values for both the basicity and the magnitude of the minimum solubility. Similar results were found for the solubilities of HfO_2 and Y_2O_3. Although, formally, the acidic solubility of CeO_2 can be described by the standard Gibbs formation energy and activity coefficient for the solute $Ce_3(VO_4)_4$, qualitatively, the more important reaction involves the conversion of metavanadate ions to orthovanadate ions. For this reason, Zhang and Rapp (*9*) implied that the acidic solubilities of most oxides would be affected similarly by the presence of metavanadate ions.

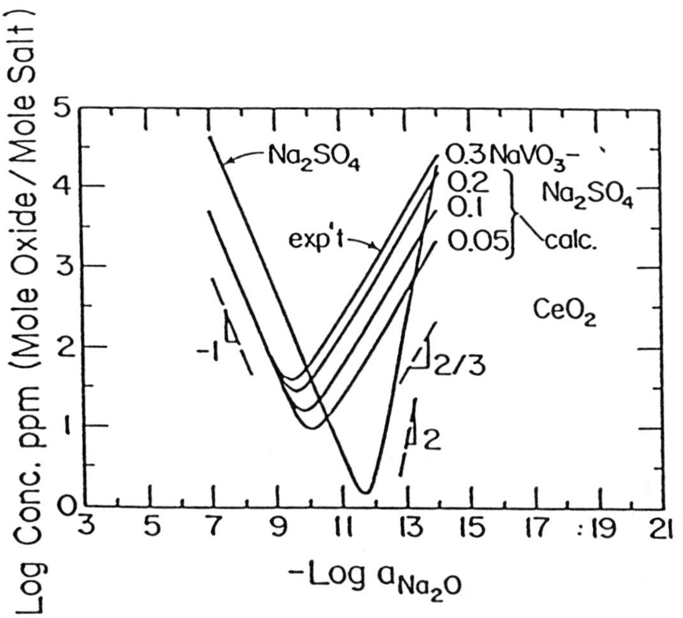

Fig. 11. Solubilities of CeO_2 in Na_2SO_4–x$NaVO_3$ solutions at 900°C in 1 atm O_2 (x = 0.3, 0.2, 0.1, and 0.05).

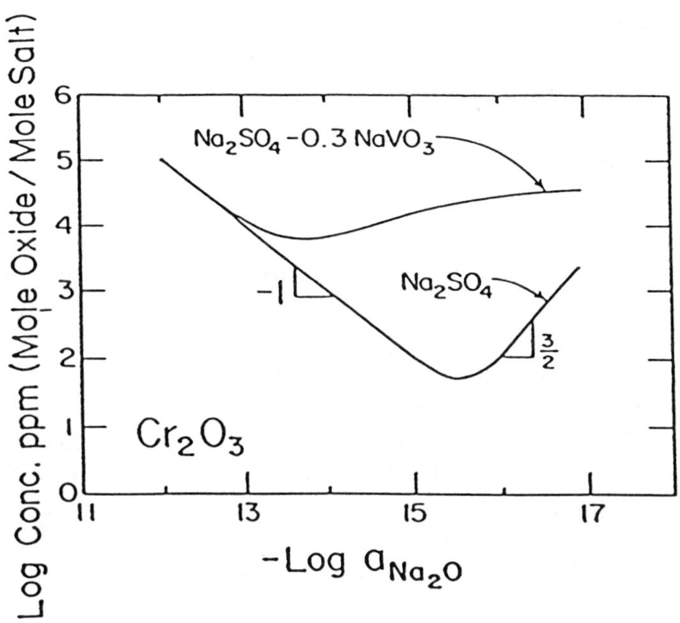

Fig. 12. Measured solubility of Cr_2O_3 in pure Na_2SO_4 and calculated solubility for 0.7 Na_2SO_4–0.3 $NaVO_3$ solution at 900°C in 1 atm O_2.

To establish the effect of 0.3 $NaVO_3$ in Na_2SO_4 on the acidic dissolution of CeO_2, the CeO_2 solubility was also measured in pure fused Na_2SO_4 at 1200 K as a function of melt basicity. Figure 9 provides a comparison for the greatly different results. The solubility minimum for CeO_2 in pure Na_2SO_4 occurred in a much more acid melt, with a much lower magnitude for the minimum solubility. In other words, the

0.3 $NaVO_3$ component of the mixed salt caused an important increase in the acidic solubility, and a similar effect must be expected for most oxides.

To extend, quantify, and generalize the stabilities of vanadates and their effect on the acidic solubilities of oxides, Hwang and Rapp (*21*) assumed an ideal solution for the 0.7 Na_2SO_4–0.3 $NaVO_3$ system to derive the equilibrium distribution of vanadate derivatives given in Fig. 10. In combination with the experimental use of the solid state electrochemical probes to measure the Na_2O activity, Fig. 10 permits a quantification of the stabilities for the vanadates, at least to the validity of the ideal solution assumption. By using the results of Fig. 9, and assuming a constant activity for the acidic solute of CeO_2 in Na_2SO_4–$NaVO_3$ solutions, the solubility of CeO_2 (or alternatively HfO_2 or Y_2O_3) can be calculated for other concentrations of $NaVO_3$ in fused Na_2SO_4, as shown in Fig. 11. Only the line for 0.3 $NaVO_3$ is a measured result; remarkably, even a low content of $NaVO_3$ in Na_2SO_4 results in an important increase in the acidic solubility of CeO_2, or by generalization, of most other oxides. By the same procedure, by assuming that the known activity coefficient for the acidic solute for Cr_2O_3 in pure Na_2SO_4 is retained in Na_2SO_4–$NaVO_3$ melts, the solubility for Cr_2O_3 in the 0.7 Na_2SO_4–0.3 $NaVO_3$ solution (where no experiments exist) can be estimated, as shown in Fig. 12. Obviously, such calculated behavior based on simplifying assumptions about the thermodynamic properties of the solutions requires experimental substantiation.

CONCLUSIONS

The solubilities of oxides and their dependencies on melt basicity and oxygen activity constitute important information to interpret hot corrosion mechanisms. Measurements of the local chemistry (reaction trace) in a thin-fused Na_2SO_4 film during the hot corrosion of preoxidized Ni indicate a basic fluxing/reprecipitation of NiO according to a negative solubility gradient criterion for hot corrosion. Synergistic dissolution of two oxides with differing basicity can occur at a solution basicity between the two minima in solubility for the oxides. The acidic solubilities of most oxides are raised by the presence of vanadate solutes in fused Na_2SO_4. Calculations have been made, involving simple (reasonable) thermodynamic assumptions, to interpret the thermochemistry of sodium sulfate-vanadate solutions and predict the associated dependencies of oxide solubilities in these solutions.

ACKNOWLEDGMENTS

This research was supported by the National Science Foundation, Metallurgy Programs of the Division of Materials Research under Grant DMR 8620311.

REFERENCES

1. Rapp, R. A., *Corrosion* **42,** 568 (1986).
2. Park, C. O. and Rapp, R. A., *J. Electrochem. Soc.* **133,** 1636 (1986).
3. Gupta, D. K. and Rapp, R. A., *J. Electrochem. Soc.* **127,** 2194, 2656 (1980).
4. Jose, P. D., Gupta, D. K., and Rapp, R. A., *J. Electrochem. Soc.* **132,** 735 (1985).
5. Zhang, Y. S., *J. Electrochem. Soc.* **133,** 655 (1986).
6. Zhang, Y. S. and Rapp, R. A., *J. Electrochem. Soc.* **132,** 734, 2498 (1985).
7. Deanhardt, M. L. and Stern, K. H., *J. Electrochem. Soc.* **129,** 2228 (1982).
8. Shi, D. Z. and Rapp, R. A., *J. Electrochem. Soc.* **133,** 849 (1986).
9. Zhang, Y. S. and Rapp, R. A., *Corrosion* **43,** 348 (1987).
10. Bornstein, N. S. and DeCrescente, M. A., *Trans. AIME* **245,** 1947 (1969).
11. Bornstein, N. S. and DeCrescente, M. A., *Met. Trans.* **2,** 2875 (1971).
12. Goebel, J. A. and Pettit, F. S., *Met. Trans.* **1,** 1943 (1970).
13. Rapp, R. A. and Goto, K. S., *Molten Salts,* Braunstein, J., Selman, J. R., eds., Electrochemical Society, Pennington, NJ, p. 81 (1981).
14. Shores, D. A., *High Temperature Corrosion,* NACE-6, Rapp, R. A., ed., National Association of Corrosion Engineers, Houston, TX, p. 493 (1983).
15. Otsuka, N. and Rapp, R. A., *Proceedings of TMS Symposium on High Temperature Corrosion* (in press), 1989.
16. Hwang, Y. S. and Rapp, R. A., *Oxidation of High-Temperature Intermetallics,* Grobstein, T. and Doychak, J., eds. TMS, Warrendale, PA, p. 257 (1988).
17. Jacobson, N. S., *Oxid. Metals* **31,** 91 (1989).
18. Goebel, J. A., Pettit, F. S., and Goward, G. W., *Met. Trans.* **4,** 261 (1973).
19. Jones, R. L., Williams, C. E., and Jones, S. R., *J. Electrochem. Soc.* **133,** 227 (1986).
20. Jones, R. L. and Williams, C. E., *Surf. and Coatings Technol.* **32,** 349 (1987).
21. Hwang, Y. S. and Rapp, R. A., *Corrosion,* in press (1989).

Oxide Acid-Base Reactions in Ceramic Corrosion

R. L. JONES

Naval Research Laboratory, Code 6179, Washington, DC 20375-5000

ABSTRACT

Because of their ability to withstand extreme temperatures, ceramics are expected to be widely used, both as coatings and integral components, in 21st century engines. Ceramics can be corroded, however, by molten deposits formed from sodium, vanadium, and sulfur contaminants occurring in engine fuel or intake air. Our research demonstrates that the molten salt corrosion of ceramics is largely controlled by Lewis type acid-base reactions between the oxide ceramics and corrodent oxides (e.g., V_2O_5, Na_2O, and SO_3). Following this hypothesis, we have identified scandia as a corrosion-resistant stabilizer for zirconia, which may allow zirconia thermal barrier coatings for engines having substantially greater resistance to molten sulfate/vanadate corrosion than the currently used yttria-stabilized zirconia coatings. This paper outlines the experiments and concepts that led to the identification of scandia and points to areas where more fundamental study is required if we are to be able to design reaction-resistant ceramics *a priori*.

Index Entries: Acid-base reaction; hot corrosion; ceramics; engines; zirconia; scandia; coatings; molten salts; vanadium; sulfur.

INTRODUCTION

By the year 2000, ceramics are expected to be widely used in engines to increase power and fuel economy by allowing higher operating temperatures and improved thermal efficiency. Ultra-high power aerospace engines are foreseen that, relying on ceramics or ceramic composites, will operate at gas temperatures of 1650–2200°C (3000–4000°F), i.e., temperatures far beyond the capabilities of metals. Ceramics are now under commercial development for use in conventional engines for such components as turbocharger rotors (already in production in Japan), pistons, gas turbine vanes, and so on, with "ceramic" diesel engines being promised for the not so distant future. Ceramic coatings, such as the heat-insulating, yttria-stabilized zirconia (YSZ) thermal barrier coatings, which can increase engine fuel efficiency or power 10–15% by permitting higher gas temperatures without overheating engine metals, have also been developed and are nearing application in some areas.

One potential problem, which has not been adequately addressed, is ceramic corrosion. All ceramics suitable for use in oxygen-rich combustion environments are oxides, or develop protective oxide layers on their surface (e.g., SiO_2 on silicon carbide (SiC) and nitride (Si_3N_4)). Oxide ceramics are highly stable and resist oxidation or reduction, but they are not chemically inert, as sometimes supposed. Yttria-stabilized zirconia thermal barrier coatings are readily attacked, for example, by the trace vanadium contaminants contained in many marine and diesel fuels, which presently limits YSZ coatings to only aviation or other "clean fuel" applications.

The fuel contaminants that have been found to attack ceramics (sodium, sulfur, vanadium, and phosphorus) are converted in combustion to oxide species (Na_2O, SO_3, V_2O_5, and P_2O_5), which all have strong acid or base characteristics. These oxides combine to form molten mixed sodium sulfate/vanadate, and so on deposits that may produce hot corrosion of ceramics in the hot section of the engine. Our research, and a review of ceramic corrosion results from the literature (*1*), suggest that the corrosion reactions between ceramic oxides and the molten salt deposits may be mostly explained as oxide acid-base reactions.

In considering oxide acid-base reactions in ceramic hot corrosion, it is useful to think both in Lux-Flood and Lewis acid-base terms. The Lux-Flood (*2*) treatment of molten salts takes the oxide anion, O^{2-}, as the fundamental basic entity, and uses the negative log of oxide activity, $-\log a_{O^{2-}}$, much as with pH in aqueous solutions, to describe the salt basicity. The sodium oxide (Na_2O) activity can be used equivalently, and it is often possible to thermodynamically calculate the Na_2O activity (basicity) of a molten deposit as, for example, with pure Na_2SO_4 at equilibrium at 900°C (*3*).

$$Na_2SO_4 = Na_2O + SO_3(g) \tag{1}$$

and then

$$\Delta G° = -RT\ln K \tag{2}$$

where $\Delta G°$ is the standard Gibbs free energy for reaction (1) at 900°C and K is the equilibrium constant

$$K = \frac{(a_{Na_2O})(p_{SO_3})}{(a_{Na_2SO_4})} \tag{3}$$

Taking $\Delta G°$ from the literature (−388.928 kJ/mol) and setting the activity of Na_2SO_4 to unity gives

$$\log (a_{Na_2O}) + \log (p_{SO_3}) = \Delta G°/2.303\ RT = -17.3 \text{ at } 900°C \quad (4)$$

The partial pressure of SO_3 in equilibrium with Na_2SO_4, therefore, fixes the Na_2O activity, or basicity of the sulfate melt.

By extending the treatment above to all possible reactions in, e.g., a sodium–metal–sulfur–oxygen system, one may derive (provided the necessary thermodynamic data are available) a phase stability diagram to predict the oxygen partial pressures and Na_2O activities at which the metal or its various oxides, sulfides, sulfates, and so on will be stable. Such phase stability diagrams are widely used in metallic hot corrosion and oxidation studies to predict or explain corrosion reactions (*3*).

When vanadium is present as a fuel contaminant, a series of sodium vanadate compounds (*4*) may be formed in the deposit, viz

$$Na_2V_{12}O_{31} \text{ (mp. 645°C)} = Na_2O + 5\ V_2O_5 \quad (5)$$

$$2\ NaV_3O_8 \text{ (decomp. 548°C)} = Na_2O + 3\ V_2O_5 \quad (6)$$

$$2\ NaVO_3 \text{ (mp. 610°C)} = Na_2O + V_2O_5 \quad (7)$$

$$Na_4V_2O_7 \text{ (decomp. 668°C)} = 2\ Na_2O + V_2O_5 \quad (8)$$

$$Na_3VO_4 \text{ (mp. 1266°C)} = 3\ Na_2O + V_2O_5 \quad (9)$$

A characteristic V_2O_5 activity (or acidity) is associated with each of these compounds, with the V_2O_5 activity decreasing as the Na_2O content increases. When Na_2O, V_2O_5, and SO_3 are generated concurrently by combustion of Na, S, and V-contaminated fuel, the V_2O_5 and SO_3 compete for reaction with Na_2O, producing a mixed sodium vanadate/sulfate engine deposit having certain equilibrium SO_3, V_2O_5, and Na_2O activities. It is these equilibrium oxide activities that are required to be known since they are the essential reagents causing corrosion of the oxide ceramics.

Few thermodynamic calculations to predict oxide activities in molten vanadate/sulfate deposits have yet been reported. Luthra and Spacil (*5*) have published calculations of the compositions of blade deposits that are likely to be formed in marine gas turbines burning fuel containing certain levels of Na, V, and S. And recently, Zhang and Rapp (*6*) have measured the solubility of CeO_2, HfO_2, and Y_2O_3 in Na_2SO_4–30 mol% $NaVO_3$ at 900°C as a function of the Na_2O activity of the melt. In general, however, there is currently little in the literature to help explain acid-base corrosion of oxide ceramics by molten salts.

Although the Lux-Flood concept provides a suitable basis for molten salt thermochemistry, the Lewis acid-base theory must be invoked to explain the relative acidic characteristics of the oxide ceramics themselves. In the Lewis theory, basicity is defined as the ability to donate electronic charge and acidity as the ability to accept electronic charge. For oxides, the oxide anion is the charge donor, and its behavior is modified by the cation, where the cation acidity increases both as the cation positive charge increases and the cationic radius decreases. For example, sodium ions (+1 charge, 0.095 nm radius) polarize the oxide anion in

Na_2O only weakly so that Na_2O retains a residual basic character; whereas, with V_2O_5, the vanadium cations (+5 charge, 0.059 nm radius) polarize the oxygen anions sufficiently that their ability to donate charge is reduced and V_2O_5 shows an acidic nature. Thus, each oxide species has a characteristic acidity (or basicity), and the tendency for reaction between oxides will depend, to a first approximation, on the difference between their relative acid or base strengths.

Lewis acid-base influences in determining oxide reactions in silicate glass and other oxide systems have been studied extensively by Duffy and coworkers (*7,8*). Also, Brewer and Wengert (*9*) have applied Lewis acid-base theory to high temperature metals and shown that transition metals with low-lying empty 4*d* or 5*d* orbitals (e.g., Zr) undergo a Lewis acid-base interaction with transition metals with six or more 4*d* or 5*d* electrons (e.g., Pt), producing highly stable intermetallics such as $ZrPt_3$.

In this paper, we recount some elementary applications of oxide acid-base theory to the probolem of molten vanadate corrosion of yttria-stabilized zirconia coatings in engines. This work has identified scandia (Sc_2O_3) as a corrosion-resistant stabilizer for zirconia that gives significantly greater resistance to molten vanadate/sulfate attack than yttria-stabilized zirconia. The paper is also intended to define areas in which new thermodynamic data and/or improved oxide acid-base theory are needed for the development of reaction-resistant ceramics for 21st century applications.

METHODS

Ceramic oxides, in the form of either pure chemical compounds, plasma-spray powders, plasma-sprayed coatings, or sintered monolithic ceramics, were mixed or coated with various mixtures of Na_3VO_4, $NaVO_3$, V_2O_5, or Na_2SO_4. The treated ceramic oxides were then exposed in platinum boats under air, or air containing equilibrated SO_2–SO_3 in controlled gas environment furnaces (*10*) at temperatures between 700 and 900°C. The ceramic/corrodent mixtures were X-rayed at various times to determine if reaction occurred and, if so, identify the products.

A major part of the investigation involved determining the rate of destabilization of various stabilized zirconia ceramics as these materials were attacked by different vanadate or vanadate/sulfate mixtures. The degree of destabilization was estimated from the relationship

$$\% \text{ destabilization} = 100 \times \frac{\text{M}}{(\text{M} + \text{T})} \tag{10}$$

where *M* is the monoclinic zirconia structure ($11\bar{1}$) peak height, and *T* is the tetragonal (111) peak height. This simple relationship of peak heights gives a nearly quantitative analysis for monoclinic and tetragonal zirconia in a mixture of the two phases (*11*). However, in our case, the X-ray specimens contain 50% salt, and there is some overlap with the salt X-ray peaks, which degrades the zirconia spectra to a certain extent. The individual destabilization data points, therefore, can be considered as only qualitative, but the trends are highly consistent and provide a valid representation of the destabilization behavior.

RESULTS AND DISCUSSION

It is generally accepted that the degradation of yttria-stabilized zirconia by molten vanadates is promoted by the reaction

$$Y_2O_3 + V_2O_5 = 2\ YVO_4 \tag{11}$$

where the formation of yttrium vanadate (mp 1810°C) leaches Y_2O_3 from solid solution in the zirconia matrix and causes destabilization from the tetragonal to monoclinic structure for zirconia. In laboratory tests, YSZ ceramics are readily attacked by thin films of molten $NaVO_3$, with lath-like YVO_4 crystals seen by scanning electron microscopy in the $NaVO_3$ layer. When molten sulfate/vanadate attack is simulated by introducing SO_3 in the furnace gas over $NaVO_3$-deposited YSZ, a molten Na_2SO_4–$Y_2(SO_4)_3$ mixed sulfate phase is formed (Fig. 1, phase marked B) from which YVO_4 crystals (marked A) are precipitated. According to Luthra and Spacil (*5*), SO_3 in the gas phase produces Na_2SO_4 and increases the V_2O_5 activity of the melt by the reaction

$$2\ NaVO_3 + SO_3 = Na_2SO_4 + V_2O_3 \tag{12}$$

To investigate the YVO_4 formation reaction, we examined the interaction of several ceramic oxides with Na_3VO_4 (low V_2O_5 activity), $NaVO_3$ (moderate V_2O_5 activity), and V_2O_5 (activity = 1). The results from this work (*12*) are summarized in Table 1, which shows classic acid-base behavior where acids react with bases, and vice versa, but no reaction occurs between compounds of comparable acid-base nature. Note that $NaVO_3$ acts as an acid with basic Y_2O_3, but as a base with acidic GeO_2 or Ta_2O_5. Introducing SO_3 into the furnace air increased the V_2O_5 activity by reaction (12), with an SO_3 partial pressure of 50 Pa at 700°C being sufficient to produce $CeVO_4$ in CeO_2–$NaVO_3$ mixtures that, under pure air, gave no $CeVO_4$ formation.

Table 1 predicts that ceria-stabilized zirconia should not be attacked by $NaVO_3$ since neither CeO_2 nor ZrO_2 alone reacts with $NaVO_3$. Ceria-stabilized zirconia, in fact, has been patented as a vanadate/sulfate resistant material (*13*). However, we found in laboratory tests (*14*) that ceria-stabilized zirconia was destabilized as rapidly as yttria-stabilized zirconia by 700°C $NaVO_3$. The ceria was leached from the zirconia matrix and segregated on the ZrO_2 surface as an oxide (apparently CeO_2 possibly containing some zirconia) and not as a vanadate, as with YVO_4 from YSZ. The reactions that could remove CeO_2 from solid solution in ZrO_2 ($a_{CeO_2} < 1$) and segregate "pure" CeO_2 ($a_{CeO_2} = 1$) on the ZrO_2 surface have yet to be explained. One possibility (*see below*) is that incorporation of CeO_2 in the zirconia matrix somehow modifies its inherent acid-base nature.

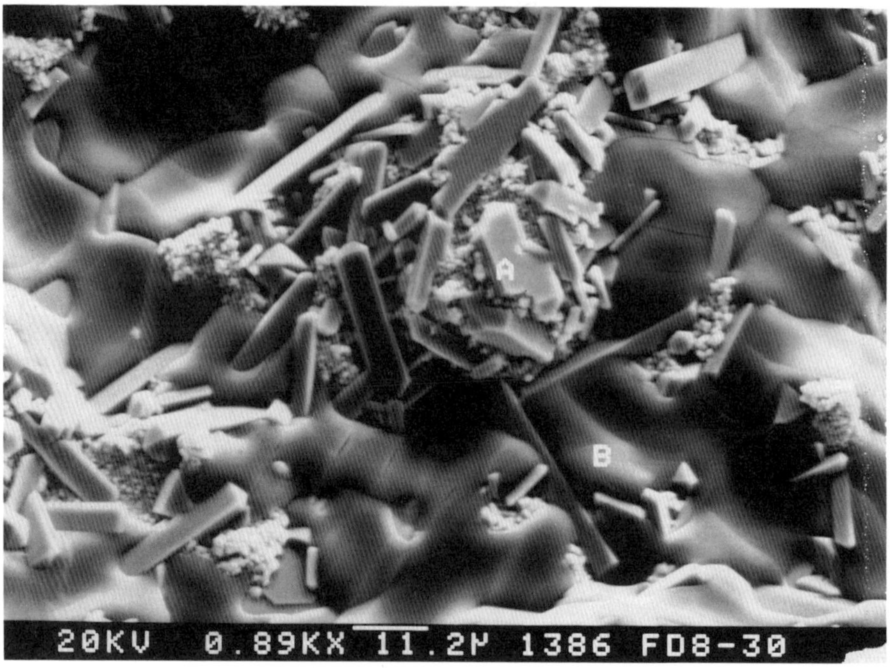

Fig. 1. Deposits on yttria-stabilized zirconia coated with $NaVO_3$ and exposed at 700°C under 9.0 Pa of SO_3; note molten mixed Na,Y sulfate phase (marked B) and YVO_4 crystals (marked A) that appear to form by precipitation from the sulfate phase.

Following the predictions of Table 1, we have recently investigated scandia (Sc_2O_3) as a corrosion-resistant stabilizer for zirconia (*15*). Scandia is the most acidic of the rare earth oxides (*16*) and an effective stabilizer of zirconia (*17*). Chemical tests confirmed that pure Sc_2O_3 was substantially more resistant than Y_2O_3 to reaction with Na_2SO_4 and SO_3 (i.e., mixed sulfate formation). Scandia was also more resistant than yttria to reaction with $NaVO_3$, producing some $ScVO_4$ at 700 but none at 900°C, whereas yttria yielded YVO_4 at both temperatures. Sintered pellets of scandia-stabilized zirconia (5 wt% Sc_2O_3) and yttria-stabilized zirconia (8 wt% Y_2O_3) were subsequently prepared and tested for resistance to molten $NaVO_3$ deposits. The scandia-stabilized zirconia (SSZ) pellets withstood molten $NaVO_3$ surface deposits for hundreds of hours at both 700 and 900°C without $ScVO_4$ formation or destabilization, whereas the YSZ pellets under the same conditions reacted strongly with the $NaVO_3$, giving rapid YVO_4 formation and destabilization at both temperatures.

We have also studied the corrosion behavior of SSZ and YSZ plasma-spray powders (obtained preliminary to the preparation of plasma-sprayed coatings). In these experiments, the corrosion resistance of a prototype SSZ plasma-spray powder (7 mol% Sc_2O_3, approx. 90% tetragonal) was compared with a commercial YSZ plasma-spray powder (5 mol% Y_2O_3, 100% tetragonal) when both were mixed with $NaVO_3$ con-

Table 1
Reaction Behavior of Ceramic Oxides with Vanadium Compounds of Increasing V_2O_5 Activity Showing Products Formed

	Na_3VO_4	$NaVO_3$	V_2O_5
Y_2O_3	NR[a]	YVO_4	YVO_4
CeO_2	NR	NR	$CeVO_4$
ZrO_2	NR	NR	ZrV_2O_7 (but slowly)
GeO_2	$Na_4Ge_9O_{20}$	$Na_4Ge_9O_{20}$[b]	NR
Ta_2O_5	$NaTaO_3$	$Na_2Ta_4O_{11}$	alpha-$TaVO_5$

[a]NR = no reaction.
[b]As ppt from H_2O solution.

taining progressively increasing mole percentages of V_2O_5 and exposed at 700 and 900°C. The results, which are essentially the same at both temperatures except for zirconia itself, are given in Table 2.

The results in Table 2 can be intrepreted as an acid-base titration, with V_2O_5 as the titrant. Yttria is sufficiently basic that YSZ forms YVO_4 and is destabilized at both 700 and 900°C by the V_2O_5 activity associated with just pure $NaVO_3$. On the other hand, scandia is more acidic, and about 15 mol% excess V_2O_5 (probably more at 900°C) must be added before SSZ begins to show $ScVO_4$ formation and destabilization. And finally, zirconia itself is the most acidic and reacts only with pure V_2O_5 and then not strongly, as evidenced by the instability of the ZrV_2O_7 product at 900°C. Note that, as with YVO_4 for YSZ destabilization, $ScVO_4$ must apparently begin to form before SSZ destabilization commences.

Corrosion of oxide ceramics in engines will involve combined molten vanadate/sulfate attack, and not simply vanadate attack, since virtually all low-quality fuels containing vanadium also contain sulfur. We have examined sulfate/vanadate attack on our SSZ and YSZ plasma-spray powders under different conditions and found these results. Neither SSZ nor YSZ was destabilized by being mixed with pure Na_2SO_4 and exposed at 800°C in air containing equilibrated SO_x ($SO_x = SO_2 + SO_3$) partial pressures as high as 150 Pa. Therefore, mixed sulfate formation is apparently not a controlling reaction in the destabilization process for SSZ or YSZ (except perhaps for very high SO_3 partial pressures). In tests with 50 mol% Na_2SO_4–$NaVO_3$ under air containing no SO_3, SSZ was not attacked, but YSZ was rapidly destabilized at temperatures up to 850°C (Fig. 2). Although addition of Na_2SO_4 reduces V_2O_5 activity by reaction (12), the V_2O_5 activity of 50 mol% Na_2SO_4–$NaVO_3$ evidently remains sufficiently high for YVO_4 formation up to 850°C.

We also sought to define the SO_3 partial pressures that could be tolerated with 50 mol% Na_2SO_4–$NaVO_3$ (Na/V ratio = 3) before scandia-stabilized zirconia would begin to destabilize. At 700 and 800°C, SSZ destabilization was observed within 24–48 h at SO_x partial pressures as low as 5 Pa (the lowest partial pressure obtainable with our gas control system). However, at 850°C (Fig. 2), SSZ withstood 25 Pa of SO_x for 100 h

Table 2
Destabilization Behavior and Products Formed for Yttria- and Scandia-Stabilized Zirconia Powders Reacting with Melts of Different $NaVO_3$–V_2O_5 Compositions

Salt composition	YSZ	SSZ
$NaVO_3$	DS,[a] YVO_4	no DS
10 mol% V_2O_5 in $NaVO_3$	–	no DS
15 mol% V_2O_5 in $NaVO_3$	–	DS starting, $ScVO_4$
20 mol% V_2O_5 in $NaVO_3$	–	DS, $ScVO_4$
NaV_3O_8 (50 mol% V_2O_5–$NaVO_3$)	DS, YVO_4	DS, $ScVO_4$
V_2O_5	DS, YVO_4, ZrV_2O_7 at 700, but not 900°C	DS, $ScVO_4$, ZrV_2O_7 at 700, but not 900°C

[a]DS = Destabilization

without any indication of destabilization and began to destabilize only after 48 h with 125 Pa of SO_x, which in our Pt-equilibrated system corresponds to about 30 Pa of SO_3. (Note: the as-received SSZ was only 90% tetragonal and simply mixing with the HT form of $NaVO_3$ perturbs the zirconia X-ray spectra so that equation (10) indicates about 15% destabilization at $t = 0$.) This performance is substantially better than yttria-stabilized zirconia, (also Fig. 2) where the ceramic was 80% destabilized after only 8 h without any SO_x in the furnace air.

Note that in molten vanadate/sulfate/SO_3 attack on YSZ (Fig. 1), it appears that yttria is taken up in a mixed Y,Na sulfate molten phase (which shows little vanadium by EDXA) from which YVO_4 is precipitated. We have observed similar behavior with SO_3/$NaVO_3$ corrosion of CoCrAlY (*18*) and ceria-stabilized zirconia (*19*). Therefore, although we find no evidence for mixed sulfate formation simply by Na_2SO_4/SO_3 reaction with YSZ or SSZ, what essentially amounts to mixed sulfate formation of the stabilizing oxide does occur when a molten Na_2SO_4–$NaVO_3$ phase is present, and this molten mixed sulfate phase apparently serves a critical intermediate role in the molten sulfate/vanadate corrosion of stabilized zirconia.

GENERAL DISCUSSION

Oxide acid-base concepts, thus, have led to the identification of scandia as a promising stabilizer to give increased molten vanadate resistance for zirconia-based high temperature coatings. The more general question, however, is how far the Lewis oxide acid-base theory can be extended to explain other ceramic corrosion, predict high temperature oxide ceramic compatibilities, or allow *a priori* design of corrosion-resistant ceramics for specific corrosion environments. By insuring that the potential reactants are of nearly the same acidic character so that the acid-base driving force for reaction is minimized, one could presumably reduce or prevent corrosion or interaction between corrodents/ceramics, or even ceramics/ceramics, in many high temperature applications. Sim-

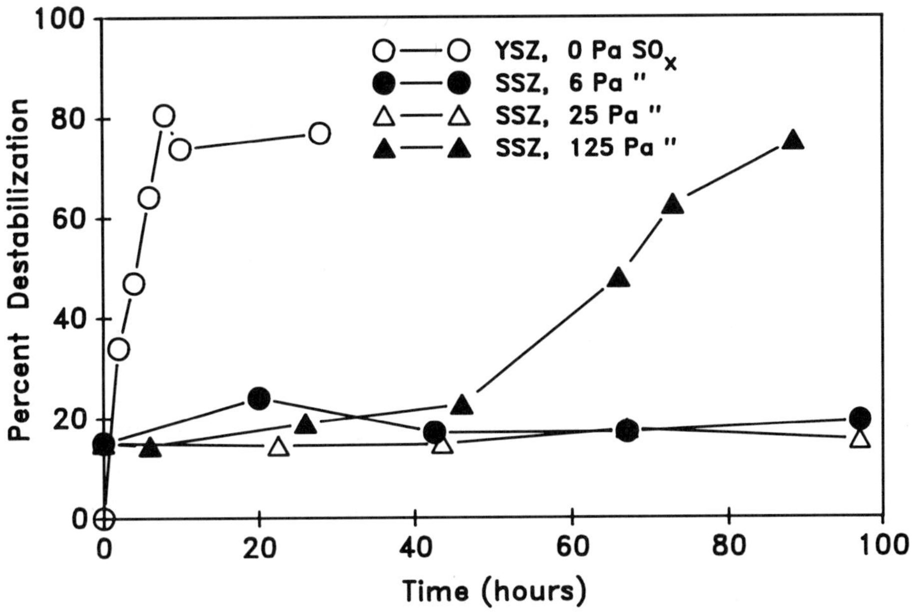

Fig. 2. Influence of SO_x ($SO_x = SO_2 + SO_3$) on destabilization of yttria-stabilized zirconia (YSZ) and scandia-stabilized zirconia (SSZ) in mixed 50 mol% Na_2SO_4–$NaVO_3$ at 850°C. Note that YSZ is destabilized quickly even without SO_x, but SSZ withstands 25 Pa and destabilizes only at 125 Pa partial pressure of SO_x.

ply using the most acidic oxide is not sufficient; for example SiC and Si_3N_4 are corroded by molten Na_2SO_4 (*20*) because the protective surface oxide, SiO_2, is so acidic that it displaces SO_3 from Na_2SO_4 by an acid-base displacement reaction

$$SiO_2 + Na_2SO_4 = Na_2SiO_3 + SO_3 \tag{14}$$

The essence of the acid-base treatment required; that is, the quantitative ranking of the Lewis acid-base strength of the various corrodent, metal and ceramic oxides, and the use of these rankings for the prediction of reaction, has been demonstrated by Dent-Glasser and Duffy (*8*) for silicates and certain other oxides. They assign an "optical basicity" (λ) (determined by UV spectroscopy measurements) to the various oxide species, e.g., SO_3 ($\lambda = 0.31$), P_2O_5 ($\lambda = 0.40$), Na_2O ($\lambda = 1.15$), where the lower values signify the greater acidity and the optical basicities are additive, viz

$$Na_2O \cdot SO_3 = \tfrac{1}{4}(1.15) + \tfrac{3}{4}(0.33) = 0.54 = \lambda \text{ for } Na_2SO_4 \tag{15}$$

Reaction can then be predicted for even complex oxysalts by considering whether a significant optical basicity difference exists between the potential reactants and whether the potential reaction would reduce the total optical basicity difference of the system. Generally, good predictive results have been found by Duffy et al. (*7*) for their model, except when the optical basicity between reactants is small and other factors, e.g., lattice energy, become dominant. In principle, V_2O_5 and ZrO_2, Sc_2O_3, Y_2O_3, and other oxide ceramics could be encompassed in the Duffy et al. optical basicity classification system. However, the oxide ceramics present experimental problems in the type of UV absorption measurement used by Duffy et al., which may be difficult or impossible to solve.

Therefore, it is desirable to develop other means to quantify Lewis acid-base character across the broad field of metal, ceramic, and corrodent oxides. Possibly other spectroscopic techniques that reflect electronic perturbations between atoms, and are thus indicative of Lewis acid-base behavior, may be found. Useful information can also be obtained by further application of competitive chemical reaction schemes as we have used (*12*), or the oxide acid-base solubility measurements developed by Zhang and Rapp (*6*). In interpreting oxide ceramic corrosion on the basis of oxide solubility measurements, however, it must be borne in mind that the corrosion morphologies (Fig. 1, also (*18,19*) so far shown suggest that molten vanadate/sulfate/SO_3 corrosion is not simply a matter of solubility, but rather of a process where the oxide is taken into solution (probably as a mixed sulfate) and then precipitated as YVO_4, $CeVO_4$, and so on as a certain V_2O_5 activity within the melt is exceeded.

Any spectroscopic or chemical reaction scheme to predict high temperature oxide behavior by quantifying the acid-base nature of the different metal, ceramic, and corrodent oxides must be capable of explaining such observations as, for example:

1. Ceria and zirconia do not react with $NaVO_3$, but yet $NaVO_3$ depletes CeO_2 from ceria-stabilized zirconia. One possibility is that the basicity of CeO_2 is increased within the ZrO_2 lattice so that $CeVO_4$ forms, but then decomposes once beyond the influence of the ZrO_2 matrix, so that CeO_2 (rather than $CeVO_4$) is deposited on the ceramic surface. Such a CeO_2–ZrO_2 matrix interaction could be akin to the localized variations in acidity that are postulated in catalysis when minority species are incorporated, or absorbed onto, solid state catalysts.
2. ZrO_2 is highly resistant to reaction with SO_3 and V_2O_5, and yet reacts strongly with P_2O_5 (yielding ZrP_2O_7) even though P_2O_5 is classified by Duffy (*7*) as being less acidic than SO_3 and is likely not more acidic than V_2O_5.
3. Rapp and coworkers found that, whereas the oxides of many high temperature metals and ceramics show sharp transitions from basic to acidic solubility behavior (*3*), SiO_2 exhibits only Na_2O activity independent, physical solubility in acidic molten Na_2SO_4 (*21*).

In conclusion, although oxide acid-base reactions are demonstrably crucial in many areas of ceramic corrosion, and certain approaches to the study of glass and ceramic acid-base behavior have been taken, many questions relating to high temperature oxide acid-base reaction are unanswered. A "need" list to answer these questions would include:

1. New thermodynamic data on the Gibbs free energies of formation for YVO_4, $ScVO_4$, and other compounds involved in vanadate/sulfate corrosion of ceramics.
2. Rigorous thermodynamic treatments of the solution chemistry of sulfate/vanadate/ceramic oxide melts to predict, e.g., the Na, V, and S fuel levels allowable when using YSZ or SSZ engine coatings.
3. Development of spectroscopic or other means to provide an internally consistent, quantitative ranking of the Lewis acid-base strength of oxides across the field of metal, ceramic, and corrodent oxides.
4. Studies to distinguish between contributions to the driving force for reaction resulting from acid-base interaction and those resulting from lattice energy or other types of bonding.
5. And (possibly related to 4) fundamental investigations of the differences between "network forming" acidic oxides, such as SiO_2 and P_2O_5, and equally or more acidic oxides, such as SO_3 or V_2O_5, which are normally not classified as network formers.

ACKNOWLEDGMENTS

The author's research has been sponsored under the NAVSEA Program for Ship Propulsion Technology, and the support is gratefully acknowledged.

REFERENCES

1. Jones, R. L., *High Temp. Technol.* **6,** 187 (1988).
2. Flood, H. and Forland, T., *Acta Chem. Scand.* **1,** 592 (1947).
3. Rapp, R. A., *Corrosion* **42,** 568 (1986).
4. Kolta, G. A., Hewaidy, I. F., and Felix, N. S., *Erzmetall* **25,** 327 (1972).
5. Luthra, K. L., and Spacil, H. S., *J. Electrochem. Soc.* **129,** 649 (1982).
6. Zhang, X. S. and Rapp, R. A., *Corrosion* **43,** 348 (1987).
7. Duffy, J. A. and Ingram, M. D., *J. Non-Crystal. Solids* **21,** 373 (1976).
8. Dent-Glasser, L. S. and Duffy, J. A., *J. Chem. Soc. Dalton Trans. 1987*, 2323 (1987).
9. Brewer, L. and Wengert, P. R., *Metal. Trans.* **4,** 83 (1973).
10. Jones, R. L., *High Temperature Corrosion: NACE-6,* Rapp, R. A., ed., NACE, Houston, TX, 1983, p. 513.
11. Schmid, H. K., *J. Am. Ceram. Soc.* **70,** 367 (1987).
12. Jones, R. L., Williams, C. E., and Jones, S. R., *J. Electrochem. Soc.* **133,** 227 (1986).
13. Siemers, P. A. and McKee, D. W., *US Patent 4328285* (May 1982).
14. Jones, R. L. and Williams, C. E., *Surface and Coatings Techn.* **32,** 349 (1987).
15. Jones, R. L., *1989 Intl. Conf. on Metallurgical Coatings,* San Diego, CA, April 17–21, 1989.
16. Sneed, M. C. and Brasted, R. C., *Comprehensive Inorganic Chemistry, vol. four,* Van Nostrand, New York, 1955, p. 168.

17. Ruh, R., Garrett, H. J., Domagala, R. F., and Patel, V. A., *Am. Ceram. Soc.* **60,** 399 (1977).
18. Jones, R. L. and Williams, C. E., *Materials Science and Eng.* **87,** 353 (1987).
19. Jones, R. L. and Williams, C. E., *Surface and Coatings Technol.* **32,** 349 (1987).
20. Jacobson, N. S., *J. Am. Ceram. Soc.* **69,** 74 (1986).
21. Shi, D. Z. and Rapp, R. A., *J. Electrochem. Soc.* **133,** 849 (1986).

High Temperature Reactions of Ceramics and Metals with Chlorine and Oxygen

N. S. JACOBSON,*[1] M. J. MCNALLAN,[2] AND E. R. KREIDLER[3]

[1]*NASA Lewis Research Center, Cleveland, Ohio 44135;*
[2]*University of Illinois at Chicago, Chicago, Illinois 60680;*
and [3]*Ohio State University, Columbus, Ohio 43210*

ABSTRACT

The reactions of chlorine and oxygen with solids are complex and quite dependent on material and reaction conditions. However, there are four reaction schemes that describe many of these reactions. These are direct reaction of chlorine with the oxide layer, penetration of the oxide layer by chlorine, paralinear behavior where oxidation and chlorination are of comparable rates, and finally, gas phase reactions where the volatile chlorides react with oxygen. Examples of chlorine and oxygen reactions with cobalt, nickel, chromium, molybdenum, and silicon carbide illustrate each of these four reaction schemes.

Index Entries: Oxidation; chlorination; high pressure mass spectrometry.

INTRODUCTION

The reactions of metals and ceramics with chlorine and oxygen mixtures at high temperatures involve the interaction of the two oxidant gases with both the solid of interest and its oxide film. A variety of reaction schemes are possible. The particular reaction scheme depends on the relative reaction rates of the solid with chlorine and oxygen and the structure of the oxide scale, as well as the experimental conditions.

Four commonly observed reaction schemes are shown in Fig. 1. Nearly all metals and nonoxide ceramics are unstable in oxygen and rely on a thin oxide film for their stability. The most obvious interaction is the

*Author to whom all correspondence and reprint orders should be addressed.

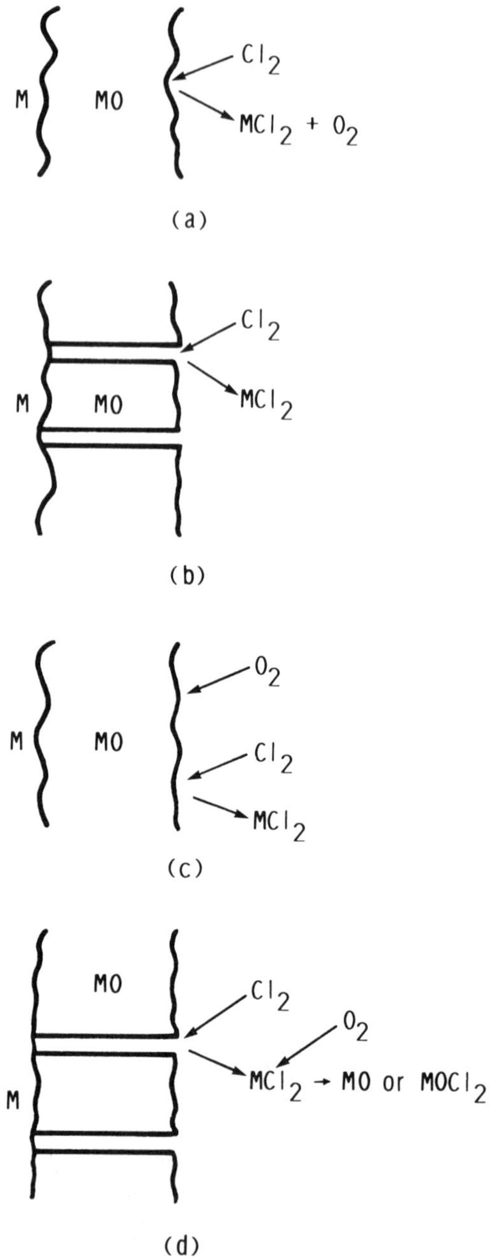

Fig. 1. Four reaction schemes observed in chlorination–oxidation reactions at high temperatures.

the direct reaction of the chlorine with the oxide. A second type of reaction involves chlorine penetration of the oxide scale and reaction with the underlying substrate. In some cases, "paralinear behavior" is observed where the overall reaction is described by a combination of the parabolic rate constant for oxidation and the linear rate constant for chlorination. Finally, gas-phase reactions may be important, wherein the volatile chloride forms via oxide penetration and then reacts with oxygen to form an oxide or oxychloride. Not all oxidation–chlorination reactions can be described by these reaction schemes, and in many cases, several reaction schemes may be simultaneously operative. However, a number of studies performed in the authors' laboratories fall into these categories and will be described below.

EXPERIMENTAL METHODS

Experimental studies of gas–solid reactions involve two primary parts: kinetic measurements and product characterization. The kinetic measurements are typically done with a recording thermobalance. In this type of apparatus, the specimen is suspended from a sensitive balance into a tube furnace. Weight changes (gains or losses) are measured as a function of time. Generally gas flow is from the top to the bottom of the furnace to avoid condensation of the volatile products on the hangwire of the balance.

Product characterization is performed with X-ray diffraction (XRD) and scanning electron microscopy (SEM) for the condensed phases on the specimen. Volatile products are identified with the mass spectrometer sampling system, shown in Fig. 2. This system allows direct, unambiguous, identification of the volatile products emerging from a process proceeding at pressures up to one atmosphere.

The mass spectrometer system has been described in detail elsewhere and will only be briefly discussed here (*1,2*). The chlorination–oxidation reaction occurs in a small tube furnace adjacent to the sampling orifice on the mass spectrometer. The volatile species enter this orifice and undergo a free jet expansion, eventually forming a molecular beam. This molecular beam passes through a series of differentially pumped vacuum chambers and finally enters a quadrupole mass spectrometer. This system preserves the chemical and dynamic integrity of the volatile products even if they are condensable (e.g., metal chlorides) or reactive.

The ions were identified from their mass-to-charge ratio and "isotopic fingerprint." For chlorine-containing molecules, the isotopes create a characteristic cluster of peaks spaced by two atomic mass units. Fragmentation patterns are determined by comparison to known metal chlorides. Ion intensities are normalized to the Cl_2^+ intensity. It should be noted that ion intensities in this instrument are related to pressure in a complex way and only give a semiquantitative measure of the amount of a species present.

RESULTS AND DISCUSSION

Reaction Scheme A: Direct Reaction with the Oxide Scale

An example of this type of reaction is the chlorination–oxidation of preoxidized cobalt at 1000 K (*3*). In this study, cobalt was preoxidized in 1% O_2/Ar at 1173 K for 40 min. This formed a thick inner layer of CoO

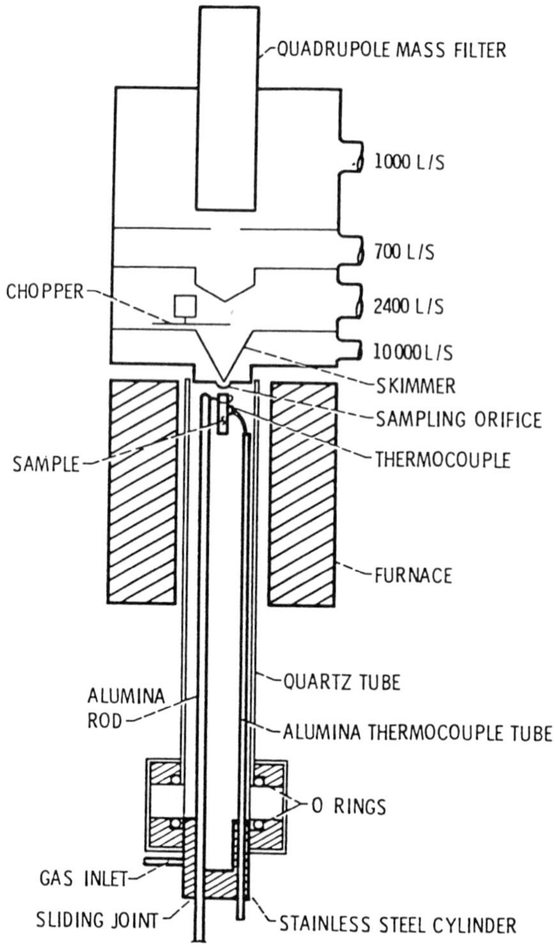

Fig. 2. Mass Spectrometer sampling system.

and a thin outer layer of Co_3O_4. The sample was then reacted with 1% Cl_2/1% O_2/Ar at 1000 K. Typical results are shown in Fig. 3. The observed linear reaction rates suggest that either a surface reaction or gas-phase diffusion is rate controlling. The approximate square root gas velocity dependence of the rate constant indicates that gas phase diffusion is rate-limiting.

The partial pressure of $CoCl_2$(g) can be calculated from the following reaction (*4*)

$$Co_3O_4(s) + 3\ Cl_2(g) = 3\ CoCl_2(g) + 2\ O_2(g) \tag{1}$$

$$K(1000\ K) = P(CoCl_2)^3 P(O_2)^2/P(Cl_2)^3 = 9.74 \times 10^{-7} \tag{2}$$

Putting 0.01 atm into Eq. (7) for $P(O_2)$ and $P(Cl_2)$ gives a $P(CoCl_2)$ of 2.14 $\times 10^{-3}$ atm. This can then be used in the equation for mass transport from a flat plate (*5*)

$$k = 0.664\ M/RT\ D^{2/3}/\nu^{1/6}\ (v/L)^{1/2}\ P(CoCl_2) \tag{3}$$

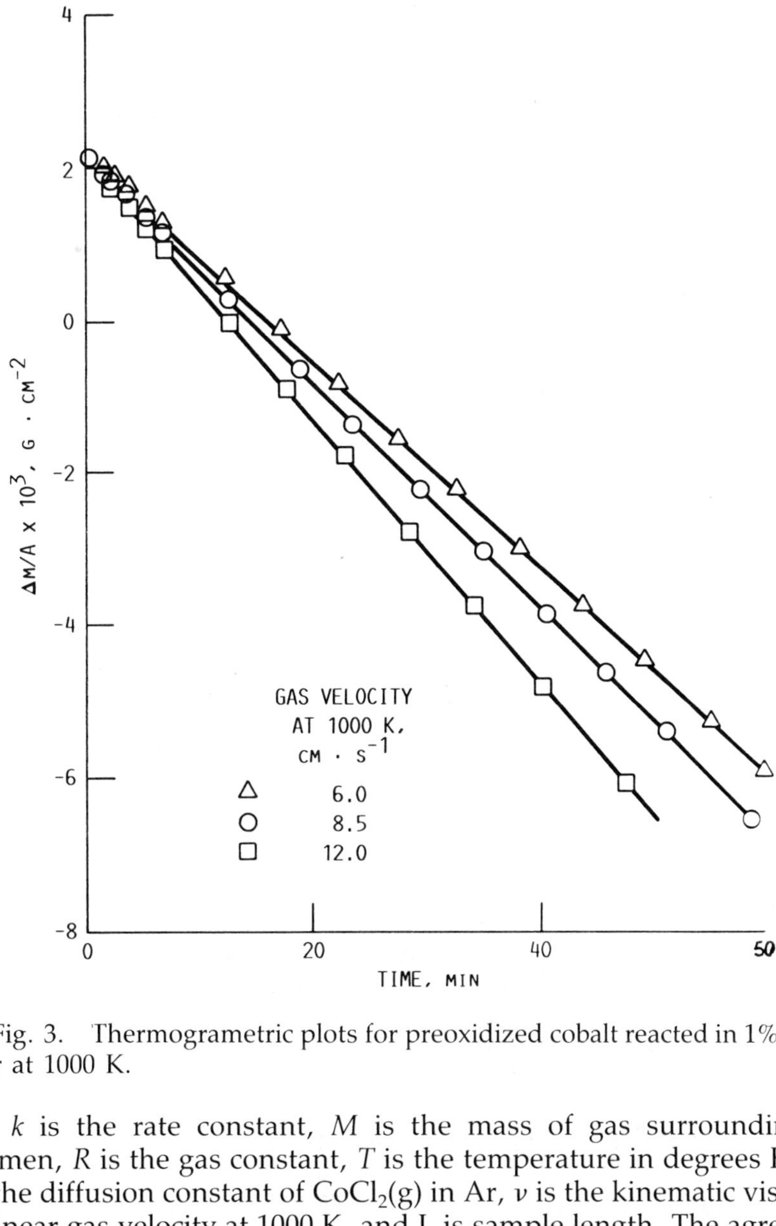

Fig. 3. Thermogrametric plots for preoxidized cobalt reacted in 1% Cl_2/1% O_2/Ar at 1000 K.

Here k is the rate constant, M is the mass of gas surrounding the specimen, R is the gas constant, T is the temperature in degrees Kelvin, D is the diffusion constant of $CoCl_2$(g) in Ar, ν is the kinematic viscosity, v is linear gas velocity at 1000 K, and L is sample length. The agreement between calculated and measured rates is reasonable, as shown in Table 1. This verifies that the major reaction is reaction (1) and the gas phase diffusion of $CoCl_2$(g) away from the sample is rate-limiting. This type of control has been observed for the reactions of pure oxides and Cl_2(g) (*6,7*).

In the case of Cr, Mo, and W, the oxychlorides are particularly stable species. Preoxidized Cr was reacted with 2% Cl_2/Ar and 1% Cl_2/Ar at 1363 K and the vapor products analyzed in the mass spectrometer (*8*). Table 2 shows that in all cases, only CrO_2Cl_2(g) was observed. Figure 4 shows the stability diagram for Cr–O–Cl at 1373 K (*9*). This is typical for

Table 1
Rate Constants for the Reaction of Co with 1% Cl_2/1% O_2/Ar at 1000°C[a]

Superficial velocity, cm/s	Measured rate constant, $mg/cm^2/h$	Calculated rate constant, $mg/cm^2/h$
8.5	8.3 ± 0.9	8.0
12	8.7	9.5

[a]Preformed oxide thickness, 15 μm.

Table 2
Principle Vapor Species for Preoxidized Cr + Cl_2/O_2 at 1363 K

Gas composition	Product	Ion intensity
1% Cl_2/Ar	CrO_2Cl_2	3.1×10^{-9}
1% Cl_2/1% O_2/Ar	CrO_2Cl_2	2.7×10^{-9}

Table 3
Principle Vapor Species for Mo + Cl_2/O_2 at 673 K

Gas composition	Products	Ion intensity
1% Cl_2/Ar	MoO_2Cl_2	3.2×10^{-8}
	MoO_3	8.9×10^{-9}
1% Cl_2/1% O_2/Ar	MoO_2Cl_2	9.4×10^{-7}

many metal–O–Cl stability diagrams. For the gas-phase compositions studied, the oxide is clearly the most stable compound. Condensed phase $CrCl_2$ could exist near the oxide/metal interface, but examination of the chromium coupons after reaction showed no clear evidence of this. Therefore, it is most likely that chlorine reacts with the Cr_2O_3 scale directly *(10)*

$$Cr_2O_3(s) + 2\ Cl_2(g) + \tfrac{1}{2}\ O_2(g) = 2\ CrO_2Cl_2(g) \tag{4}$$

$$K(1400\ K) = P(CrO_2Cl_2)^2\ /P(O_2)^{1/2}\ P(Cl_2)^2 = 2.9 \times 10^{-3} \tag{5}$$

It should be noted that even in the case of the 2% Cl_2 reactant, sufficient O_2 (estimated to be 100 ppm) exists to maintain a Cr_2O_3 scale. The amount of $CrO_2Cl_2(g)$ appears to be independent of $P(O_2)$, which was not expected. This may be related to the dense protective Cr_2O_3 scale, which only forms in small quantities regardless of the $P(O_2)$.

At low temperatures, the reactions of Mo with chlorine and oxygen appear to be similar to those of Cr. Table 3 shows some mass spectrometer results for these reactions. A stability diagram for Mo–O–Cl *(9)* at 673 K (Fig. 5) shows that $MoO_2Cl_2(g)$ is the most stable product at pressures close to the operating conditions. This suggests that the following reaction occurs at or near the oxide surface *(4)*

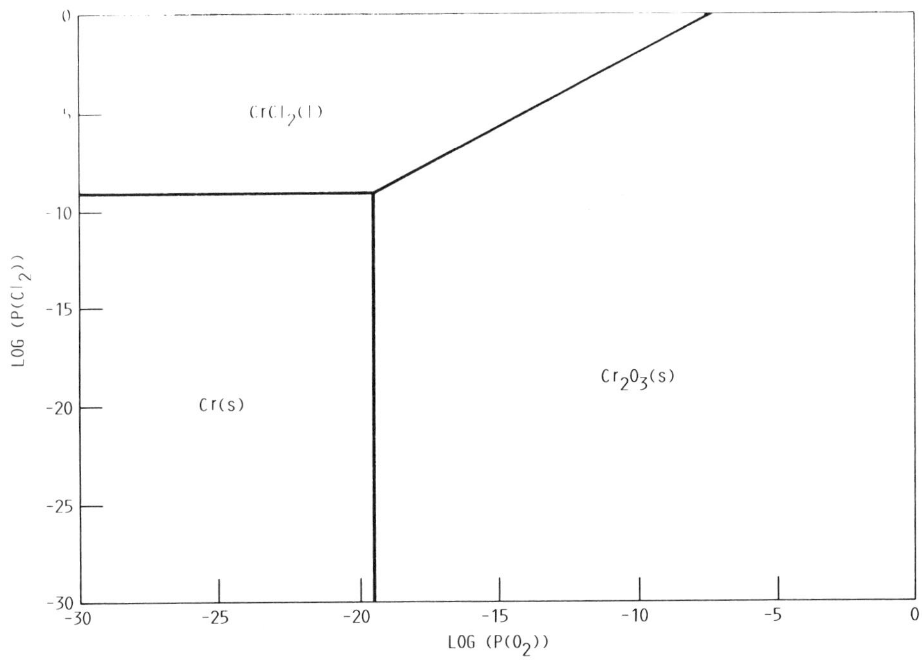

Fig. 4. Cr–O–Cl stability diagram at 1373 K (*9*).

$$MoO_3(s) + Cl_2(g) = MoO_2Cl_2(g) + \tfrac{1}{2} O_2(g) \qquad (6)$$

$$K(700\ K) = P(MoO_2Cl_2)\ P(O_2)^{1/2}/P(Cl_2) = 5.7 \times 10^{-2} \qquad (7)$$

The larger amount of reaction in the O_2-containing gas is very likely owing to the larger amount of MoO_3 formed. Under these conditions, a thick oxide layer was observed. At temperatures higher than 973 K, the reaction mechanism for Mo appears to change, and oxide penetration may be important (*8*).

Reaction Scheme B: Penetration Through the Oxide Scale

At lower temperatures, presumably owing to the formation of thinner oxide scales, the penetration of the oxide scale on cobalt by chlorine becomes a dominant reaction mechanism. In a study of the reaction of cobalt at 923 K with chlorine–oxygen mixtures (*11*), evidence strongly supported the formation of a condensed phase chloride below the oxide. Figure 6 shows an electron microprobe image and associated elemental maps of a cobalt sample surface after a 15 min exposure to 1% Cl_2/1% O_2/Ar. A large crack exposes a product layer below the oxide. Note that this layer is quite rich in cobalt and chlorine, but contains very little oxygen. Thus, it appears that a layer of $CoCl_2(s)$ forms below the oxide layer. For this gas mixture, the vapor pressure of $CoCl_2(g)$ is generated by the following reaction

$$CoCl_2(c) = CoCl_2(g) \qquad (8)$$

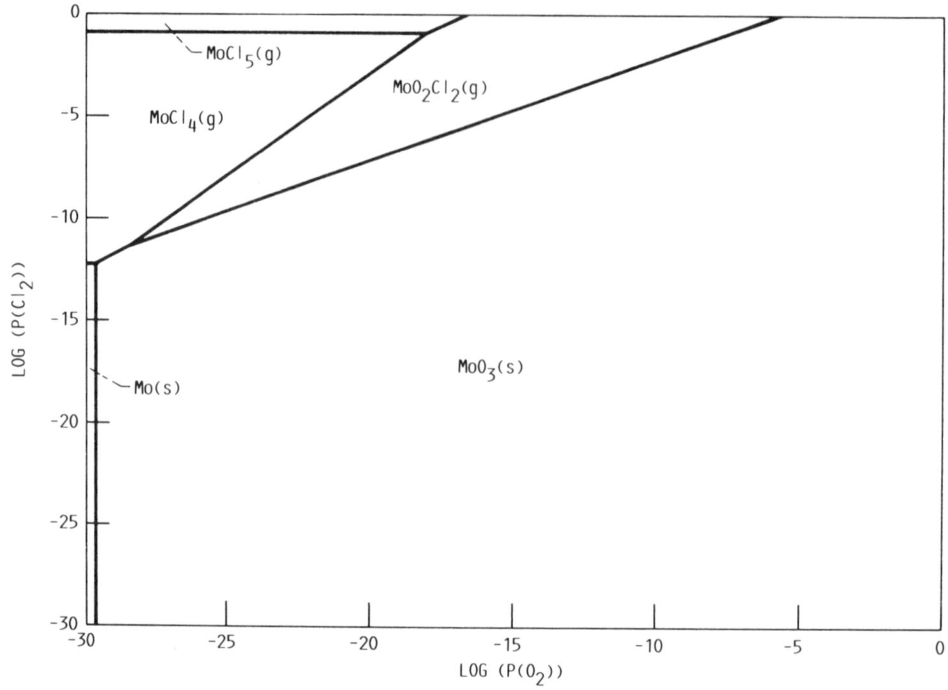

Fig. 5. Mo–O–Cl stability diagram at 673 K (*9*).

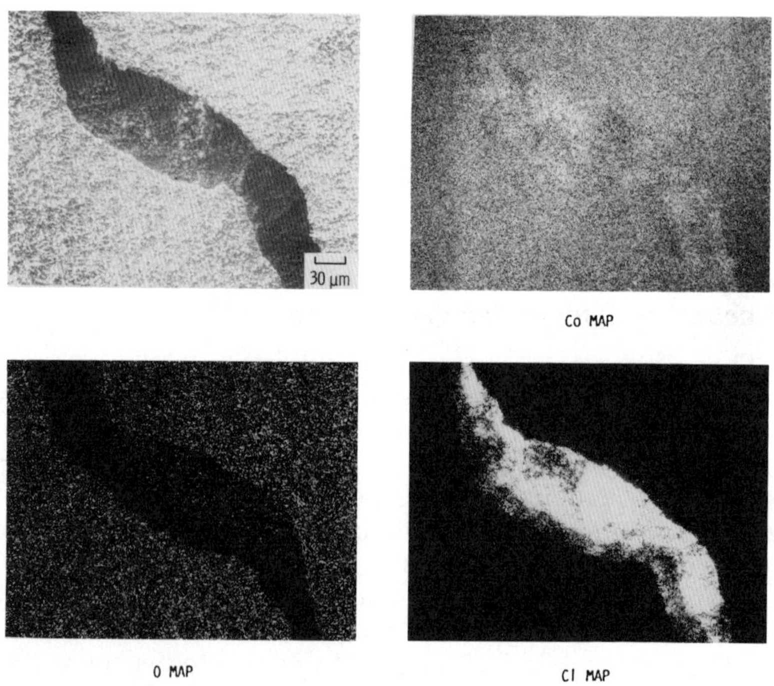

Fig. 6. Cobalt specimen surface and associated element maps after 15 min exposure to 1% Cl_2/1% O_2/Ar at 923 K.

Using the vapor pressure from this reaction, a rate based on gas diffusion control of $CoCl_2(g)$ away from the sample can be calculated using Eq. (3). The measured (-2.0 mg/cm^2/h) and calculated (-1.7 mg/cm^2/h) rates show good agreement despite the presence of a porous oxide scale above the $CoCl_2(c)$. This reaction scheme has been observed in the case of pure metals with chlorine and no oxygen (*12*).

A somewhat similar reaction scheme has been observed in the case of SiC reacting with chlorine–oxygen mixtures (*13*). Coupons of various types of SiC are reacted with 2% Cl_2/Ar and 1% Cl_2/1–20% O_2/Ar at 1223 K. Although the results are somewhat more complex than in the above cobalt case, scale penetration appears to be the dominant reaction mechanism. Mass spectrometer measurements determined $SiCl_4(g)$ was the primary vapor specie. However, direct reaction with the SiO_2 scale is not really feasible

$$SiO_2(s) + 2\ Cl_2(g) = SiCl_4(g) + O_2(g) \tag{9}$$

$$K(1273\ K) = P(SiCl_4)\ P(O_2)/P(Cl_2)^2 = 2 \times 10^{-8} \tag{10}$$

For $P(O_2) = P(Cl_2) = 0.01$ atm, one can calculate a $P(SiCl_4)$ of 2×10^{-10}, when put into Eq. (3), gives a rate much slower than observed. Thus, reaction to form $SiCl_4(g)$ must be occurring by scale penetration and direct reaction with the SiC

$$SiC(s) + 2Cl_2(g) = SiCl_4(g) + C \tag{11}$$

Oxide growth at these temperatures is very slow and the oxide scale is formed in only limited quantities. The reaction rates were flow-rate dependent, which suggests that either Cl_2 gas diffusion to or $SiCl_4$ gas diffusion away from the sample is rate-controlling. Putting $P(Cl_2) = 0.01$ atm into Eq. (3) gives a reaction rate much higher than measured. Thus, $SiCl_4$ gas diffusion away from the sample is very likely rate-limiting.

Reaction Scheme C: Paralinear Kinetics

When the chlorination and oxidation reaction rates are similar, the reaction is best described by a combined parabolic rate constant and linear rate constant. The rate of change of scale thickness is thus described as

$$dx/dt = k_p/x - k_l \tag{12}$$

A study with pure Co and x% Cl_2/50% O_2/Ar at 900–1200 K illustrates this reaction scheme (*14*). At these temperatures, the scale consists primarily of CoO, and the rate of the following oxidation reaction in a chlorine free environment was measured

$$Co(s) + \tfrac{1}{2}\ O_2(g) = CoO(s) \tag{13}$$

The rate of chlorination was calculated from the following reaction:

$$CoO(s) + Cl_2(g) = CoCl_2(g) + \tfrac{1}{2}\ O_2(g) \tag{14}$$

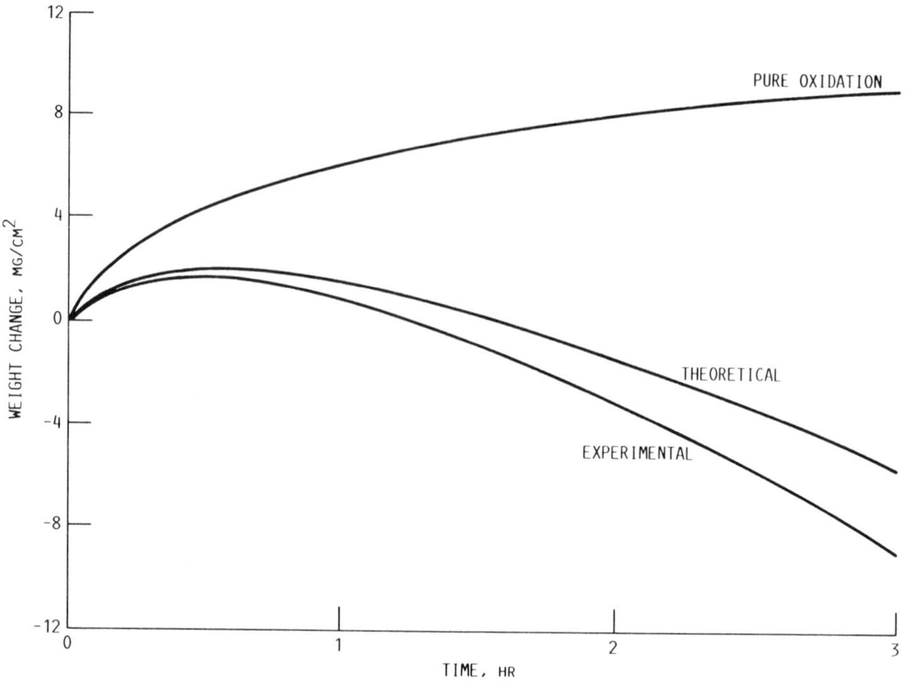

Fig. 7. Calculated and experimental thermogravimetric plots for cobalt reacted in 0.5% Cl_2/50% O_2/Ar at 1200 K.

These were put into the integrated form of Eq. (12):

$$t = \frac{K_p/2}{\alpha^2 K_v^2}\left[\frac{-\alpha K_v \dfrac{\Delta M}{A}}{K_p/2} - \ln\left(1 - \frac{\alpha K_v \dfrac{\Delta M}{A}}{K_p/2}\right)\right] \tag{15}$$

Here, α is the weight fraction of oxygen in the scale. At 1200 K, the approximation is good for a range of chlorine contents ranging from 0.25 to 1.0% and 50% oxygen. A representative curve is shown in Fig. 7. The initial weight gain and bending over is typical of paralinear behavior. At lower temperatures the paralinear approximation does not hold for cobalt. Chlorine may induce cracks and fissures in oxide scale, which leads to accelerated oxidation. Formation of Co_3O_4 and oxide penetration by chlorine may also be more important at the lower temperatures, so reaction (14) does not fully describe the chlorination process.

Reaction Scheme D: Gas Phase Reactions

The presence of chlorine can induce a drastic increase in the rate of oxidation of nickel. One way in which this occurs is via the gas-phase reaction of $NiCl_2$(g) with oxygen (*15*). This mechanism occurs as follows. Chlorine penetrates the oxide layer to form $NiCl_2$(g) at the oxide–metal interface where the oxygen potential is low. Then $NiCl_2$(g) diffuses out through the oxide to the gas–oxide interface where it is converted to oxide by the following gas phase reaction

$$NiCl_2(g) + \tfrac{1}{2} O_2(g) = NiO(s) + Cl_2(g) \quad (16)$$

The oxide produced in this manner has a porous and nonprotective morphology. The rate of oxidation by this mechanism is several orders of magnitude higher than the rate of oxidation of nickel in pure oxygen. Under some conditions, this rate can be so high that the heat of reaction causes the temperature of the corroding specimen to rise above the furnace temperature. The particular temperatures and gas mixtures that lead to these rapid reaction rates are well-defined and have been determined (*15*).

Gas-phase reactions may also lead to the formation of oxychloride molecules. The reaction of SiC and chlorine–oxygen mixtures leads to small amounts of $Si_2OCl_6(g)$ (*16*). These are approximately $\frac{1}{100}$ of the amount of $SiCl_4(g)$ and proposed to form as follows

$$2SiCl_4(g) + \tfrac{1}{2}O_2(g) = Si_2OCl_6(g) + Cl_2(g) \quad (17)$$

Passing oxygen gas over $SiCl_4(\ell)$ at room temperature also forms $Si_2OCl_6(g)$, which supports the above mechanism.

SUMMARY AND CONCLUSIONS

The reactions of chlorine and oxygen with metals and ceramics are dependent on the thermodynamic stabilities of the products, the individual reaction rates for oxidation and chlorination, and the reaction conditions. However, much of the experimental results can be categorized according to four reaction schemes. Examples of each have been given. These involve reaction of chlorine directly with the oxide, penetration of the oxide, simultaneous oxide growth and chlorination at similar rates, and finally, gas-phase reactions.

REFERENCES

1. Stearns, C. A., Kohl, F. J., Fryburg, G. C., and Miller, R. A., *Characterization of High Temperature Vapors and Gases*, NBS-SP-561/1, National Bureau of Standards, Washington, DC, 1978, pp. 303–355.
2. Jacobson, N. S., *Application of an Atmospheric Pressure Sampling Mass Spectrometer to Chlorination Reactions*, NASA TM-87270, NASA, Washington, DC, 1986.
3. McNallan, M. J., Kang, C. T., and Liang, W. W., *Metall. Trans. A* **15,** 403 (1984).
4. Chase, M. W., Jr., Davies, C. A., Downey, J. R., Jr., Frurip, D. J., McDonald, R. A., and Syverud, A. N., *JANAF Thermochemical Data*, third ed., American Institute of Physics, Inc., New York, 1986.
5. Skelland, A. H. P., *Diffusional Mass Transfer*, Wiley, New York, 1974, p. 117.
6. Fruehan, R. J. and Martonik, L. J., *Metall. Trans.* **4,** 2793 (1973).
7. McNallan, M. J. and Liang, W. W., *J. Am. Ceram. Soc.* **64,** 302 (1981).
8. Kreidler, E. R. and Jacobson, N. S., (in preparation).
9. Thompson, W. T., Pelton, A. D., and Bale, C. W., *F*A*C*T* (Facility for the Analysis of Chemical Thermodynamics), Thermfact, Ltd., Quebec, Canada, 1985.

10. Sano, N. and Belton, G. R., *Metall. Trans.* **5,** 2151 (1974).
11. Jacobson, N. S., McNallan, M. J., and Lee, Y. Y., *Metall. Trans. A* **17,** 1223 (1986).
12. Fruehan, R. J., *Metall. Trans.* **3,** 2585 (1972).
13. Marra, J. E., Kreidler, E. R., Jacobson, N. S., and Fox, D. S., *J. Am. Ceram. Soc.* **71,** 1067 (1988).
14. Maloney, M. J. and McNallan, M. J., *Metall. Trans. B* **16,** 751 (1985).
15. Lee, Y. Y. and McNallan, M. J., *Metall. Trans. A* **18,** 1099 (1987).
16. Marra, J. E., Kreidler, E. R., Jacobson, N. S. and Fox, D. S., *J. Electrochem. Soc.* **135,** 1571 (1988).

Erosion and Corrosion of Nickel in Sulfurous Atmospheres

D. M. RISHEL, F. S. PETTIT, AND N. BIRKS*

Department of Materials Science and Engineering, Rm. 848, Benedum Hall, University of Pittsburgh, Pittsburgh, PA 15261

ABSTRACT

It is shown that the attack of metals by erosion and corrosion simultaneously at high temperatures can be characterized and placed in several distinct regimes. Such regimes of interaction range from erosion-dominated to corrosion-dominated behavior. In all cases studied, the interaction of erosion and oxidation leads to degradation that exceeds that owing to oxidation alone or erosion alone. Mechanisms are proposed to account for the observed phenomena and the current paper will emphasize how these mechanisms are modified by the presence of sulfur dioxide and sulfur trioxide in the aggressive atmosphere.

It will be shown that the sulfurous species penetrate the growing oxide scale, the presence of sulfides in the scale modifies its physical properties and that the presence of the erosion component increases the permeability of scales formed to sulfurous gas species. A further regime of degradation involving spalling of the scale under erosion-corrosion attack is also identified.

Index Entries: erosion-corrosion; mixed oxidants; nickel, spallation; sulfidation/oxidation.

INTRODUCTION

In the evolution of advanced energy conversion technologies, it is recognized that material selection, along with plant design and operation criteria, is dictated to a large degree by the effects of combined high temperature erosion-corrosion (*1*). For instance, erosion-corrosion prob-

*Author to whom all correspondence and reprint requests should be addressed.

lems have been identified in gas turbines (*2*), fluidized bed coal combustors (*3*), and coal gasifiers (*4*).

Erosion-corrosion is the material degradation owing to simultaneous erosion and corrosion. As frequently observed, however (*5–8*), the concomitant interaction is much more complex than an additive response and simple scale removal. Instead, a complex interaction exists between the erosion and corrosion processes. For example, it has been found (*5*) that combined hot corrosion-erosion is more severe than the sum of the erosion and hot corrosion processes acting independently. Adding to the complexity of the erosion-corrosion process in energy conversion processes is a myriad of variables (*9*). As typified by coal gasifiers, these variables may include, but are not limited to, multiple oxidant environments (i.e., oxygen, sulfur, and carbon) and reactive erodents (i.e., coal, coke, or char) with a variety of sizes, loadings, velocities, and impingement angles. In addition, temperature and pressure must be taken into account.

The response of materials, in part, may be described in terms of interaction regimes. In this regard, Kang et al. (*7*) describe that the erosion-corrosion response of pure metals exposed to a high temperature atmosphere of air falls into three interaction regimes; (1) pure erosion, (2) erosion-enhanced oxidation, and (3) oxidation-affected erosion. These regimes are summarized in Fig. 1, which details scale thickness in terms of erosion rate vs corrosion rate. In general terms, the erosion rate may be taken to be a function of particle velocity, particle loading, particle size/type, and impingement angle; similarly, the corrosion rate is a function of temperature and gas composition.

The regime of pure erosion of oxide encompasses the situation where the oxidation rate is rapid and scale damage or removal occurs along lines similar to the erosion of ductile and brittle materials at room temperature (*10*). The pure erosion of oxide regime may dominate for erosion-corrosion of an oxide at high temperatures, whereas pure erosion of metal represents erosion-corrosion of a pure metal when exposed to low temperatures and/or nonoxidizing environments.

The erosion-enhanced oxidation regime dominates under conditions when a metal sample is subjected to a high temperature, erodent-laden, oxidizing environment. In this situation, the oxide scale develops under two opposing processes; growth owing to oxidation and removal owing to erosion. Under certain circumstances, and particularly at the start of the erosion-corrosion process, if the corrosion rate exceeds the erosion rate, the scale will grow to a certain thickness. As the scale thickens, however, its rate of growth decreases, and a point is reached whereby the growth and erosion rates are equal. When this occurs, the scale has achieved a "steady state" thickness.

When a metal is subjected to a high temperature, erodent-laden, oxidizing environment, but where either the scale growth rate is relatively low or the erosion rate is high, the attack of the specimen enters the oxidation-affected erosion regime. the existence of this regime was shown by Kang et al. (*7*) to occur when a slowly oxidizing metal, such as

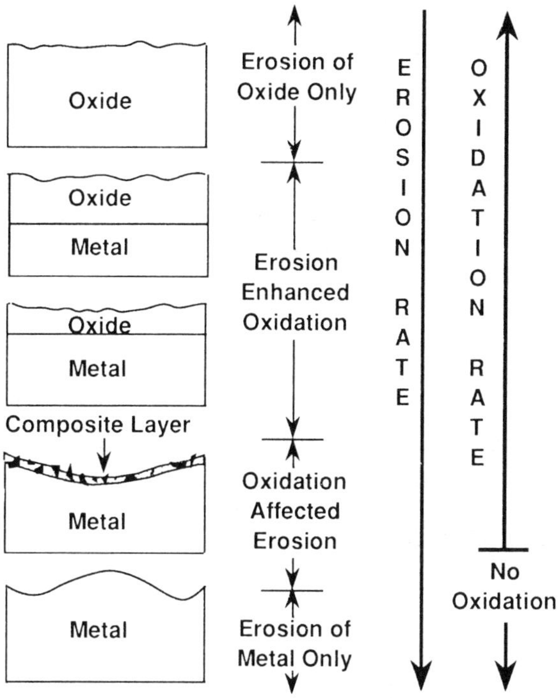

Fig. 1. Suggested interation regimes describing erosion-corrosion behavior.

pure Ni, was exposed to an erodent-laden airstream at 800°C and 140 m/s. In this situation, the erosive stream thinned and removed the slowly growing oxide scale to the point where substantial deformation of the substrate resulted. When this occurred, the formation of a series of undulations or moguls was noted. Additionally, erodent particles penetrated the thin oxide and embedded into the metal substrate.

The investigation by Hogmark et al. (*11*) proposed erosion-corrosion degradation regimes that are in many ways similar to those proposed by Kang et al. (*7*). Erosion-corrosion of a metal, parameterized in terms of corrosive vs erosive aggressiveness, is a process that can be divided into six regimes. However, unlike Kang (*7*) where spallation was not observed for the experimental conditions used, Hogmark (*11*) considered flaking or spallation of the oxide scale to play an important role. In their first regime, pure corrosion, material removal is considered to proceed in the form of flaking as a result of either a combination of poor adhesion between the oxide layer and the substrate or internal stresses. In the second regime, erosion-affected corrosion, material removal by flaking is influenced or enhanced by a relatively mild erosion component. In regime number three, erosion of the corrosive film without flaking occurs either by brittle or ductile mechanisms. Regime number four is entered when local flaking of the oxide scale at individual impact sites is observed. The dimensions of the individual craters are of the same order of magnitude as the instantaneous layer thickness. In the fifth regime, simultaneous erosion of the oxide layer and metallic substrate occur. This is owing to the fact that, when erosion dominates over corrosion, impinging particles penetrate into the substrate and detach oxide fragments, as well as displacing the metal. The last regime, pure erosion, is restricted to conditions of no corrosion or in situations when the oxide thickness is much less than the average depth of particle penetration and when the mechanisms of material removal, depending upon the material, occur in a ductile or brittle manner.

Of the six regimes discussed, a dramatic increase in the wear rate is expected in the erosive-corrosive conditions of regimes four and five. This is particularly true if the adhesion of the oxide layer is poor.

In the present work, the major emphasis was to examine the effects of SO_2/SO_3 on the erosion-corrosion behavior of nickel. It will be shown that, depending upon the experimental conditions, the observed behavior generally falls into one of the regimes proposed by Kang (*7*) and Hogmark (*11*). However, it may be necessary to redefine or expand the definition of some of these regimes.

METHODS

Ni-270 (Ni-99.98, C-0.01, Mn-0.003, balance < 0.007) plate was fabricated into coupons nominally $8 \times 8 \times 2.5$ mm in size. A 1-mm diameter hole (~5 mm in depth) was drilled into one edge of the specimen, providing a means of monitoring temperature using a K-type thermocouple. An aluminide coating of high oxidation resistance was applied to the five sides of the specimen not directly exposed to the erosive stream in order to minimize their weight gain owing to corrosion. The remaining surface, exposed to erosion-corrosion, was ground down to a 600 grit finish with SiC paper and cleansed in methanol.

Samples were exposed in the apparatus previously described by Kang et al. (*7*) and modified to incorporate the use of sulfurous gases. Briefly, the apparatus delivers a high velocity (~30–170 m/s), high temperature (600–850°C), particle laden corrosive gas (containing mixtures of air, SO_2, and SO_3). The specimen is positioned by a holder that can be swiveled to permit a variety of impingement angles and is placed at the exit of a 9-mm diameter nozzle. The specimen is heated by impingement of the incident hot gas stream. A laser doppler velocimeter is used to measure the velocities of erodent emerging from the nozzle. Determination of the actual SO_2 and SO_3 concentrations is accomplished by the analytical technique described in Appendix A of ref. *12*.

A Sylco MK-IX powder dispenser is used to introduce 20 μm angular alumina particles for entrainment into the main air stream used for specimen erosion; the amount of powder actually entrained into the acceleration tube is measured. In the work to be described, a number of conditions were varied in a systematic manner and categorized into sets. However, in all cases, the specimen material was Ni-270, the exposure temperature is 600°C, the erodent particle was 20 μm Al_2O_3, impingement was at 90° to the specimen surface, and the average measured velocity was 70 m/s. For each specific experimental set, a number designation is assigned. The experiments conducted are listed in Table 1, along with the pertinent conditions. In the body of this discussion, reference to a particular experimental condition will be with its designated number. To minimize referral to Table 1, after the appearance of the words "Set #1," for example, additional descriptors such as (6-⅔-L) will appear. In this case, the 6 refers to 600°C, the ⅔ refers to the atmosphere composition of air + SO_2/SO_3, and L refers to low particle loading rates. This nomenclature is summarized in Table 2.

In all experimental sets, reaction kinetics were established from specimen exposures of 15, 41, 67, 93, 120, 270, and 420 min. Each data point on the weight change vs time plots represents a single specimen. With the exception of experimental Set #3(6-A-L), all curves were derived from data obtained through duplicate exposures. Exposed specimens were weighed using an analytical balance with an accuracy of 0.1 mg and examined using optical metallography, SEM-EDS and X-Ray diffraction.

DISCUSSION

It was found that the presence of sulfurous atmospheres, compared to exposures in air only, significantly influenced the erosion-corrosion behavior of nickel. This is reflected in Figs. 2 and 3, where the weight change behavior for experimental conditions Sets #1 (6-⅔-L), 4 (6-⅔-H), 3 (6-A-L), and 6 (6-A-H) is displayed. One important fact to consider, however, when analyzing these plots is the effect of corrosion. For instance, under erosion-corrosion in atmospheres of air + SO_2/SO_3 (Fig. 2), significant corrosion occurs and the "true" weight loss owing to erosion is somewhat masked by the weight gain owing to corrosion. On the other hand, for erosion-corrosion in air (Fig. 3), the oxide formed is very thin and results in a negligible weight gain component. The importance of the corrosion product layer regarding erosion is reflected in the surface micrographs of Figs. 4 and 5 for specimens exposed to Sets #3 (6-A-L) and 1 (6-⅔-L), respectively. In atmospheres of air, erodent particles impinging normally can easily penetrate the thin oxide layer formed and extensively deform the substrate metal, resulting in a surface characterized by moguls (7). The weight loss in this case is owing to the erosive removal of both metal and oxide. When the corrosion rate is high, as in atmospheres of air + SO_2/SO_3, the resulting thick scale shields the underlying nickel substrate, and the weight loss that occurs is owing to erosion of the corrosion product. As defined by Kang et al. (7) erosion-corrosion of nickel in atmospheres of air falls into the oxidation affected erosion regime, whereas in atmospheres of air + SO_2/SO_3, particularly under "low" particle loading rate conditions, the erosion-corrosion response is in the erosion-enhanced oxidation regime.

Table 1
Experimental Conditions to Measure Erosion-Corrosion[a]

			Gas Composition			
Set No.	T°C	Vel. (m/s)	Air	SO_2 (ppm)	SO_3 (ppm)	Part. Load. Rate (mg/min.)
1	600	70	Bal.	2256†	68†	174.5 ± 52.4 (low)[b]
2	600	70	Bal.	2256†	68†	0 (no erosion)
3	600	70	Bal.	0†	0†	174.5 ± 52.4 (low)
4	600	70	Bal.	2256†	68†	584.0 ± 81.8 (high)[c]
5	600	70	Bal.	1960‡	0‡	0 (no erosion)
6	600	70	Bal.	0	0	584.0 ± 81.8 (high)

[a]To minimize referral to this table, after the appearance of the words "Set #1," for example, will appear a few additional descriptors. For instance, after Set #1, appears (6-2/3-L). The 6 refers to 600°C, the 2/3 refers to the atmosphere composition of air + SO_2/SO_3, and L refers to low particle loading rates. This nomenclature is summarized in Table 2.

[b]658 ± 198 ppm.

[c]2,200 ± 309 ppm.

†Average values for standard settings.

‡Uncatalyzed.

Table 2
Nomenclature of Descriptors Used in Table 1

Set #1	⇔	(6-2/3-L)	⇔	600°C, air + SO_2/SO_3, low particle loading
Set #2	⇔	(6-2/3-NE)	⇔	600°C, air + SO_2/SO_3, no erosion
Set #3	⇔	(6-A-L)	⇔	600°C, air, low particle loading
Set #4	⇔	(6-2/3-H)	⇔	600°C, air + SO_2/SO_3, high particle loading
Set #5	⇔	(6-2-NE)	⇔	600°C, air + SO_2, no erosion
Set #6	⇔	(6-A-H)	⇔	600°C, air, high particle loading

To understand erosion-corrosion in atmospheres containing SO_2 and SO_3, a knowledge of the mechanisms responsible for corrosion are required. To this end, experiments under Sets #2 (6-⅔-NE) and 5 (6-2-NE) were conducted. Plots of the weight change behavior for corrosion only, in atmospheres of Air + SO_2/SO_3 and Air + SO_2 (i.e., Sets #2 (6-⅔-NE) and 5 (6-2-NE)), are shown in Fig. 6, illustrating that the weight gain for corrosion in Air + SO_2/SO_3 is an order of magnitude higher than in Air + SO_2. As a means of comparison, the kinetic data for specimens exposed to corrosion only, Set #2 (6-⅔-NE), and erosion-corrosion, Sets #1 (6-⅔-L) and 4 (6-⅔-H), are provided in Fig. 2. For the case of corrosion only, it was noted for samples exposed to Set #2 (6-⅔-NE) conditions, that the surfaces had a combined shiny and matt black appearance, whereas specimens exposed under Set #5 (6-2-NE) parameters were dark

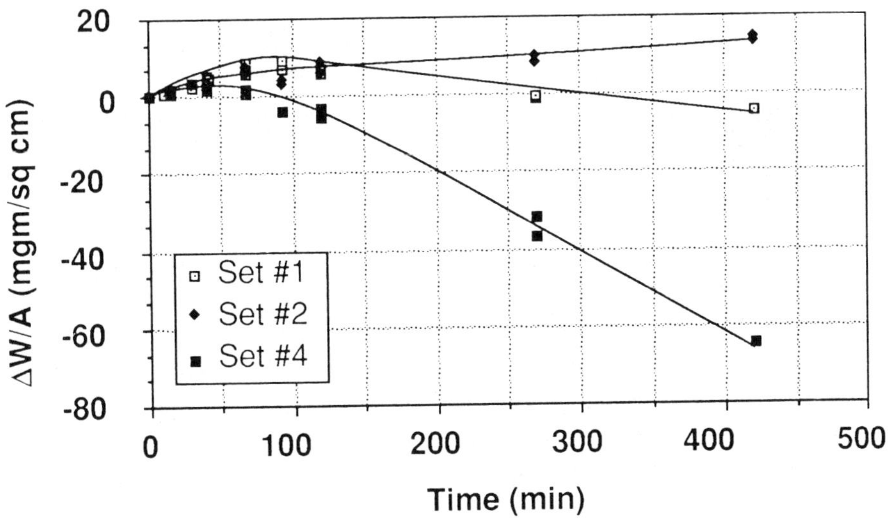

Fig. 2. Comparison of erosion-corrosion and corrosion kinetics of Ni-270 in Air + SO_2/SO_3 under: (1) Set #1 (6-⅔-L), (2) Set #2 (6-⅔-NE), and (3) Set #4 (6-⅔-H) at 600°C for a flow velocity of 70 m/s.

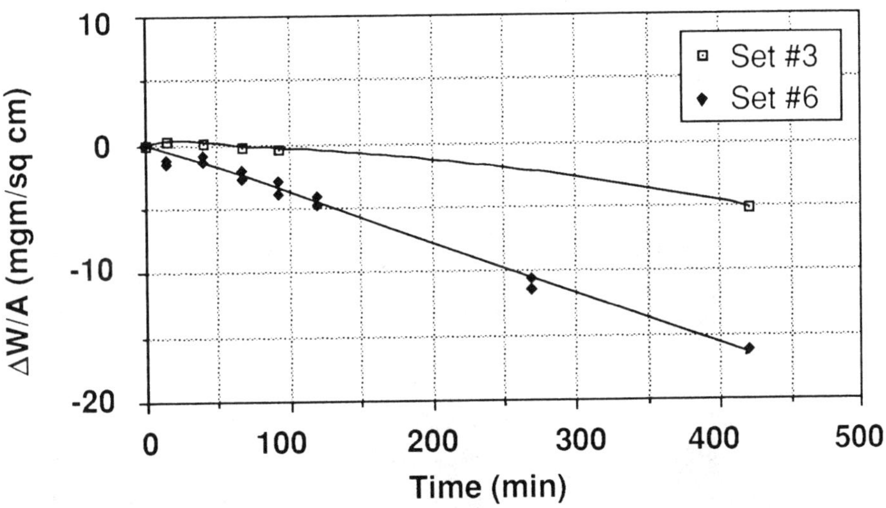

Fig. 3. Comparison of erosion-corrosion kinetics of Ni-270 in (1) Air (Set #3 (6-A-L) under "low" particle loading rates and (2) erosion-corrosion kinetics in Air (Set #6 (6-A-H)) under "high" particle-loading rates at 600°C for a flow velocity of 70 m/s.

gray. The natures of the scales developed under these two conditions are also reflected in cross-sectional micrographs. Figs. 7(a and b) contrast the scales developed during Condition Sets #2 (6-⅔-NE) and 5 (6-2-NE), respectively. Both a thick overall scale and a Ni_3S_2 layer forms in atmospheres containing SO_3, whereas only a thin scale with no detectable Ni_3S_2 layer forms in atmospheres without SO_3.

Fig. 4. Surface micrograph of specimen subjected to erosion-corrosion under Set #3 (6-A-L) conditions for 420 min.

Fig. 5. Surface micrograph of specimen subjected to erosion-corrosion under Set #1 (6-⅔-L) conditions for 420 min.

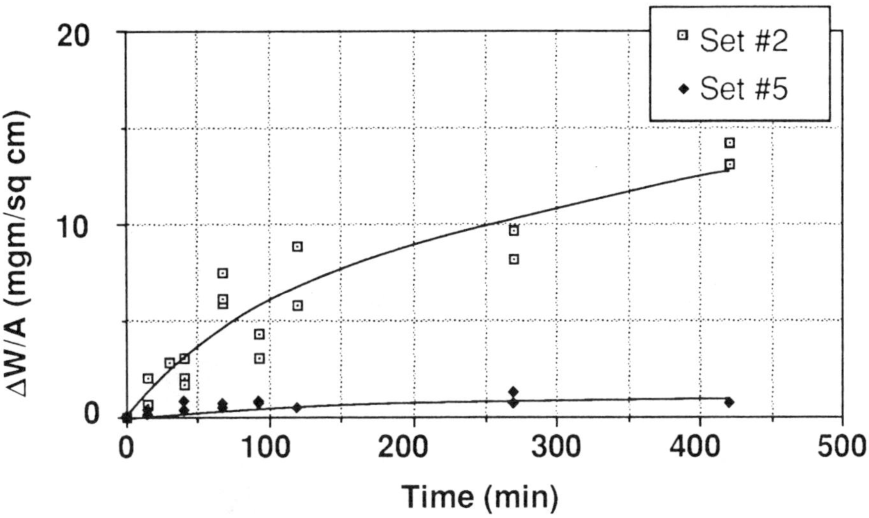

Fig. 6. Comparison of corrosion kinetics of Ni-270 in (1) air + SO_2/SO_3 (Set #2 (6-⅔-NE)) and (2) air + SO_2 (Set #5 (6-2-NE)).

A thermochemical stability diagram for nickel, constructed at 600°C is shown in Fig. 8. Included in this figure are points delineating the bulk gas composition (with respect to SO_3), SO_2 isobars, and the possible reaction path between points 1, 2, and 3. The strong effect of the presence of SO_3 is believed to be connected with the tendency toward $NiSO_4$ formation. Although the average SO_2 pressures employed were high enough to favor the reaction

$$NiO + SO_2 + \tfrac{1}{2}O_2 = NiSO_4 \qquad (1)$$

a significant increase in the corrosion rate was not observed until the average SO_3 pressure (6.8×10^{-5} atm), as produced by unequilibrated catalysis in the reaction

$$SO_2 + \tfrac{1}{2}O_2 = SO_3 \qquad (2)$$

was higher than the equilibrium SO_3 pressure (3.5×10^{-5} atm) for the coexistance of NiO and $NiSO_4$ according to (*13*)

$$NiO + SO_3 = NiSO_4 \qquad (3)$$

The increase in the corrosion rate of Ni in $O_2/SO_2/SO_3$ atmospheres, as related to $NiSO_4$ formation, is described in detail elsewhere (*14–16*). Briefly, the corrosion mechanism is associated with $NiSO_4$ formation at the scale-gas interface with conversion to NiO and Ni_3S_2 at the inner surface according to the reactions

$$4\ Ni + NiSO_4 = NiS + 4\ NiO \qquad (4)$$

$$Ni + 2\ NiS = Ni_3S_2 \qquad (5)$$

or overall as

$$9\ Ni + 2\ NiSO_4 = 8\ NiO + Ni_3S_2 \qquad (6)$$

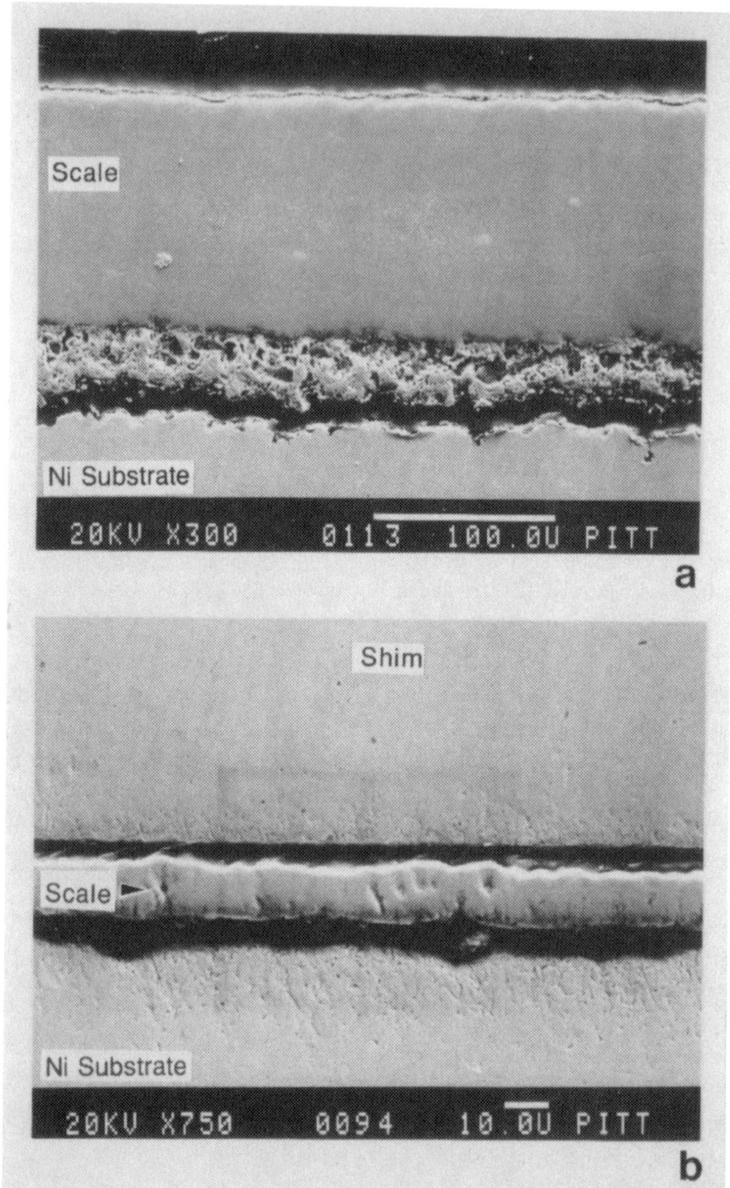

Fig. 7. (a) Cross-section of specimen exposed to Set #2 (6-⅔-NE) conditions; (b) Cross-section of specimen exposed to Set #5 (6-2-NE) for 420 minutes each.

If the rate of this decomposition reaction is rapid and the rate of $NiSO_4$ formation at the scale gas interface is slow, only a thin $NiSO_4$ layer would be expected to form. In fact, it has been shown (*16,17*) that, if the corrosion of nickel involves sulfate formation, only a very thin layer (on the order of 100 Å) of $NiSO_4$ typically forms on the scale surface and that a dedicated surface analytical technique must be employed to determine its presence. In previous research, it was found (*12*) through ESCA analysis that the surfaces of samples exposed to atmospheres containing

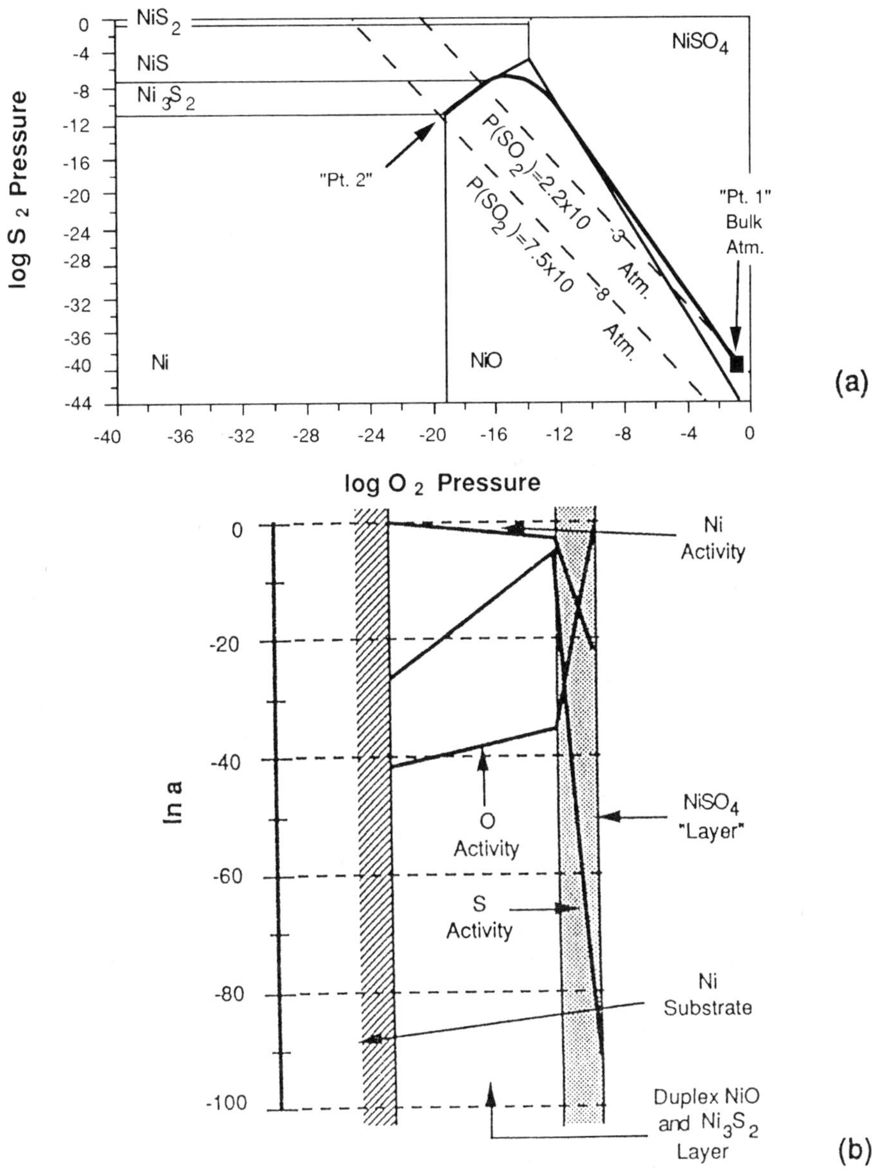

Fig. 8. (a) Ni–S–O Phase Stability Diagram for 600°C; (b) Scale Ni, S, O activity profile (not drawn to scale).

SO_2/SO_3 host Ni configured in a oxidic- and sulfate-binding state, whereas sulfur is present also in a sulfate-binding state. This strongly suggests that the surface of the specimen consists of, or is host to both NiO and $NiSO_4$.

Once formed, the presence of Ni_3S_2 has been found to be especially deleterious (*14–16*). Because of the fact that Ni_3S_2 is heavily defected (*17*), enhancement of the outward migration of Ni cations (*18*) and the inward diffusion of sulfur anions (*19*) occurs. In sum, in atmospheres capable of supporting the formation of $NiSO_4$, substantial amounts of Ni_3S_2 form just within the scale surface.

The predominant mechanism by which corrosion in atmospheres of Air + SO_2 without SO_3 (Set #5 (6-2-NE)), may occur is expressed in the following equation

$$7\ \mathrm{Ni} + 2\ \mathrm{SO_2} = 4\ \mathrm{NiO} + \mathrm{Ni_3S_2} \quad (7)$$

As explained elsewhere (*20*), although Ni_3S_2 is unstable in such a gas phase, it is possible for it to form beneath the outer NiO layer. In contrast to the mechanism previously discussed above, the inward transport of sulfur may occur, as described by Pope and Birks (*21*). The inward transport of sulfur (from the scale-gas to the scale-metal interface) may result from either bulk diffusion of sulfur or transport of SO_2 molecules down microcracks. Thus, when the SO_2 partial pressure in the bulk gas is greater than that corresponding to the SO_2 isobar defined by the Ni–NiO–Ni_3S_2 triple point, the formation of a second phase beneath the external oxide by dissociation of SO_2 is feasible. As shown in Fig 8(a) by the dashed lines, the SO_2 partial pressure used in these experiments, ($1.96–2.3 \times 10^{-3}$ atm) is greater than required for equilibrium between Ni, NiO, and Ni_3S_2 (7.5×10^{-8} atm). Thus, such a mechanism is theoretically feasible.

The influence of particle-loading rate is reflected in two ways. First, as seen in Fig. 2, for exposures less than 100 min, a higher weight gain occurred for specimens exposed under Condition Set #1 (6-2/3-L) compared to Condition Set #2 (6-2/3-NE). The apparent higher corrosion rate, owing to "low" particle loadings, is also reflected in thicker corrosion product scales and, thus, suggests that the corrosion process is enhanced in a manner unrelated to oxide thinning, as described by Kang et al. (*7*).

The second aspect of the particle loading rate to be noted is the difference in the weight change behavior for erosion-corrosion (in atmospheres of Air + SO_2/SO_3) under low and high particle-loading rate conditions (i.e., Sets #1 (6-⅔-L) and 4 (6-⅔-H), respectively). As shown in Fig. 2, a threefold increase in the particle loading rate (174.5 vs 584.0 mg/min) produces a weight loss that is approximately 13 times higher. On the other hand, as shown in Fig. 3, a threefold increase in the particle-loading rate causes a corresponding threefold increase in the measured weight loss when the atmosphere is air only (Condition Sets #3 (6-A-L) and #6 (6-A-H)). This strongly suggests that this synergistic behavior is clearly related to the presence of SO_2 and SO_3.

Figures 9–11 form a sequential series of surface micrographs for specimens exposed under Condition Set #4 (6-2/3-H) parameters. It can be seen that, starting at approximately 67 min of exposure, in Fig. 9, is the appearance and subsequent growth of an intensely eroded center region. High magnification micrographs, for instance Fig. 11(a), show that within these zones numerous spalls occur that are usually associated with what appear to be mounds. Although the presence of sulfur is suspected of inducing spallation (*22*), it is not clear if these are regions where the oxide scale is locally thick or where the scale separated or delaminated from the underlying substrate. The spalls, typically

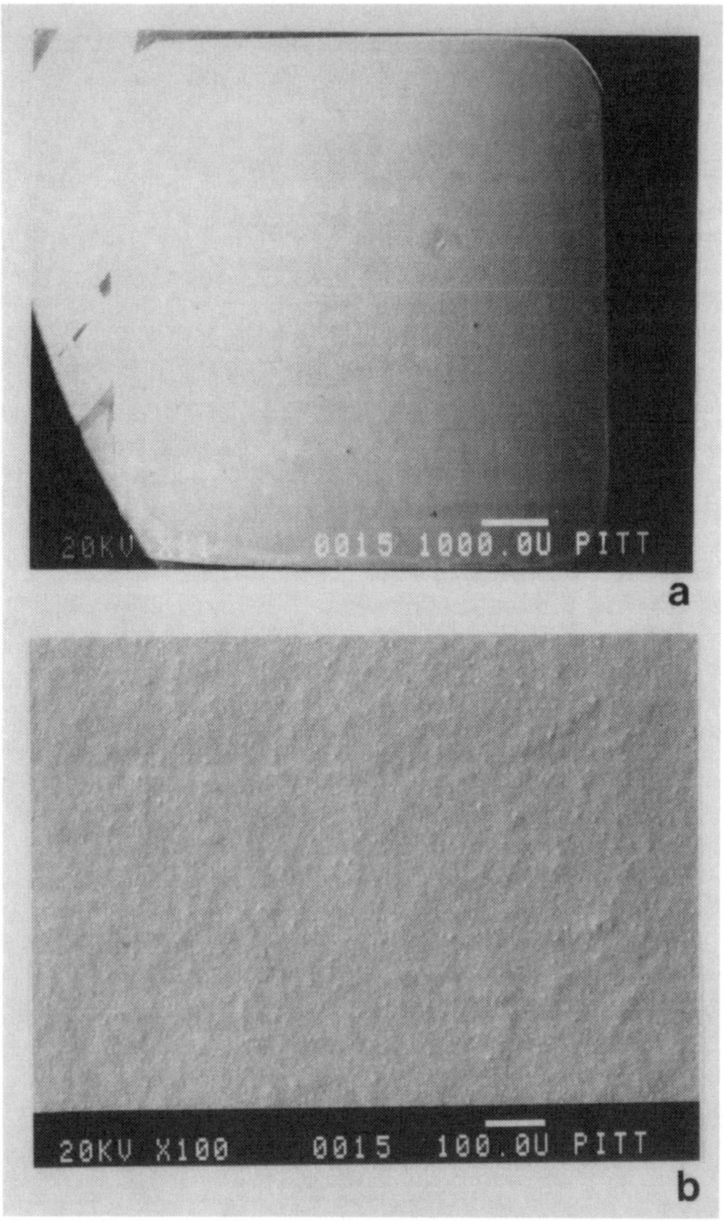

Fig. 9. (a) Surface micrograph of specimen subjected to erosion-corrosion in atmospheres of air + SO_2/SO_3 under "high" particle-loading rates at 600°C and 70 m/s for 15 min. (Set #4 (6-⅔-H)); (b) Magnification of (a).

10–30 μm in diameter, are the same approximate size as the alumina erodent and were noted to occur where the scale is about 2–3 μm thick. However, these spalls are not seen in the outer periphery, where the oxide scale is thicker. The occurence of spallation in regions where the scale thickness is relatively thin was also noted in earlier work (*12*). There, erosion-corrosion of Ni-270 was studied under similar conditions, with the exception that the particle velocity was 84 m/s. However, unlike

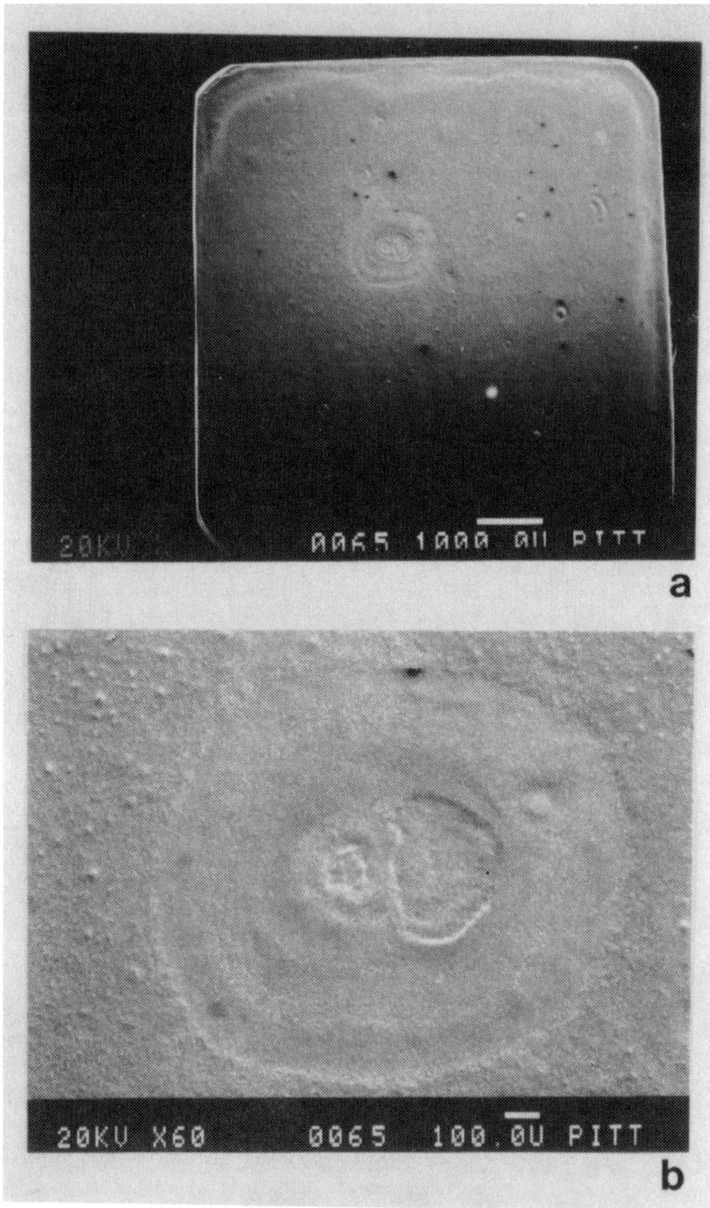

Fig. 10. (a) Surface micrograph of specimen subjected to erosion-corrosion in atmospheres of air + SO_2/SO_3 under "high" particle-loading rates at 600°C and 70 m/s for 67 min. (Set #4 (6-⅔-H)); (b) Magnification of (a).

the situation presented in this paper, the intensely eroded "thinner" scale and, thus, spalling were present from the onset of specimen exposure.

The spallation phenomenon may also be related to impingement angle in that it is more likely to occur when the impingement angle is close to 90° in the specimen center and less likely to occur near the specimen edge, where the impacts may be oblique. From Fig. 12(b), it appears that the elevated portion of the oxide scale "shields" the corrosion product to its right from particles impacting at nonnormal impinge-

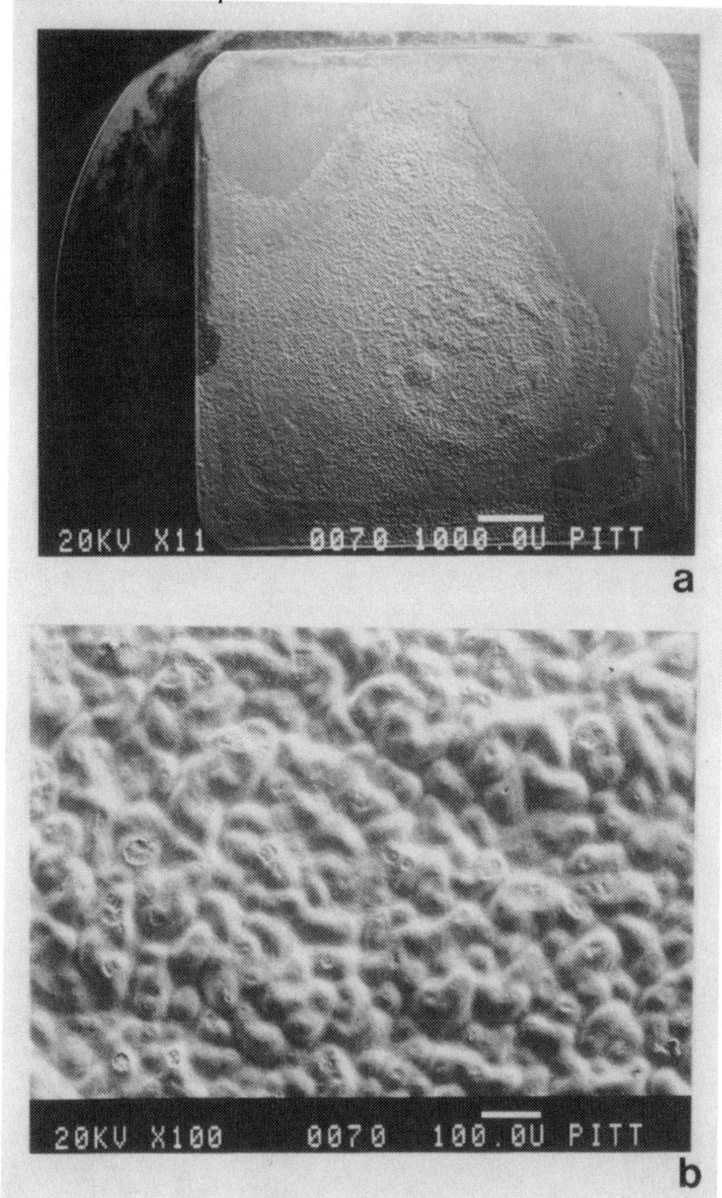

Fig. 11. (a) Surface micrograph of specimen subjected to erosion-corrosion in atmospheres of air + SO_2/SO_3 under "high" particle-loading rates at 600°C and 70 m/s for 420 minutes (Set #4 (6-⅔-H)); (b) Magnification of (a).

ment angles. This observation strongly suggests that the impinging erodent particles are aerodynamically influenced (*23*) or, more likely, deflected by rebounding particles and/or debris.

Thus, it is concluded that the occurrence of spallation is responsible for the increase in the rate of weight loss (*see* Fig. 2) for erosion-corrosion under "high" particle loading rate conditions, compared to "low" loading rate conditions in atmospheres of Air + SO_2/SO_3. Specimens exhibiting the type of spallation discussed appear to fall into regime number four, as suggested by Hogmark et al. (*11*).

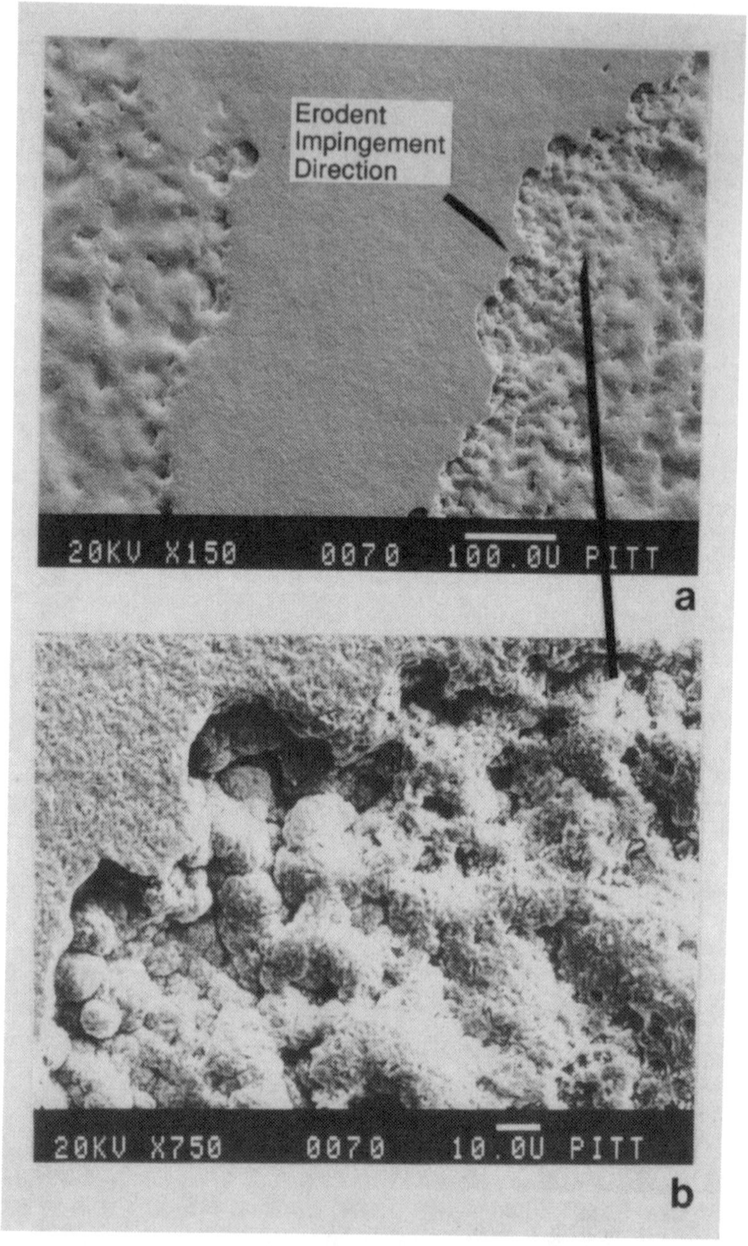

Fig. 12. (a) Surface micrograph of specimen exposed to Set #4 (6-⅔-H) parameters for 420 minutes, illustrating artifact owing to impingement angle; (b) Magnification of (a).

SUMMARY

1. The presence of a sulfurous atmosphere has been found to influence the erosion-corrosion behavior of nickel significantly, compared to erosion-corrosion response in air.
2. Compared to atmospheres containing Air + SO_2, specimens

exposed to atmospheres of Air + SO_2 and SO_3 experience significantly greater degrees of corrosion.

3. The rate of weight loss for erosion-corrosion in atmospheres containing air was directly related to the particle-loading rate. That is, since the particle loading was tripled, the rate of weight loss also tripled. However, for erosion-corrosion in atmospheres containing Air + SO_2/SO_3, the rate of weight loss increased by a factor of 12–13 when the particle loading rate was tripled, suggesting a synergistic response to particle loading rates.
4. It appears that increased amount of weight loss under "high" particle loading rate conditions is related to the occurrence of spallation.

ACKNOWLEDGMENTS

This work was carried out under a grant provided by the Electric Power Research Institute, which also provided support for D. M. Rishel.

REFERENCES

1. Wright, I. G., "High Temperature Erosion in Coal Combustion and Conversion Processes: A Review", *Materials Science and Engineering*, (1987), pp. 261–271.
2. Brunetaud, R., Coutsouradis, D., Gibbons, T. B., Lindbloom, Y., Meadowcroft, D. B., and Strickler, R., eds., *High Temperature Alloys for Gas Turbines 1982*, "High Temperature Erosion and Erosion-Hot Corrosion of Superalloys and Coatings, by J. C. Galsworthy, J. E. Restall and G. C. Booth" (Dordrecht, Holland, D. Reidel Publishing Co., 1982), pp. 207–235.
3. Krause, H. H., Vaughn, D., Wright, I. G., Carlton, H. E., and Hall, A. M., "Corrosion and Erosion of High Temperature Tube Alloys in a Fluidized Bed Coal Combustor", *Materials Performance*, Vol. 21, No. 11, (1982), pp. 52–58.
4. Wire, G. L., Vesely, E. J., and Agarwal, S., "Erosion-Corrosion of Metals in Coal Gasification Atmospheres," *Journal of Materials for Energy Systems*, Vol. 8, No. 2, (Sept. 1986), pp. 150–167.
5. Barkalow, R. H., Goebel, J. A., and Pettit, F. S., *Material Problems in Fluidized-Bed Combustion Systems: High-Temperature Erosion-Corrosion by High Velocity (200 m/s) Particles*, Pratt and Whitney Aircraft Group, May 1980, EPRI CS-1448, Project 979-4, Final Report.
6. *Symposium on High Temperature Materials Chemistry-III*, Las Vegas, October, 1985, "Erosion-Corrosion in Elevated Temperture Sulfidizing Gas-Solids Environments, by G. Sorell and M. A. H. Jones" (Pennington, NJ: The Electrochemical Society, 1986), pp. 100–128.
7. Kang, C. T., Pettit, F. S., and Birks, N., "Mechanisms in the Simultaneous Erosion-Oxidation Attack of Nickel and Cobalt at High Temperatures," *Met. Trans. A.*, Vol. 18, (1987), pp. 1785.
8. Wright, I. G., Nagarajan, V., and Merz, W. E., *Erosion-Corrosion of Metals and Alloys at High Temperatures*, Battelle, Columbus Laboratories, June 1984, EPRI CS-3504, Project 979-8, Interim Report.
9. Agarwal, S. C. and Howes, M. A. H., "Erosion-Corrosion of Materials in High-Temperature Environments: Inpingement Angle Effects in Alloys 310

and 6B under Simulated Coal Gasification Atmosphere," *J. Materials for Energy Systems*, Vol. 7, No. 4, (March 1986), pp. 370–381.

10. *Proceedings of the 6th International Conference on Erosion by Liquid and Solid Impact*, "Erosion Mechanisms in Ductile and Brittle Materials, by Alan V. Levy," (University of Cambridge, 1983, pp. 39-1 through 39-7).
11. *Proceedings of the 6th International Conference on Erosion by Liquid and Solid Impact*, "On the Combined Effects of Corrosion and Erosion, by S. Hogmark, A. Hammersten and S. Soderberg," (University of Cambridge, 1983, pp. 37-1 through 37-8).
12. Rishel, D. M., "The Erosion-Corrosion Behavior of Ni at 600°C in Mixed Oxidant Atmospheres," (unpublished M. S. Thesis, Department of Materials Science and Engineering, University of Pittsburgh, 1988).
13. Kellog, H. H., "A Critical Review of Sulfation Equilibria," *Transactions of the Metallurgical Society of AIME*, Vol. 230, (December 1964), pp. 1622–1634.
14. Haflan, B. and Kofstad, P. "The Reaction of Nickel with SO_2 + O_2/SO_3 at 500–900°C," *Corrosion Science*, Vol. 23, No. 12, (1981), pp. 1333–1352.
15. Alcock, C. B., Hocking, M. G., and Zador, S., "The Corrosion of Ni in O_2 + SO_2 Atmospheres in the Temperature Range 500–700°C," *Corrosion Science*, Vol. 9, (1969), pp. 111–122.
16. Lillerud, K. P., Haflan, B., and Kofstad, P., "On the Reaction Mechanism of Nickel with SO_2 + O_2/SO_3," *Oxidation of Metals*, Vol. 21, Nos. 3/4, (1984), pp. 119–134.
17. Stoklosa, A. and Stringer, J., "Defect Structure and Chemical Diffusion in Nickel Sulfide β-Ni_3S_2," *Oxidation of Metals*, Vol. 11, No. 5, (1977), pp. 277–288.
18. Mrowec, S. and Przybylski, K., "Transport Properties of Sulfide Scales and Sulfidation of Metals and Alloys," *Oxidation of Metals*, Vol. 23, Nos. 2/3, (1985), pp. 107–139.
19. Hocking, M. G. and Sidky, P. S., "The Hot Corrosion of Nickel-Based Ternary Alloys and Superalloys for Gas Turbine Applications-II. The Mechanism of Corrosion in SO_2/O_2 Atmospheres," *Corrosion Science*, Vol. 27, No. 2, (1987), pp. 205–214.
20. Pettit, F. S., Goebel, J. A., and Goward, G. W., "Thermodynamic Analysis of the Simultaneous Attack of Some Metals and Alloys by Two Oxidants," *Corrosion Science*, Vol. 9, (1969), pp. 903–913.
21. Pope, M. C. and Birks, N., "The Penetration by Sulfur of NiO Scales Growing on Nickel," *Oxidation of Metals*, Vol. 12, No. 2, (1978), pp. 173–181.
22. Smialek, James L., "Adherent Al_2O_3 Scales Formed on Undoped NiCrAl Alloys," *Metallurgical Transactions A*, Vol. 18A, (January 1987), pp. 164–167.
23. Laitone, J. A., "Aerodynamic Effects in the Erosion Process," *Wear*, Vol. 56, (1979), pp. 239–246.

Laser-Pulse Vaporization of Refractory Materials

DONALD R. OLANDER

Materials and Chemical Sciences Division, Lawrence Berkeley Laboratory and the Department of Nuclear Engineering, University of California, Berkeley, CA 94720

ABSTRACT

Recent experimental and theoretical developments concerning the interaction of laser light with refractory solids are reviewed, with emphasis on long, low-power density pulses. Compared to short, high-power density irradiation, long pulses produce principally thermally-driven evaporation that can be related to the thermochemical properties of the heated surface. Surface temperature measurement and calculation methods are outlined and the use of mass spectrometry to analyze the vapor is reviewed.

Index Entries: Laser light; refractory solids; low-power density pulses; high power density irradiation.

INTRODUCTION

The incidence of short nanosecond- to millisecond-wide pulses of laser light on solids serves two functions. First, it is used to produce desirable changes in the substrate, including welding, annealing, and alloying (*1,2*). Second, laser irradiation can remove solid, either by a thermally-activated evaporation process or an explosive ejection of material sometimes called ablation. The second application is the subject of this review.

In addition to the well-established laser machining operations (cutting, welding, and drilling), two examples of the practical application of laser vaporization are production of clusters of many materials and production of thin films of high-T_c superconducting ceramics. In the former application, the laser-generated vapor is condensed in an inert carrier gas and transported to detection instrumentation (*3*). In the latter application, targets of bulk material are vaporized in vacuum and deposited as superconducting films on suitable substrates (*4*). In this application, congruency of vaporization, meaning that the composition of the vapor is equal to that of the solid, is critical. Finally, laser vaporization (or ablation) is the basis of a commercial ion source for mass spectrometric analysis of solids.

Laser heating can also be used as a convenient surface heat source for fundamental studies that simulate vaporization processes occurring in systems that are normally heated by other means. For example, the very high temperatures that can occur in a nuclear reactor core during a severe accident are produced by a combination of nuclear heating from decay of fission products and chemical heating from steam-oxidation of the metallic components of the fuel element. Temperatures as high as 3000°C can be attained, and vaporization of the UO_2 fuel and refractory fission products is an important process in such an accident. In another example, disruptions of the thermonuclear plasma in tokamak-type fusion reactors results in rapid energy deposition on plasma-facing components, which are constructed of or lined with refractories, such as graphite, silicon carbide, or tungsten. In this case, heating is caused by a combination of high-energy ions and electrons and X-rays. Vaporization of the wall materials can adversely affect the power balance in the plasma and must be controlled. As a final example, aerodynamic frictional heating during reentry of space vehicles is accommodated by heat shields fabricated of graphite composites, which are designed to vaporize in a controlled fashion at the peak temperatures achieved.

In these three cases of diverse modes of heating of a solid, substitution of pulsed-laser heating provides a useful method for studying the kinetic and thermodynamic aspects of the vaporization process that cannot be achieved using the actual heat source. The laser can be focused onto a spot of small diameter (typically several millimeters) on the surface, which, as a result, is driven briefly to very high but controllable temperatures (4000–5000 K). The laser provides a localized, mainly surface heat source so that the target serves as its own crucible. Conventional steady-state means for investigating vaporization processes, such as Knudsen cell or transpiration methods, require heating of bulk structures to a high temperature. These methods are limited to maximum temperatures of about 3000°C by the lack of container materials to hold specimens at higher temperatures. Usually, chemical reaction between specimen and container dictate a lower usable maximum temperature than determined solely by container melting. Both of these restrictions are absent in laser heating.

Except where an ambient gas atmosphere is required for a particular application, most laser evaporation experiments are conducted in vacuum. This environment aids in maintaining surface cleanliness, allows use of *in situ* detectors, such as mass spectrometers and surface analyzers, and avoids vapor interactions with another gaseous species.

Measurement of vaporization by laser heating is not without severe drawbacks. Principal of these is the difficulty in measuring the temperature of the surface during the heating transient. This is possible only for the relatively long pulses (order of milliseconds) characteristic of conventional-mode Nd-glass lasers. Conventional optical pyrometry is not feasible for nanosecond pulses from an excimer laser. With ceramic targets, the thermal shock engendered by the laser pulse heating may fragment the surface. This problem can be avoided if the target can be preheated to a temperature at which it is plastic.

Another difficulty is the variety of physical processes that can occur during the interaction of the laser with the solid. The surface of the target usually melts, and because of the recoil pressure of the vaporizing material, radial flow of the liquid can occur. The molecules in the vapor may collide with one another, changing the blowoff from a free-molecule flow to a gasdynamic expansion. At sufficiently high power densities, the laser may interact with the plume, thereby creating a plasma and shielding the surface from the incoming laser light. The multitude of processes that can occur and their dependence on laser power, pulse length, wave length, and material properties, such as thermal conductivity and vapor pressure, renders clean-cut physical interpretations of the results problematic. Despite these difficulties, pulse-laser heating of solids has become an important tool in studying the vaporization process at temperatures unattainable by conventional means.

REGIMES OF LASER–SOLID INTERACTIONS

The absorbed power density is a convenient, although not unique, measure for categorizing the nature of the laser–solid interaction. This quantity, together with the pulse duration and reflectivity of the surface, determines the absorbed energy density. The consequences of energy absorption depend on a wide variety of physical and chemical properties of the target material, chief among which are the thermal diffusivity, vapor pressure, and optical absorbance. These properties govern the depths of penetration of the laser light and the thermal front, and consequently of the maximum surface temperature and density of the vapor just above the surface.

In this review, only laser power densities that produce significant vaporization will be considered. At power densities from approximately 10^3 to 10^4 W/cm^2 only modifications of the target, such as melting or annealing, occur.

Low Laser Power Density-Long Pulse Time

Power densities less than about 10^5 W/cm^2 lasting for approximately a millisecond are sufficient to heat the surface of poorly-conducting materials, such as oxides, to temperatures in the vicinity of 4000 K. To reach comparable temperatures on a metal surface, longer pulses or higher power densities are required to compensate for the greater heat loss by conduction into the bulk solid. Under these conditions, the main mechanism of material removal is ordinary vaporization into a vacuum. Langmuir analyzed the kinetics of this process using the principle of detailed balancing. This principle states that at equilibrium, the rates of all forward and reverse processes must be equal. Thus, the rate of evaporation of a solid or liquid in contact with its equilibrium vapor is equal to the rate of impingement of molecules from the gas on the surface, which is calculable from gas kinetic theory, times the probability that the molecules stick, or condense. Langmuir then assumed that the vaporization rate remains unchanged when the equilbrium vapor is replaced by a vacuum. The resulting Hertz-Langmuir formula

$$F = \frac{\alpha P(T_S)}{\sqrt{2\pi m k T_S}} \tag{1}$$

is the basis for measurement of the vapor pressure–temperature relationship, $P(T_s)$, for refractory metals such as tungsten. In Eq. (1), F is the vaporization flux from the surface, T_s is the surface temperature, which is in general time-dependent, m is the mass of the evaporating molecule, and k is the Boltzmann constant. The parameter α is the evaporation (or condensation) coefficient, which accounts for imperfect sticking of impinging molecules on the surface. It is usually close to unity unless the vapor species has an appreciably different molecular structure from that in the condensed phase. Crystalline arsenic, for example, has a rhombohedral crystal lattice structure, but the vapor consists of tetrahedral As_4 molecules. The sluggish rate of atomic rearrangement prior to vaporization is manifest as a very low evaporation coefficient for this element (5). Because of the strong covalent bond, molecular oxygen has a low sticking probability on the surfaces of refractory oxides.

If the solid contains more than one element, say A and B, and evaporates as molecules A_iB_j, the flux of a particular species characterized by the stoichiometric numbers i and j is given by

$$F_{ij} = \frac{\alpha_{ij} P_{ij}(T_S, x_s)}{\sqrt{2\pi m_{ij} k T_S}} \tag{2}$$

where P_{ij} is the partial pressure of $A_1B_j(g)$ in equilibrium with the condensed phase at the surface of temperature T_s and composition x_s(expressed as the atom fraction of A, for example). The dependence of the vaporization rate of each molecular species on the surface A:B ratio means that all vaporization fluxes depend on the rate at which A or B can be supplied to the surface by diffusion from the bulk in order to replenish the loss by vaporization (6).

The less-volatile constituent of the condensed phase becomes enriched at the surface with respect to the bulk composition. This is associated with incongruent vaporization, in which the composition of the vapor is different from that of the bulk-condensed phase. For example, the vapor flux from uranium carbide is rich in U(g), with smaller quantities of UC_2(g) and C_3(g); vacuum vaporization of the superconducting oxide $YBa_2Cu_3O_7$ first produces essentially pure molecular oxygen, followed by the oxides and atoms of copper, then the vapor species of the more refractory oxides of barium and yttrium.

Incongruency occurs only in the initial stages of vacuum vaporization. If the surface is maintained at a constant temperature (greater than that of the bulk), the total vaporization rate, and hence the surface recession speed, also approach steady-state values. When this state has been achieved, a simple mass balance over the moving surface requires that the compositions of the solid and vapor be equal. This condition of "forced congruent vaporization" (*7*) is achieved at times greater than the characteristic time D/u^2, where D is the mutual diffusion coefficient in the binary A–B condensed phase (which is usually liquid for significant vaporization to occur) and u is the speed of surface recession. The latter is related to the vaporization fluxes by (*6*)

$$u = (M_A \sum iF_{ij} + M_B \sum jF_{ij})/\rho \tag{3}$$

where ρ is the mass density of the condensed phase at the surface, assumed to be independent of the A/B ratio, and M_A and M_B are the atomic weights of elements A and B, respectively. Forced congruent vaporization can also be achieved in periodic pulsed evaporation.

In a single pulse, the constituents of the solid may vaporize at different times during the transient. The more volatile components evaporate first, followed by the less volatile constituents. At any instant in time, vaporization may be highly incongruent. However, if sufficient material is removed from the target during the pulse, the time integrals of the component fluxes approach congruency with the bulk solid. If the vapor phase condensation coefficients are near unity, the stoichiometry of the target material is reproduced in films of material deposited on substrates exposed to the irradiated surface.

Thus, equilibrium vacuum vaporization of multicomponent solids by laser or other heat sources involves a complex coupling of: heat and mass transport in the near-surface regions of the condensed phase, which governs the surface temperature and surface composition; vaporization kinetics according to the Hertz-Langmuir theory; and the thermochemistry of the material, by which the equilibrium pressures in Eq. (2) are determined.

Effect of Molecular Collisions in the Gas Phase

If the laser power density is such that only modest surface temperatures are reached and the vapor pressure of the material is low at these temperatures, the molecules emitted from the target do not undergo collisions in the vapor phase. This is the free-molecule flow regime. Far from the surface, a density-sensitive detector should observe a cosine angular distribution, a speed distribution corresponding to a Maxwell distribution characteristic of the surface temperature, and a total evaporation rate given by Eq. (1).

At input power densities roughly between 10^5 and 10^6 W/cm^2, collisional effects in the vapor become significant. The transition between the free-molecule and the gasdynamic vapor blowoff regimes can be studied by Monte Carlo methods (*8*) or by the simpler test-particle method (*9*). In the latter method, the number-density distribution of vapor molecules above the evaporating surface is first calculated assuming collisionless flow. For a finite source, this distribution varies from the density corresponding to one half of the equilibrium vapor pressure at the surface temperature to zero far downstream from the surface. Then a test particle is emitted from the center of the evaporating surface along the normal. Knowing the atomic scattering cross-section of the vapor molecules, the probability that the test particle escapes from the cloud without undergoing a collision is determined as a function of surface temperature (or equivalently of the vapor pressure). Figure 1 shows the results of such a calculation for steady-state evaporation of uranium dioxide from an 8-mm diameter heated spot. The collision cross-section was taken to be 0.43 nm^2. The curves correspond to various levels of rigor in the analysis; curves c and d probably bracket the exact result. At low temperatures, the escape probability is unity, which signifies collisionless flow. At high temperature, the probability of escape is zero, meaning that the flow is of the gasdynamic type similar to that of a gas jet expanding from a high pressure reservoir through an orifice into a vacuum. Over a 400 K range, the escape probability drops from 1 to 0. The transition from free-molecule to gasdynamic flow is defined to occur when the escape probability is one-half. This condition is achieved for UO_2 at about 2600 K.

The calculation shown in Fig. 1 is for steady evaporation. A similar analysis has been performed for the temperature transient produced by a heat pulse that has a triangular temporal shape. For a pulse of a full width of 0.2 ms, the 50% escape probability occurs at a temperature about 50 K higher than for the steady-state case.

The analysis leads to the following rule of thumb for approximating the 50% escape probability temperature for other conditions or other materials: the dimensionless product of the equilibrium vapor number density at the surface temperature, the collision cross-section, and the radius of the heated spot is approximately 2.5 (*9*).

The effect of intermolecular collisions in the vapor in the gasdynamic regime is to transform random thermal energy of the vapor molecules into translational kinetic energy along the normal to the surface. This phenomenon occurs in both expansion of a high-pressure gas from an orifice (*10*) or the blowoff from a laser-heated spot on a solid or liquid surface (*11,12*). However, the analogy between these two situations is not exact because, in the latter, a thin transition zone known as the Knudsen layer separates the surface and the beginning of the adiabatic expansion (Fig. 2). In this layer, the translational distribution along the normal (x direction) changes from a half-Maxwellian

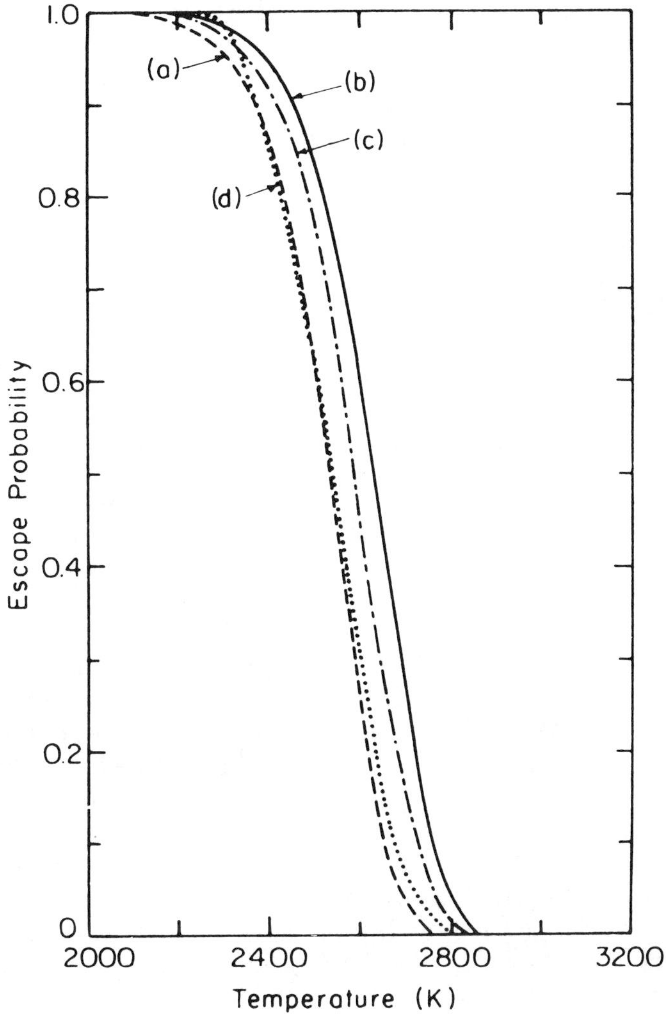

Fig. 1. Escape probability as a function of surface temperature for steady vaporization of uranium dioxide from an 8 mm diameter heated surface into vacuum. The curves labeled a–d represent various approximations in the calculation (9).

$$f_{FM} \propto \exp\left(-\frac{mv_x^2}{2kT_S}\right) \qquad (v_x \geq 0) \tag{4}$$

at the surface to the full Maxwellian

$$f_{GD} \propto \exp\left(-\frac{m(v_x - v_K)^2}{2kT_K}\right) \qquad (-\infty < v_x < \infty) \tag{5}$$

on the downstream side of the Knudsen layer. In Eq. (5), k is Boltzmann's constant, v_x is the molecular speed component in the x direction, v_k is the

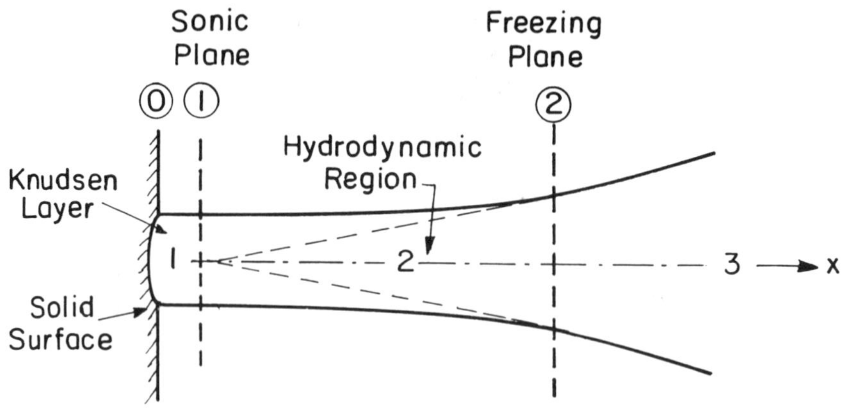

Fig. 2. Flow regimes in a gasdynamic blowoff.

flow speed about which the distribution is centered, the T_k is the temperature characterizing width of the velocity distribution along the normal to the surface.

The adiabatic expansion following the Knudsen layer narrows the distribution somewhat, but does not appreciably affect the mean flow speed (*13*). The number density of the vapor decreases roughly as the inverse square of the distance from the target surface. The flow becomes supersonic in the sense that the flow velocity exceeds the local sound velocity. In addition, the flow velocity is greater than the mean speed of the Maxwellian distribution at the surface temperature by a factor ranging from 1.26 to 1.64 (*13*). The gasdynamic flow is "cold" because $T_k < T_s$ as a result of transformation of thermal motion to directed flow. The angular distribution of the flow is more sharply peaked about the normal than in a collisionless vapor; the distribution varies as $\cos^n\Theta$, where Θ is the angle from the surface normal an $n > 2$. At the freezing plane in Fig. 2, the gas density becomes so low that intermolecular collisions cease, the gas continues to expand in free-molecule flow, and its density continues to decrease according to the inverse square of the distance from the target.

With respect to a detector downstream of the surface, the principal effect of the modifications of the flow owing to the combined effects of the Knudsen layer and the gasdynamic expansion is to speed up the gas by 26 to 64% (*13*). The gas density sensed by the detector, however, is not greatly affected. In both free-molecule and gasdynamic expansions, molecules leave the surface at a rate given by Eq. (1). In the latter case, three effects must be considered. First, recondensation on the surface that occurs by backscattering of molecules emitted from the surface on the outer edge of the Knudsen layer reduces the vaporization flux by 18–25% Second, for a fixed evaporation rate, the higher flow speed proportionately reduces the number density to which the detector responds. Finally, the sharper-than-cosine angular distribution provides more molecules along the surface normal where the detector is usually located.

The combination of these three effects leads to very little change in the measured number density from the gasdynamic expansion compared to what would be observed for a collisionless flow (*14*). Thus, the principal observable difference between a gasdynamic expansion and a free-molecule flow vaporization process is the higher mean speed of the former. This conclusion ignores the possibility that the gasdynamic expansion may result in a highly supersaturated vapor in which nucleation and growth of clusters may occur.

High Laser Power Density-Short Pulse Times

Although the principal concern here is the nature of the laser–solid interaction during low power density pulses of millisecond lengths, a brief review of the type of interaction when the power density exceeds 10^7 to 10^8 W/cm^2 is in order.

At low power densities, the interaction has an essentially thermal character, in the sense that the material removal rate follows the vapor pressure according to Eq. (1). For higher power densities delivered in pulse times on the order of tens of nanoseconds (typical of excimer lasers) a variety of other, nonequilibrium, mechanisms assume greater importance than thermally-driven emission. These are generally of the "explosive" type involving emission of large masses of material rather than individual molecules.

One of the earliest of these mechanisms to be proposed is the recoil pressure model proposed by Ready (*15*) and others (*16*). In this process, the rapid, initially thermally-driven vaporization flux creates a back pressure on the surface in much the same way that the flow from a hose produces a force on the structure (or person) holding the hose. The back pressure created by the departing vapor is about half of the vapor pressure at the surface temperature (*17*). This recoil pressure can contain the hot liquid surface and prevent further evaporation, thereby producing a superheated liquid. As the pulse ends, the restraining vapor escapes, and rapid nucleation of vapor bubbles in the superheated liquid leads to explosive release in the form of ejected droplets.

A modification of this model involves subsurface heating of the liquid (*18,19*). If the condensed phase does not strongly absorb the incident laser light, in-depth heating rather than purely surface heating occurs. The energy deposited decreases exponentially with depth in the condensed phase with a characteristic length equal to the reciprocal of the optical absorption coefficient. The combination of in-depth energy deposition and cooling of the surface owing to the necessity of supplying the heat of vaporization to sustain the initially thermal evaporation results in a maximum temperature below rather than at the surface. The high vapor pressure of the material at the location of the temperature maximum drives nucleation and growth of vapor bubbles, thereby leading to an explosive release of both the gas and entrained liquid.

Droplet emission has been clearly demonstrated by scanning electron micrographs of the surface of laser-irradiated gold (*20,22*). The model proposed involves heating and thermal expansion of small protuberances or asperities on the surface which, because of thermal expansion, tend to be driven away from the surface. If the kinetic energy of the minute liquid hills exceeds the energy required to create additional surface area to form detached drops, the latter are ejected from the heated surface.

Emission of chunks of solid can occur in metals that do not even reach the melting point during the laser pulse (*20,21,23*). In this process, the steep temperature gradients in the solid generated thermal stresses that exceed the fracture stress. The result is creation of flakes of solid that become detached from the surface. Tungsten and alumina exhibit this type of behavior (*20*).

In addition to the athermal material removal processes, absorption of the incident laser light by the vapor or a superheated liquid at the surface can also occur at high laser power densities. The hot, high-density vapor plume produced by thermal vaporization can become ionized by a variety of mechanisms and ultimately create a full-fledged plasma that strongly absorbs incident laser light. Free electrons and ions are produced in any hot gas by the Saha equilibrium

$$A(g) = A^{+}(g) + e$$

which has a Boltzmann-factor temperature dependence involving the ionization energy of the neutral atom or molecule A. In addition, electrons and ions can be emitted directly from the condensed phase at a rate governed by Richardson's equation

$$J = BT^2 e^{-W/kT}$$

where W is the electron or ion work function of the solid or liquid and B is a constant (*24*).

Mechanisms by which laser light is absorbed by the vapor include an inverse bremsstrahlung process, whereby free electrons in the vapor interact directly with the laser photons (*25*); bound-free absorption, in which bound electrons are pumped to the ionization level by absorption of one or more photons (2); and bound-bound transitions driven by laser photon absorption in the atoms or molecules of the vapor (*26*). Ignition of plasmas in atmospheric–pressure air occurs at power densities as low as 10^6 W/cm^2 (*27*) so comparable thresholds would be expected in vapor plumes that attain similar number densities. For most systems, the laser power density at which vapor absorption of laser light begins is probably closer to 10^8 W/cm^2 (*25*). In short-pulse laser irradiation, vapor-absorption processes can so efficiently shield the surface from the laser light that the material removal rate per laser pulse actually decreases as the power density increases (*25*). However, target heating, even with strong vapor absorption, can continue by thermal conduction or radiation from the plasma to the surface. In metals, ultraviolet reradiation from the plasma can even enhance energy absorption by the surface if the laser is in the infrared where absorbance is low (*27*).

EXPERIMENTAL ASPECTS

Low Power Densities

A large number of diagnostic tools have been employed for characterizing the nature and extent of laser-evaporated plumes. These include:

1. Measurement of the depth of the depression in the target material after a number of identical laser shots. This information, along with measured surface temperature during the transient permits determination of the vapor pressure by use of Eq. (1) (*24*). Alternatively, if the vapor pressure–temperature relation is known for the target material, the crater depth serves as a measure of the temperature (*20,28*). The amount of material removed is an integral measure of the consequence of the laser–solid interaction.
2. Vapor condensation on collection disks surrounding the target. This technique is also an integral measure of the interaction. In conjunction with neutron activation analysis, it is useful for measuring the total quantity vaporized, as well as the angular distribution of the plume (*29*). The condensate on the disks can also be examined by scanning electron microscopy or replica transmission electron microscopy to determine if visible (micron size) droplets were emitted from the surface or formed by condensation in the gasdynamic expansion (*29,30*).
3. *In situ* mass spectrometry. This technique is useful in assessing whether the ratios of molecular species emitted during the pulse conform to thermochemical expectations (*14,24,29,31,32*).
4. Recoil momentum measurements by ballistic pendulum. Both the target and a collection disk can be suspended by fine wires and their deflection following a laser pulse on the target measured. This technique can be used to provide an average "temperature" and vapor pressure during the pulse (*33*).
5. Charged particle (ions, electrons) detection (*24*) or deflection (*29*). Faraday probes, deflecting magnets and grids and the mass spectrometer proper can be used to assess the contributions of ions and neutral species in the vapor.
6. Laser-induced fluorescence analysis of the vapor plume using a separate pulsed dye laser probe that traverses the plume, but does not strike the surface (*28*). This technique also requires a spectrophotometer to analyze the light emitted by the excited vapor molecules. The technique is useful for determining the rotational and vibrational temperatures of species in the vapor.
7. Time-of-flight techniques using a quadrupole mass spectrometer (*29*) or spectroscopic methods (*28*). Measurement of the time required for the vapor molecules to traverse a known distance in vacuum permits determination of their average

speed. It is particularly interesting to know whether this speed is characteristic of a Maxwellian distribution at the surface temperature, which would indicate free-molecule flow, or whether the flow has been accelerated by the gasdynamic expansion and changes occurring as the Knudsen layer is crossed (*13,29*).

8. Use of a time-of-flight mass spectrometer (not to be confused with the time-of-flight technique described in the preceding paragraph). This instrument differs from a quadrupole mass spectrometer because separation of the masses is effected by drift down a long tube. It is especially useful in analyzing very high molecular weight molecules and is probably the only instrument that can detect clusters owing to condensation in gasdynamic vaporization. It has been used for this purpose with supersonic molecular beams for some time (*3*).

A typical experimental setup using quadrupole mass spectrometric detection is shown in Fig. 3. Its features and the type of measurements it permits are discussed below.

Surface Temperature Measurement and Calculation

As shown in Fig. 3, an optical pyrometer is aimed at the same spot on the target that is irradiated by the laser beam. Either specially-made pyrometers (*34*) or commercial automatic optical pyrometers (*29*) can be used for this purpose. Direct temperature measurement by optical pyrometry is applicable only to slow transients (i.e., pulse widths of about 100 μs or greater). Accurate temperature measurement requires careful attention to a number of potential problems:

1. Appropriate preamplification of the output signal from the electron multiplier of the pyrometer is necessary to avoid signal distortion by cables leading to the recording device, which is usually an analog-to-digital recorder.
2. Means of extending the calibration of the pyrometer to temperatures well beyond those available in standard NBS-calibrated sources. Blackbody sources are available up to 3000°C; calibration to higher temperatures can be accomplished using Planck's law.
3. The emissivity of the surface (in both the solid and liquid states) must be known or estimated in order to convert the brightness temperature to the true temperature.
4. Pulse-power densities must be low enough to avoid significant emission of radiation from the hot plume at the wavelength of the optical pyrometer. Also, the plume should be optically transparent to the thermal radiation originating from the surface (*35*). This requirement effectively limits the maximum temperatures that can be accurately determined by optical pyrometer to somewhat over 4000°C for refractory targets that have relatively

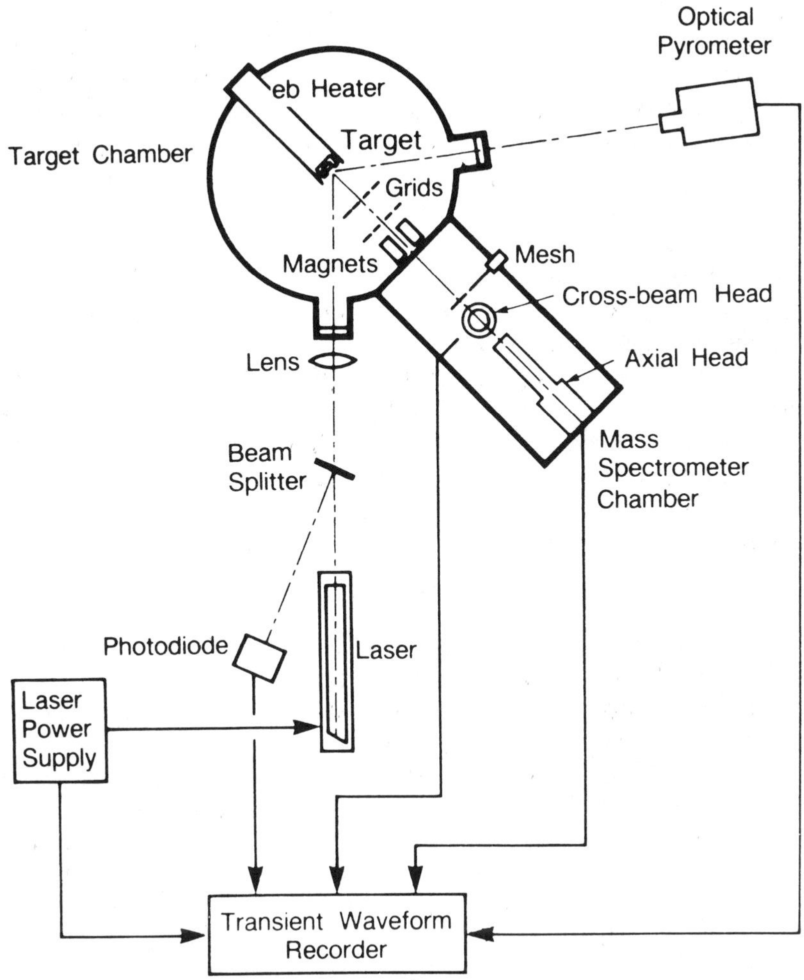

Fig. 3. Sketch of a laser-pulsing apparatus with pyrometric temperature measurement and mass spectrometric detection.

low vapor pressures (*26*). The limiting temperature would be considerably lower for aluminum, for example.

5. The radial power density distribution in the laser beam should be flat enough that the pyrometer view spot intercepts a circle that is uniformly heated. Pyrometer view spots are on the order of a millimeter so that the temperature of targets irradiated by highly focused laser beams cannot be measured. Conventional laser beams (e.g., from a Nd-glass laser operated in the conventional mode) produce a pulse that is uniform, over about 7–9 mm.

One method of assessing the accuracy of the surface temperature measurement is to compute the transient from a first-principles, no-adjustable-parameter heat transfer calculation. Such a calculation is tractable if the experiment satisfies a few simple restraints.

First, the temperature should depend only on time and depth beneath the surface, but not on radial position in the solid. The depth of the laser-heated zone is of the order of $(\kappa t_p)^{1/2}$, where κ is the thermal diffusivity of the condensed phase and t_p is the width of the pulse. For millisecond-scale pulses in insulators, the thermal penetration depth is of the order of a few tens of microns. If the radius of the uniformly-irradiated spot is much larger than this depth, radial heat conduction can be neglected in the analysis. In addition, a large, uniformly-illuminated spot tends to suppress radial movement of the liquid (*23,35*). This permits the radial convection term in the energy equation to be neglected and justifies the assumption of a flat surface.

Second, if the thermal penetration depth is significantly larger than the optical absorption length (as it usually is for long pulses), the laser can be considered to be a surface rather than a volume heat source.

Third, the reflectivity (R) of the surface at the wavelength of the laser light must be known in order to convert the specifiable incident laser power density on the surface, q_{inc}, to the absorbed power density,

$$q_{abs} = (1 - R)q_{inc} \tag{5}$$

which drives the heat transport process. The reflectivity of metals and ceramics generally decrease with increasing temperature (*37,38*).

Fourth, the temporal shape of the laser pulse must be known. The average power density trace is measured by a photodiode. However, conventional-mode lasers actually produce a series of very narrow triangular pulses about one-half microsecond wide and 1.5 μs apart. This heat input results in a surface temperature that oscillates about the value that a smooth pulse of the same power density as the average of the spikes would produce (*39*). The amplitude of the oscillations depends on the physical properties chosen for the calculation, but appears to be from 50 to 100 K for most materials; in effect, the surface has a considerable amount of thermal inertia, and unless a special effort is made, neither the pyrometer nor the mass spectrometer responds to these oscillations.

When strong vaporization and high temperatures are involved, the nonlinearities inherent in the heating process preclude application of standard analytical solutions of the heat conduction equation. Only in the case of surface melting or annealing are such solutions sufficiently accurate. Nonlinear behavior of the governing equations appears in temperature-dependent thermal properties and in the necessity of accounting for the heat effect of vaporization at the surface and melting at the liquid–solid interface.

With the above restrictions, the energy equation in the condensed phase is given by (*6*):

$$\rho \overline{C}_p \left(\frac{\partial T}{\partial t} - u \frac{\partial T}{\partial z} \right) = \frac{\partial}{\partial z} \left(\kappa \frac{\partial T}{\partial z} \right) \tag{6}$$

where ρ is the density of the solid or liquid at the surface, κ is the thermal conductivity and $\overline{C}_p$ is an effective heat capacity. The latter provides a reasonably accurate method for accommodating the heat effect of fusion without requiring the more detailed treatment of conduction in a moving two-phase system (*40*). The function

$$\overline{C}_p = C_p + \frac{\Delta H_F}{\sqrt{\pi}\,\Delta T_F}\exp\left(-\left(\frac{T - T_F}{\Delta T_F}\right)^2\right) \tag{7}$$

spreads the melting discontinuity over a small, but nonzero temperature range about the melting point T_F, while correctly accounting for the enthalpy of fusion, ΔH_F. The true heat capacity of the condensed phase is C_p. The temperature range over which the heat of fusion is spread, ΔT_F, is chosen so that the partially-melted zone occupies a small spatial region compared with the depth of penetration of the thermal disturbance $(\kappa t_P)^{1/2}$, but large enough to avoid instabilities in the numerical solution of Eq. (6). This method is acceptable for the present application because the fraction of the laser input energy required to melt the upper layer of the solid is small. The accuracy of this approximate method of treating melting has been tested by comparing the results with standard solutions to problems of conduction with phase change (*41*).

The temperature dependences of the thermal properties should be considered for large surface temperature changes. The second term in the parentheses on the left hand side of Eq. (6) represents recession of the surface owing to evaporation, given by Eq. (3) (if only one evaporating species is considered). The backscattering factor in a gasdynamic blowoff can be included in this equation.

The surface boundary condition for Eq. (6) is given by the energy balance

$$q_{abs} = -\kappa\left(\frac{\partial T}{\partial z}\right)_0 + F\Delta H_V + \varepsilon\sigma(T_S^4 - T_{amb}^4) \tag{8}$$

where the absorbed power density is given by Eq. (5) and the vaporization flux by Eq. (1). ΔH_v is the heat of vaporization, ϵ is the total hemispherical emissivity of the surface, σ is the Stephan-Boltzmann constant, and T_{amb} is the temperature of the surroundings of the system (300 K). The surface temperature is denoted by T_S. The kinetic energy carried off by the departing vapor molecules has been neglected in Eq. (8). However, their momentum is the source of the recoil pressure discussed earlier.

The remaining conditions for Eq. (6) are the initial condition

$$T(z,0) = T_o \tag{9}$$

where T_o is the initial temperature of the solid. The condition representing a semi-infinite medium is

$$T(\infty,t) = T_o \tag{10}$$

The above set of equations can be solved numerically by standard partial differential equation solvers in commercial software libraries. The principal output is the time variation of the surface temperature, although subsurface temperatures and the location of the melting front can be extracted if desired.

Figure 4 shows the measured and calculated temperature histories following a laser pulse with a peak power density of 10^4 W/cm^2 on uranium dioxide at an initial temperature of 1600 K. The comparison is based on the temperature rise normalized to the temperature rise at the peak temperature in order to illustrate two aspects of the results. First, the agreement between the measured and predicted times of the maximum surface temperature is good. The temperature reaches a maximum at 354 μs following onset of laser irradiation, whereas the laser power density peaks at 320 μs. Second, the shape of the measured transient is in good accord with the predicted shape. The principal uncertainty in comparing the experimental and theoretical absolute values of the maximum surface temperature T_{smax} is in assignment of the surface reflectivity. Fortunately, this quantity has been measured for UO_2 (*38*). For pulses that produced measured maximum surface temperatures from 2380 to 3670 K, the calculated values average about 130 K lower than the observed values (*35*).

The generally good agreement between the observed and computed temperature transients for laser pulsing of uranium dioxide suggests that measuring surface temperatures by optical pyrometry is reasonably accurate and can be generally applied for millisecond-scale laser pulses on most materials, at least those for which the high-temperature reflectivity is either known or so close to unity that errors in this parameter do not significantly affect the measurement. The limit imposed by the onset of plasma emission or shielding of the surface must also be respected.

A potentially serious temperature-measurement problem is emission from the hot vapor of light at the wavelength accepted by the optical pyrometer. This was investigated by measuring the apparent temperatures with the pyrometer viewing direction parallel to the hot surface (*35*). The results indicated that the energy emitted from the vapor in the pyrometer bandpass is two or three orders of magnitude lower than emitted from the hot surface.

Species Detection by Mass Spectrometry

One of the principal applications of mass spectrometric detection of the vapor emitted from a laser-heated target is measurement of the thermochemical properties of refractory substances at temperatures beyond the reach of conventional methods. Both the total vapor pressure and the various molecular species in the vapor are of particular interest. In addition, mass spectrometers can be used as time-of-flight detectors to determine the energies of the species emitted from the surface.

Use of a mass spectrometer for this purpose requires extensive calibrations of instrumental sensitivity to the molecules detected and

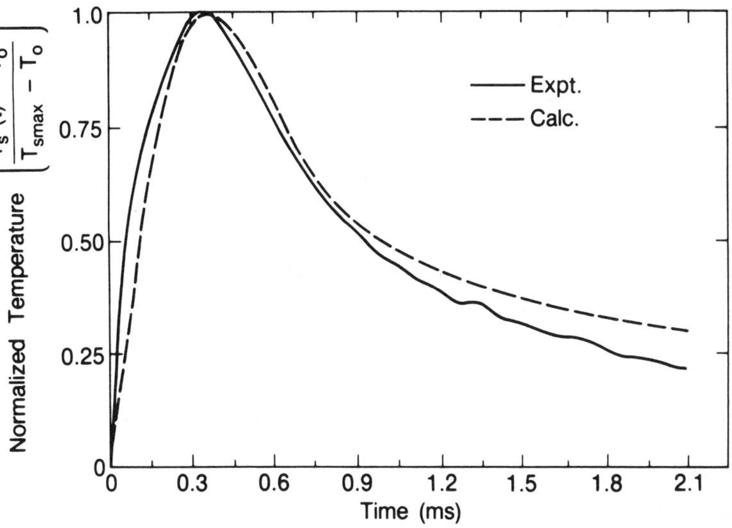

Fig. 4. Measured and calculated normalized surface temperature rises for a laser pulse with a peak power density of 10^4 W/cm^2 incident on a UO_2 surface of initial temperature 1600 K. The peak surface temperature measured in this experiment was 2400 K. (after Ref. 35)

estimation of their fragmentation patterns. These problems are the same as those that have to be dealt with in conventional Knudsen-cell mass spectrometric vaporization studies.

A convenient method of comparing theory and experiment is by calculation and measurement of the number density of particular species in the ionizer. Since mass spectrometers are density-sensitive devices, the instantaneous output signal, V(t), is proportional to the instantaneous number density in the ionizer, n(t), or

$$(n(t))_{\text{exp}} = KV(t) \tag{11}$$

where K is an instrumental constant that depends on the geometries of the flight path between the target and the mass spectrometer, the ionizing electron current, the fragmentation pattern, various mass spectrometer efficiency factors, and the gain of the electron multiplier. Calibration can be performed by steady-state vaporization of the material or by using laser pulses of sufficiently low energy density that the maximum temperature of the surface is in the range in which the vapor pressure is known.

As explained earlier, the number density in the vapor downstream of the target is not greatly dependent on the extent of intermolecular collisions in the blowoff (provided that only thermally-driven evaporation occurs). For free-molecule vaporization, the number density in the vapor at a distance, L, from the target is given by a convolution of the vapor pressure (driven by the surface temperature transient $T_s(t)$) and the Maxwellian speed distribution of the emitted molecules (*42*)

$$n(t)=\frac{A_s L}{k}\left(\frac{m}{2\pi k}\right)^{3/2}\int_0^t \frac{P\{T_S(\tau)\}}{\{T_S(\tau)\}^{5/2}(t-\tau)^4}\exp\left(-\frac{mL^2}{2kT_S(\tau)(t-\tau)^2}\right)d\tau \qquad (12)$$

Here A_s is the area of the irradiated spot on the target and $P(T_s)$ is the vapor pressure of the material at the temperature of the surface (only one component is assumed). Equation (12) is a theoretical time-of-arrival curve, which can be computed if the vapor pressure is known. It is strictly speaking valid only for free-molecule flow, and indeed it does not correctly predict the arrival time in a gasdynamic blowoff. However, if only the quantity of material removed is of interest, the maximum value of n (or the integral of n over the entire pulse) is nearly equal whether in the free-molecule or the gasdynamic evaporation regime. In the latter case, however, the right-hand side of Eq. (12) must be reduced by the backscattering factor discussed earlier.

RESULTS

Uranium Dioxide

Figure 5 shows an experimental trace of the UO_2^+ signal [Eq. (11)] for a relatively weak laser pulse on solid UO_2 (wavy line) and the predicted time-of-arrival curve from Eq. (12) (smooth curve). The maximum surface temperature attained in this pulse was 2400 K, where the vapor pressure is too low to initiate a gasdynamic blowoff. The theoretical and experimental curves in Fig. 5 have been normalized at their peak values (the normalization factor is the temperature-independent calibration constant for this species). The excellent fit of the two curves, particularly the time at which the maximum signal occurs and the overall shape, demonstrates that the evaporation mechanism is classical free-molecule flow with the emission rate given by the Hertz-Langmuir equation.

Figure 6 shows a similar comparison for a laser shot that produced a peak surface temperature just below the melting point of UO_2. The maximum temperature for this pulse is greater than the critical temperature at which the flow regime is calculated to switch from free-molecular to gasdynamic (which is about 2600 K). The experimental trace shows distinctly nonMaxwellian behavior. Two peaks are observed. One (labeled "fast" in Fig. 6) arrives earlier than expected for free-molecule vaporization and may be owing to gasdynamic acceleration of the flow. The larger peak is slower than expected from free molecule vaporization. Neither peak contains any ions, as was verified by turning off the electron current in the ionizer (which caused all signals to disappear) and also by placing magnetic and electric deflection plates around the vapor plume between the target and mass spectrometer (as shown in Fig. 3), which had no effect on the signal.

In addition to the clear departure of the shape of the time-of-arrival curve from that expected for equilibrium collisionless vaporization, the

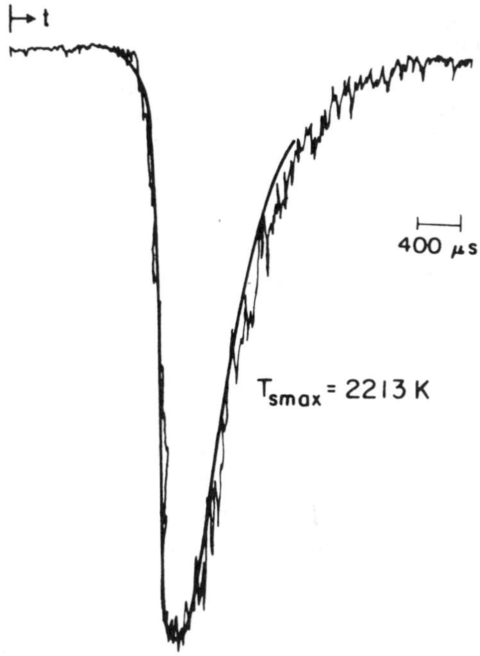

Fig. 5. Mass spectrometer signals (wavy trace) for UO_2^+ following a laser pulse that resulted in a maximum surface temperature of 2213 K. The smooth curve is calculated from Eq. (12). The laser was fired at the time represented by the vertical line in the upper left-hand corner of the diagram. (after Ref. 29)

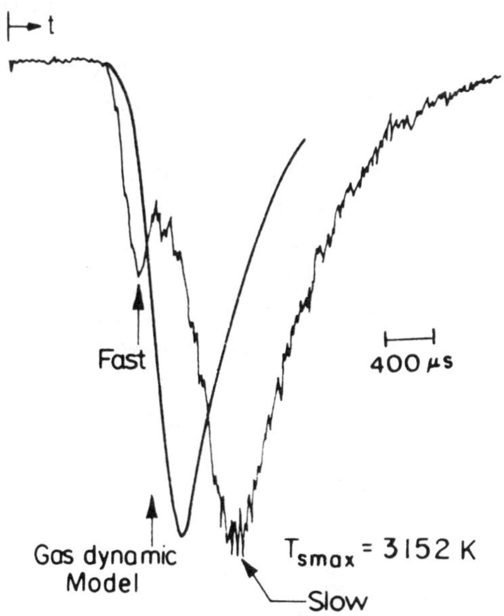

Fig. 6. Experimental signals and free-molecule flow prediction similar to those shown in Fig. 5, but for a pulse giving a maximum surface temperature of 3152 K. (after Ref. 29)

quantity of vapor emitted, as measured by the maximum signal, is a factor of nearly 100 smaller than predicted by Eq. (12). This phenomenon is shown in Fig. 7, wherein the maximum signal is plotted as a function of various measures of the strength of the laser pulse. Up to peak surface temperature of about 2400 K, the measurement conforms to equilibrium vacuum vaporization. The slope of a line through the data points in this regime is unity and the level of the signal gives the calibration factor. At higher temperatures, however, the signal levels off and practically ceases to increase with laser pulse energy. It should have continued to increase along the extrapolated line of the low-temperature results.

A number of possible experimental artifacts were examined and discarded as explanations of this discrepancy (*29*). This behavior is also observed with ZrO_2 but not chromium. The loss of signal at high surface temperatures does not mean that the quantity vaporized is smaller than expected theroretically; it simply may mean that the vapor is not principally UO_2. In fact, neutron activation analysis of the deposit on collector plates near the target was consistent with vaporization according to Eq. (1).

The best explanation of the phenomenon seen in Fig. 7 is that UO_2 nucleates and grows clusters during the gasdynamic expansion. By virtue of their larger mass, the clusters are not detected as UO_2^+ but as some higher parent ions or fragments. This explanation cannot be tested for UO_2 because even the dimer $(UO_2)_2$ is beyond the mass range of the instrument. However, laser pulsing of MgO at temperatures above 1700 K showed the presence of masses corresponding to $(MgO)_n$, with $n = 2,3,6$, and 7. The "slow" peak in Fig. 6 is believed to be owing to fragmentation of slowly-moving $(UO_2)_2$ clusters to produce small quantities of UO_2^+.

A possible reason that metals, such as chromium, do not show the anomalous leveling-off seen in Fig. 6, whereas oxides such as UO_2 and ZrO_2 do, is the difficulty of nucleating clusters from atomic monomers. Small clusters formed from single-particle monomers have fewer degrees of vibrational and rotational freedom to absorb the binding energy of cluster formation compared to clusters formed from molecular monomer units. Gaponov et. al. (*30*) offer a different explanation. The reflectivity of metals for the 1.06 μm radiation from a Nd laser can be as large as 70–90%, whereas ceramic oxides have much lower reflectivies (for UO_2 the value is approximately 15%). The laser light reflected by the metal surface can be absorbed by clusters and cause them to revaporize. Even for low-reflectivity targets, a similar effect can be obtained by reducing the angle between the laser axis and the surface normal, which will cause the laser light to intercept more of the expanding plume than if it impinged on the target at an oblique angle. Gaponov et. al. (*30*) proposed a criterion for condensation in the expanding gasdynamic plume to occur; the product of the vapor density corresponding to the vapor pressure at the surface temperature and the radius of the irradiated spot should be greater than 10^{17}–10^{18} cm^{-2}. This criterion does not involve the pulse duration, but

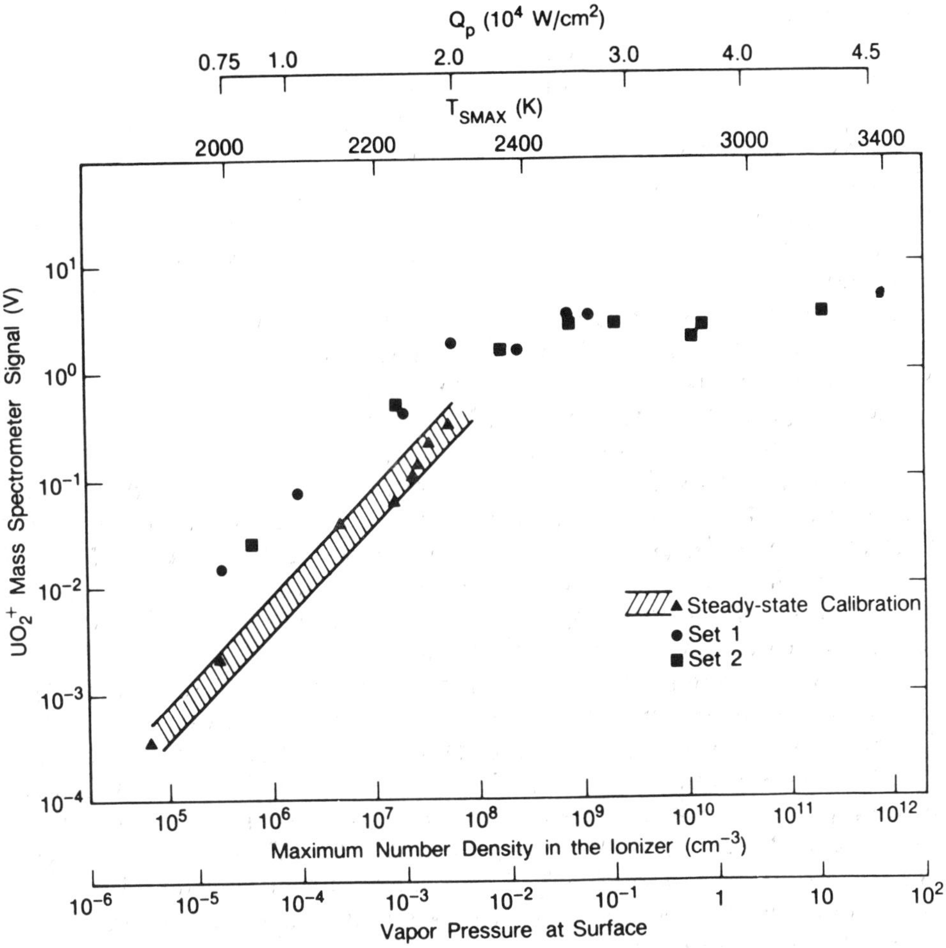

Fig. 7. Maximum UO_2^+ signals from the mass spectrometer for laser pulses of varying strength. Q_p is the peak absorbed power density, and T_{Smax} is the measured maximum surface temperature in the pulse. The scale designating the maximum number density in the ionizer of the mass spectrometer was calculated from Eq. (12) and the vapor pressure is of UO_2 at the peak surface temperature. The hatched area represents the range of results of the steady-state calibrations. (after Ref. 29)

applies to millisecond-length pulses. When this criterion was met for normal-mode laser evaporation of bismuth, submicron-size spherical particles were observed on collector plates a few centimeters downstream of the target. A detailed discussion of condensation in laser-driven gasdynamic flows is given by Anisimov et al. (*43*).

Condensation in a gasdynamic plume is particularly bothersome because it occurs when the surface is at high temperatures where the thermochemical information is needed. There are several options for dealing with this problem. First, nothing may need to be done if only ratios of vapor species concentrations is desired rather than absolute values of the equilibrium partial pressures (or vaporization rates). There is evidence from laser-pulsing of UO_2 and UC (*14,29*) that the ratios of the vapor species do not change drastically as the blowoff changes from free-molecule to gasdynamic in nature. The high temperature equilibium thermochemistry at the surface conditions may be preserved despite clustering and cooling during the expansion. Figure 8 compares the predicted and measured ratios UC_2/U and C_3/C as functions of the maximum surface temperature reached in a series of shots of different laser energy density. Fragmentation patterns of the neutral species for 70-eV ionizing electrons were determined separately by steady vaporization tests under conditions of known partial pressures. During the laser pulsing experiments at much higher temperatures, the ratios of the uranium-bearing species were found to agree within an order of magnitude with a thermochemical model based on low temperature data. However, no C_2 was detected although this species was predicted thermochemically to be present at partial pressures comparable to those of C_3. The work of Hastie and Bonnell, on the other hand, suggests that the relative amounts of the C_n species emitted from laser-pulsed graphite is in accord with equilibrium expectations (*32*).

Second, it may be possible by experimental means to prevent clustering during the expansion. Two such methods have been suggested by Gaponov et at. (*30*), namely using small irradiated spots and tilting the incident laser beam as close to the normal as possible. Third, it may also be possible to evaporate clusters using a second laser crossing the expanding plume, as has been demonstrated for SF_6 clusters in a supersonic jet (*44*). Even more simply, illuminating the vapor plume by a laser flash lamp coiled around the vapor plume may deliver enough energy to dissipate the condensate.

In any case, it is clear that more work is needed to elucidate the nature of formation of clusters by condensation of vapor in the gasdynamic expansion of a plume formed by laser irradiation of solid targets. The distribution of cluster sizes in the plume is information of prime importance. This should be accessible by using the reflectron time-of-flight mass spectrometer (*45*), which provides an accessible mass range of tens of thousands of amu.

LASER VAPORIZATION OF HIGH-T_c SUPERCONDUCTING CERAMICS

Long-pulse vaporization of targets of the superconducting ceramic $YBa_2Cu_3O_7$ exhibits several of the characteristics discussed above for UO_2. However, because of the greater number of elements involved, the vapor is more complex. With the multicomponent oxide, a substantial number of vapor species, including ternary oxides containing two of the metal constituents of the target, were detected. Figure 9 shows the

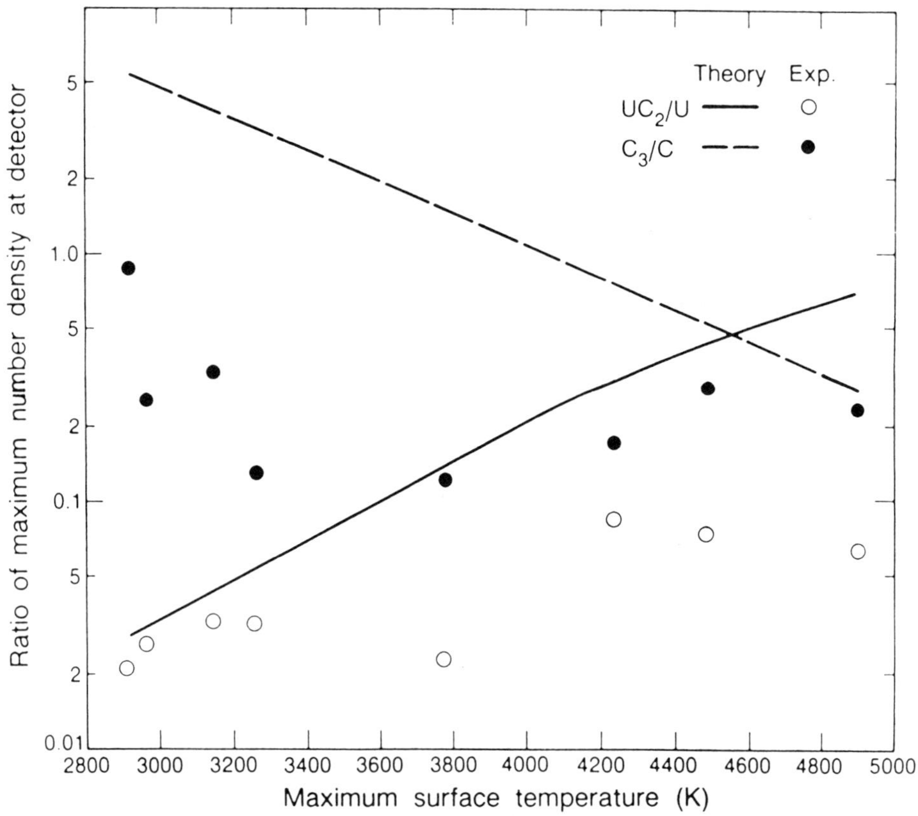

Fig. 8. Measured and theoretical ratios of uranium and carbon species owing to laser-pulsing of uranium carbide. The maximum surface temperature is measured. (after Ref. 14)

surface temperature transient caused by a 45-J laser pulse of 0.5 ms width and the time-of-arrival curves for three species produced by evaporation. The peak power density was $\sim 10^6$ W/cm^2, which resulted in a maximum surface temperature of at least 4000 K (the emissivity is not known, so blackbody behavior was assumed).

At any time, vaporization is highly incongruent, with the initial vapor being copper-rich and subsequent emission containing the bulk of the yttrium and barium. Despite this temporal incongruency, films deposited from this vapor were stoichiometric to within 5% because the temporal integrals of the time-of-arrival curves contain the metallic elements in nearly the same proportions as in the target material (*46*). The oxygen content of the as-deposited films, however, is below stoichiometry. This is because the molecular oxygen evolved from the target has a low sticking probability on the substrate or the film growing thereon. However, subsequent annealing of the films in oxygen gas produced

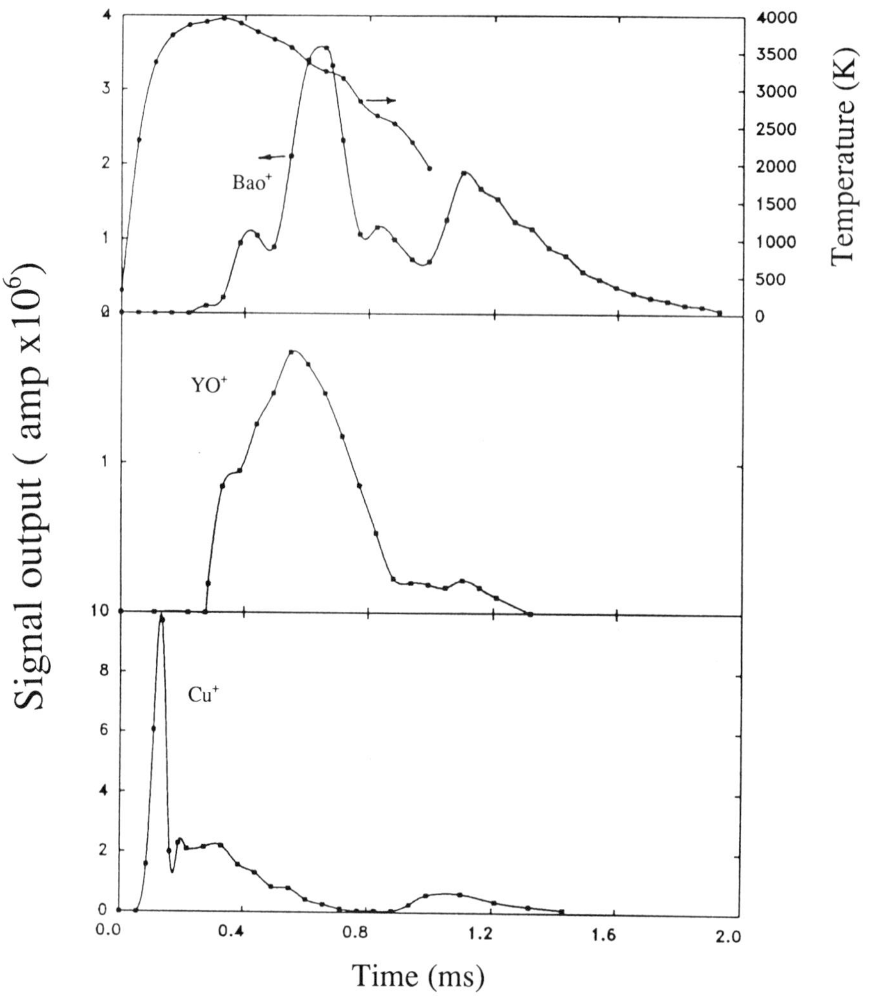

Fig. 9. Time-of-arrival curves for copper and the monoxides of yttrium in barium following laser pulsing of the yttrium–barium–copper superconducting oxide.

films with transition from normal to superconductivity between 78 and 62 K.

Figure 10 shows a scanning electron micrograph of the surface of the evaporated film. The large spherical object embedded in the film is probably a large cluster nucleated and grown from the supersaturated vapor during transit from the target to the substrate on which the film was deposited.

CONCLUSIONS

Laser irradiation with millisecond-width pulses is a useful technique for studying the thermochemistry of refractory solids at very high temperatures. For power densities less than about 10^6 W/cm^2, the vaporiza-

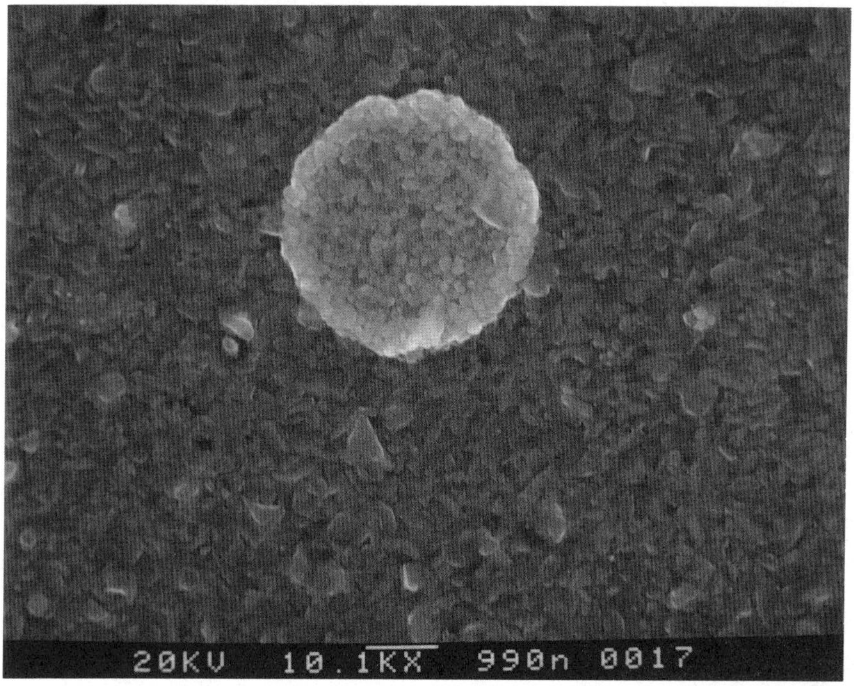

Fig. 10. Scanning electron micrograph of a large cluster probably produced by condensation in the expanding plume from a laser-irradiated target of superconducting oxide. (after Ref. 46)

tion rate is closely represented by the Hertz-Langmuir formula. Within this regime, two types of vapor blowoff have been identified, namely free-molecule and gasdynamic expansions. The latter regime may be accompanied by nucleation and growth of weakly-bound clusters, but the relative concentrations of the various molecular species in the vapor appears to be characteristic of thermochemical equilibrium at the temperature and composition of the emitting surface.

In contrast to short (nanosecond scale) laser pulses, the surface temperature transients created by long laser pulses can be measured with reasonable accuracy by optical pyrometry, which is essential for thermochemical investigations. Plasma formation downstream of the surface does not occur in long-pulse vaporization because of the relatively low power densities involved.

The main unresolved questions connected with thermally-driven vaporization by long laser pulses are the extent of ionization of the vapor and the conditions under which cluster formation occurs. Recently-developed mass spectrometers and novel experimental techniques should lead to a better understanding of clustering and to development of methods of avoiding this phenomenon.

Pulsed-laser vaporization is also a promising method for producing thin films of refractory materials for use in integrated circuits and high-T_c ceramic superconductors. Recent work in this field has concentrated on short-time, high-power density laser pulsing, but the use of long-pulse laser processing may have some advantages. Chief among these are avoidance of chunk or droplet ejection from the irradiated spot and elimination of high-energy ion emission, both of which can lead to less-than-optimum characteristics of the deposited films. The large quantity of material vaporized in each pulse results in films that approach the metal composition of the target material, but postirradiation annealing in oxygen is required to restore the proper overall stoichiometry of the deposit. Finally, a technique must be found to reduce condensation in the plume that results in micron-size spheres of the target material embedded in the deposited film.

ACKNOWLEDGMENT

This work was supported by the Director of the Office of Energy Research of the Office of Basic Energy Science, Material Science Division of the US Department of Energy under contract number DE-AC03-76sf00098.

REFERENCES

1. Proceedings of the *American Society for Metals*, Metzbower, E. A., ed., Washington, DC, (1979).
2. von Allmen, M., *Laser Beam Interactions with Materials*, Springer-Verlag, (1986).
3. Woste, L., *Contribution of Cluster Physics to Materials Science and Technology*, Davenas, J. and Rabette P. M., eds., pp. 1–50.
4. Inam, A., Wu, X. D., Venkatesan, T., Ogale, S. B., Chang, C. C., and Dijkkamp, D., *Appl. Phys. Lett.* **51,** 1112 (1987).
5. Rosenblatt, G. M., *Accounts of Chem. Research* **9,** 169 (1976).
6. Tsai, C. H. and Olander, D. R., *J. Nucl. Mater.* **137,** 279 (1986).
7. Breitung, W., *Kernforschungszentrum Karlsruhe Report No. KfK-2240* (1976).
8. Murakami, M. and Oshima, K. K., *Rarefied Gas Dynamics*, Becker, M. and Fiebig, M., eds., (DFVLR, Porz-Wahn, W. Germany, 1974).
9. Tsai, C. H. and Olander, D. R., *Phys. Fluids* **30,** 386 (1987).
10. Anderson, J. B., *Molecular Beams and Low-Density Gasdynamics*, chapter 1, Wegener, P. P., ed., Marcel Dekker, New York (1974).
11. Anisimov, S. E., *Sov. Phys. JETP* **27,** 182 (1968).
12. Cerciganani, C., *Rarefied Gas Dynamics*, Fisher, S., ed., AIAA, New York, 9181, p. 305.
13. Kelly, R. and Dreyfus R. W., *Surf. Sci.* **198,** 263 (1988).
14. Tehranian, F. and Olander, D. R., *High Temp. Sci.* **25,** 55 (1988).
15. Ready, J. F., *J. Appl. Phys.* **36,** 462 (1965).
16. Gagliano F. P., and Paek, U. C., *Appl. Opt.* **13,** 274 (1974).
17. Mazhukin, V. I. and Samokhin, A. A., *Sov. J. Quant. Electron.* **14,** 1608 (1984).
18. Dabby, F. W. and Paek, U. C., *IEEE J. of Quant. Electron.* **QE-8,** 106 (1972).
19. Kozlov, B. M., Krynetskii, B. B., and Samokhin, A. A., *Sov. J. Quant. Electron.* **4,** 1303 (1975).

20. Kelly, R., Cuomo, J. J., Leary, P. A., Rothenberg, J. E., Braren, B. E., and Aliotta, C. F., *Nucl. Instr. and Meth.* **B9,** 329 (1985).
21. Rothenberg, J. E. and Kelly, R., *Nucl. Instr. and Meth.* **B1,** 291 (1984).
22. Kelly, R. and Rothenberg, J. E., *Nucl. Instr. and Meth.* **B7/8,** 755 (1985).
23. Magill, J., Ronchi, C., Babelot, J.-F., Long, K. A., and Ohse, R. W., *High Temp.-High Press.* **12,** 503 (1980).
24. Ohse, R. W., Babelot, J-F., Cercignani, C., Hiernaut, J.-P., Hoch, M., Hyland, G. J., and Magill, J., *J. Nucl. Mater.* **130,** 165 (1985).
25. Ready, J. F., *Industrial Applications of Lasers,* chapter 13, Academic, 1978.
26. Bober, M. and Karow, H. U., *7th Symp. on Thermophysical Properties,* NBS, Washington, DC (1977)
27. McKay, J. A. and Schriempf, J. T., Washington, DC, p. 55.
28. Dreyfus, R. W., Kelly, R., and Walkup, R. E., *Nucl. Instr. and Meth.* **B23,** 557 (1987).
29. Olander, D. R., Yagnik, S. K., and Tsai, C. H., *J. Appl. Phys.* **64,** 2680 (1988).
30. Gaponov, S. V., Gudkov, A. A., and Fraerman, A. A., *Sov. Phys. Tech. Phys.* **27,** 1130 (1983).
31. Olstad, R. A. and Olander, D. R., *J. Appl. Phys.* **46,** 1509 (1975).
32. Hastie, J. W., Bonnell, D. W., and Schenk, P. K., *High Temp.-High Press.* **20,** 73 (1988).
33. Bober, M., Karow, H. U., and Schretzmann, K., *Nucl. Technol.* **26,** 237 (1975).
34. Ohse, R. W., Babelot, J.-F., Frezzotti, A., Long, A., and Magill, J., *High Temp-High Press.* **12,** 537 (1980).
35. Yagnik, S. K. and Olander, D. R., *J. Nucl. Mater.* **154,** 253 (1988).
36. Zhiryakov, B. M., Korotchenko, A. I., Popov, N. I., and Samokhin, A. A., *Sov. J. Quant. Electron.* **13,** 763 (1983).
37. Park, K. and Walter, W. T., Washington, DC, p. 21.
38. Karow, H. U. and Bober, M., *Proceedings of Symp. on the Thermodynamics of Nuclear Materials,* vol. 1, Julich, IAEA, 1979, p. 155.
39. Olstad, R. A., *PhD Thesis,* LBL-1177 (1972).
40. Hassanein, A. M., Kulcinski, G. L., and Wolfer, W. G., *J. Nucl. Mater.* **103/104,** 321 (1981).
41. Olander, D. R. and Kim, Y. S., Internat. J. Heat and Mass Transfer, submitted.
42. Olstad, R. A. and Olander, D. R., *J. Appl. Phys.* **46,** 1499 (1975).
43. Anisimov, S. I., Imas, Y. A., Romanov, G. S., and Khodyko, Y. U., *Effect of High-Power Radiation on Metals,* National Tech. Info. Service, Springfield, VA (1970).
44. Rechsteiner, R., Monot, R., Woste, L., Zellweger, J. M., and van den Bergh, H., *Helv. Phys. Acta* **54,** 282 (1981).
45. Mamyrin, B. A., Karataev, V. I., Shmikk, D. V., and Zagulin, V. A., *Sov. Phys. JETP* **37,** 45 (1973).
46. M. Balooch, D. R. Olander and R. Russo, *Appl. Phys. Lett.* **55,** 197 (1989).

Direct Laser–Materials Interaction

Laser Ablation of Superconductor Materials and Laser Welding

D. A. CREMERS,* R. D. DIXON,[1] R. C. ESTLER,[2] G. K. LEWIS,[1] J. L. LYMAN,[1] R. E. MUENCHAUSEN,[1] N. S. NOGAR,[1] AND M. S. PILTCH[1]

[1]Los Alamos National Laboratory, Los Alamos, NM 87545; and [2]Department of Chemistry, Fort Lewis College, Durango, CO 81301

ABSTRACT

This paper considers two applications of direct laser–materials interaction: laser ablation of superconductor materials and Nd–YAG laser welding. The laser ablation experiments with 355 nm, 10 ns optical pulses have demonstrated uniform, thin superconductor films. The film uniformity was improved significantly by rotating both the laser target and film support and by vertically dithering the laser focus. The film stoichiometry does vary with laser fluence, but it is within 10% of the target material at high fluence. Annealing of the thallium-containing films restores lost oxygen. For the welding application, we have investigated the conditions that influence the melt depth. High-speed photography has shown the plume to consist of a series of regenerated pulses that move in a direction normal to the surface. The delay time for plume initiation and the laser power threshold for plume formation are dependent on the substrate material. The plume temperatures depended on the welding atmosphere. Temperatures and melt depths are higher for the reactive gases, air, O_2 and SF_6, than for He, Ar, and N_2.

Index Entries: Laser ablation; superconductor; thallium; perovskite; laser welding; aluminum; stainless steel.

*Author to whom all correspondence and reprint orders should be addressed.

INTRODUCTION

Processes as diverse as welding and the production of superconductor films can take advantage of the unique properties of laser radiation. This paper reports the results of research in each of these two areas: (1) the production of thallium-containing, high-temperature superconducting thin films by laser ablation and (2) laser welding. The first application requires short laser pulses (~10 ns) and the second much longer pulses (~8 ms). Consequently, the conditions under which the irradiated material absorbs the laser radiation are quite different. For the ablation process, the local pressure during and shortly after the laser pulse may greatly exceed ambient. This results in the near explosive removal of the material in the irradiated region. The welding laser, on the other hand, heats the material slowly enough that acoustical processes keep the pressure in the irradiated region close to ambient. The welding laser does form an evaporative plume, but it results from slow evaporation, not an explosive expansion.

Laser welding is a well-developed technology. The objective of the research at Los Alamos is to develop sufficient understanding of the laser interaction process to bring more versatility to the technique. Laser ablation of superconductor materials is only a recent development. Again, our goal is to increase understanding of the process. We wish to find conditions where the stoichiometry is preserved and the deposited films are of high quality.

LASER ABLATION OF SUPERCONDUCTOR MATERIALS

Applications of the new metal–oxide based perovskite high-temperature superconductors (HTSC) are dependent, in many instances, upon the ability to fabricate superconducting thin films. A number of approaches, including sputtering, coevaporation and ion-assisted deposition, employing both elemental and compound sources, have been used to produce high-quality HTSC thin films (1). One difficulty with elemental sources is that the relative evaporation rates for each source must be kept constant throughout the deposition by independently varying the temperatures and/or sputter rates. On the other hand, compound sources often suffer from the fact that the various metals and their oxides have significantly different vapor pressures and ion and electron sputter yields. The final composition of the film will also depend on process variables, such as background pressure, deposition rate, and so on.

Laser deposition can reduce or circumvent a number of these difficulties (2). A single source may be used, obviating the need for control of multiple deposition rates. In addition, the rapid energy deposition possible with laser sources results in temperature jumps so rapid ($\geqslant 10^{11}$ K/s) that congruent evaporation can occur without significant elemental fractionation. In addition, laser evaporation can be used under a variety of processing conditions: deposition pressure may vary from UHV to several torr of inert or reactive gas; deposition temperatures are restricted only by the physical properties of the target and substrate materials used.

A number of groups have demonstrated deposition of metal-oxide thin films (*3*) using Nd:YAG (*4*), excimer (*5–7*), and CO_2 lasers (*8*). The majority of this work has been done with YBaCuO systems (*4–7*) though more recently some work has been done on Bi–Sr–Ca–Cu oxide systems (*8*).

We have recently demonstrated (*9*) the production of superconducting $Y_1Ba_2Cu_3O_{7-x}$ films, using the Q-switched, third-harmonic (355 nm) output from a Nd:YAG laser. Third harmonic radiation was used, in preference to 1064 or 532 nm, because of the large absorption coefficient, shallow penetration depth, and relative immunity from both slow thermal processes and particulate formation in the film at this wavelength (*3*). More recently, we have turned our attention to the deposition of thallium-containing thin films, because of the more desirable chemical, physical, and electrical properties of these superconductors (*10*). We will describe recent results on deposition conditions, plume diagnostics, and thin film properties resulting from deposition at 355 nm from both 2122 and 2223 thallium–calcium–barium–copper oxide targets.

EXPERIMENTAL

Films were deposited in a 70 mm id stainless-steel conflat six-cross, with an additional port on one diagonal for optical access, as shown in Fig. 1. Pressures during deposition were typically maintained at a few μtorr (1.3×10^{-4} Pa) by an oil-free mechanical pump. Bulk samples ~25 mm in diameter were mounted on a vacuum feedthrough and rotated at 0.5 rpm. In order to assure even deposition, the substrate, located ~3 cm from the target, was also rotated at a comparable rate in the opposite direction. The substrate can be heated by a quartz lamp mounted on a vertical port, up to a temperature of 773 K. This also heats the target to a maximum temperature of 623 K. The deposition temperature used in these experiments was typically between 298 and 523 K to avoid excessive loss of thallium from the film or target.

Laser pulses of 15–25 mJ, at 355 nm, were generated by frequency tripling the Q-switched output (~10 ns) from a Nd:YAG laser, equipped with filled-beam optics. Mild focusing produced a fluence at the sample surface of 1–15 J/cm^2, and a confocal parameter much larger than the etch depth in these experiments. The laser was operated at 10 Hz, and deposition times were from 15–90 min. The final turning prism was dithered vertically, at 0.1 Hz, in order to etch the target surface evenly.

As-deposited films were insulating and required a postdeposition anneal at 850 K for 15 min in a Tl-enriched crucible containing the bulk starting material to compensate for Tl loss from the film. Subsequent annealing at 750 K in an O_2 atmosphere for 2–8 h gave shiny, black films with room temperature resistances between 100–500 Ω, and T_c near 90 K.

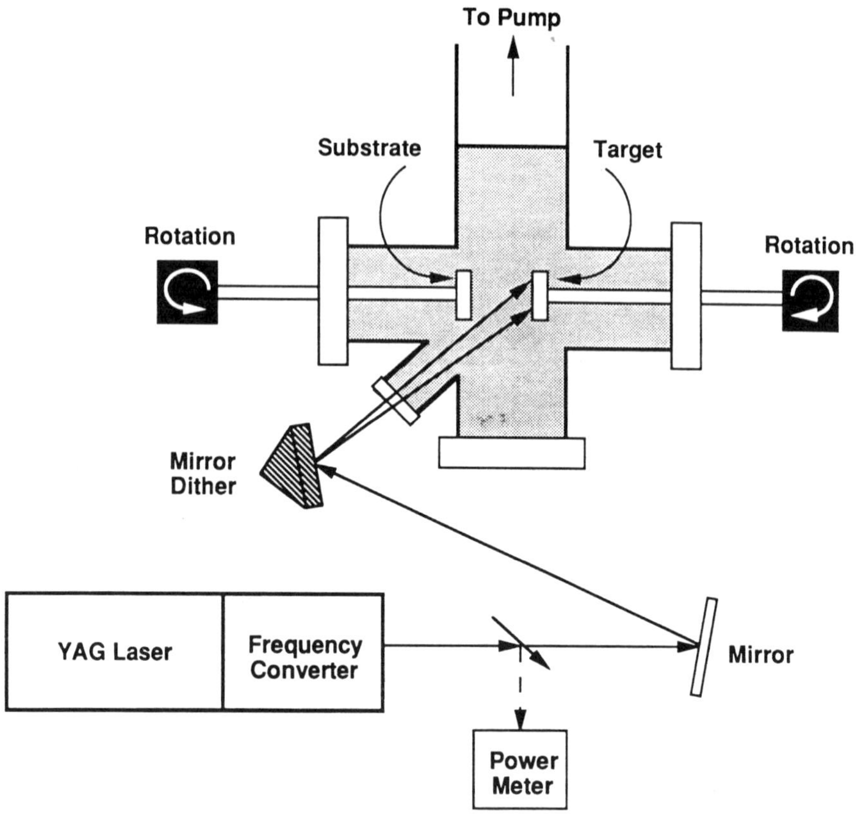

Fig. 1. A schematic of the experimental apparatus. The target and substrate are rotated at ~0.5 rpm, whereas the laser beam is dithered at 0.1 Hz. Typical pressures in the deposition chamber were $\leq 10^{-5}$ torr.

RESULTS

Most evaporations utilized a laser fluence in the range 4–7 J/cm^2. Dithering the steering prism resulted in a circular etch pattern, covering most of the target surface, as seen in Fig. 2. This proved to be far superior to etching a single track at fixed radial distance, which resulted in severe spatial distortion of the emitted plume, as well as thickness and compositional variations in the films. Consequently, a more reproducible deposition rate was obtained as evidenced by Fig. 3, which shows a plot of film thickness as a function of deposition time for subsequent runs. The film thickness for the rotated substrates, as measured with a stylus profilometer, was uniform, to ± 10%, over the area coated, ~0.5 cm^2. This was a factor of two better than obtained for unrotated substrates.

During the deposition process, 200 ng/pulse was typically removed from the bulk target, and only ~1 ng/pulse deposited on the substrate. This corresponds to a deposition rate between 0.1–0.5 μm/s. The difference between the etch rate and the deposition rate was presumably

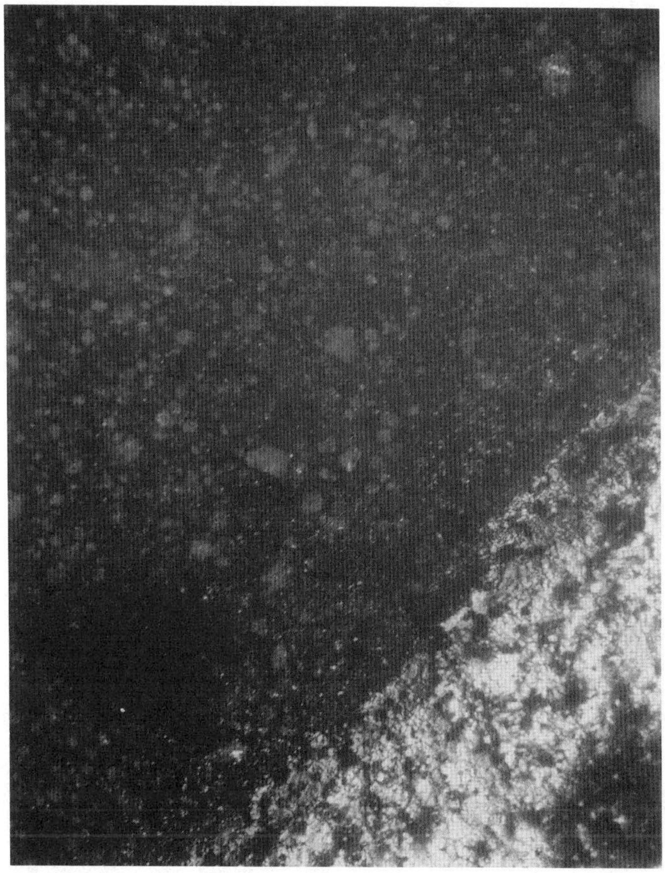

Fig. 2. A micrograph (50×) of the laser-etched (dark), and unexposed (bright) parts of the target pellet. The grains observable on the etched part of the pellet are characteristics of the starting material. The unexposed area is brighter owing to a crust formed during final sintering of the pellet.

owing to a combination of imperfect overlap of the ablation plume with the substrate, nonunity sticking coefficient and, perhaps, gas phase reactions occurring prior to impact on the substrate. Empirically, the deposition rate was found to vary with target composition, grain size, sintering process, length of previous exposure, and length of time between target synthesis and exposure to the laser.

From our unrotated film thickness profiles and exposure geometry, we deduce a $\cos^n(\Theta)$ distribution, $n \geq 8$, which is consistent with previous reports (*11*). In addition, there was very little lateral composition variation over the surface of the film, as monitored by Rutherford backscattering spectrometry (RBS), using a lateral resolution of ~2 mm.

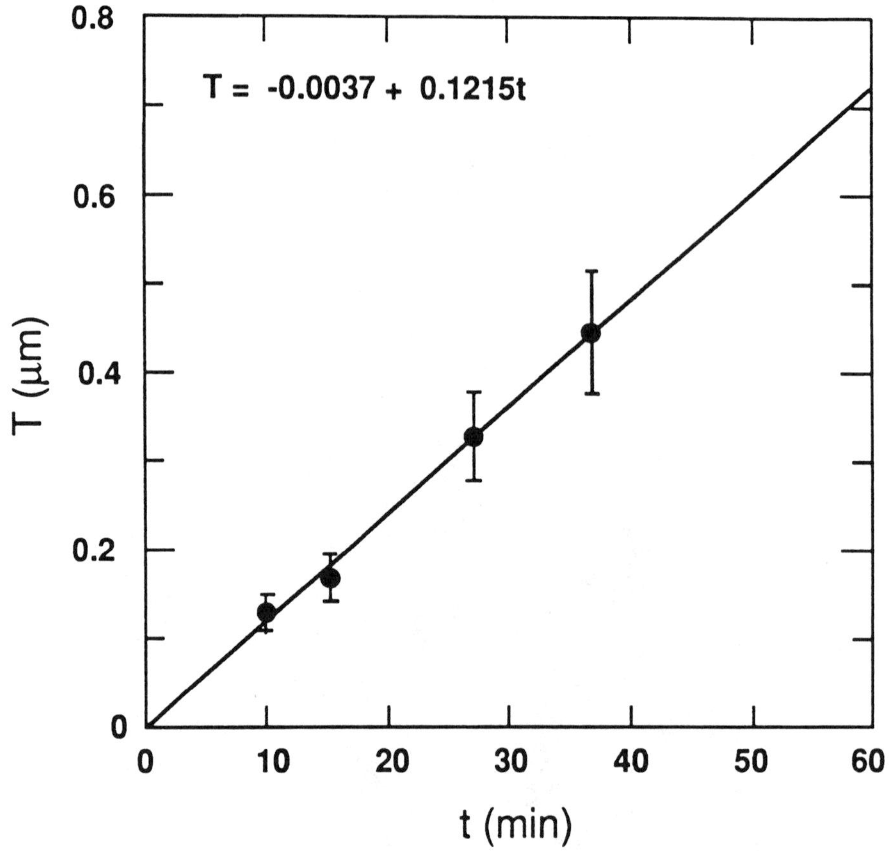

Fig. 3. Deposition thickness of the thin film, in μm, vs deposition time at constant fluence and repetition rate. The data indicate that a constant amount of material was deposited with each laser shot.

However, composition was seen to vary as a function of laser fluence for irradiations at both 532 and 355 nm, as shown in Fig. 4. For both wavelengths, the films were found to be deficient in calcium and enriched in Tl, relative to the starting material, near the threshold fluence (minimum fluence for deposit formation). Although this result cannot be explained in terms of simple volatility or reactivity arguments, it is consistent with recent observations on CO_2 laser evaporation of the similar Bi–Ca–Sr–Cu–O films. In our experiments, the stoichiometry approached that of the starting material as the fluence was increased. At five times the threshold fluence, the film deposition was stoichiometric, within ±10%, to the target material. At higher fluences, the composition is then relatively independent of fluence over a broad range. An upper limit on useable fluence of ~10 J/cm^2 is imposed by the deposition of relatively large particles, >10μm, as shown by scanning electron microscopy (SEM).

The as-deposited Tl–Ca–Ba–Cu–O films were invariably deficient in oxygen and required a postdeposition anneal as described above. In

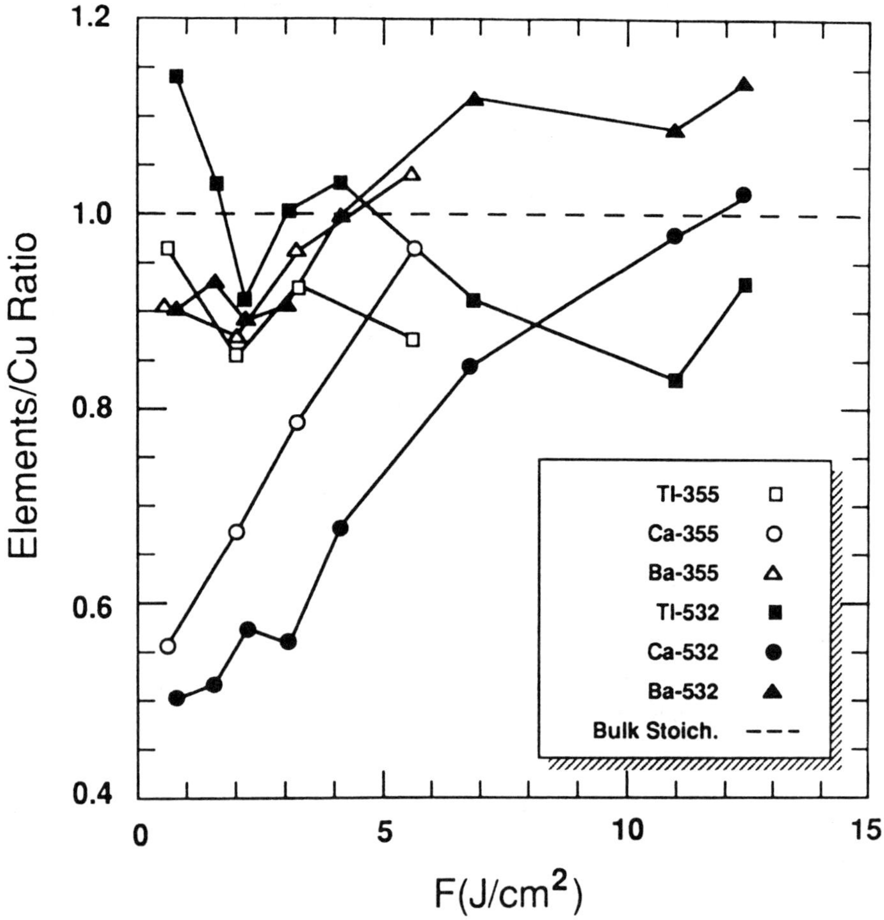

Fig. 4. Thin film composition, determined by Rutherford backscattering spectrometry, as a function of fluence (F) for 355 nm irradiation (open symbols) and 532 nm irradiation (filled symbols). The normalized element/Cu ratio is the element/Cu ratio in the film divided by that ratio in the original sample.

order to develop an *in situ* deposition process, we are investigating a number of diagnostics on the laser generated plume, including optical emission and mass spectroscopy. Representative mass spectrometry results are shown in Fig. 5, which shows both the dynamic range available in our measurements and some interesting aspects of the evaporation process. In particular, no BaO ions are visible in this spectrum. We were also unable to detect any oxide ions containing Ca, Cu, Ba, or Tl in the plume. On the other hand, we were able to detect large quantities of both neutral and ionic oxygen atoms and molecules. This suggests that the evaporation process is sufficiently violent that the metal–oxide bonds are dissociated. This higher volatility and reactivity of the oxygen species, relative to metal atoms and ions, suggests an explanation for the lack of oxygen in the as-deposited films relative to the target.

The laser-produced films are superconductive after annealing with the bulk starting material and oxygen. Figure 6 shows an example of the

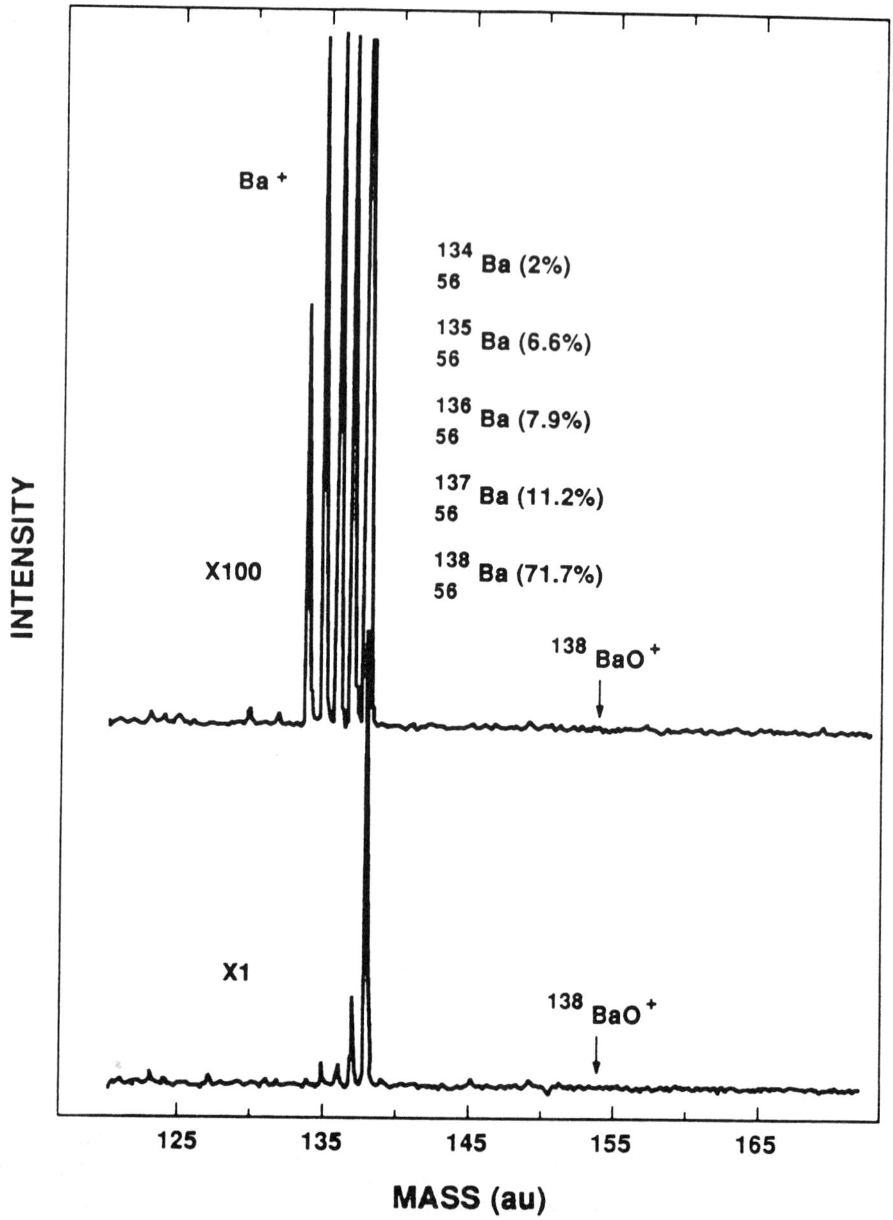

Fig. 5. Mass spectrum of the ablated plume, showing dynamic range and sensitivity of the measurement for barium, and the lack of any detectable metal oxide. The wavelength was 447 nm and fluence 5.7 J/cm^2.

measurement of resistance vs temperature for one film. The film was ablated from $Tl_2Ca_2Ba_2Cu_3O_{10}$ target material (2223) and was 500 nm thick. The deposition was at room temperature, and the postannealing was as described above. X-ray diffraction showed that the film was unoriented mixed phase 2122 and 2223 material. The morphology of the films, as seen by scanning electron microscopy, showed micron-sized, randomly-oriented crystalites.

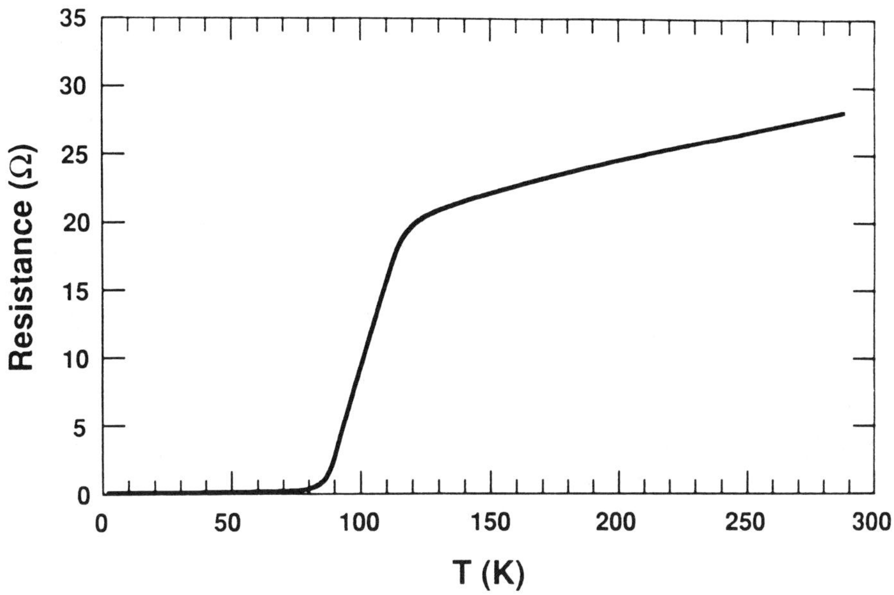

Fig. 6. DC resistance vs temperature for laser-deposited 500 nm film ablated from $Tl_2Ca_2Ba_2Cu_3O_{10}$ target material.

CONCLUSIONS

Laser evaporation at 355 nm has been demonstrated to produce superconducting thin films from the bulk under mild processing conditions. Rastering the laser beam across the target surface, in combination with rotation of the target, produces a much more uniform film than does simple target rotation. In addition, dithering of the laser beam external to the target chamber is more easily implemented than placing optics within the deposition chamber (*3*), and it significantly improves the deposition rate reproducibility. *In situ* deposition of thallium-containing thin films is more difficult than $Y_1Ba_2Cu_3O_{7-x}$ films owing to the high volatility of thallium and its oxides. Future efforts in this laboratory will center on implementation of an activated oxygen source in an attempt to lower the processing and/or annealing temperatures required.

LASER WELDING

In the laser welding process, the laser is used as a source of energy that heats the substrate material above its melting point to produce a fusion weld. However, the process is complex because interactions between the laser beam and the material may either increase or decrease the coupling of energy into the weld (*12–15*). During the melting process, a fraction of the incident laser energy is absorbed by the substrate. The remainder is reflected or absorbed above the surface by the plume, which forms if the surface temperature is sufficiently high. The plume consists of evaporated metal, hot gases, and reaction products. At low laser power, the bulk metal is heated by thermal conduction from the laser-heated surface (conduction mode). At higher laser powers, a vapor depression is formed in the melt pool, which greatly enhances both the energy coupling and the melt depth. This process is called keyhole formation (*16*) (keyhole mode). Other processes that influence the energy coupling include: absorption of laser radiation by the plasma or plume, convection in the weld pool, reflection from the solid or molten metal, chemical reactions in the plume or melt pool, scattering from particles in the plume, and defocusing effects caused by the changing index of refraction in the plume.

Control over the process depends on understanding the laser material interactions and the energy balance involved among these interactions. With this knowledge, process diagnostics and enhancement techniques can be developed.

Most Nd–YAG welding lasers are pulsed. (Most CO_2 welding lasers are continuous wave.) The time scales of events (melting, evaporation, and thermal conduction) during the Nd–YAG pulse are generally shorter than the laser pulse length (~8 ms). From an energy coupling point of view, even the pulsed laser may be considered continuous wave. From the metallurgical viewpoint, however, abrupt heating, melting, cooling, and solidification occurring with each pulse may produce different physical and mechanical properties compared to a continuous-wave weld.

The major emphases in the Nd–YAG laser welding research have been: (1) to determine the conditions that influence melt depth and (2) to develop real-time welding diagnostics. The experimental observations were integrated light emission from the plume and the plume temperature. The temperature was determined by atomic emission spectrometry and the melt depth by metallography. The experimental conditions evaluated include laser pulse energy, gas pressure, and gas reactivity. The melt depth was correlated with gas composition and oscillations in the plume height.

EXPERIMENTAL

The Nd–YAG laser used in the welding experiments had a maximum pulse energy of 40 J at 1064 nm. The pulse rate for all experiments was 10 Hz and the pulse length was 7.8 ms. Careful measurements of the spot size showed that it varied somewhat with the pulse energy. At low pulse energy (10 J), it was 0.75 mm diameter, and at 40 J it was 0.5 mm diameter. For some of the experiments, we report the average laser power and for others the power density at the target during the pulse. For the former, the pulse energy is obtained by dividing the average power by 10.

The high speed photography experiments were performed with a Spin Physics video camera that operated at 2000 frames per second and 6 images per frame. Consequently, the time between individual images was 83 μs.

Figure 7 shows a diagram of the experimental layout for the spectroscopic experiments. The light from the weld plume was imaged onto the entrance slit of a 0.5-meter spectrograph. The spectrally-dispersed light was detected by a time-gated photodiode array. The array was gated to monitor events occurring over an interval down to 20 ns in duration. Light from the plume and the laser pulse temporal profile were monitored using fiber-optic cables and photomultiplier detectors. The output of the detectors was recorded on a digital oscilloscope.

WELDING PLUME ANALYSIS

The characteristics of the plume produced by the laser are influenced by several properties of the target. These include: the fraction of the incident laser beam absorbed (emissivity), thermal conductivity, enthalpy of vaporization, and boiling point. The most important of these properties is probably the emissivity, which is 0.08–0.15 for aluminum (*17*) and 0.25–0.47 for stainless steel (*17*). The emissivity depends strongly on the extent of oxidation and the surface roughness. The thermal conductivity determines the rate of conduction of the absorbed laser energy to the bulk metal. The thermal conductivity of aluminum (3 W/cm − K) exceeds stainless steel (0.3 W/cm − K) by a factor of 10 (*18*). Melting significantly lowers the thermal conductivity (*17*).

Figure 8 shows high speed photographs of the plume formation above individual spot welds made on 304 stainless steel (304SS) and commercially pure aluminum (1100Al). These two materials were chosen to be representative of materials that highly reflect (Al) and highly absorb (SS) the 1.06 μm light from the Nd–YAG laser. Each horizontal panel shows a succession of individual photographs of the plume, which expands vertically above the horizontal sample surface. Each panel corresponds to a weld made by one laser pulse. The time between successive photographs is 83 μs. The sample surface does not show in the photographs because the camera f-stop was adjusted for proper exposure of the plume, which was much brighter. A fiducial mark shown as a white block to the left of each photograph sequence marks the start of the laser pulse. The powers listed at the left are the average laser powers measured at a pulse rate of 10 Hz.

The temperatures of the weld plumes on 1100Al and 304SS were measured using neutral ironlines in the spectral region 360–390 nm. The concentration of iron in 1100 Al is below 1%, so these lines are suitable to construct a Boltzmann plot to determine temperature. Figure 9 shows the temperature, as determined by atomic emission spectroscopy, at various times during the laser pulse. The laser pulse energy is the same (21.4 J) for both materials. The more strongly absorbing 304SS target gives a plume temperature that is several hundred degrees higher than the aluminum target.

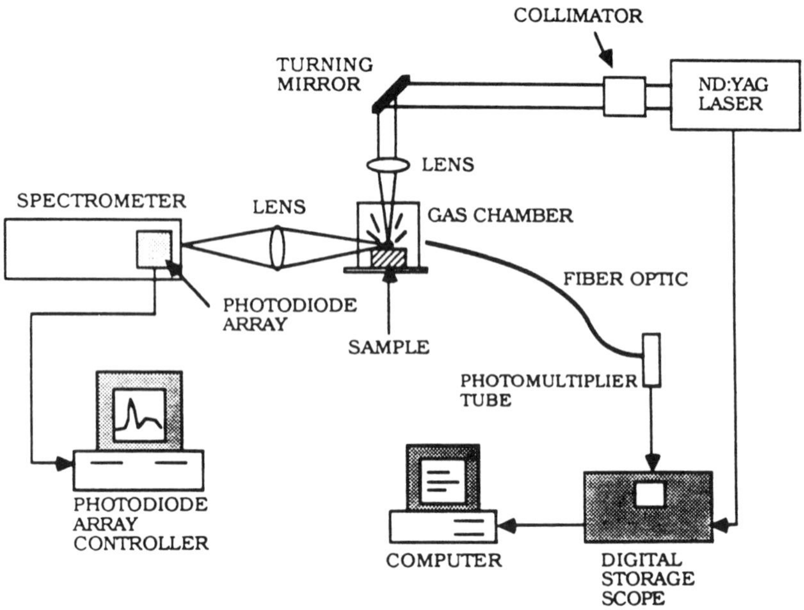

Fig. 7. Apparatus diagram for the laser welding experiments.

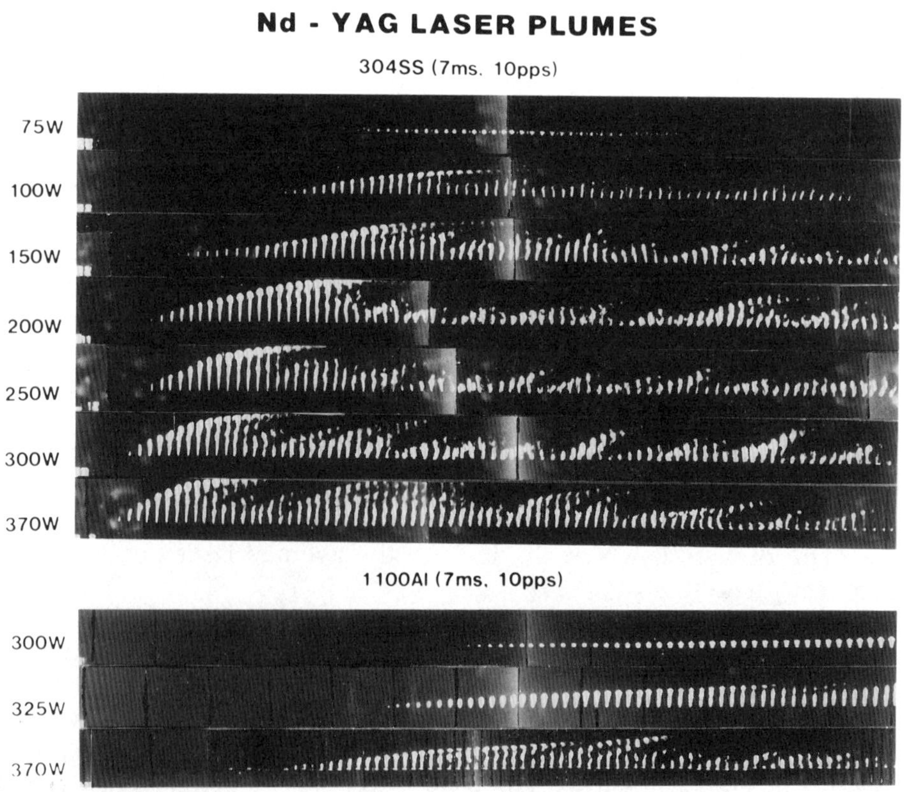

Fig. 8. High-speed photographs of the rising plume during a single Nd–YAG laser pulse. Time increases to the right, and the time between images is 83 μs. The label on the left margin is the average laser power (10 Hz). The greatest plume heights are 1.75 cm for 304SS and 1.25 cm for 1100Al.

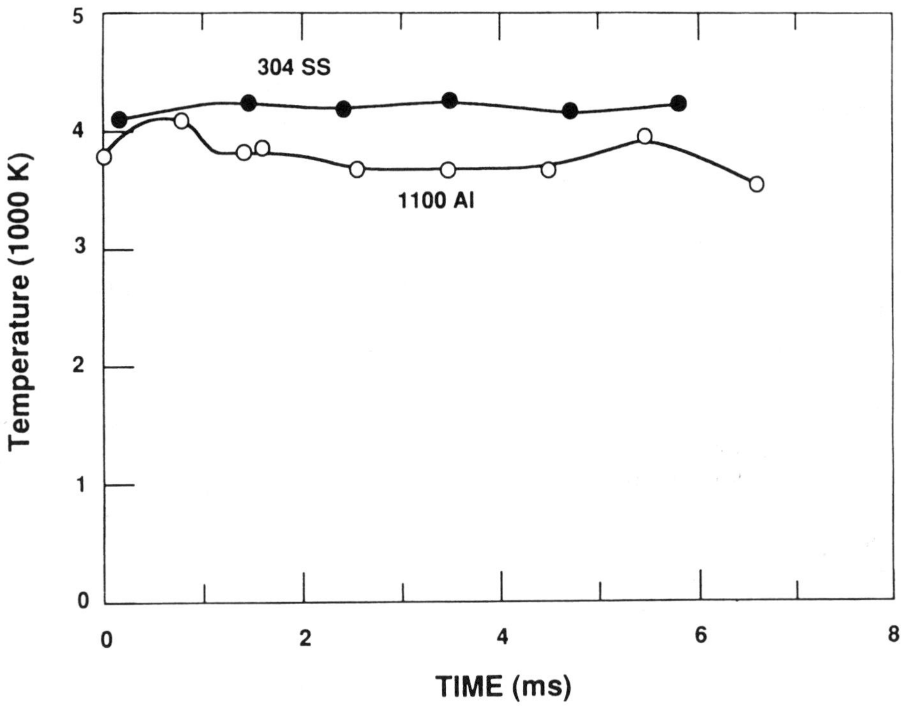

Fig. 9. Temperature (from atomic emission) vs time during a 7.8 ms laser pulse (21.4 J) for stainless steel and aluminum.

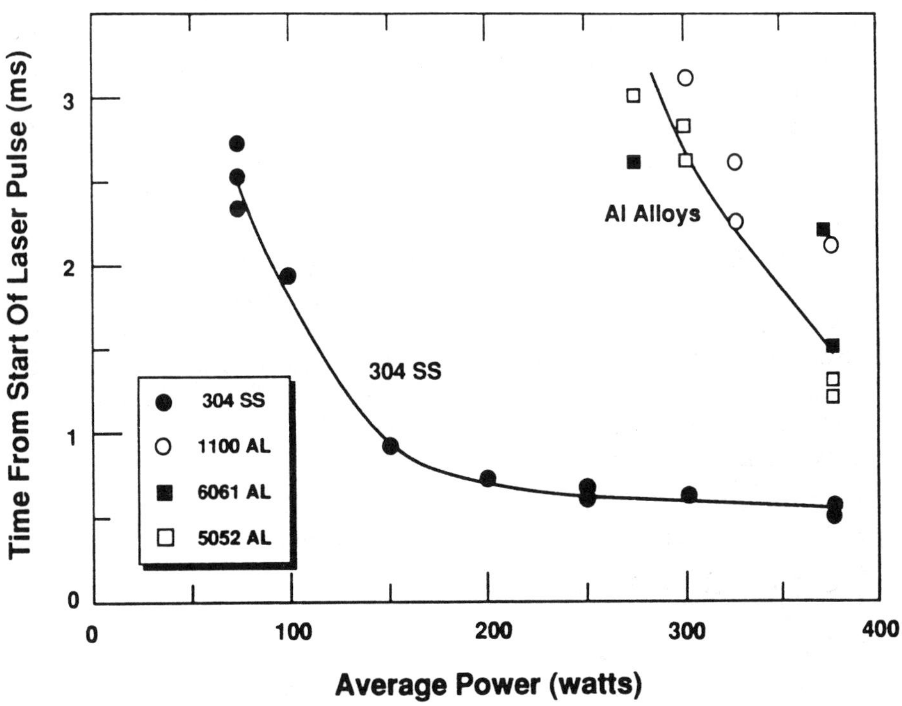

Fig. 10. Time delay to laser plume initiation vs average laser power for 304SS and several aluminum alloys.

The time to the initiation of the plume is shorter for 304SS than 1100Al because of the greater absorption of the steel sample. Figure 10 shows the initiation time, measured from the photograph sequences, plotted vs average power. Plumes initiate at 75 W (7.5 J) average power on the 304SS sample compared to 300 W (30 J) on the 1100Al sample. Initiation times for plume formation drop from 2.7 to 0.5 ms in changing the power from 75 to 370 W on 304SS. On the 1100Al sample, plume initiation occurred after 3.0 ms at 275 W (27.5 J) and dropped to 1.3 ms at 370 W (37 J). The power density required for plume initiation is constant for a given surface condition, material, and cover gas.

The rate of rise of the plume will be influenced by the same material properties that determine other plume characteristics. For an average laser power of 370 W, the initial stainless steel plume rises at a rate of 2200 cm/s, whereas the initial aluminum plume for the same average laser power rises at a rate of 590 cm/s. Both of these rates are only a small fraction of the sonic velocity. For example, the rise rate for 304SS is 6.4% of the room-temperature sonic velocity in air. The ratio of the rates for 304SS and 1100Al (3.7) is about the same as the ratio of emissivities (1/4) for these two materials.

At high powers, the welding plumes for both materials tend to rise, expire, and reinitiate during the pulse. The occurrence of these oscillations appears to have an energy threshold and the number of pulses increases quickly for both materials with increasing power. Up to 22 plume pulses were counted on the 304SS sample and 18 on the 1100Al sample. Experiments with 5052 and 6061 aluminum showed similar effects. The threshold pulse energies appear to be the same for the three aluminum alloys, but the number of plume pulses for 5052 Al was about 2/3 of the other alloys. A significant increase in melt depth occurs above the threshold energy for these plume pulses. The occurrence of the plume oscillations has been attributed (*12*) to successive opening and closing of the keyhole depression, that is, to oscillation of the melt pool.

Keyhole formation results in increased absorption because of multiple reflections of the incident laser radiation. In addition to the dramatic increase in melt depth, we also see that the rate of rise of successive plumes is greater than for the initial plumes. This is particularly true for 1100Al. For 370 W average power, the rate of rise of later plumes is three times that for the initial plume. We conclude, then, that the keyhole formation increases the fraction of the laser radiation absorbed by aluminum by about a factor of three.

EFFECT OF THE WELDING ATMOSPHERE ON MELT DEPTH

The gas above the melt pool is known to affect the welding process. For example, gases such as N_2 and H_2 in the cover gas during welding can be incorporated into the weld and lead to embrittlement. To study the effect of different gases on the weld depth, the workpiece was placed in a sealed chamber that was purged by different gases (Fig. 7). The melt depths obtained as a function of laser power densities using various purge gases are shown in Figs. 11 and 12 for continuous welds on 1100Al and 304SS. The welds were made at 10 Hz with 75–80% overlap between adjacent welds. The divergence of the laser beam was kept as small as possible during these experiments by adjusting the collimator located at the laser output each time the laser pulse energy was changed. Even with this adjustment, the spot size of the focused beam on the workpiece still decreased with increasing laser power.

Two distinct regions are evident in Figs. 11 and 12: (1) At power densities below about 1.3 MW/cm^2 there was a small, but gradual increase in the melt depth with increased power density and (2) above 1.3 mW/cm^2, there was 5–10-fold increase in the melt depth with a small increase in power density. Comparison of the figures in the low power regime shows that the melt depths in 304SS were about three times the melt depths in 1100Al at the same power densities. Also, for 304SS, the slopes of the curves at low power density are about five times greater than those obtained for 1100Al. Both these effects may be indicative of the greater thermal conductivity and reflectivity of aluminum compared to stainless steel. Inspection of the figures also shows that at a specific power density, the welds made in air, O_2, and SF_6 were deeper than those obtained in the other gases. The greatest effects were for the 304SS welds in which melt depths in O_2 and SF_6 were 5.4 and 2.9 times, respectively, the depth obtained using He at a density of 1 MW/cm^2. In addition, the rates of change of melt depth with power density were greatest for the more reactive gases. These differences are attributed mainly to the different reactivities of the gases. As stated above, temperatures in the welding plume can reach 3500 K, which is sufficient to generate free oxygen and fluorine atoms that can react with aluminum or iron to produce heat. The reaction $2Al + 3O = Al_2O_3$, for example, is very exothermic with $\Delta H = -2414$ kJ/mole. The energy required to melt the volume of the largest melt produced in 1100Al (2 mm D $\times$ 1.8 mm W) is 3.23 J. This energy is generated if only 2% of the aluminum in the melt volume reacts to form the oxide. Similar results apply to the formation of iron oxides and fluorides. Measurements of the temperature of the plumes produced in the different gases is presented in Fig. 13. These data show that the temperature is greater for the reactive gases, suggesting that energy is being released via a reactive mechanism.

The increase in melt depth obtained with the reactive gases could also be owing to greater coupling of the laser energy into the metal by formation of the oxides. Aluminum oxide, for example, exhibits much stronger absorption at 1064 nm than pure aluminum metal. Other gas properties, such as thermal conductivity, may have a role in determining the melt depth, but their effect is probably small. The thermal conductivities of O_2 and N_2, for example, are almost identical, but significantly deeper melt depths were achieved in O_2 (Fig. 11). It is also possible that

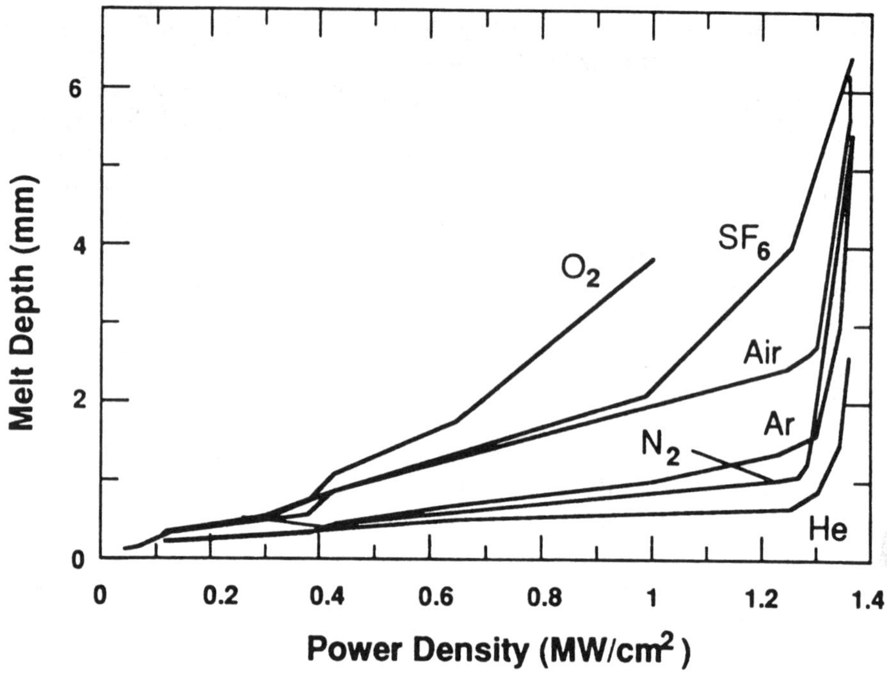

Fig. 11. Melt depth in 304SS vs laser power density for six different cover gases (590 torr, 78.7 kPa) from a series of overlapping welds.

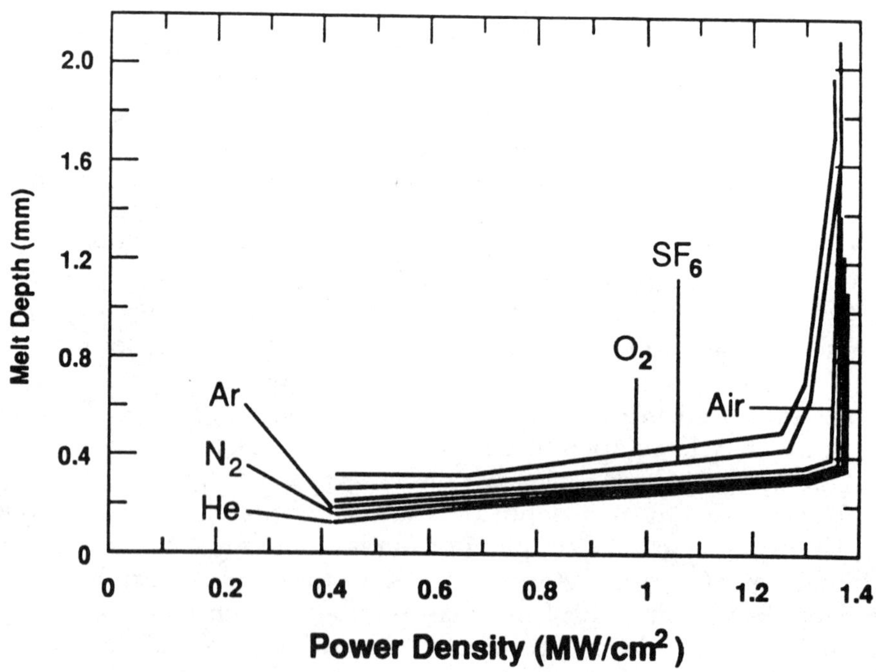

Fig. 12. Melt depth in 1100Al vs laser power density for six different cover gases (590 torr, 78.7 kPa) from a series of overlapping welds.

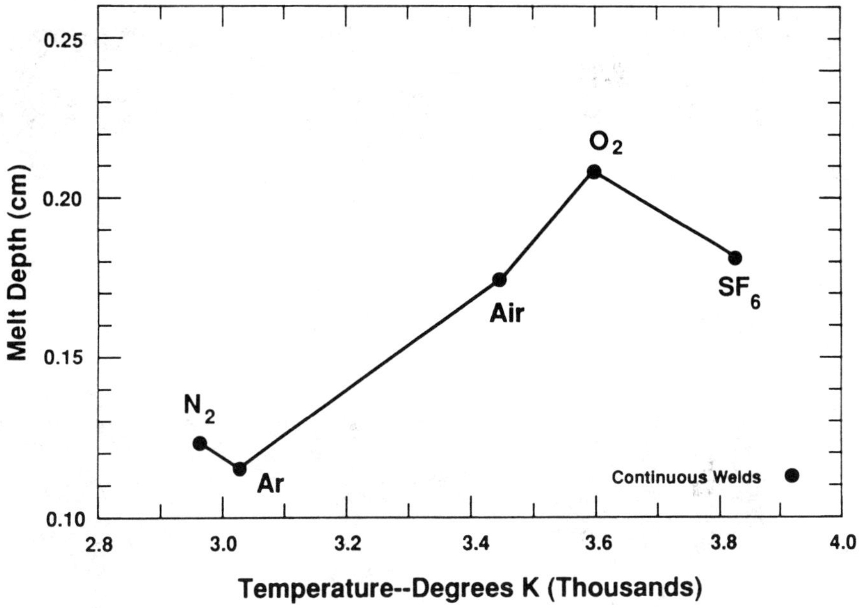

Fig. 13. Melt depth for continuous welds in 1100Al vs temperature of the laser plume. The laser power was the same for all points. The temperature variation was produced by varying the gas over the weld.

some as yet undetermined parameter of the welding process, for example, changes in the spot size owing to thermal lensing in the different gases may account for the observations reported here, but these effects are expected to be small.

The curves in Figs. 11 and 12 show that, at powers above 1.3 MW/cm^2, a new mechanism becomes operative in each gas that produces a significant increase in the melt depth. This dramatic increase in the weld depth at higher powers is probably owing to a change in the welding mode from conduction to keyholing in which the laser beam is effectively channeled into a vapor cavity within the melt via multiple reflections. In the keyholing mode, as noted in the previous section, the fraction of laser energy coupled into the metal increases significantly over that obtained in the conduction mode. The transition from conduction to keyhole mode is demonstrated in Fig. 14, which shows cross-sections of welds made on 1100Al in argon and air at different laser powers. At 300 W, which corresponds to a power density of about 1.34 MW/cm^2, the weld profile corresponds to the conduction mode. At 350 W (1.36 MW/cm^2), the onset of keyholing is evident with a significant increase in the depth-to-width ratio of the weld (see Fig. 14). The main effect of the reactive gases over the transition region appears to be a slight reduction in the power density at which the transition from conduction to keyholing occurs. Unfortunately, it was not possible to increase the power density above 1.36 MW/cm^2 to observe whether the large increase in melt depth with power continues at higher powers.

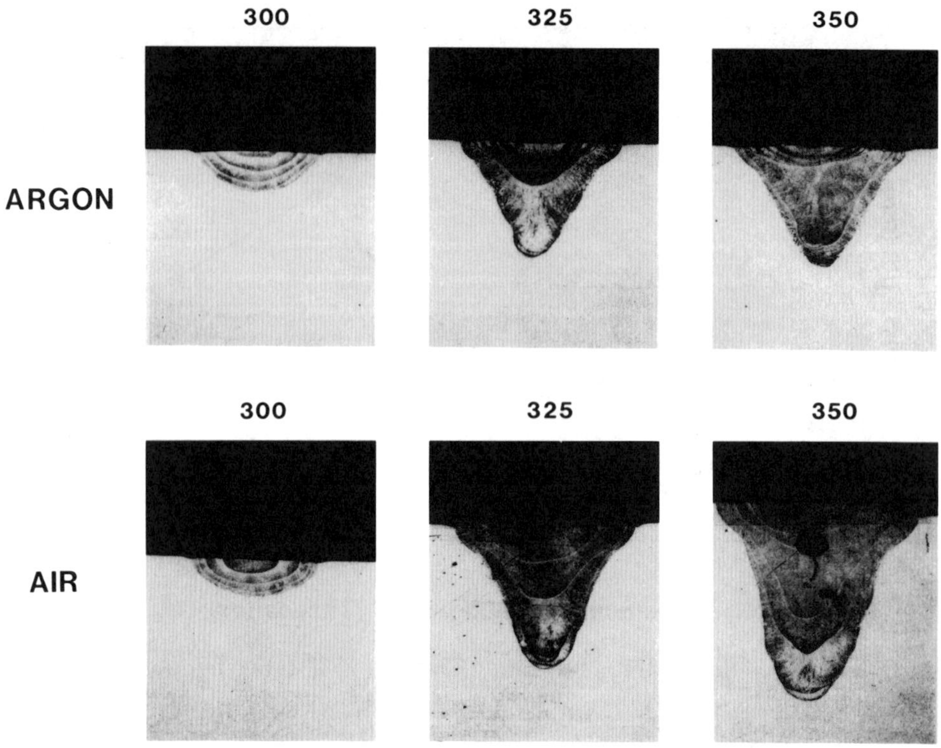

Fig. 14. Weld profiles in 1100Al for two gases and three laser powers that span onset of keyhole formation.

CONCLUSION

The properties of the welding plume correlate with the characteristics of a laser weld. A prompt initiation and rapid rise of the plume indicates effective coupling of the laser radiation to the workpiece. The plume oscillations are also indicative of operation in the keyhole mode. We now have sufficient data and understanding to begin the development of diagnostics based on plume properties that can be good indicators of weld characteristics. The studies with different cover gases show the importance of gas/metal vapor reaction in determining the heat input to the weld. This gives another degree of freedom that one can use in controlling the weld properties. As this research continues, the development of a computer model of the welding process will become a high priority. The development of improved heat deposition diagnostics and the establishment of a more quantitative correlation between plume characteristics and weld properties is continuing.

ACKNOWLEDGMENTS

This work was performed under the auspices of the United States Department of Energy. The authors appreciate the support of the management of the Los Alamos National Laboratory during the course of this research.

REFERENCES

1. Wolf, S. A. and Kresin, V. Z., eds., *Novel Superconductivity*, Plenum, New York, 1987.
2. Wu, X. D. and Venkatesan, T., *Chemistry of Oxide Superconductors*, Rao, C. N. R., ed., Blackwell Scientific Publications, Oxford, England 1988.
3. Sankur, H. and Cheung, J. T., *Appl. Phys.* **A47,** 271 (1988).
4. Kwok, H. S., Mattocks, P., Shi, L., Wang, X. W., Witanachchi, S., Ying, Q. Y., Zheng, J. P., and Shaw, D. T., *App. Phys. Lett.* **52,** 1825 (1988).
5. Wu, X. D., Dijkkamp, D., Ogale, S. B., Inam, A., Chase, E. W., Miceli, P. F., Chang, C. C., Tarascon, J. M., and Venkatesan, T., *Appl. Phys, Lett.* **51,** 861 (1987).
6. Dijkkamp, D., Venkatesan, T., Wu, X. D., Saheen, S. A., Jisrawi, N., Min-Lee, Y. H., Mclean, W. L., and Croft, M., *Appl. Phys. Lett.* **51,** 619 (1987).
7. Roas, B., Schultz, L., and Endres, G., *Appl. Phys. Lett.* **53,** 1557 (1988).
8. Jaggi, N. K., Meskoob, M., Wahid, S. F., and Rollins, C. J., *Appl. Phys. Lett.* **53,** 1551 (1988).
9. Garcia, A. R., Muenchausen, R. E., and Nogar, N. S., unpublished results.
10. Sleight, A. W., *Science* **242,** 1519 (1988).
11. Venkatesan, T., Wu, X. D., Inam, A., and Wachtman, J. B., *Appl. Phys. Lett.* **52,** 1193 (1988).
12. Herziger, G., *The Industrial Laser Annual Handbook*, Belforte, D. and Levitt, M., eds. Pennwell Books, Tulsa, OK, 1986, p. 108.
13. Pirri, A. N., Root, R. G., and Wu, P. K. S., *AIAA Journal* **16,** 12 (1978).
14. J. F. Lancaster, ed. *The Physics of Welding*, Pergamon, Oxford, England 1984, p. 9.
15. Cambel, A. B., *Plasma Physics and Magneto-Fluidmechanics*, McGraw-Hill, New York 1963.
16. *Metals Handbook, 9th ed.*, vol. 6, American Society for Metals, pp. 655–657, published in 1983.
17. *Thermophysical Properties of Matter*, vol. 7, p. 249 published by: IFI Plenum in 1970.
18. *Handbook of Chemistry and Physics, 69th edition*, Weast, R. C., ed., The Chemical Rubber Co., Cleveland, OH, 1988/1989.

Laser Probe Absorption Spectroscopy Measurements on Laser-Induced Plumes

TERRY D. KUNZ,[1,2] RICHARD F. MENEFEE,[1,2] BRENDAN D. KRENEK,[1,2] LEIF G. FREDIN,[1,2] AND MICHAEL J. BERRY*,[1,2,3]

[1]Antropix Corporation, 30 Meadowfair Court, The Woodlands, TX 77381; [2]Laser Applications Research Center, HARC, 4802 Research Forest Drive, The Woodlands, TX 77381; and [3]Department of Chemistry and Rice Quantum Institute, Rice University, P. O. Box 1892, Houston, TX 77251

ABSTRACT

A computer-controlled laser absorption spectroscopy (CCLAS) system has been developed to study high temperature plume species (atoms, molecules, and ions) generated by laser/materials interactions. Direct temporally/spatially resolved measurements on carbon plume/plasma species have been obtained during continuous wave (cw) and repetitively pulsed (rp) laser irradiation experiments. The measured plume characteristics (species, abundances, energy contents, and optical properties) have been interpreted to provide mechanistic understanding of laser/materials interaction phenomenology and effects.

Index Entries: Plume; laser/materials interaction; ablation; laser probe spectroscopy; laser effects; high temperature materials; diagnostic instrumentation; optical properties; plasma generation; carbon.

INTRODUCTION

Laser effects on materials depend strongly on laser and interaction parameters (wavelength, irradiance, pulse waveform, irradiated spot size, angle of incidence, and so on), material characteristics (composition, optical and thermal properties, pyrolysis kinetics, and so on),

*Author to whom all correspondence and reprint orders should be addressed.

plume characteristics (composition, geometric structure, optical properties, and so on), and environmental conditions (pressure, flowfield, acceleration, and so on). Laser interactions with materials (*1–3*) are governed by: the value of the wavelength-dependent coupling coefficient α_λ and its origins owing to optical properties of the target and plume, and the value of the enthalpy of mass removal H_{MR} and its connection with plume composition. Plume species (atoms, molecules, particles, and/or plasmas) may absorb and/or scatter the incident laser beam; these beam/plume interactions decrease α_λ from its transparent plume value. Beam/plume interactions dominate overall energy transport to the target surface if a dense, high opacity plasma is formed within the plume; then, plasma absorption of the incident laser radiation and subsequent plasma reradiation determine the irradiance distribution at the target surface (*4*). Currently, phenomena associated with plasma initiation and growth are poorly understood and poorly modeled; therefore, direct experimental measurements on model plasma systems are needed to stimulate the development of useful theoretical models.

The present paper describes:

1. Plume diagnostic instrumentation systems developed to obtain temporally-, spatially-, and spectrally-resolved plume optical properties (absorption spectra, emission spectra, plasma opacity, and so on) during irradiation by single pulse (sp), repetitively pulsed (rp), and continuous wave (cw) laser sources;
2. Selected measurements and analyses on laser-induced carbon plumes;
3. Mechanistic implications of the results.

EXPERIMENTAL METHODS

The plume diagnostic instrumentation system, schematically shown in Fig. 1, includes a computer-controlled laser absorption spectroscopy (CCLAS) system to probe ablation plumes and an optical multichannel analyzer (OMA) system to measure plume emission spectra. Both continuous wave (cw) carbon dioxide (CO_2) and repetitively pulsed (rp) hydrogen fluoride (HF) damage lasers are used to irradiate target materials, such as carbon, with controlled and well characterized laser beams. The ablation plume expands into vacuum or into air, N_2, or He at specified pressures. The CCLAS system generates a tunable cw dye laser probe beam with a small spot size (typically, 0.5 mm diameter) that is directed through the plume at a fixed, but adjustable distance (typically, 0.5–4 mm) from the target surface. The transmitted probe beam is detected and ratioed to a reference beam. Transient digitizers (or other detection electronics) and computer-based data acquisition instrumentation systems are used to record, store, and process ratioed signals for quantitative spectroscopic analysis of plume species (atoms, molecules, and ions).

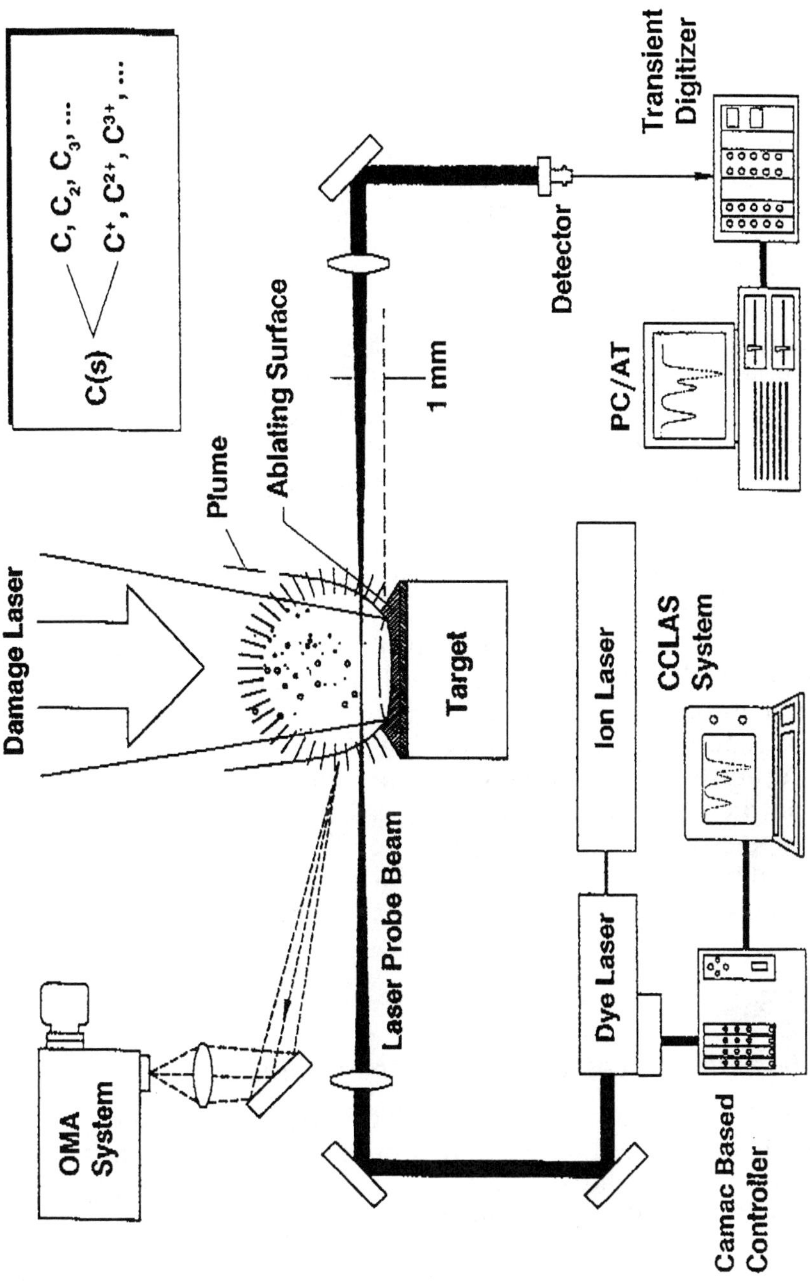

Fig. 1. Plume diagnostic instrumentation system.

GraphNOL(G), a fine-grained polycrystalline pure carbon with well-known nearly isotropic thermal properties, was used as the target material to produce laser-induced carbon plumes. G samples were machined into small specimens (typically, 1.5–5 mm diameter × 0.5–2 mm thickness disks or 20 mm long × 5 mm wide × 3 mm thick strips) for laser experimentation.

In cw experiments, G targets were ablated in air using the Houston Area Research Center (HARC) cw CO_2 Laser Test Resource (*5*) to deliver ca. 600 W power on target at $\lambda = 10.6$ μm incident wavelength. The cw beam was focused by a ZnSe lens to produce peak irradiances on the target surface in the range 7–400 kW/cm^2 and peak front surface temperatures in the range 4000–4600 K. Over this temperature range, assuming equilibrium vaporization, triatomic carbon (C_3) is the predominant (greater than 0.5 mole fraction) vapor species in the plume, followed by substantial amounts of monatomic carbon (C; ca. 0.02–0.04 mole fraction), diatomic carbon (C_2; ca. 0.05–0.06 mole fraction), and larger carbon clusters C_n, where $n > 4$ (*6*). Only neutral species, rather than ions, are significant under these conditions.

In rp experiments, G targets were ablated in vacuum using the HARC rp HF Chemical Laser Test Resource (*7*) to deliver ca. 200 mJ/pulse multiline (λ = ca. 2.7–3.1 μm) pulse energy on target at low pulse repetition frequency (0.5–1 Hz PRF). The rp beam was focused by a CaF_2 lens to produce peak fluences on the target surface in the range 8–165 J/cm^2. Under these conditions, peak front surface temperatures are not known well (owing to complex fluence-dependent and time-dependent absorption in the plume, as well as the target front surface). However, significant ionization and high plasma temperatures are generated in the ablated plume at peak pulse fluences greater than ca. 30 J/cm^2, leading to the observation of intense plasma emissions owing to ionic carbon vapor species (C^+, C^{2+}, and so on). Neutral carbon clusters are also present both prior to plasma formation and following recombination of plasma species (ions and electrons) after the incident laser pulse is completed.

RESULTS

CW Carbon Dioxide Laser/Materials Experiments

CW CO_2 laser irradiation of graphNOL (G) targets generates a blue plume with an emission intensity that depends on background gas and pressure (*8*). In air, the plume is much brighter than in N_2, He, or vacuum because of the chemiluminescent reaction (*9*)

$$C_3 + O_2 \longrightarrow C_2^*(d^3\Pi_g) + CO_2 \qquad [1]$$

Figure 2 shows a survey plume emission spectrum in the λ = 330–670 nm spectral region obtained by ablating G in air at a peak irradiance of 137 kW/cm^2. The OMA system was used in low resolution (ca. 0.5 nm/detector element) to view the ablation plume perpendicular to its propa-

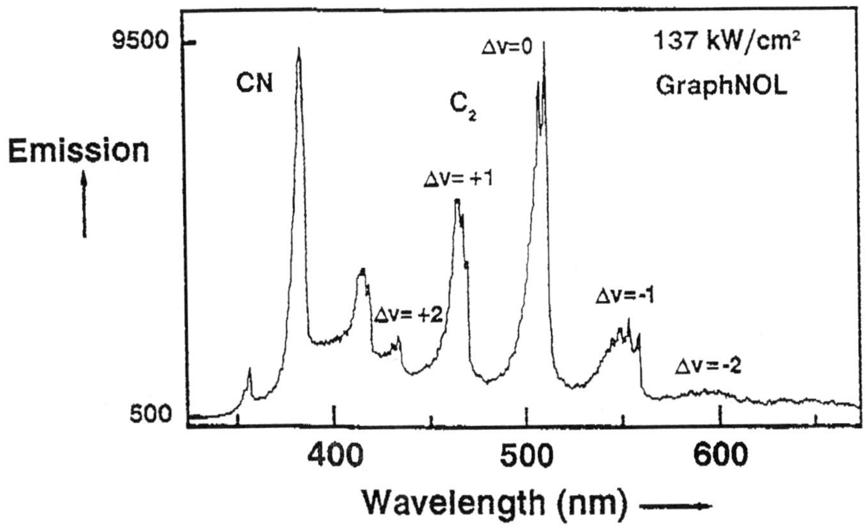

Fig. 2. Survey plume emission spectrum.

gation at a distance ca. 3 mm from the initial target surface. Prominent plume emission features are owing to diatomic band spectra [the CN $B^2\Sigma^+ \longrightarrow X^2\Pi_i$ bands (*10*), with peak band emissions in the 360–420 nm spectral region and the C_2 $d^3\Pi_g \longrightarrow a^3\Pi_u$ bands (*11,12*), with peak band emissions in the 430–610 nm spectral region]. Several C_2 band sequences are labeled in the figure. Higher resolution plume emission spectra were also recorded to confirm detailed C_2 assignments (*8*).

Figure 3 (top panel) shows a portion of the experimental C_2 plume absorption spectrum, recorded with the CCLAS system (resolution: ca. 0.02 cm^{-1}) during G ablation in air, using a peak cw CO_2 laser irradiance of 7 kW/cm^2; individual rotational transitions of the 0,0 band of the Swan system (*11,12*) are identified in the figure. Similar spectra were also obtained at other irradiance levels (*8*). The bottom panel of Fig. 3 shows a spectral simulation based on known molecular constants (*12*), with an assumed rotational temperature of 1400 K and a Doppler width of 0.23 cm^{-1} for each transition (corresponding to a translational temperature of 6600 K). Spectral simulations of high resolution emission spectra yield "best fits" with vibrational temperatures of 2300 K. All three (translational, rotational, and vibrational) temperatures are probably produced by complex reactions and/or relaxation processes rather than by the vaporization process itself.

RP Hydrogen Fluoride Chemical Laser/Materials Experiments

RP HF chemical laser irradiation of graphNOL (G) targets in vacuum generates relatively faint plumes at low peak fluences (F_p less than 10 J/cm^2). However, highly luminous plumes are observed at high peak fluences (F_p greater than ca. 30 J/cm^2). Figure 4 shows survey plume emission spectra in the λ = 350–650 nm spectral region obtained by

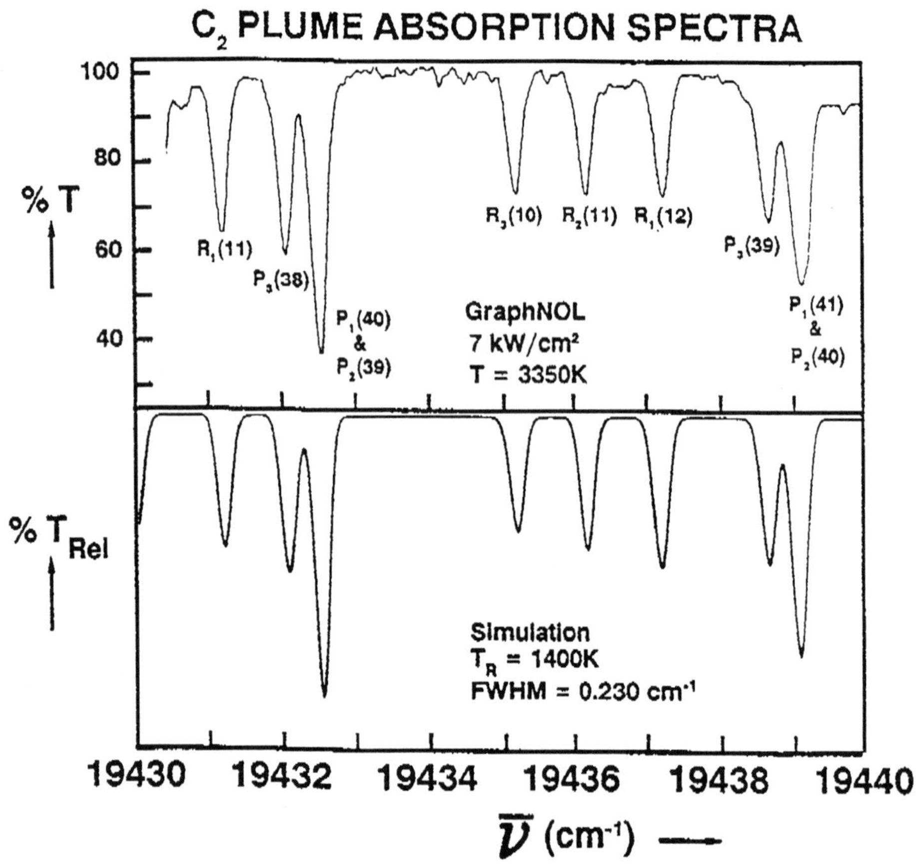

Fig. 3. Experimental (top) and simulated (bottom) C_2 plume absorption spectra.

ablating G in vacuum at peak fluences of 32 and 124 J/cm^2 (bottom and top panels, respectively). The OMA system was used in low resolution (ca. 0.5 nm/detector element) to view the ablation plume perpendicular to its propagation at a distance ca. 3 mm from the initial target surface. Prominent plume emission features are owing to ionic carbon transitions (*13,14*), such as λ = 426 nm: C^+ (C II) $^2F^o_{5/2,7/2} \longrightarrow {}^2D_{3/2,5/2}$ and λ = 465 nm; C^{2+} (C III) $^3P^o_{0,1,2} \longrightarrow {}^3S_1$. These particular transitions dominate the plume emission spectra observed in the visible spectral region, but it is likely that many other intense transitions are present in the ultraviolet and vacuum ultraviolet spectral regions.

Figure 5 shows plasma emission spectra generated by rp HF chemical laser irradiation of G in vacuum at a peak fluence of 50 J/cm^2 as a function of delay time (0–200 ns) from the damage laser pulse peak (which occurs ca. 50 ns after pulse initiation). The OMA system collected emission from a 1 mm diameter spot positioned ca. 2.5 mm in front of the target surface and was used with a gate width of 100 ns for each spectrum. Note the temporal evolution of prominent emission features near λ = 426 and 465 nm wavelengths; the doublet components near

$\lambda = 465$ nm have different time dependences with respect to each other and the apparent singlet (actually a blended triplet) near $\lambda = 426$ nm. This temporal evolution indicates that there are different formation and decay processes and timescales for particular ionic carbon states.

Figure 6 shows the spatial dependence of plasma emission spectra generated by similar rp HF chemical laser irradiation at a peak fluence of 75 J/cm^2 and at a fixed time (ca. 50 ns; OMA gate width: 100 ns) after pulse initiation. Spectra observed at distances $D_{Surf} = 300$ μm, 800 μm, 1.4 mm, and 2.5 mm from the G surface are shown. The spectral width variation of the $\lambda = 465$ nm emission feature is caused by Stark broadening, which is a function of electron density (*8,14,15*).

Figure 7 shows an expanded region of part of Fig. 5, together with a spectral simulation based on transition wavelengths (*13*) and spontaneous emission coefficients (*16*) for C II $^2F^o \longrightarrow$ 2D multiplet transitions near $\lambda = 427$ nm wavelength, together with a 25 cm^{-1} Stark broadening linewidth for each of the three transitions within the multiplet. This Stark

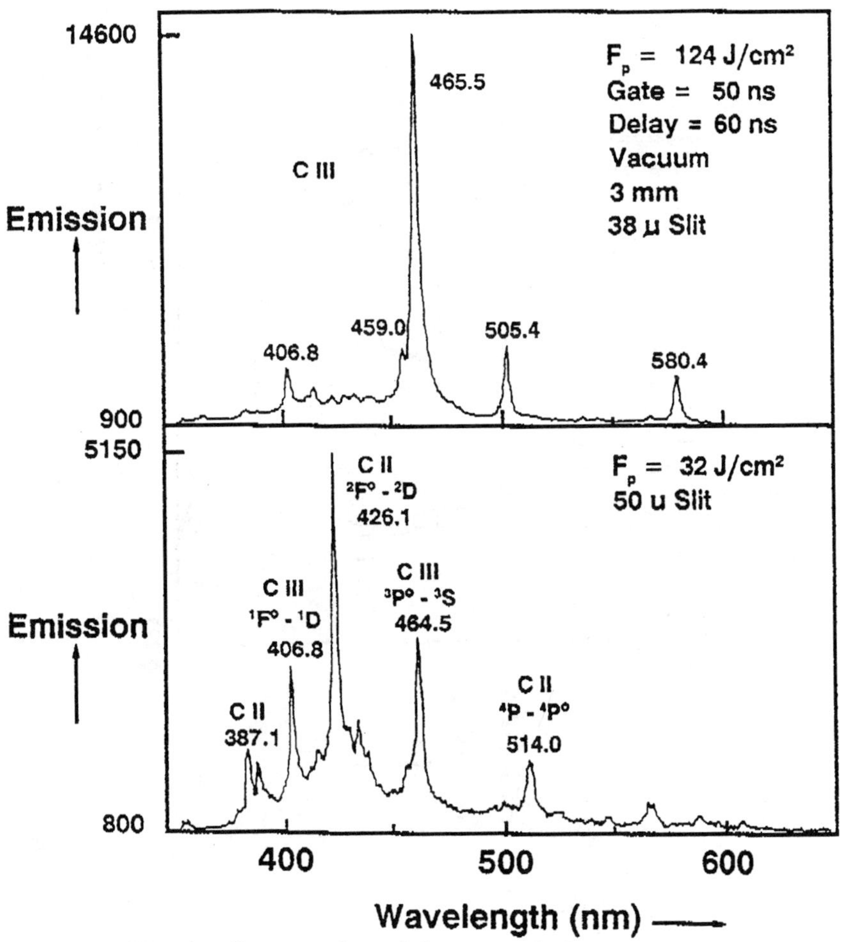

Fig. 4. Survey plume/plasma emission spectra.

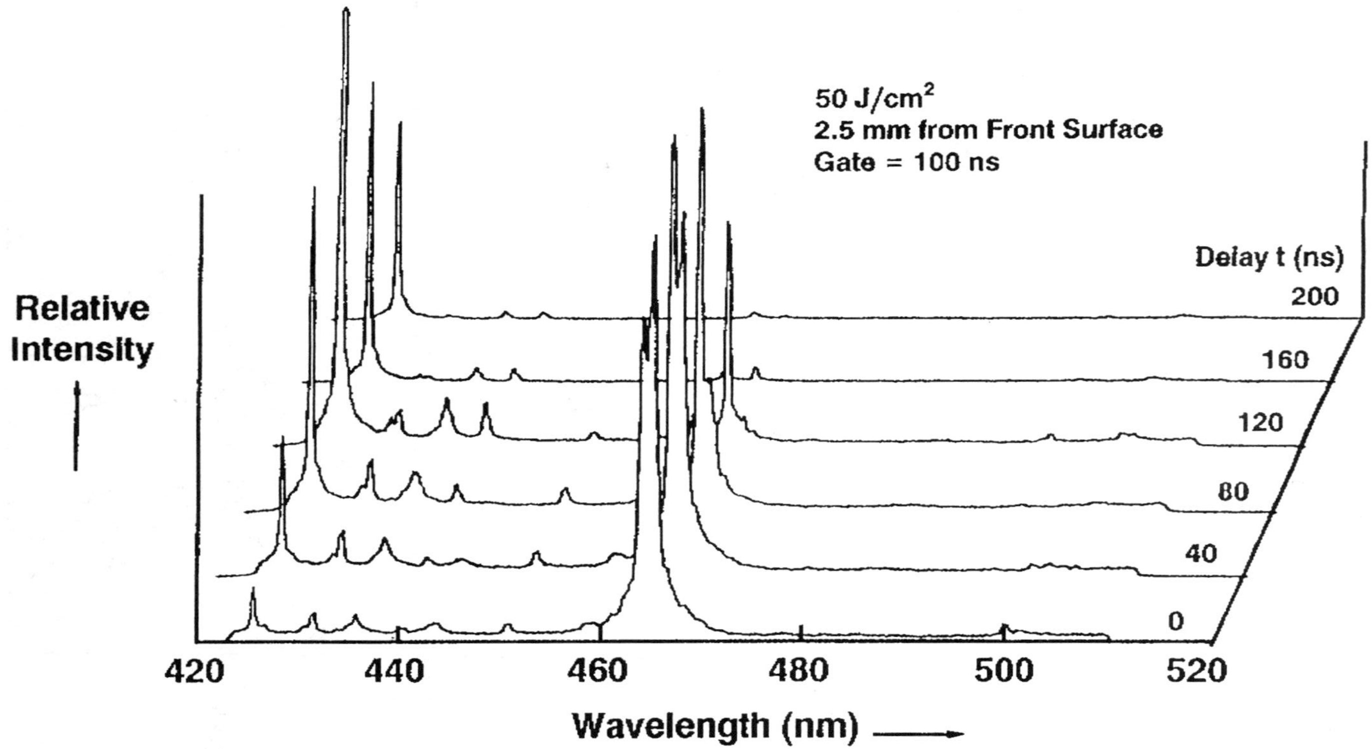

Fig. 5. Plasma emission spectra as a function of delay time from peak irradiation.

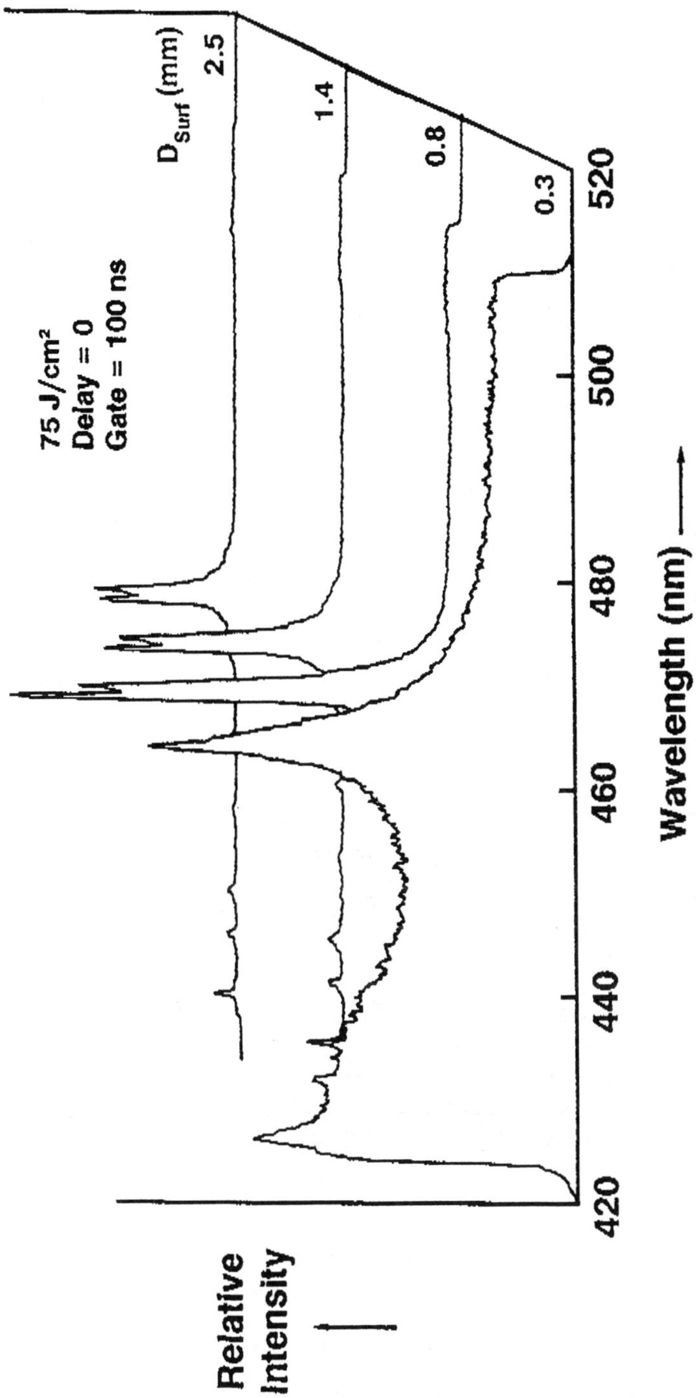

Fig. 6. Plasma emission spectra as a function of distance from the target surface.

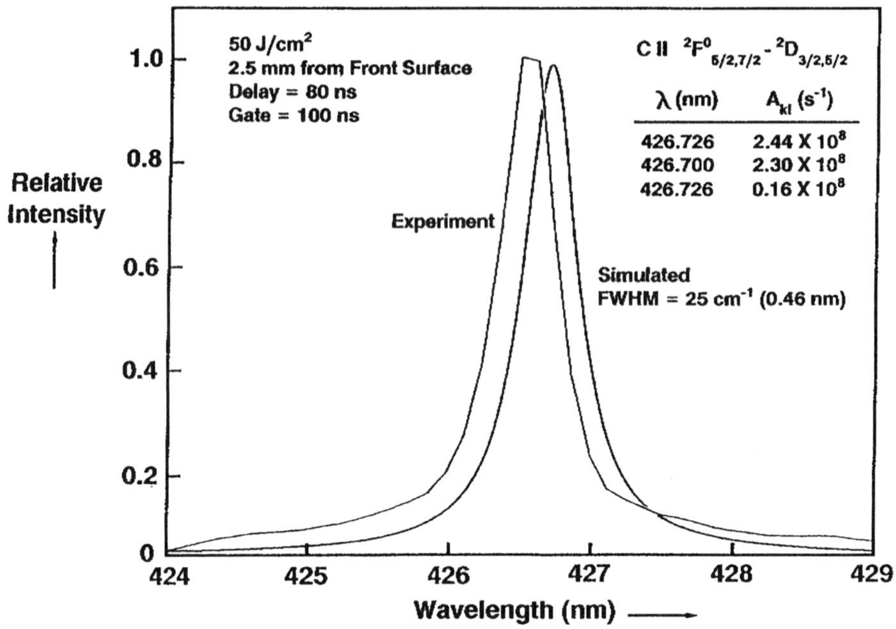

Fig. 7. Experimental and simulated C II plasma emission spectra.

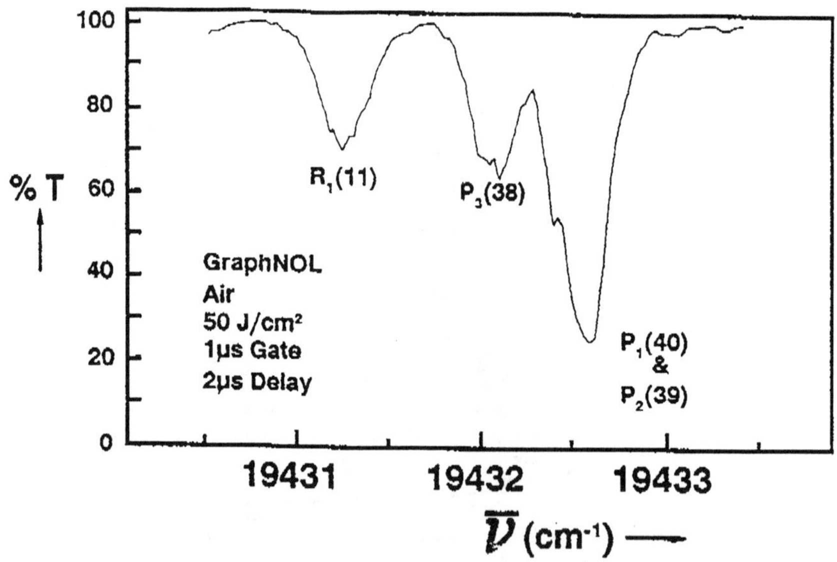

Fig. 8. C_2 plume absorption spectrum obtained during RP HF laser ablation of G.

broadening linewidth corresponds to an electron density of ca. 3×10^{16} cm^{-3} (*8*).

Absorption spectra of ionic carbon vapor species were not obtained. However, Fig. 8 shows an absorption spectrum of C_2 recorded at relatively long delay times (2 µs after the peak of the incident rp HF chemical laser pulse) following rp HF chemical laser irradiation of G in air at a peak fluence of 50 J/cm^2. Spectral simulation of these transitions indicates that the Doppler width is ca. 0.27 cm^{-1}, corresponding to a translational temperature of ca. 9000 K. This very high temperature is probably produced during plasma recombination. The plasma temperature during peak plasma emission is at least 20,000 K (*14*).

Instrumentation Capabilities and Extensions

In addition to the carbon ablation experiments reported above, the computer-controlled laser absorption spectroscopy (CCLAS) system has been used to observe atomic aluminum (Al I) in plumes generated by cw CO_2 laser ablation of pure aluminum in vacuum (*8*); atomic titanium (Ti I) and singly-ionized titanium (Ti II) in plumes generated by rp HF chemical laser ablation of pure titanium in vacuum (*8*); silicon monoxide (SiO) as a reaction product in plumes generated by cw HF chemical laser irradiation of silicon carbide (SiC) in an oxygen (O_2) atmosphere (*8*); and aluminum oxide (AlO) in plumes generated by cw CO_2 laser ablation of alumina (Al_2O_3) (*17*). Frequency-doubled outputs of the CCLAS system were used to obtain tunable probe beams near λ = 308 and 235 nm wavelengths for Al and SiO detection, respectively.

On the basis of laser probe absorption spectroscopy (as well as OMA emission spectroscopy), experiments on plume species performed to date, the following capabilities of the instrumentation have been demonstrated:

1. Both nascent (*i.e.*, directly ablated) and chemically-formed (owing to plume species reactions and/or plume/atmosphere reactions) vapor phase species can be monitored sensitively (to number densities as low as 10^{12} cm^{-3}) and measured quantitatively with excellent temporal/spatial/spectral resolution; these measurements can be made even in the presence of highly luminous plumes/plasmas over a wide range of laser interaction parameters (wavelength, irradiance, waveform, and so on) for all irradiated materials (including the most refractory elements and compounds);
2. The kinetics of chemical reactions on target surfaces and/or in plumes can be obtained as a function of laser parameters, atmospheric composition, distance from the irradiated surface, and so on;
3. The identities, abundances, energy contents, and optical properties of plume species, as well as plasma parameters (electron density and temperature, plasma opacity, and so on), can be determined by measurements on relative absorption intensities and spectral linewidths.

Laser probe absorption spectroscopy permits quantitative measurements on plume species, but is only moderately sensitive. Recently, laser induced fluorescence (LIF) detection of SiO, using the CCLAS system probe beam for excitation of fluorescence, has been performed (*18*); at least two orders-of-magnitude detection sensitivity improvement has been obtained, compared to the typical absorption experiment. LIF detection is preferable when plume emissions are not brighter than the fluorescence signals. Further extensions of the CCLAS system capabilities to vacuum ultraviolet spectral region experiments are planned.

Carbon Ablation Experiments

Under cw laser ablation conditions, much smaller absorption and emission signals of diatomic carbon (C_2) are observed when graphNOL (G) samples are irradiated in a vacuum, rather than an air, environment. In addition, the C_2 spectra obtained in air are practically independent of irradiance level over a broad range of irradiances (7–400 kW/cm^2). Both observations indicate that most of the C_2 is not a nascent ablation product; rather, it is formed mainly by reaction of triatomic carbon (C_3) with oxygen in the air, according to reaction (1) (*see* Results section). This C_2 reaction product has nonequilibrium energy distributions ($T_{translational}$ = 6600 K, $T_{vibrational}$ = 2300 K, and $T_{rotational}$ = 1400 K). Work in progress is directed toward laser-induced fluorescence detection of the nascent C_2 product, as well as other nascent carbon vapor species.

Additional experiments (*1,3,19*) on cw laser ablation of G can be summarized as follows:

1. The enthalpy of mass removal is consistent with thermochemical ablation of carbon to yield an equilibrium distribution of carbon vapor species (*1,3*);
2. No discrete mass removal is observed when fixed frequency probe laser beams are used to identify possible particles with diameters larger than 10 μm (*19*);
3. The thermal coupling coefficient α_λ is equal to 0.8–0.85 at λ = 3.8 and 10.6 μm wavelengths during laser ablation by a cw DF chemical laser using irradiances of 12–46 kW/cm^2 (*1*) and a cw CO_2 laser using irradiances of 9–100 kW/cm^2 (*19*); ablation plumes are transparent at both wavelengths, so these coupling coefficients are caused solely by target optical properties.

Under rp laser ablation conditions, very different phenomena and effects are observed below and above the plasma "threshold" (at peak fluences F_p greater than ca. 30 J/cm^2 for HF chemical laser irradiation with ca. 200 ns FWHM pulsewidth). Highly luminous plumes and specific ionic carbon transitions are observed above this "threshold"; the most intense emissions switch from singly ionized carbon (C II) near threshold to doubly ionized carbon (C III) at fluences well above threshold (up to peak fluences of 165 J/cm^2).

Effective enthalpies of ablation Q^* and apparent total reflectances R_λ at Λ = 2.9 μm have also been determined as a function of peak fluence F_p (*20*). Q^* values increase from ca. 80 kJ/g at low peak fluence (8 J/cm^2) to a maximum value of ca. 800 kJ/g at the highest peak fluence (165 J/cm^2). Similarly, apparent total reflectances $R_{2.9}$ drop from 0.15 to less than 0.04

as a function of increasing peak fluence. Our interpretation of spectral emission data and Q^* and $R_{2.9}$ measurements is that the incident HF chemical laser radiation is absorbed more and more strongly in the plume/plasma as a function of increasing peak fluence. This leads to a higher temperature plasma (and a higher degree of ionization) and greater decoupling from the target surface. Q^* rises as less and less of the incident HF chemical laser radiation reaches the target surface. The apparent total reflectance also drops from a value characteristic of the optical properties of G surfaces (and a transparent plume) to values characteristic of plasma absorption with increasing plasma opacity.

Further experiments will be performed to measure plasma opacity at $\lambda = 2.9\ \mu m$ directly by use of a cw laser probe beam directed through the ablation plume. In addition, further emission and absorption spectra will be obtained over an extended spectral region to observe the formation and decay of particular carbon vapor species and their states of excitation.

Spectroscopic techniques (OMA emission spectroscopy for surveys plus laser probe absorption/LIF spectroscopy for high temporal/spatial/spectral resolution measurements) are valuable for *in situ* nonintrusive characterization of high temperature plume/plasma species. They are complementary to mass spectrometric techniques that are more suitable for survey identification of laser ablated species without specification of energy contents.

Carbon vapor species generated by laser irradiation are strongly dependent on laser parameters and plume properties; above the threshold for plasma formation, formation of carbon ions and free electrons leads to strong plasma absorption of incident laser radiation that dominates the overall energy balance.

ACKNOWLEDGMENTS

This work was sponsored by the Strategic Defense Initiative Organization (SDIO/T/IS) under Contract DNA001-87-C-0271 managed by the Defense Nuclear Agency, by the Air Force Office of Scientific Research under Grant AFOSR-85-0365, and by the Robert A. Welch Foundation under Grants C-0002 and C-0812.

REFERENCES

1. Klein, C. A., Menefee, R. F., Krenek, B. D., and Berry, M. J. (1987), *J. Appl. Phys.* **61,** 1701.
2. Berry, M. J. (1987), *Proceedings of the Sixth DoD Conference on DEW Vulnerability, Survivability and Effects,* vol. 1, Office of the Chief of Naval Research, Arlington, VA, pp. 333–342.
3. Klein, C. A., Berry, M. J., and Miles, P. A. (1989), *J. Appl. Phys.* **65,** 3425.
4. Dingus, R. D. (1989), *Report LA-10113,* Los Alamos National Laboratory, Los Alamos, NM.

5. Berry, M. J. (1988), *The HARC cw Carbon Dioxide Laser Test Resource,* Houston Area Research Center, Woodlands, TX.
6. Leider, H. R., Krikorian, O. H., and Young, D. A. (1973), *Carbon* **11,** 555.
7. Berry, M. J. (1988), *The HARC rp HF/DF Chemical Laser Test Resource,* Houston Area Research Center, Woodlands, TX.
8. Kunz, T. D., Fredin, L. G., Menefee, R. F., and Berry, M. J. (1988) *Final Technical Report,* Contract DNA001-87-C-0271, Defense Nuclear Agency, Washington, DC.
9. Wicke, B. G. (1985), *Carbon* **23,** 131.
10. Brocklehurst, B., Hébert, G. R., Innanen, S. H., Seel, R. M., and Nicholls, R. W. (1972), *Identification Atlas of Molecular Spectra #9, The CN Violet System,* York University, Toronto, Canada.
11. Tyte, D. C., Innanen, S. H., and Nicholls, R. W. (1967), *Identification Atlas of Molecular Spectra, #5, The C_2 Swan System,* York University, Toronto, Canada.
12. Phillips, J. G. and Davis, S. P. (1968), *The Swan System of the C_2 Molecule: Berkeley Analyses of Molecular Spectra,* vol. 2, University of California, Berkeley.
13. Bashkin, S. and Stoner, J. O., Jr. (1975), *Atomic Energy Levels and Grotrian Diagrams,* Vol. 1, North-Holland, Amsterdam, pp. 55–98.
14. Glumb, R. J., Hilyard, R. C., Perry, R. B., Petach, M. B., Rubin, I., and Shackleford, W. L. (1988), *AIAA Paper No. 88-1078,* American Institute of Aeronautics and Astronautics, New York.
15. Knudtson, J. T., Green, W. B., and Sutton, D. G. (1987), *J. Appl. Phys.* **61,** 4771.
16. Wiese, W. L., Smith, M. W., and Glennon, B. M. (1966), *Atomic Transition Probabilities,* vol. 1, NSRDS-NBS 4, US Government Printing Office, Washington, DC.
17. Kunz, T. D., Fredin, L. G., Menefee, R. F., Krenek, B. D., and Berry, M. J. (1989), in progress.
18. Kunz, T. D., Menefee, R. F., Krenek, B. D., and Berry, M. J. (1989), in progress.
19. Kunz, T. D., Menefee, R. F., Krenek, B. D., Fredin, L. G., and Berry, M. J. (1989), *Final Technical Report,* Contract DNA001-88-C-0163, Defense Nuclear Agency, Washington, DC.
20. Fredin, L. G., Menefee, R. F., Krenek, B. D., and Berry, M. J. (1989), in progress.

Space/Time Resolved Spectroscopic Analysis of High Power Laser–Materials Interaction

A. KASUYA AND Y. NISHINA*

Institute for Materials Research, Tohoku University, Sendai 980, Japan

ABSTRACT

Dynamic aspects of the laser ablation process have been analyzed by space/time resolved spectrometry by using a streak camera combined with an optical monochromator. The result shows that the transient process in the desorption of surface atoms into vapor is important in understanding the ablation mechanism of solid surface. Our measurement demonstrates that this instrument is capable of analyzing the physical, as well as the chemical process of transient phenomena through optical detection.

Index Entries: Laser–materials interaction; laser desorption; microcluster, fragmentation; Group IV elements; transient analysis.

INTRODUCTION

The microscopic mechanism of laser beam–solid surface interaction has been investigated extensively since pulsed high power lasers became available in the 1960's. Early work is concerned with the excitation process at the surface by analyzing the emission characteristics of electrons, neutral atoms, and ions ejected from the irradiated surface (*1*). The results have demonstrated the feasibility of using lasers as a tool for materials processing. This technique recently has marked significant progress in its application to many fields of basic research and materials technology, such as production of microclusters, growth of thin films, annealing, etching, or chemical modification of surface layers in the

*Author to whom all correspondence and reprint orders should be addressed.

fabrication of microelectronic devices (*2*). At an extremely high level of energy fluence, the laser irradiation produces a highly excited plasma that can be used for initiating the controlled nuclear fusion process (*3*).

Despite the vast varieties of applications of this technique to new materials technology, our physical understanding of its microscopic process is rather limited to a narrow range of irradiation conditions and to a few kinds of target materials. The process for desorption of particles from the irradiated surface may be classified into three consecutive steps of transformation: (1) excitation of solid target by laser beam, (2) desorption of particles from the target surface (transition from condensed matter to vapor), and (3) expansion of vapors into vacuum. The dominant process for step (1) is the electronic excitation in metals and insulators having energy gaps smaller than the photon energy of laser beam. The energy absorbed in the electronic system is transferred to the lattice in a time typically less than 1 ps, during which the thermal equilibrium is established in the solid (*4*). Then, the surface atoms evaporate into vacuum as a thermal process at the temperature of a surface heated by the laser irradiation. This is the so-called thermal model of laser desorption, which may be applied to the case of low power laser excitations where the irradiated surface does not change its structure and composition significantly before atoms are ejected from their lattice sites (*1,4*).

For higher levels of excitation, a very high density of atoms desorb from the irradiated surface. These atoms may be highly ionized and act as plasma plume. This plasma can be formed in the early part of a laser pulse and continue to absorb the latter part of the same pulse through inverse bremsstrahlung process while it expands into vacuum in step (3) (*1*). This process of excitation has been investigated extensively in relation to the inertial confinement of laser-produced plasma in fusion studies (*3*). The inverse bremsstrahlung process causes desorbed particles to be excited to a state of high kinetic energy with considerable degree of ionization, in agreement with many experimental observations at very high levels of laser fluence (*1,3*). There must be a lower limit in laser intensity and pulse duration above which the inverse bremsstrahlung process becomes significant. A simple estimate by Archbold et al. (*5*) shows that for a 100 ns pulse and atoms with an ionization energy of 8 eV, the initial temperature of plasma before expansion must be above 5000°C in order for the plasma to absorb a significant amount of energy during its expansion. This minimum value of temperature increases with decreasing pulse width and increasing ionization energy of the target atom. This estimate indicates that there may be a range of laser fluence that is too high to apply the model of thermal evaporation processes, but is too low to apply that of the inverse bremsstrahlung process since boiling points of most of the elements and compounds are below 5000°C. It is questionable whether the thermal model can be applied to cases when the surface temperature exceeds the boiling point of the target material. There are many experimental observations in this range of laser fluences where desorbed atoms exhibit those values of kinetic energy and an ion density much higher than the corresponding quantities predicted by the thermal model (*6*).

One may note in the above discussion that neither of the two models of laser desorption takes explicit account of step (2), during which atoms are actually ejected out from the surface. The dominant step to absorb laser energy is in step (1) for the lower excitation region, and it is in step (3) for the higher excitation region. In such cases, the contribution from step (2) may be neglected to a good approximation in calculating the kinetic energy and degree of ionization of desorbed particles. This seems to be the reason why simple theoretical models agree rather well with experiments in these two ranges of excitation despite the fact that these models do not quantify the most fundamental part of the desorption process (step 2).

In the intermediate region of excitation, the final state of desorbed atoms may depend critically on the transient process of step (2), during which a significant part of the laser energy is absorbed. It is necessary, therefore, to identify the process of transformation of surface atoms from their condensed phase into vapor. This paper presents space/time resolved analysis of light emission spectra obtained from the plume of vaporized materials using a streak camera system. Similar measurements were conducted during the 1960's (*7*), but were either in higher regions of excitation, or not detailed enough to reveal many peculiar features of the streak pattern observed in our measurements (*8–10*). Our system resolves the desorption processes in time, space, and wavelength in a one shot measurement. The result indicates that the surface atoms leave the target in the form of various sizes of clusters that subsequently decompose into atoms and their ions in a time scale longer than 10 ns, in agreement with our previous time-of-flight analysis of desorbed ions (*11*). The clusters have high absorption coefficients comparable to the solid. Hence, they continue to absorb the latter part of a laser pulse. This absorption process can account for the high ion yield and translational energies observed by many of the previous investigators (*1,6*). It should be emphasized that this mechanism of energy transfer from laser beam to the cluster system has never been conceived nor proposed by those workers who have presented either the thermal model for lower excitations (*4*) or the inverse bremsstrahlung process for much higher levels of excitation (*1,3*).

EXPERIMENTAL RESULTS AND DISCUSSION

Our optical system records the transient process of light emission from the plume of desorbed atoms from a solid target by projecting its light intensity in a coordinate system of (1) distance in millimeters in any two choices of the x, y, and z directions, (2) time, t, in nanoseconds, and (3) wavelength, λ, ranging from 200 to 800 nm. The streak camera displays a cross-sectional intensity distribution of any of the two choices of the coordinates. Figures 1 and 2 show results on graphite under KrF laser irradiation with its photon energy of 5.0 eV. The patterns represent the time integrated intensity distribution in the cross-section of the y–z plane with the laser beam incident normal (z-direction) to the sample surface (Fig. 1) and 45° off normal toward tangential (y-direction) (Fig. 2). The light emission is viewed along the x-direction. Figure 3 shows the z–t pattern, which represents the time evolution of the intensity distribution along the z-axis in Fig. 1. These figures, together with our previous investigations in group IV elements (*8–10*), show the following characteristics of the intensity pattern: (1) the maximum intensity of light emission from the plume takes place in an area of a few tenths of a millimeter off the surface toward the z-direction and at a time a few nanoseconds after the tail end of 10 ns laser pulse; and (2) the surface atoms expand in vacuum into two different components. These two components are in marked contrast with each other in respective characteristics described as follows: (2a) one moves faster with wider distribution of velocity both in magnitudes and directions. The other moves slower with narrower distribution; (2b) the area of maximum intensity in Figs. 1 and 2 corresponds to the slower component, and the outer area of weaker intensity the faster component; and (2c) the light emission from the faster component begins immediately after the tail end of the laser pulse and the slower component is delayed by a few nanoseconds. This delay is more significant in the case of Si, where the maximum emission takes place after the end of the laser pulse (*9*). Furthermore, the ratio of the intensity of a faster component to a slower one is larger for Si than graphite (*8*). Characteristics (2a–c) show that the plume expands into vacuum in a way much more complicated than one would expect from the time-integrated pattern shown in Fig. 1. The pattern of Fig. 1 gives us an impression as if the plume would expand from the spot of the maximum intensity toward the outer area (*1*). Careful comparison of this time-integrated pattern with the time-resolved one of Fig. 3 verifies that the maximum intensity pattern is dominated by the slower component, whereas the outer region of weak intensity in Fig. 1 is owing to the faster. (2d) The intensity ratio of continuum to discrete emission is larger for the faster component than the slower one between 200 and 800 nm. This conclusion is drawn from our experimental results showing that the intensity of the faster component is much weaker than the slower one in the z–t pattern if a time-resolved spectrum is measured with a monochromator placed in front of the streak camera and set to any line of strong emission. (2e) For the off-normal incidence of laser beam in Fig. 2, the intensity distribution of the faster component does not change much from the normal incidence in Fig. 1, whereas the slower component changes appreciably and points back toward the direction of incident laser beam. These peculiar features cannot be explained easily in terms of existing models for expansion of micro-plasma of desorbed gas produced by laser excitation.

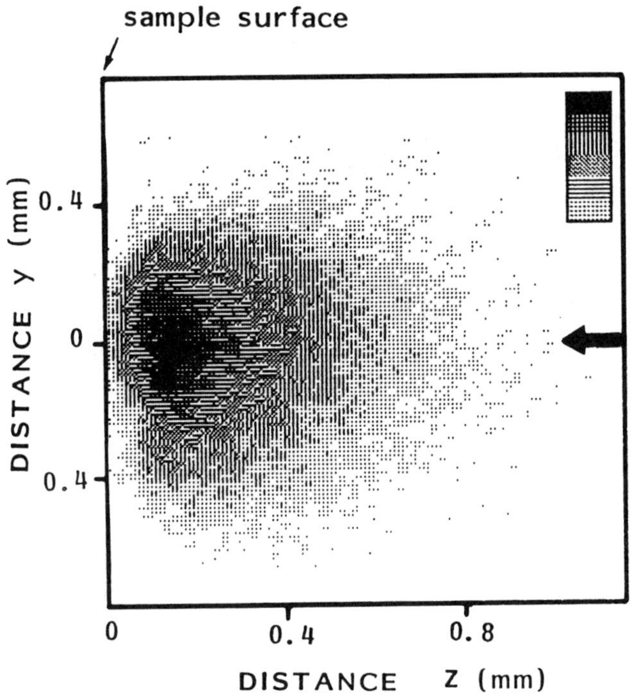

Fig. 1. Time-integrated image of the plume emitted from a graphite surface excited by a KrF laser pulse of its peak intensity at 30 MW/cm^2. The distances y and z are measured along the tangential and normal directions with respect to the irradiated surface of graphite, respectively. The relative scale of intensity is quantized linearly by dot patterns shown in the inset. The direction of laser beam incidence is indicated by the arrow from the right.

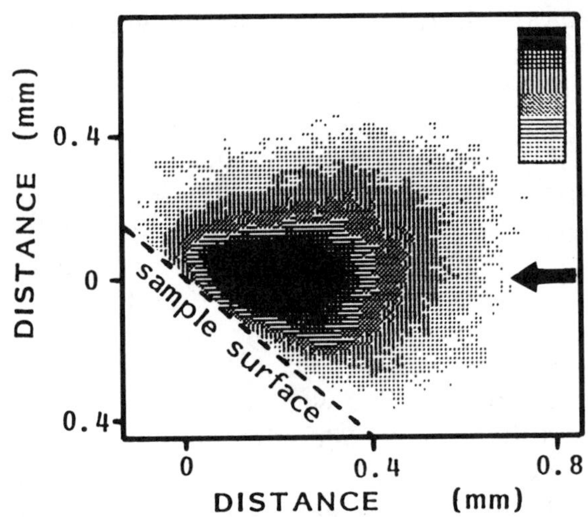

Fig. 2. Time-integrated image of the plume emitted from a graphite surface with its surface tilted 45° with respect to the KrF laser beam. The incident direction of laser beam is indicated by the arrow from the right.

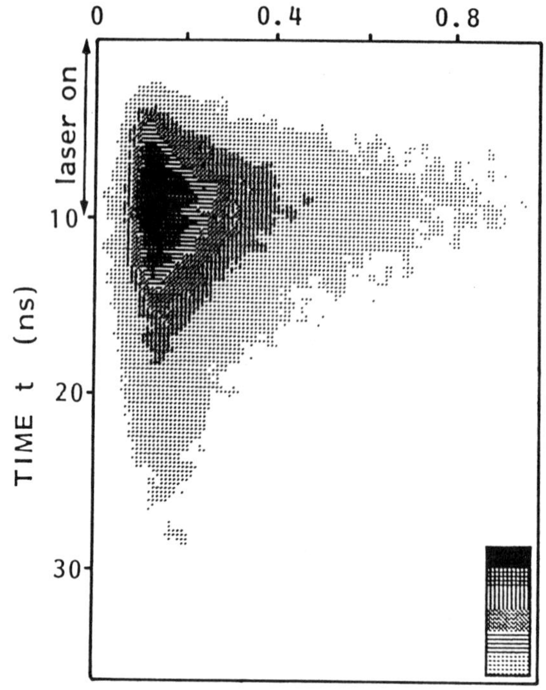

Fig. 3. Time evolution of the intensity of light emitted from the plume of laser desorbed particles toward the z direction normal to the surface of graphite.

Origin of the Faster Velocity Component

The velocity of the faster component exceeds 10^6 cm/s, which corresponds to kinetic energy of 6 eV for a carbon atom. On the other hand, weakness in the intensity of visible line emission from the faster component indicates that the desorbed atoms carry lower energy in their internal electronic system than in their translational energy. This fact implies that the desorbed atoms are not excited by the inverse bremsstrahlung process. Otherwise, the process would excite the system in both degrees of freedom in energy in comparable magnitudes. This result is consistent with another experimental fact (described in 2e) that the faster component does not travel toward the direction of the incident laser beam, which is tilted at 45° off-normal to the sample surface, but expands rather isotropically. If the faster component absorbed laser energy during its expansion, it would exhibit its maximum emission intensity in the time and space domains of the maximum laser intensity. The results in Figs. 2 and 3 are evidently incompatible with this argument.

The apparent difference between the internal energy and expansion energy of desorbed atoms may be explained if one assumes a transient process of a desorption and subsequent expansion. The process is analogous to a rapid release of compressed gas into vacuum. The energy stored in the compressed gas is converted into translational energy of expansion. The free expansion, for example, of compressed gas into vacuum in one dimension yields its asymptotic expansion velocity

$V_{as} = 2v_s/(\gamma - 1)$, where v_s is the velocity of sound in the initial gas and γ the adiabatic exponent. This value can be higher than the thermal velocity by a factor of 3.9 for monoatomic gases. This asymptotic velocity, however, is still much smaller than the observed high velocity of expansion of over 10^6 cm/s. Gurevich et al. have treated desorbed atoms as a collection of partially ionized gas in a state of plasma. They take into account the Coulomb attraction between ions and fast moving electrons and show that ions may be accelerated to a velocity comparable to the thermal velocity of electrons during their expansion (*12*). This model, therefore, for the free expansion of the plasma predicts an expansion velocity comparable to our experimental values.

The Origin of the Slower Velocity Component

The slower velocity component is characterized by its delay of emitting visible light with respect to the laser pulse. In addition, the luminous plume tends to direct back toward the incident direction of the laser beam rather than point toward the surface normal as the laser beam is tilted from the normal toward the tangential. These characteristics indicate that this component of desorbed particles absorbs beam energy as they move (expand) away from the sample surface so that the intensity of luminescence is higher in the beam area of higher laser intensity. The delay of luminescence has been observed by many investigators (*1*) under irradiation conditions at much higher powers (over 10 GW/cm^2) or for longer laser pulses (over 10 ns) than ours. In these cases, desorbed particles (atoms) are excited by the inverse bremsstrahlung process into multiply-charged ions. Then, light emission takes place initially from ions of highest charge multiplicity and subsequently from those of lower ones down to neutral atoms at later times as they recombine with electrons during the process of relaxation and expansion. Our spectrum, on the other hand, shows lines dominantly from neutral and singly-ionized atoms with their delay of luminescence in comparable magnitudes. No lines from multiply-charged atoms are detected. Since there are many allowed transitions in the multiply-charged atoms in the visible region, they should be detected if their density is appreciable. The absence of lines from multiply-charged ions is consistent with our intermediate level of excitation where the inverse bremsstrahlung process is negligible. It is emphasized, however, that the desorbed particles still absorb a laser beam in our excitation condition and exhibit a significant length of delay in luminescence both in the forms of ions and neutrals comparable to the period of laser pulse. It is also noted in the z–t pattern of luminescence in Fig. 3 that the slower component does not represent a simple form of expansion of excited atomic gas (laser-produced plasma) into vacuum. The pattern rather indicates that the desorbed particles are confined in a small volume in space and luminesce strongly only at the onset of expansion. The z-λ pattern also exhibits peculiar spectral features unexpected for the laser-produced plasma. Figure 4 shows the spectra for graphite under ArF laser excitation at the photon energy of 6.4 eV. The

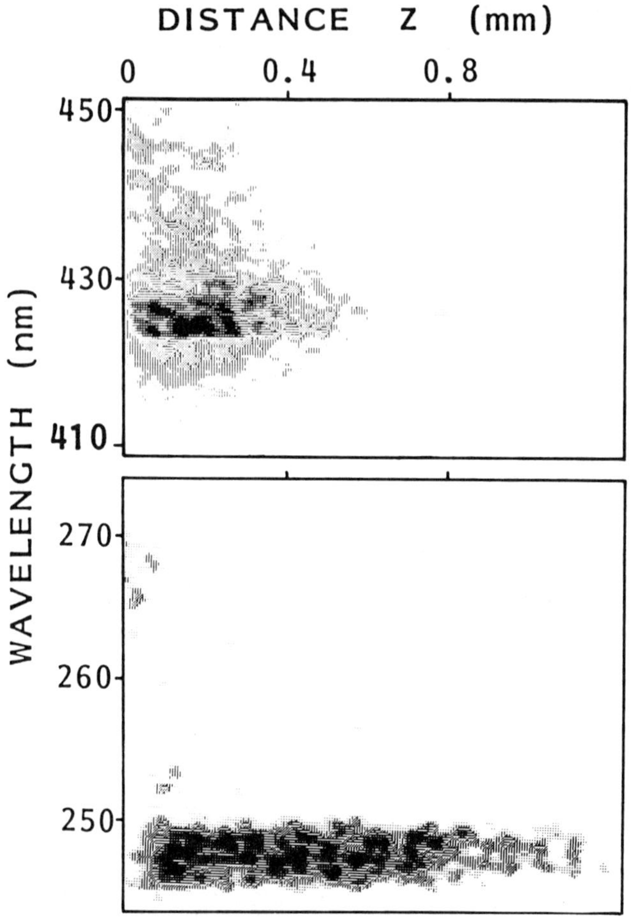

Fig. 4. Emission spectrum by ArF laser (193 nm) excitation in the vicinity of 427 nm (ionic) and 248 nm (neutral) lines of the plume along the z direction normal to the surface of graphite.

spectral linewidth of ions at 427 nm (transition from 4f $^2F^o$ to 3d ^2D) is found to be much larger than neutrals at 248 nm (from 3p ^1S to $2p^3$ $^5S^o$). These features are found common to all other ionic and neutral lines from C, as well as from Si, Ge, and Sn (*8*). The broadening of the ion lines is too large to be accounted for by Doppler and collision broadenings for a gaseous temperature below 10^6 K and density below 10^{23}/cc. The Stark broadening in many cases yields larger linewidths for neutral atoms than ions. These broadening mechanisms, therefore, cannot account for our experimental results. All of the above features of our experimental observations cannot be explained in terms of any hydrodynamical process in plasma previously proposed (*1*).

The discrepancies found between experiment and the model of plasma excitation and expansion through the inverse bremsstrahlung process can be explained in a simple picture of the desorption process by assuming that particles are emitted in the form of various sizes of clusters. If the surface is excited and its temperature rapidly exceeds the boiling point, it can no longer remain in the solid. If the excitation is so rapid and inhomogeneous, it cannot evaporate layer by layer from the top surface. It is most likely that atoms within the skin depth leave the surface simultaneously in the form of various sizes of clusters. These clusters continue to absorb laser light, because they have absorption coefficients comparable to the bulk solid. Furthermore, they are excited much more efficiently than in the solid because they are isolated from the bulk and, hence, they do not lose their absorbed energies through heat conduction and charge neutralization to the bulk solid. These two effects give rise to a highly-excited layer of surface as soon as clustering takes place within the skin-depth. A cluster can keep a large amount of energy, while distributing it into high degrees of freedom through many internal nonradiative transitions. This channel of relaxation is much more probable than prompt radiative decay in the visible during and immediately after laser excitation. If the desorbed cluster is highly excited, it eventually decomposes into smaller fragments. Atoms in highly-excited states may be ejected in the process of fragmentation and subsequently relax through radiative decays that contribute to the delayed luminescence. The ejection of atoms owing to fragmentation of a cluster takes place in various local environments. Such a situation can give rise to spectral line profiles for neutral atoms different from those of ions.

The above mentioned desorption process of clusters and their subsequent fragmentation into highly excited atoms and ions can account for all the emission characteristics of the slower component observed in the experiments. If the slower component consists of microclusters of the target elements, they can absorb laser energy after desorption from the surface, but do not expand immediately until their fragmentation takes place at a later time. During fragmentation, they expand and emit light from atoms and ions ejected in highly excited states from the parent fragments. The specific properties of microclusters, such as a high optical absorption coefficient, high degrees of internal freedom are recognized in our experimental observations for the slower component. It is emphasized that the absorption of laser beam by clusters is significant in the intermediate levels of excitation where the inverse bremsstrahlung process is not appreciable (surface excitation below 10^4 K in Si (*5*)). At higher levels of excitation, the clusters are produced in higher density and excited states. These clusters decompose more rapidly into monoatoms/ions within an early period of a laser pulse so that the absorption of laser energy by desorbed clusters becomes less significant. The density (temperature) of decomposed atoms, on the other hand, becomes higher and, hence, the inverse bremsstrahlung process begins to be increasingly dominant and can account for many of experimental observations on desorbed particles at the high levels of excitation (*1,3*).

CONCLUSION

The transient process of laser ablation has been analyzed by space/time resolved spectrometry with a streak camera system combined with optical monochromator. For an intermediate level of laser energy fluence, the streak patterns in the z–y, z–t, and z–λ planes show many peculiar features that have been overlooked in previous investigations. These features can be explained in terms of desorption of surface atoms in the form of clusters, but not of the expansion of plasma of vaporized monoatoms or small molecules into vacuum. The desorption of surface atoms in a cluster form may be regarded as a transient intermediate step of transition for surface atoms in a solid to atomic vapor. This process, together with thermal model of surface excitation and the inverse bremsstrahlung process of postdesorbed excitation, provides a unified picture of laser desorption in the entire region from low to extremely high intensity of excitation. This space/time spectroscopic technique can be applied to many fields of interest where a transient analysis is essential in understanding their physical, as well as chemical processes.

ACKNOWLEDGMENT

The authors would like to thank C. Horie and T. Goto of Tohoku University for valuable discussions. This work was supported, in part, by a scientific grant from the Mitsubishi Foundation.

REFERENCES

1. Ready, J. F., *Effects of High Power Laser Radiation*, chapter 4, Academic, New York, 1971.
2. *Materials Research Society Symposia Proceedings*, vol. 129, Johnson A. W., Sigmon, T. W., and Loper, G. L., eds., 1989.
3. Nuckolls, J. H., *Phys. Today* **35,** 24 (1982).
4. *Laser Annealing of Semiconductors*, chapters 1–4, Poate, J. M. and Mayer, J. W., eds., Academic, New York, 1982.
5. Archbold, E., Harper, D. W., and Hughes, T. P., *Brit. J. Appl. Phys.* **15,** 1321 (1964).
6. Furstenau, N. and Hillenkamp, F., *Int. J. Mass Spectr. Ion Phys.* **37,** 135 (1981).
7. Basov, N. G., Boiko, V. A., Dement'ev, V. A., Krokhin, O. N., and Sklizkov, G. V., *Soviet Phys. JETP* **24,** 659 (1967.).
8. Kasuya, A. and Nishina, Y., *4th Int. Symp. on Small Particles and Inorganic Clusters*, Aix-En-Provence *Zeitschrift fur Physik D*, **12,** 493 (1989).
9. Kasuya, A. and Nishina, Y., *Proc. of 19th Int. Conf. on Physics of Semicond.*, Warsaw, ed. W. Zawadzki (Inst. Phys. PAN, Warsaw 1988) p. 725.
10. Kasuya, A. and Nishina, Y., *Materials Research Society Symposia Proc.*, vol. 129 (in press) (1989).
11. Kasuya, A. and Nishina, Y., *Phys. Rev. Lett.* **57,** 755 (1986).
12. Gurevich, A. V., Pariiskaya, L. V., and Pitaevskii, L. P., *Sov. Phys. JETP* **22,** 499 (1966).

In Situ Analysis of Laser-Induced Vapor Plumes

PETER K. SCHENCK, DAVID W. BONNELL, AND JOHN W. HASTIE*

Metallurgy Division, National Institute of Standards and Technology, Gaithersburg, Maryland 20899

ABSTRACT

Laser-induced vapor plumes are becoming increasingly important in the processing of advanced materials and as a medium for fundamental high temperature chemistry experiments. These aspects have been discussed by us in detail previously (*1–5*) and by others elsewhere in this volume. However, the detailed molecular and thermal structure of laser-induced vapor plumes and plasmas is still largely unknown. State-of-the-art spectroscopic and spectrometric techniques are required for the spatial and temporal analysis of such plumes. This chapter outlines recent developments in our laboratory for the analysis of laser-induced vapor plumes, generated by the interaction of high power laser pulses with refractory targets in vacuum. The principal techniques used are high pressure sampling molecular beam mass spectrometry, and optical multichannel emission spectroscopy.

Time-resolved mass spectra of C_n (n = 1–9) neutral vapor species have been obtained from graphite targets using various laser-surface interaction orientations and vapor plume sampling geometries. Ionic species C_1^+, C_2^+, and C_3^+ have also been observed in the laser-induced plume, but their abundance is strongly dependent on the plume sampling orientation. Mass spectra obtained from superconducting $YBa_2Cu_3O_x$ (x = 6.5–7.0) targets also showed a diversity of species present in the laser-induced plumes, including both neutral and ionic Y, Ba, and Cu. In addition, molecular species, such as BaO, CuO^+, YO, and bimetallics (BaCu and YCu) were observed.

*Author to whom all correspondence and reprint orders should be addressed.

For similar target materials, but in a separate experimental system, time and spatially-resolved optical emission spectra have been obtained with an optical multichannel analyzer (OMA). With graphite targets, intense emission was observed from C, C^+, and C^{++} in the near UV. Molecular C_2 emission was also observed in the vicinity of the surface/plume interface. The emission spectra obtained from the superconducting $YBa_2Cu_3O_x$ targets included intense lines from both neutral and ionic Y, Ba, and Cu, as well as YO. With repetitive laser pulses at a fixed target location, the spectral emission intensities of Y and Ba decreased more rapidly than for Cu with increasing number of laser pulses. This observation is consistent with surface analysis results where films deposited with higher energy are Cu-deficient relative to Ba and Y.

Index Entries: Graphite; $YBa_2Cu_3O_x$ laser vaporization; mass spectrometry; emission spectroscopy; refractory materials; superconductivity; high temperature; thermochemistry; temperature measurement.

INTRODUCTION

Reliable information on materials behavior at very high temperatures, particularly beyond 2000 K, is generally lacking. Measurement problems associated with this temperature regime have prompted us to investigate the applicability of laser vaporization, coupled with high pressure molecular beam sampling mass spectrometry (HPMS) and optical emission spectroscopy, as a new tool for thermochemical and kinetics studies. Initial studies on this problem have been reported by us elsewhere for the boron nitride and graphite systems (*1–3*).

The vapor plumes produced by this approach also have practical application as a material transport medium for the deposition of thin films. The formation of superconducting thin films is one of many such examples of current interest (*5,6*). Recent observations by Moorjani et al. (*6*) indicate that laser vaporization or ablation of the superconducting phase, $YBa_2Cu_3O_x$ ($x = 6.5–7.0$) in vacuum and onto a room temperature substrate can directly form thin films with superconducting regions without oxygen annealing. In order to understand this process, an *in situ* analysis of the laser-generated plumes from $YBa_2Cu_3O_x$ targets is required. Preliminary results, obtained using both HPMS and optical emission spectroscopy, are reported here.

For typical laser heating experiments, a gas-dynamically stabilized vapor plume is generated by a 10 ns pulsed Nd/YAG laser beam focused onto a target. The generated plume has instantaneous pressures (~1 atm; 1 atm = 101325 Pa) much greater than those required by classical free molecular flow sampling techniques ($\sim 10^{-4}$ atm). At high pressures a rapid isentropic expansion occurs to generate a nonperturbed sample beam for mass spectrometric analysis. Ideally, for thermochemical studies, the laser should serve as a heat source without perturbing the vaporization, plume formation, or analysis processes. Also, it is desirable to establish laser heating conditions such that the vapor species and target surface are effectively at thermodynamic equilibrium. Establishment of these conditions may depend on such factors as: the optical and thermal properties of the target material, surface and system geometry, and laser energy, wavelength, and pulse width. Earlier laser vaporization studies on boron nitride and graphite materials indicated that it is possible to establish near-thermodynamic local equilibrium conditions for laser conditions similar to those reported here (*2,3*). The present study addresses the dynamic, nonequilibrium features of laser-induced vapor plumes, particularly with respect to species formation and decay and their thermal state.

The materials used for these studies are as follows. The graphite targets were made from high density/high purity graphite. This material has a bulk density of 1.75 gm/cm^3 and a relatively small grain size. Specimens 0.24 cm thick were parted from a 5 cm diameter rod with carefully cleaned steel cutting tools. The $YBa_2Cu_3O_x$ targets were prepared by sintering appropriate amounts of $BaCO_3$, Y_2O_3, and CuO followed by hydraulic pressing into disks (*6*).

METHODS

Mass Spectrometry

The basic apparatus, shown schematically in Fig. 1, is similar to that described in detail elsewhere (*2*) with the exception of a modified plume sampling orientation described here. An important feature of the sample-mount region is the very high pumping conductance, exceeding 2000 ℓ/s in the plume formation region. This allows good vacuum conditions to be maintained with minimal interference to the vapor plume and molecular beam-forming processes. No evidence was found in our earlier laser vaporization studies for a significant perturbation of species identities and partial pressure distributions during the plume and molecular beam-forming processes (*2,3*). The mass filter (Fig. 1) is a quadrupole system with a mass range of 1–1000 u (standard unified mass units) and typically better than unit resolution. An electron impact ion source, used to produce positive ions from neutral beam species, operates at adjustable electron energies, typically from 10–100 eV. The ion source allows an unrestricted entrance and exit of the molecular beam.

The laser used for the present studies is a Nd/YAG 20 Hz pulsed laser with a typical pulse width of 10 ns. Since the pulse duration of the laser is short, the sample cools nearly to the ambient temperature between laser shots. Discrete wavelengths of 355, 532, and 1064 nm, with energies up to 200 mJ/pulse at 1064 nm, are available. For the present studies, the 532 nm wavelength has been used, with energies of 10–40 mJ focused to a 250 μm beam spot diameter. A radiometer is used to directly measure the laser energy. These measurements are made periodically during experimental runs to confirm constant laser power conditions. Alignment of the laser impact point with the mass spectrometer (MS)

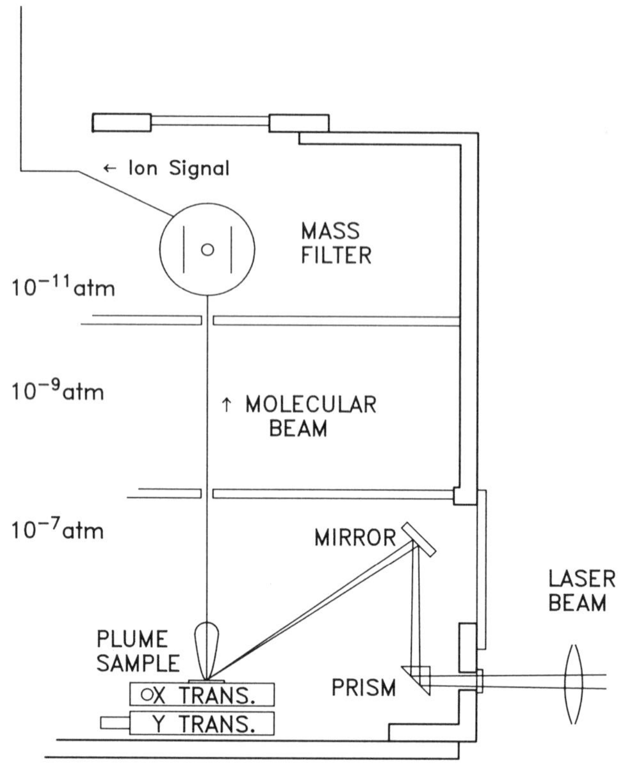

Fig. 1. Schematic of laser-induced vaporization mass spectrometric facility showing laser-sample molecular beam geometry for on-axis plume sampling.

beam-acceptance axis is monitored using a photodiode located just above the mass filter ion source. Initial alignment is obtained by adjusting the laser mirror (*see* Fig. 1) to maximize the intensity of the laser light reflected from the sample surface, as monitored from an observation port at the top of the beam path. The sample targets are mounted on a two-axis translation stage. This allows samples to be positioned or rastered, as desired, between laser shots. Raster rates up to 200 μm/s are possible (100 μm/s is typical) with a 95% heated area overlap and a ~250 μm beam spot diameter. The sample can be mounted with the translation stage flat, as shown in Fig. 1, or the stages can be mounted on a wedge to vary the sampling angle of the mass spectrometer relative to the sample surface. The periscope, shown in Fig. 1, allows for plume sampling normal to the target surface. For off-axis sampling, the laser beam bypasses the periscope and is focused directly onto the sample.

Two signal detection schemes are used. The first approach provides time resolved time-of-arrival (TOA) ion intensity signals for individual masses by tuning the mass spectrometer to a specific mass value and directly signal averaging the detector output. The time scales for the laser pulse and ion transport between the mass filter ion source and the detector are negligibly short compared with those of the TOA signals.

Hence, the TOA profiles very closely approximate the flight times of the neutral or charged plume species between the sample target and mass filter (*1–3,7*). These profiles contain important velocity distribution and species assignment information, as discussed by us in the section below and elsewhere (*2,7*).

The second signal detection scheme provides time-gated mass spectral scans by integrating the time-resolved information with a boxcar integrator. The ramp generator directly sweeps the mass pass band of the mass spectrometer via an external command input of the quadrupole control unit. This ramp generator also triggers the signal averager used to record and average the data runs. The time base of the boxcar averager is synchronized with the laser. The boxcar gate, or integration time, is positioned in time to collect the ion signal synchronous with the laser. Since the arrival time of the different ions are mass dependent, the gate width is set wide enough, typically 500 μs, to collect signals for all masses of interest. The signal averager is connected to a laboratory computer for data storage and subsequent analysis.

Plume Temperature Analysis

One of the objectives of this work is to establish the degree of chemical and thermal equilibration present in the plume and also in the extracted molecular beam. In our earlier studies, evidence was presented for a frozen chemical equilibration of the plume species even though the translational temperature and surface temperature conditions differed markedly (*1–3*). Indirect determinations of surface temperature were also made in these initial studies. At the present time, no direct pyrometric, or other approach, has been developed for the very short pulse-time (ns) conditions used. We anticipate that the OMA technique described here can be adapted to multiwavelength pyrometry in future studies. For much longer pulse times (ms), pyrometric approaches have been used successfully, as discussed by Olander elsewhere in this volume.

TOA Molecular Weight Correlation

Where more than a single molecular weight species is present, as is the case for the C_n (n = 1–9) system, a translational temperature can also be determined for a thermally equilibrated system from the expression:

$$\Delta t = L/v = k \cdot L(M/T)^{1/2} \quad (1)$$

where Δt is the arrival time, *L* the molecular beam flight distance (26 cm), *v* the species beam velocity, *k* a constant characteristic of the beam formation dynamics, *M* the neutral species molecular weight, and *T* the temperature. Thus, a linear plot of Δt vs $M^{1/2}$ should indicate that the species-specific ion signals are assigned to the correct neutral precursors and that the species are formed, and sampled, with a common thermal history. This relationship was, in fact, observed in both the graphite and BN systems, as presented elsewhere (*2,3,7*).

TOA Deconvolution

The mass and time-resolved TOA species profiles can be deconvoluted to give velocity distribution, and hence translational temperature information, for individual species as discussed below. Time-of-arrival ion intensity signals $I^+(t)$ can be calculated from the convolution integral of the creation function, speed distribution, and detector response, as described by Young (*8*)

$$I^+(t) = K \int_0^t g_d(t-\lambda) \int_0^\lambda s^2 f(s)(\gamma/L)\, A(\tau)\, d\tau\, d\lambda \tag{2}$$

where $A(\tau)$ is the creation function, L the distance, γ the most probable speed, $s = v/\gamma = L/\gamma(\lambda-\tau)$ the speed ratio, $g_d(t-\lambda)$ the detector response, K the calibration constant of the detection electronics, and $f(s)$ the speed distribution. λ and τ are formal integration time variables yielding t, the laboratory time. It can be shown that the time scale of the creation function, i.e., laser heating pulse and subsequent cooling, is short compared with plume formation, beam transit, and detection. For the purpose of initial application of the deconvolution, $A(\tau)$ may be approximated by a delta function at $\tau = 0$. The integral then becomes

$$I^+(t) = K \int_0^t g_d(t-\lambda)\, s^2 f(s)(\gamma/L)\, d\lambda \tag{3}$$

where now $s = v/\gamma = L/\gamma\lambda$. Our detector response is not fast on the time scale of the experiments. The general response function for an RC type detector is

$$g_d(t) = (1/\tau_e) \exp(-t/\tau_e) \tag{4}$$

where τ_e is the time constant of the detector electronics, determined in our case to be ~120 μs. Through the use of Laplace transforms (*8*) or by standard differentiation by parts (of the definite integral), using this response function, we obtain

$$J(t) = I^+(t) + \tau_e\, d[I^+(t)]/dt \tag{5}$$

where $J(t)$ is now the deconvoluted TOA signal with the response time delay removed. To clarify the relation of f(s) to J(t), it is important to note that the $s^2f(s)(\gamma/L)$ term in the integral Eq. (2) is just the conversion of the speed distribution f(s)ds to an arrival time distribution f(t)dt since $ds/dt = -(\gamma/L)s^2$. The Maxwell velocity distribution in a cavity is given by

$$P_m(v) = 4\pi\, (M/2\pi kT)^{3/2}\, v^2 \exp(-Mv^2/2kT) \tag{6}$$

where M is the mass, k Boltzmann's constant, and T the temperature. The root, mean-square velocity for this distribution is

$$v_{rms} = (3kT/M)^{1/2} \tag{7}$$

and the most probable velocity is given by

$$v_{mp} = (2kT/M)^{1/2} \tag{8}$$

Now the conditions in the HPMS produce a molecular beam that is similar to that produced by molecules effusing from a hole in a cavity (box) into a vacuum. In this case, the velocity distribution in the beam is skewed toward higher velocities, i.e., the faster moving molecules are more likely to escape through the "hole," having a new distribution, P_b, related to the source distribution, P_m, by:

$$P_b(v) \propto v\, P_m(v) \tag{9}$$

where both $P_b(v)$ and $P_m(v)$ have the appropriate normalization. In order to analyze the TOA signals, these distributions must be converted from velocity space to time space. As noted earlier in the discussion of the convolution integral, the P(v)'s are velocity distributions and the conversion to time space is not only given by setting $t = L/v$, but also involves the conversion of P(v)dv to P(t)dt. Defining $t_{mp} = L/v_{mp}$ we obtain

$$P_m(t) = \kappa_m(t_{mp}^3/t^4)\exp[-(t_{mp}/t)^2] \tag{10}$$

where κ_m is the normalization factor for the cavity distribution and

$$P_b(t) = \kappa_b(t_{mp}^4/t^5)\exp[-(t_{mp}/t)^2] \tag{11}$$

is the arrival time distribution in a beam. Setting $dP/dt = 0$ for both of these distributions reveals that $t_{mp} = L/v_{mp}$ is not the observed most probable arrival time but rather

$$t_{max} = t_{mp}/\sqrt{2} \tag{12}$$

is the time of the maximum TOA signal for the Maxwell cavity distribution and that

$$t_{max} = t_{mp}/\sqrt{2.5} \tag{13}$$

is the time of the maximum TOA signal for the beam distribution.

Now the mass spectrometer ion source produces a signal that is proportional to the gas number density in the active ionizer volume, rather than to an instantaneous flux in the ionizer. Therefore, it is necessary to note that the average residence time of a beam element in the ion source is proportional to 1/v, compensating precisely for the factor, v, relating P_b to P_m (Eq. 9). Thus, although the flux distribution in the beam, P_b, is different (Eq. 9), the number density in the beam is in fact the same as that obtained from the cavity distribution. Eq. 12 gives the appropriate t_{max} for the present mass spectrometer electron impact ionization detector, whereas Eq. 13 is appropriate for a flux detector, such as a surface ionizing detector (conversion dynode phototube or similar detector).

Using the above model, nonlinear least-squares fits to the observed TOA signals have been made. The fitting program was designed to either include the instrumental response with a supplied value for the instrument RC time constant, or to consider the time constant as one of the fit parameters. From these fits beam species temperatures, corresponding to a directed Maxwell distribution of velocities are obtained, as indicated below.

Optical Spectroscopy

A separate multiported reaction chamber is used for optical spectroscopic observations of the laser-generated plumes. In future studies, the same chamber will be used for simultaneous optical spectroscopic and mass spectrometric analysis. The reaction chamber is continuously evacuated to 10^{-7} atm. Sample targets are mounted on a continuously rotatable shaft to provide a fresh sample surface to each laser pulse, and minimize cratering effects. The Nd/YAG laser is focused through an optical port onto the target using a 20 cm focal length lens. Side ports on the chamber allow for several line-of-sight optical access paths perpendicular to the laser-plume axis. Other side ports on the chamber are available for future access, such as electrical probing of the plume ions (*9*) and optical probing of the plume particulates (*10*).

An optical multichannel analyzer, consisting of a 0.275 m monochromator with a gatable intensified photodiode array, is used to record the emission spectra of the plume, synchronous with the laser. A fused silica 10 cm focal length lens is used to obtain one-to-one imaging of the plume on the entrance slit of the OMA. By aperturing the slits and moving the imaging lens, the emission spectra from the plume can be spatially resolved.

Two OMA detection time schemes are used. In the first approach, the detector gate is 100 ns long and fixed in time after the laser pulse to avoid saturating the detector array. In the second approach, the gate is 10 ns wide and is swept under computer control from before the laser onset to several hundred nanoseconds after the laser firing in order to time-resolve the plume emission spectra. The OMA controller averages the detector array output over 100–1000 laser pulses and subtracts a previously stored dark current signal from the emission spectrum prior to transfer to the laboratory computer.

RESULTS AND DISCUSSION

Graphite Targets

Mass Spectrometric Analysis

With the perpendicular on-axis sampling geometry (depicted in Fig. 1), the molecular beam pattern clearly coincided with the mass spectrometer beam axis, even though the periscoped laser beam was directed at the surface at approximately a 15° angle from the normal. In contrast with earlier off-axis (15° tilted specimen stage) sampling experiments (3), the on-axis survey spectra (via the gated sweep method described above) showed significant laser-synchronous mass spectral peaks at mass positions not identified with C_n species. The mass spectral data also indicated that the major C_n (n = 1–5) species were present at similar concentrations to those observed off-axis.

Typical time-gated mass spectral scans are shown in Fig. 2. Spectrum A represents the background signal, integrated during the boxcar gate time. The relatively high background observed (17, 18, 28, 32, 39, and 44 u) was enhanced by the absence of effective cryopumping, normally used in the source region, which resulted in apparently synchronous signals at the permanent gas mass positions. Under normal cryopumping conditions, these background peaks were not detectable and were typically minor peaks in the time-resolved spectra even with saturated cryopump conditions. Spectrum B shows a laser-induced increase for the background signals noted in spectrum A, in addition to new features at 12, 23, and 36 u. The difference between the spectra B and A represents the laser-induced mass spectrum. The 12 and 36 u peaks are identified as C_1 and C_3, respectively. In Fig. 2C, the signal consists only of surface or plume ions formed synchronously with the laser. Using time-of-arrival single peak detection, mass peaks corresponding to C_1^+, C_2^+, C_3^+, and C_5^+ were observed under the conditions of Fig. 2C. That is, these ions were produced directly by the laser rather than by electron impact of the arriving molecular beam in the mass spectrometer ion source. Comparison of the three spectra in Fig. 2 indicates that the signals at 23, 39, and 42 u result mainly from impurity ions formed during the laser-induced vaporization or ejection process. These peaks correspond to Na and the two K isotopes in their correct natural abundance.

Time-of-arrival measurements of the alkali impurity peaks indicated a bimodal peak shape, with the faster component being dominant. With the mass spectrometer electron impact ion source off, only the fast component of Na and K TOA profiles was observed, which indicates that this component results, as expected, from laser-produced ions, and the slower component represents neutral plume species. The C_1^+, C_2^+, C_3^+, and C_5^+ TOA profiles were also bimodal, and probably even trimodal, in marked contrast to the off-axis profiles reported elsewhere (3), which showed single peaks. Figure 3 shows examples of this multimodal behavior. Similar data were obtained for C_5 ions and neutrals (not shown in Fig. 3). The broad, slower peaks correspond well with those obtained for the gas-dynamically cooled neutral plume species, as observed in the earlier off-axis results (see (3) and the off-axis curves in Figs. 4 and 5). The on-axis fast, relatively broad peaks were not present with the ionizer off, which establishes their origin as laser-produced neutrals. The fastest, quite narrow, peaks were also present with the ionizer off; hence, these peaks correspond to laser-produced ions. Both the TOA peak location

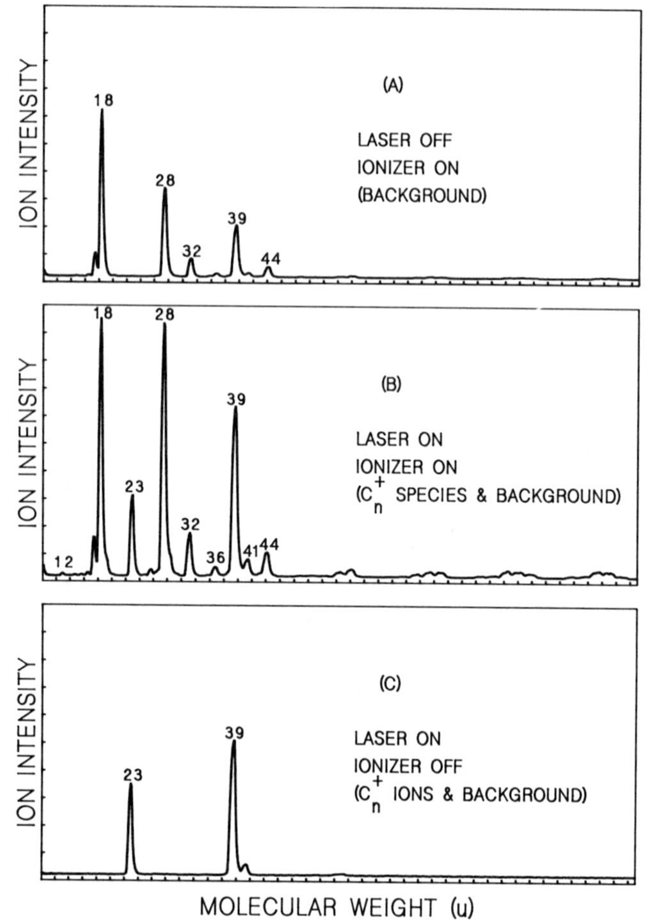

Fig. 2. Gated (500 μs width) mass spectral scans for mass range 10–120 u. Spectrum A was obtained with the laser off and ionizer on; B, with the laser and ionizer both on, and C with the laser on and ionizer off.

and width indicates these ions to be translationally very hot (~8000 K). It should be noted that the mass spectrometer system is intrinsically three to four orders of magnitude more sensitive to ions than neutrals, because of the low efficiency of electron impact ionization. The TOA peak locations for the ions, and the hot and cool neutrals, correspond to temperatures (directional Maxwell distribution) of 8000 ± 1000 K, 2500 ± 500 K, and 300 ± 200 K, respectively. These values are not unexpected for the gas dynamic conditions present. From the earlier off-axis studies, the plume density and individual species concentrations are consistent with a surface temperature of 4000 ± 200 K (3).

An example of fitting the TOA data to the Maxwell velocity distribution model (Eqs. 10 and 12) is given in Fig. 4 for the relatively simple off-axis sampling case. Note that it is not possible to completely fit the TOA data by a simple distribution characterized solely by a single tempera-

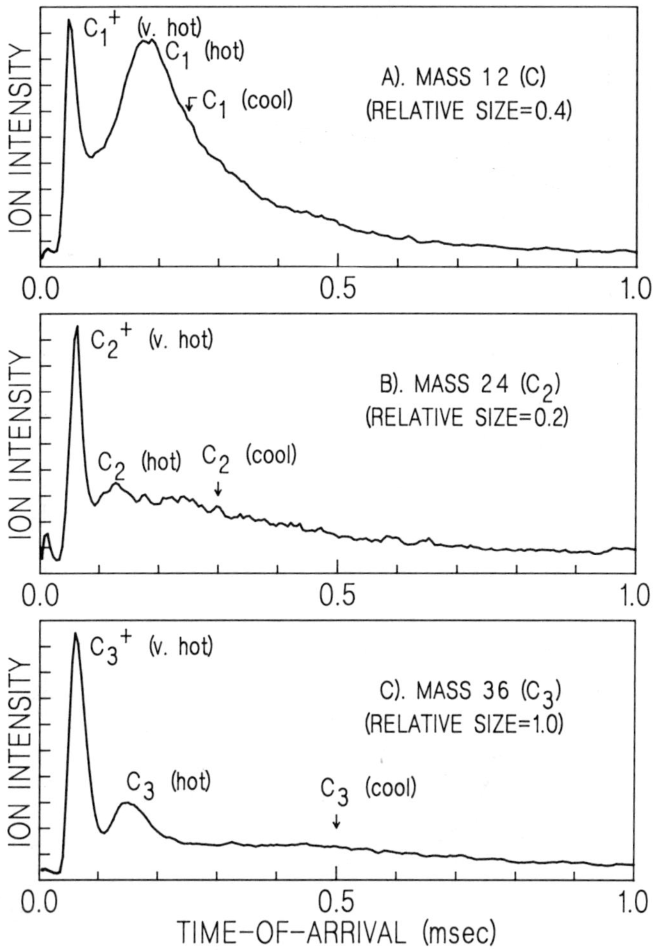

Fig. 3. Time-of-arrival ion intensity profiles (arbitrary units) for C_1, C_2, and C_3 (Figs. A, B, and C, respectively) obtained with both the laser and ionizer on and with on-axis plume sampling.

ture. The good fit obtained for the early portions of the peak correspond to a temperature of 567 ± 6 K, which is attributable to the major fraction of C_1 species. The fact that there is excess intensity in the tail region, relative to the model fit, suggests the existence of an even lower temperature region (the TOA Maxwellian curve is wider in time at lower temperatures even though its velocity distribution is narrower), which persists to a later time. A more accurate fit of the observed and model TOA signals will need to include a surface time/temperature distribution that models the cooling and plume formation processes. Such models have been considered by Olander for millisecond laser pulses, as discussed elsewhere in this volume. The main features of such models include detailed consideration of thermal conduction, mass vaporization, and radiation heat losses as the major cooling mechanisms. On the Nd/YAG

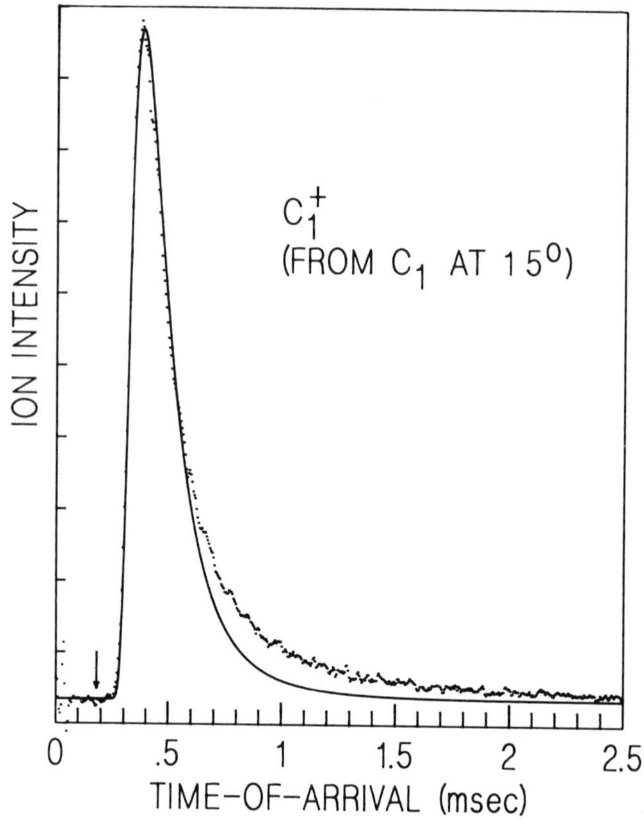

Fig. 4. Comparison of model (solid curve) and experimental (data points) time-of-arrival ion intensity profiles. The arrow indicates the laser on/off time position.

laser pulse heating time scale, heat conduction is the major energy loss mechanism. For the very fast heating rates of Nd/YAG pulses, relative to the TOA time resolution (currently 5 μs), the model can ignore the heating cycle, including laser pulse time structure, modeling only the cooling curve. We have ignored physical loss mechanisms (chunking), based on SEM examination of smooth (~1 μm surface roughness) graphite surfaces after only a few shots. Also, as very little surface is removed in a single shot, relative to long pulse (>1 ms) lasers, it is unnecessary to consider surface recession during a single pulse. However, it is necessary to consider the details of energy deposition by the laser. An iterative model is presently under development, which consists of calculating the instantaneous energy transport via conduction, heat of vaporization, and radiation.

Figure 5 shows the effect of varying the beam sampling to surface geometry. Note that the very fast ion and fast neutral peaks are relatively much more prominent for on-axis plume sampling. The plume ion concentration is estimated to be equivalent to a pressure of $\sim 10^{-4}$ atm, as compared with the ~1 atm pressure of the fully-developed plume. Since

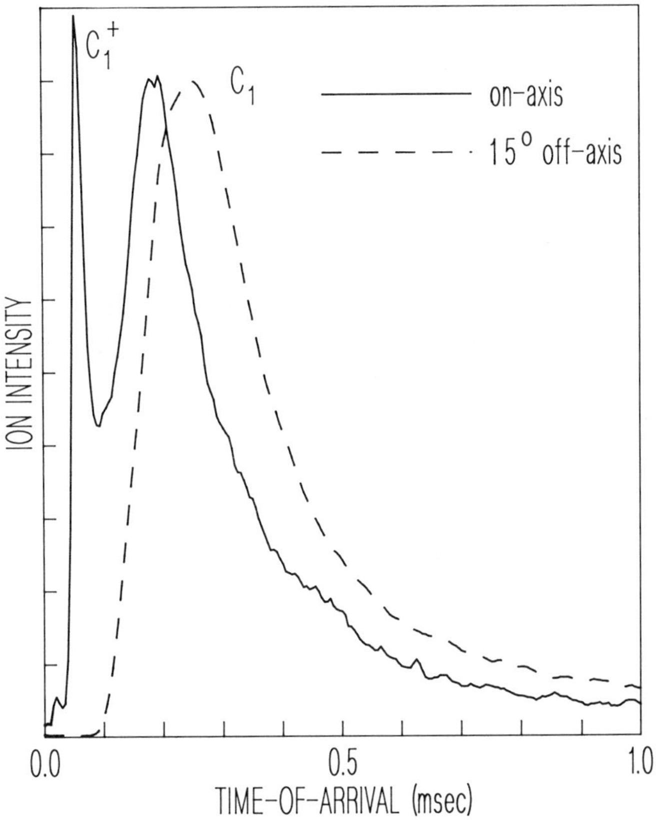

Fig. 5. Time-of-arrival ion intensity profiles (arbitrary units) for mass 12 u (C_1 and C_1^+) for on-axis and 15° off-axis plume sampling.

the observed ions are most likely accelerated out of the plume by the other ions present in the plume center, the observed ion temperature does not reflect the ion kinetic energy on leaving the surface. Possible mechanisms for ion formation are, multiphoton ionization of C during the laser pulse and dissociative ionization of larger C_n groups in the hotter center of the plume. It should be noted that the measured ion concentration is, perhaps fortuitously, of the same order of magnitude as expected from a Saha-Langmuir calculation of the thermal concentration of ions present in equilibrium with neutrals at the expected surface temperature of 4000 K.

Optical Emission Analysis

The optical emission spectra generated from graphite targets, depended dramatically upon the focus conditions of the laser spot on the sample. With a tight focus, and hence high laser fluence, the plume emission was dominated by line spectra, primarily owing to C^+ and C^{++}. Defocusing the laser, and thereby reducing the laser fluence, resulted in spectra dominated by C_2 emission. Figure 6A shows a low resolution spectrum of the C_2 plume emission obtained with a 100 ns

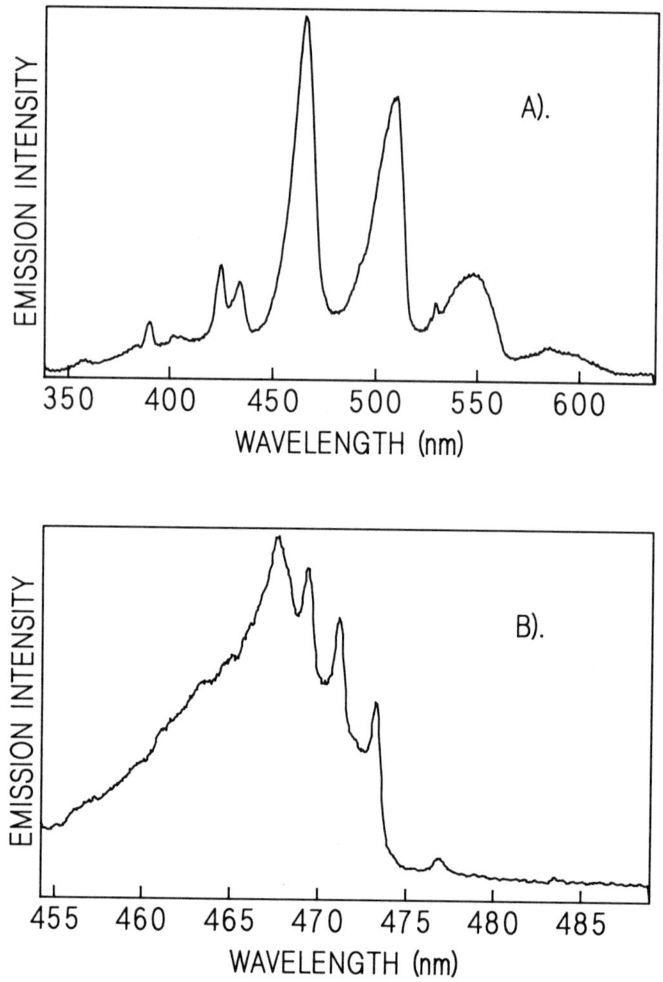

Fig. 6. OMA spectra of the C_2 Swan bands in the laser-generated graphite plume. Fig. A was obtained with low spectral resolution and Fig. B was obtained with high spectral resolution.

OMA gate set just after the laser pulse. The well-known Swan system ($A^3\Pi_g$ – $X'^3\Pi_u$) is apparent in Fig. 6A. The $\Delta\nu = -2, -1, 0, +1, +2$ vibrational bands at 436.5, 473.7, 516.5, 563.5, and 619.1 nm, respectively, dominate the spectrum. Additional features at 386.7 and 426.7 nm are the result of a strong emission from C^+. A small amount of residual laser scattering is also observable at 532 nm. Fig. 6B is a higher resolution spectrum of the 1→0, 2→1, 3→2 and 4→3 transitions (running red to blue). The intensity ratios of these vibrational transitions should provide information on the vibrational temperature of the C_2 molecule in the plume, assuming local equilibration of vibrational states. Preliminary analysis of this spectrum indicates vibrational temperatures of 6000–8000 K. The rotational temperatures are substantially lower, 3000–4000 K, owing to the expected more rapid rotational cooling of the C_2 molecules in the expanding plume. This emission spectrum is neither time nor spatially resolved and it is unlikely that a single rotational or vibrational temperature can be assigned since both rotational and vibrational cooling will occur during the plume expansion. These vibrational and rotational temperatures are also lower than the kinetic temperatures derived from the TOA analysis of the mass spectrometric data, most likely owing to the rapid adiabatic expansion of the plume.

Time and spatially resolved emission spectra can be obtained by using a 10 ns gate on the OMA and aperturing the entrance slit of the OMA's monchomator, as described above. Using a 1 mm aperture and imaging the target surface at the aperture, we obtained the time and spatially-resolved emission spectra shown in Fig. 7. Normally, the spectra are accumulated every 10 ns from just prior to the laser pulse to 400 ns after the laser pulse, but for clarity in Fig. 7, only the spectra taken every 20 ns from 70 to 190 ns after the laser are shown. The strong emission line from C^+ at 426.7 nm rapidly decreases in time owing to the short radiative lifetime of the transition (~5 ns) and the fact that ion production within the plume falls off sharply 50–100 ns after the laser pulse, as may also be seen in the mass spectral TOA profile in Fig. 5. The C_2 Swan band emission intensity falls off more slowly owing to its longer radiative lifetime (170 ns) and possible creation in the plume after the laser pulse has ended. The relative heights and widths of these bands depends on both the C_2 vibrational and rotational temperatures; however, the relative heights are more sensitive to the vibrational temperature and the relative widths to the rotational temperature. The ratio of the $\Delta\nu = -1$ vibrational band at 473.7 nm to the $\Delta\nu = 0$ band at 516.5 nm can be used to provide a C_2 vibrational temperature. Likewise, the width of these bands provides a measure of the C_2 rotational temperature. Both the rotational and vibrational temperatures of C_2 decrease with time, as one would expect.

$YBa_2Cu_3O_x$ Targets

Mass Spectrometric Analysis

The major plume species observed from the high temperature superconductor target, $YBa_2Cu_3O_x$, were Y, Ba, Cu, Cu^+, CuO^+, O_2, and O_2^+. In addition to these species, mass spectrometric signals identified as H_2O, CO_2, BaO, YO, Na^+, K^+ and bimetallics (CuBa and YCu) were also observed in the plume. Species assignments were confirmed by isotopic ratio measurements. Oxygen atoms and ions, which are known to be present from the optical emission spectra, were not detectable in the mass spectrometer.

Typical TOA profiles for the major mass spectral species are shown in Fig. 8. TOA profiles for the metal atom species Y, Ba, and Cu are bimodal, as shown in Fig. 8A, for mass 63 u (Cu). In the case of graphite, the fast component arises from ionic plume species and the slower component from cooler neutral plume species. The signal at mass 79 u, identified as CuO in Fig. 8B, consisted only of ionic plume species. Fig. 8C shows the signal from mass 32 u (O_2), which is also bimodal. Laser

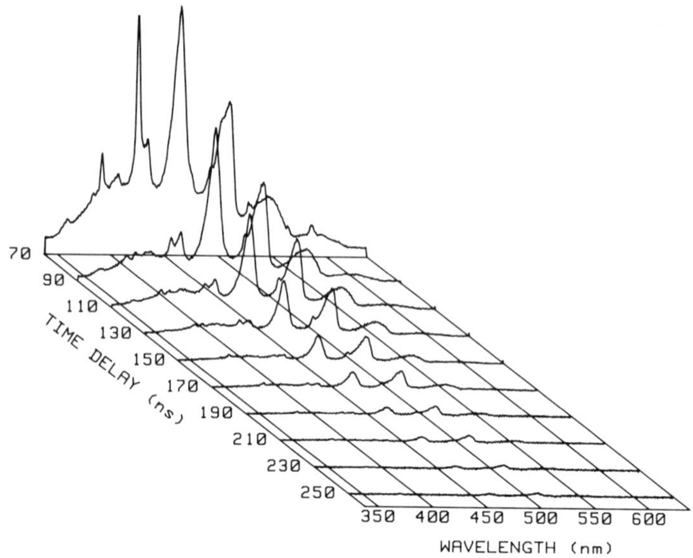

Fig. 7. Time and spatially resolved OMA spectra of the C_2 Swan bands in the laser-generated graphite plume.

deposited films from $YBa_2Cu_3O_x$ targets are usually O-deficient (*6,11*) and normally require postdeposition, high temperature annealing to restore the stoichiometry of the superconducting phase. Molecular O_2 is less likely than O atoms to be deposited as part of the film. The presence of O_2 in the plume is probably an indication of O-atom recombination in the plume as a cause of the oxygen-deficient film deposits.

Optical Emission Analysis

Figure 9 shows two representative OMA emission spectra obtained from the $YBa_2Cu_3O_x$ target plume. In Fig. 9, the sample position was static, and the spectrum in Fig. 9A was obtained with a fresh target, whereas Fig. 9B shows the spectrum after cratering of the surface had occurred from repeated laser impact. Note the dependence of relative intensities to the number of accumulated laser shots. The more volatile Y and Ba species have been depleted, relative to Cu, after 5000 laser shots. This result is consistent with the observation (*11*) that films deposited with higher energy are deficient in Cu and rich in Ba and Y.

SUMMARY

The present results indicate that the vaporization and ejection of vapor species by laser heating and excitation can produce both equilibrium and nonequilibrium effects. Vapor plume formation is nonhomogeneous, both in space and time. Fast, very hot, nonequilibrated ions result from ions produced, preferentially, on-axis early in the laser–surface interaction phase. These ions are accelerated by the laser driven

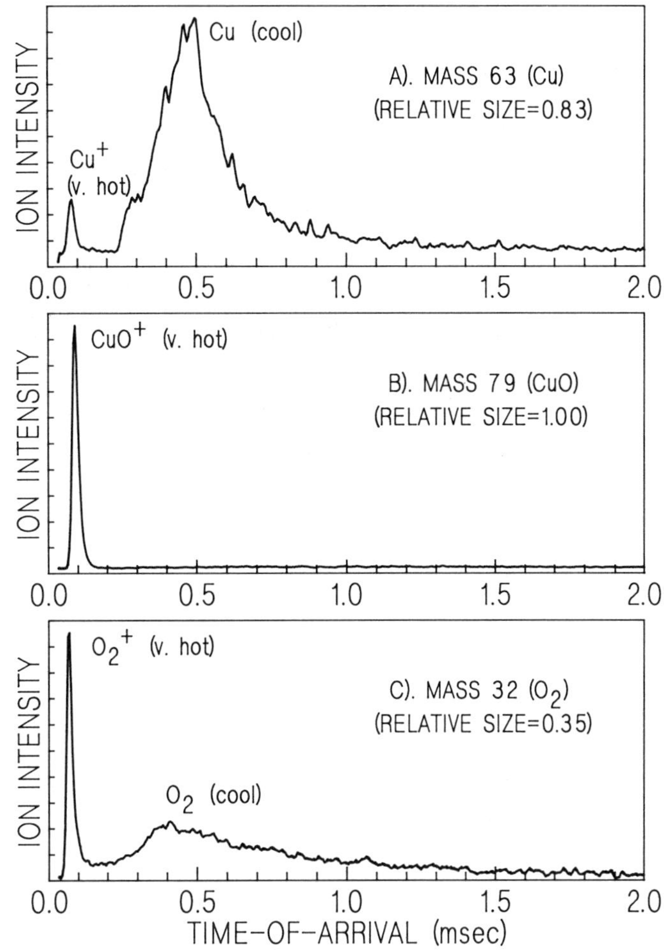

Fig. 8. Time-of-arrival ion intensity profiles (arbitrary units) for plume species at 63 u (Cu, Cu^+), 79 u (CuO^+), and 32 u (O_2, O_2^+) (Figs. A, B, and C, respectively) obtained with both the laser and ionizer on and with on-axis plume sampling.

plasma formed just above the surface (*see* R. W. Dreyfus and also Y. W. Kim, elsewhere in this volume). On-axis hot neutrals are also formed prior to the full development of an expansion-cooled plume jet. At later sampling times, and with off-axis sampling, only cool neutrals are observed. Under these conditions, as discussed elsewhere, the species intensities appear to follow those expected for an equilibrium vaporization process (*2,3*).

The techniques of time-resolved mass spectrometry and optical emission spectroscopy complement each other in the *in situ* analysis of laser-induced vapor plumes. Optical emission spectroscopy is well suited to monitoring the evolution of highly excited species formed during plume development, whereas mass spectrometry is more suited to analysis of the cooler postplume expansion species. Together, the two tech-

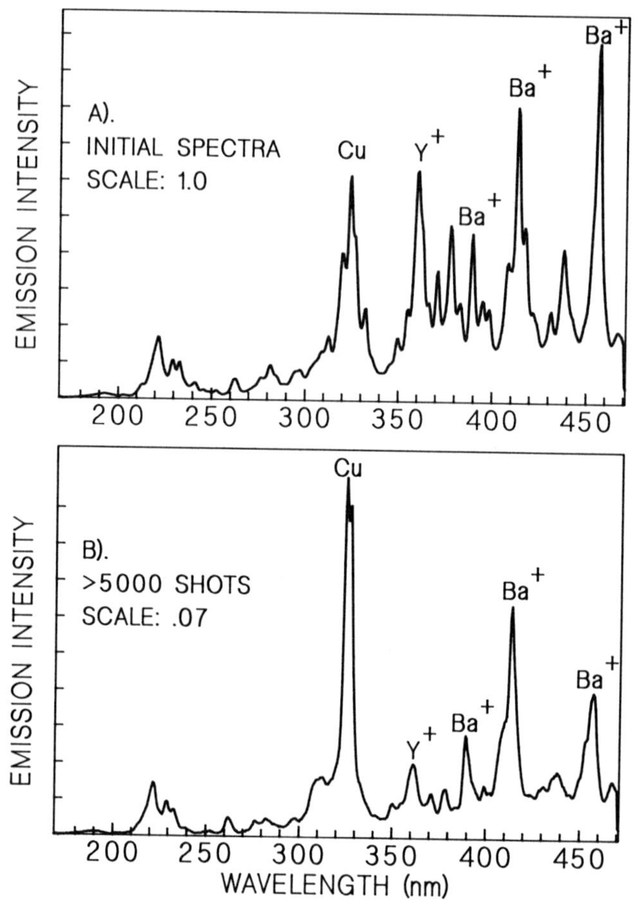

Fig. 9. Optical emission spectra from the laser-generated plume integrated over 100 ns after the laser pulse. Fig. A shows the average spectra from the first 1000 shots, and Fig. B shows the average spectra of 1000 shots averaged after 5000 shots impacted the target.

niques provide a more complete picture of plume composition, temperature, and dynamics than either one alone.

In future studies, methods will be developed and applied to time-resolved optical measurement of surface and plume temperatures. The results will be used to test models, under development, of the laser heating and plume formation process.

ACKNOWLEDGMENTS

The authors would like to acknowledge the assistance of L. P. Cook and C. K. Chiang in obtaining $YBa_2Cu_3O_x$ targets, Jianrong Zhao in the analysis of the OMA spectra, and W. G. Mallard for computer simulation of the C_2 spectra. This work was supported, in part, by the Air Force Office of Scientific Research.

REFERENCES

1. Hastie, J. W., *Pure and Applied Chem.* **56,** 1583 (1984).
2. Hastie, J. W., Bonnell, D. B., and Schenck, P. K., *High Temp. Sci.* **25,** 117 (1988).
3. Hastie, D. W., Bonnell, D. W., and Schenck, P. K., *High Temp.-High Press.* **20,** 251 (1988).
4. Schenck, P. K., Bonnell, D. W., and Hastie, J. W., *J. Am. Vac. Soc.*, in press (1988).
5. Cook, L. P., Schenck, P. K., Zhao, J., Farabaugh, E. N., and Chiang, C. K., *Amer. Cer. Soc.* in press (1989).
6. Moorjani, K., Bohandy, J., Adrian, F. J., Kim, B. F., Schull, R. D., Chiang, C. K., Swartzendruber, L. J., and Bennett, L. H., *Phys. Rev. B* **36,** 4036 (1987).
7. Bonnell, D. W., Schenck, P. K., and Hastie, J. W., *Proc. Intl. Symp. Laser Processes for Microelectronic Applications; 172nd Electrochemical Soc. Meeting,* Honolulu, Hawaii, Oct. 23–27, 1987.
8. Young, W. S., *Rev. Sci. Inst.* **44,** 715 (1975).
9. Krehl, P., Schwirke, F., and Cooper, A. W., *J. Appl. Phys.* **46,** 4400 (1975).
10. Chen, G. and Yeung, E. S., *Anal. Chem.* **60,** 864 (1988).
11. Sudarsan, U., Cody, N. W., Bozack, M. J., and Solanki, R., *J. Mater. Res.* **3,** 825 (1988).

Interactive Effects in Excimer Laser Photoablation

R. W. DREYFUS

IBM Research Division, T. J. Watson Research Center, Yorktown Heights, NY 10598

ABSTRACT

Laser irradiation of surfaces leads to the rapid ejection of atomic and larger species. An ablative process based on (classical thermal) vaporization is often contrasted to photochemical etching, i.e., material removal owing to bond breaking. Collisional sputtering, electrical breakdown (plasma formation), and mechanical disintegration are also possible. This talk will summarize interactions which occur when removing $<<$10 monolayers in $\lesssim$10 ns and how these interactions affect the etch product, i.e., plume material.

Our experiments primarily utilize laser-induced fluorescence, LIF, to measure the density and energy of specific species in the plume. Energy information is particularly useful as one then has information about the surface temperature. The energy is derived both from densities, and rotational/vibrational state distributions (of diatomic radicals), *see* Fig. 1. At low fluences both C_2 and Cu (from the resp. substrates) are vaporization products. The electronic energy distributions are also being followed currently; this includes *simultaneously* following Cu° atoms, Cu^+ (metastable) ions and a few Cu_2 radicals above laser-etched copper metal. Initial generation of the Cu^+ ions appears to be a photon→vapor interaction. With higher fluences ($>4 \times 10^8 W/cm^2$) plasma effects become dominant.

The overall conclusion is that while direct thermal effects are present in nsec laser ablation, other interactions can be quantitatively followed by laser probing techniques, primarily LIF. Specific examples (sapphire) exist for lasers producing photochemical ablation when the excimer beam interacts with sensitive sites. Also, secondary interactions between photons or radicals and other radicals, ions and electrons are common, and indeed, amenable to quantitative evaluation and understanding. One example of this is that plasma physics concepts and equations explain many of the effects seen at slightly higher fluences.

Index Entries: Photoablation; lasers; vaporization.

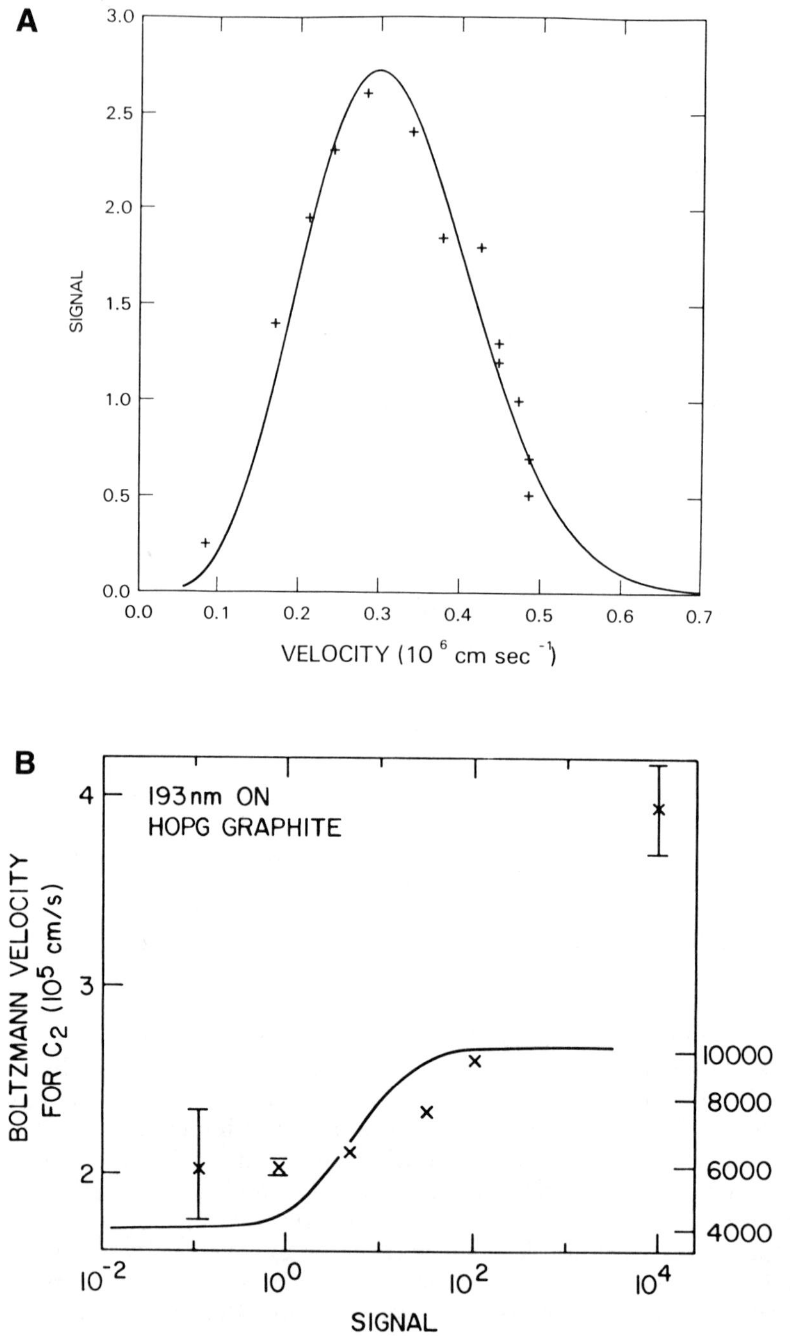

Fig. 1. (**A**)Typical LIF results showing the signal amplitude vs. the velocity perpendicular to the surface, as determined by time-of-flight. The theoretical curve represents Boltzmann-type distributions; note that the fit is reasonable. (**B**) The Boltzmann velocity ($V_B^2 = 2kT/m$) vs the peak LIF signal. The latter quantity is used to bring out the dependence of velocity on the amount of material removed per pulse.

INTRODUCTION

Understanding the physics of ablation on a microscopic scale is primarily a matter of understanding the various atomic interactions, particularly the atom-surface interactions. Historically this extends back to determining the velocity distribution of bismuth vapor (*1*) and even earlier. More recently, the behavior of laser vaporized sodium atoms has been followed by probing with a second laser (*2*). The area of the present work is to explore even more dynamic regions; i.e., UV excimer laser vaporization of monolayers of material in nanoseconds. The objective is to evaluate the various interactions that transpire between the laser beam, the surface and the vaporized material. By removing less than a monolayer of surface per laser pulse it will be seen that a number of the interesting interactions cannot only be observed but also put into a quantitative framework. In other words, while the initial species densities, velocities and angular dependences are altered in nanoseconds by events in the etch plume (whose density goes through many decades during expansion) these alterations are still interpretable in terms of simple physics and chemistry concepts.

The fact that *interactions* are emphasized here is a different but yet complementary approach to isolated atom removal, where the energy pathways are primarily considered. Besides thermal energy, in both cases one often considers the possibility of surface atom ejection as arising from several other energy pathways; e.g., desorption induced by electronic transitions [vacuum UV or fast electron activated, i.e., DIET processes (*3*)], or momentum transfer sputtering (owing to fast atom bombardment) (*4*). DIET-type processes can result from multiphoton ($n\,h\nu$) events, although the laser intensities required sometimes bring the surface near the point of thermal vaporization (*5*) or gross photochemical ablation (*6*). The reason for the lack of interactions between the vaporized components in the DIET and sputtering cases is that energy deposition commonly involves a single isolated site, hence collision or gross surface feature alteration are not directly involved. On the other hand, this same low rate of atom removal means that only ultra high vacuum equipment is suitable as surface perfection and purity becomes primary questions. Many experiments are aimed at bridging the gaps in knowledge between the etching particles and removal rates, so that a cohesive, comprehensive view evolves. The current study of interactions is thus just one of the series of endeavors to form these bridges.

The presence of interaction in the present experiments is directly linked to the fact that our ablative energy comes from pulsed (10 ns) excimer lasers, and, furthermore, that the surface removal is typically in the $\leqslant 1$ monolayer region. A simple reaction rate calculation shows why this is the region for the onset of interactions between species. For the removal of 0.3 monolayers ($\sim 3 \times 10^{14}$ atoms/cm^2) we are dealing with an above surface vapor density of $\sim 10^{17}$ atom/cm^3 since plume expansion velocities are $\sim 3 \times 10^5$ cm/s. Taking a gasdynamic reaction rate constant of 3×10^{-10}cm^3/s (i.e. a 10Å^2 cross section) we have a reaction rate of 3×10^7s^{-1} for the initial 10 ns period; i.e., the reaction probability will be ~0.3. Continuing to allow the vapor to expand and to react for t→100 ns, we find that the majority of the species have suffered a collision and possibly undergone a reaction (*7*).

Considering the free electron induced reactions, we note that the electrons travel much faster (~4×10^7 cm/s for 1 V), but they are present in somewhat lower densities and usually have slightly lower reaction cross sections. In summary, monolayer ablation rates are quite often a transition region for going from isolated atom removal to the onset of interactions between the vaporized species. In exploring the interactions, we shall first outline our experimental approach, and then illustrate the nature of this onset by a few examples.

EXPERIMENTAL TECHNIQUES

A Lambda Physics Model 102 excimer laser is the UV energy source for the pulsed ablation. The wavelength is changed between 193, 248, and 351 nm by changing the noble gas. The apertured beam (to aid uniformity) is focussed onto the sample by a quartz lens to give an irradiated area of ~0.8×3.2 mm^2. The energy per pulse is ascertained by using a quartz beam splitter plate and a Scientech Powermeter. The etching and laser probing take place in a vacuum chamber at ~3×10^{-6}T.

The plume of ablated material is interrogated by the laser-induced fluorescence (LIF) technique. Both this technique and the above etching arrangement have been described in detail earlier (*5,6*). Basically, LIF excites atoms/diatomics starting from specific (usually ground or low-lying) energy states. Some nanoseconds after excitation, the excited electronic state emits light at the same or a nearby wavelength which is then collected and detected by a conventional spectrometer/photomultiplier arrangement. The advantage of this technique is that relative values of specific species density in the plume are determined accurately.

The ability of LIF to determine relative densities opens up the possibility of determining species' energies through three different approaches. The simplest approach is to determine kinetic energies through time-of-flight (TOF), since both the etch and dye laser pulses are only ~10 ns, whereas 1 to 3 cm flight paths provide time intervals of ≳1 μs. The second approach is to determine transverse kinetic energies by the Doppler width of LIF signal as will be seen later for Cu° atoms. The third technique is to determine occupancy of various internal energy states (rotational and vibrational) and relate this to a temperature (presuming a Boltzmann distribution) (*8*).

Interaction effects between various plume species can thus be recorded as changes in both their relative densities and energies.

Primary attention will be paid to the species derived from graphite and copper substrates. Graphite surfaces primarily produce C, C_2, and

C_3. The C_2 is detected on the (~230 nm) Mulliken Bands (*9*). C_3 is detected by excitation near 388 or 405 nm and fluorescences near 407 nm (*10*). LIF also monitors copper species (Cu°, Cu^{+}, and Cu_2) in the etch plume formed by UV excimer laser etching of Cu metal. A unique feature of this work is that both ground state Cu atoms ($4p'^4D^o_{1/2} \leftarrow 4s^2S_{1/2}$ excitation at 222.6 nm) and the lowest metastable ion state ($4p^3P^o_2 \leftarrow 4_s{}^3D_3$ transition at 224.7 nm) are monitored interchangeably under a specific set of etching conditions. Not only are the above wavelengths adjacent, but Cu_2 radicals can be excited at 454.6 nm using the (1-0) band of the $B^1\Pi_u \leftarrow X^1\Sigma_g^+$ transition, i.e., one uses the same laser dye but does not utilize the frequency doubler. With this LIF approach, the results for the various species can be directly compared with minimal experimental uncertainty.

RESULTS

Graphite

Sublimation of graphite is a good starting point for understanding LIF results from laser etch plumes. The requisite fluences of $\leqslant$1J/cm^2 do not melt the surface (a scanning electron microscopy, SEM, result) nor create an intense free electron plasma over the surface (of the type discussed later for Cu) (*11*). The fact that the surface does not show overall melting indicates that the surface temperature remains at <4600K (*12*) for moderate (<1 J/cm^2 fluence) and monolayer ablation. The amounts of material sublimed from the surface are in agreement with being below this peak temperature, and furthermore establish a lower temperature boundary of 3500 K.

Figure 1(a) shows a typical velocity distribution for graphite, the theoretical curve represents a Boltzmann type distribution. The velocities derived by such fits for various fluences have been collected into Fig. 1(b). In the latter figure, we see that the kinetic energies for removal of small amounts of graphite (i.e., at minimal etch fluence) are ~6000 K. When the fluence is increased (large LIF signal) this energy increases. This is interpreted as interactions (momentum transfer collisions) between the ablated carbon species. Gas dynamic concepts can be applied to such an interacting ablated layer (Knudsen layer) in order to give a theoretical correction to the kinetic energy (*13*). The solid line shows the result of such corrections. The break point has no adjustable parameters; it is set by the collision cross sections and measured atom removal rates. If one makes the quite reasonable assumption that the surface does not ablate uniformly, but rather there are comparable microscopic yields from areas ablating at ⅓ and 3 times higher densities, then the theoretical curve is smoothed out and gives a quite reasonable representation of the momentum transfer interaction between plume species. The theory does not cover the very highest density point, however; an approach equivalent to supersonic flow from a pulsed nozzle is required. There may also be some small free electron/plasma effects accelerating the ions and then neutrals. Nevertheless, we do conclude that the gasdynamic interaction is the dominant effect and its magnitude can be calculated.

In contrast to gasdynamic cross sections (for momentum transfer collisions) the transition probabilities for rotational and vibrational interchange are quite small, especially for the C_2 diatomic. Two reasons for this are: first, the homonuclear C_2 molecule has no electric dipole moment for such transitions, and, second, since the nuclear spin is zero, alternate rotational (J) states are absent (*8*). Thus, any transition must be of the form $\Delta J = 2,4. \ldots$ Because of the insular nature of vibrational and rotational states, one believes that their behavior accurately represents the state of C_2 as it left the graphite surface. In fact, the internal energies are nearly unchanged even when removing from submonolayers up to ~40 Å of surface atoms. Fig. 2(A,B) also supports this belief. The distribution of C_2 among the various internal energy states indeed follows a Boltzmann distribution in the 3600–3800 K temperature range.

A different type of interactive effect is observed when one uses LIF to follow the density of C_3 radicals. On a freshly cleaved surface of graphite, the LIF signal from C_3 is quite large, despite its large degree of spectroscopic splitting. However, after about 50 etching pulses the C_3 signal drops by nearly 50 x in intensity. At present, we believe the C_3 originates from cleavage steps on the surface and that these are smoothed out as the ablation process proceeds.

Copper Ablation

While the vaporization of copper metal has some similarities to graphite, there exists some distinct differences. The similarities start with the point that the kinetic energy of Cu (for removal of $\sim 6 \times 10^{-3}$ monolayers) give values in quite reasonable agreement to the estimated surface temperature, e.g., we find the Cu atom Boltzmann energy distribution to be fit by an energy of 2180 K, *see* Fig. 3. Such a value is in reasonable agreement with removing 6×10^{-3} monolayers in 10 ns. Figure 4 shows the slope of the Cu° signal vs the reciprocal of the fluence. Plotting the data in this way is based on the assumption that thermal vaporization is furnishing the Cu° atoms, i.e.,

$$\log (\text{LIF signal}) \propto \log (\text{max. vapor pressure}) \propto -E_{vap}/kT \propto [\text{fluence}]^{-1}$$

The heat of vaporization, E_{vap}, thus determines the slope of the line in Fig. 4. It is seen that the observed yield of Cu° fits the known vapor pressures over a range of 10^5. This then extends the demonstration of classical vapor pressure from the earlier gentle case of sodium vaporization to the more dynamic situations with graphite and with copper.

With respect to Cu_2 we note that there is a maximum in the Cu_2 density and then a significant decrease with the presence of an intense luminous plasma above the Cu surface. Such behavior had been noted earlier with the case of AlO diatomics derived from sapphire photoabla-

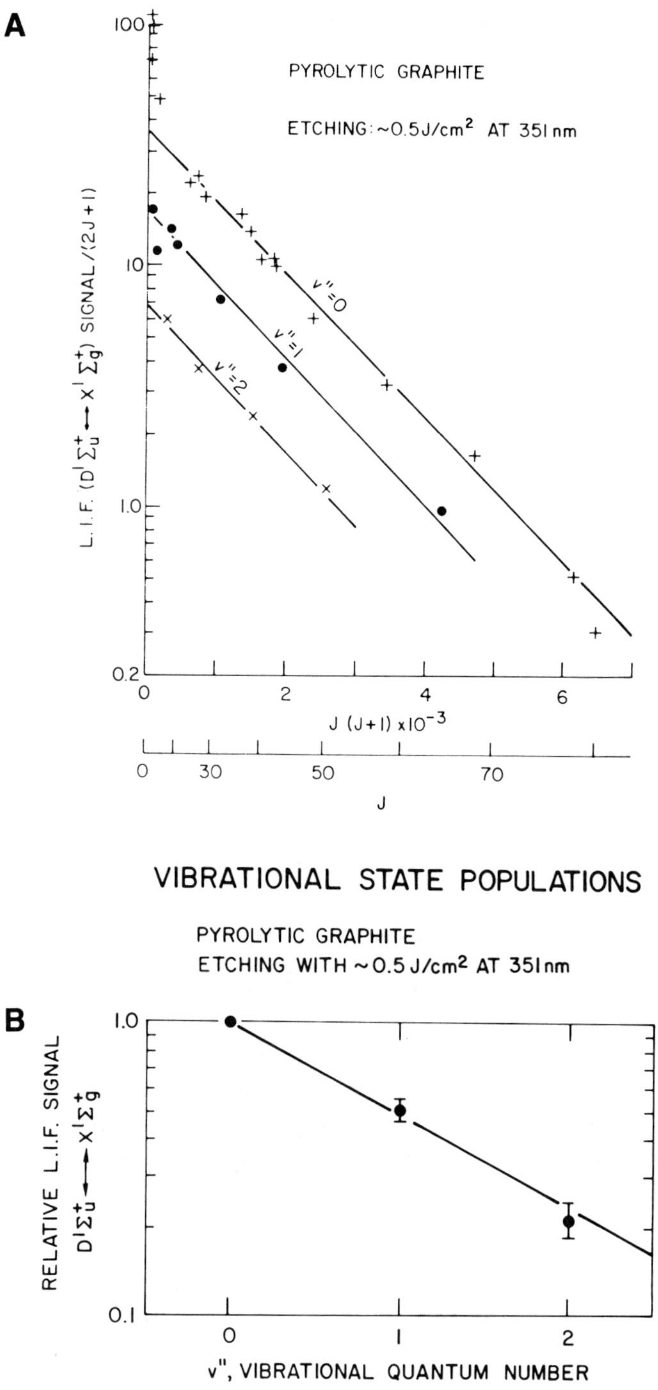

Fig. 2. (**A**) Magnitude of the LIF signal from C_2 (divided by a degeneracy factor) vs the rotational energy. Note that the straight line fit is in agreement with a Boltzmann-type activation energy. The data are shown for the lowest three vibrational states. (**B**) Similar to (**A**) but for the vibrational energies. Note that in both cases the slopes of the curves are reasonably close to 3600 K, the estimated sublimation temperature.

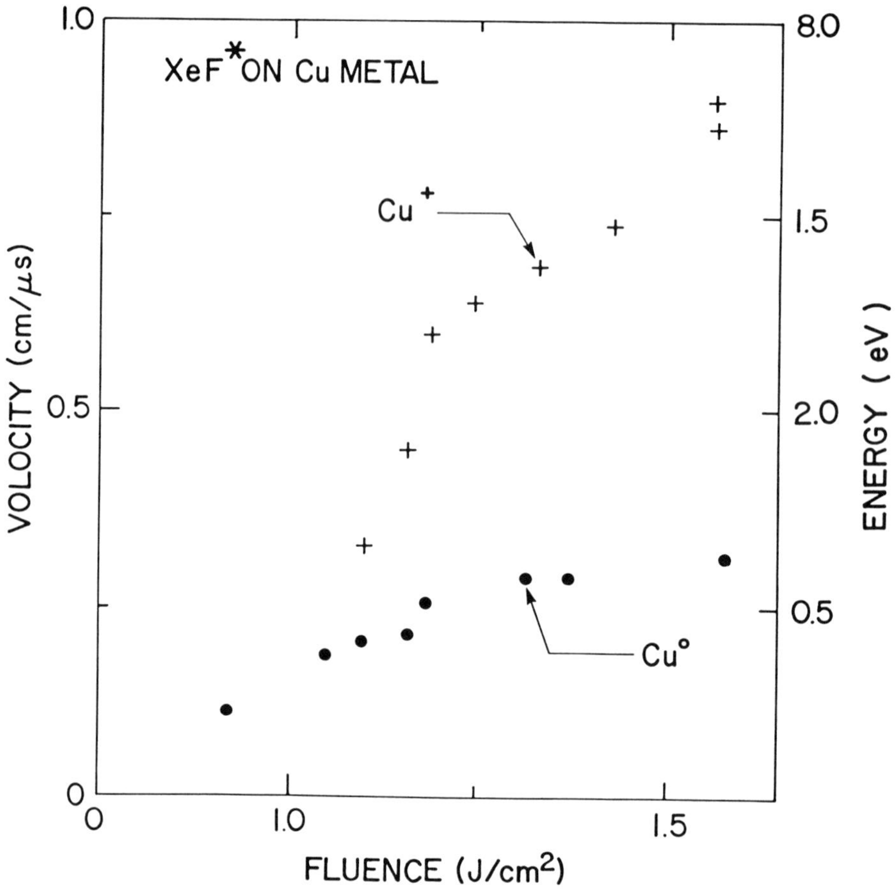

Fig. 3. (**A**) The LIF signal vs velocity similar to 1 (**B**), but now for Cu° from the bulk metal surface.

tion (*14*). The conclusion is that plasma electrons (with eVs of energy) excite the diatomics into dissociation electronic states. It has not yet been possible to determine a rotational/vibrational temperature for the Cu_2 because of the high degree of splitting of the spectroscopic states due primarily to the isotopic makeup of Cu and also electronic multiplicity. Besides plasma electrons dissociating the Cu_2, initial observations indicate that the 193 nm etching wavelength also reduces the Cu_2 density by greater than one order of magnitude as compared to the 351 nm ablation. While the 193 nm radiation has $h\nu = 6.4$ eV and this is slightly below the ionization energy (7.4 eV) of Cu_2, there appears to be either a dissociative channel accessed by the 6.4 eV photons or an allowed multistep dissociative or ionizing excitation.

As the fluence increases about 20% above the thermal vaporization threshold, the signal from Cu^+ (metastable) ions grows rapidly. The rapid increase in Cu^+ is particularly notable with 351 nm etching and is in rough agreement with a 3 $h\nu$ process producing ionization of the Cu° vapor, *see* Fig. 4. With 193 nm etching, the ratio of the Cu° and Cu^+ signals remains nearly constant. The explanation for this constant ratio is

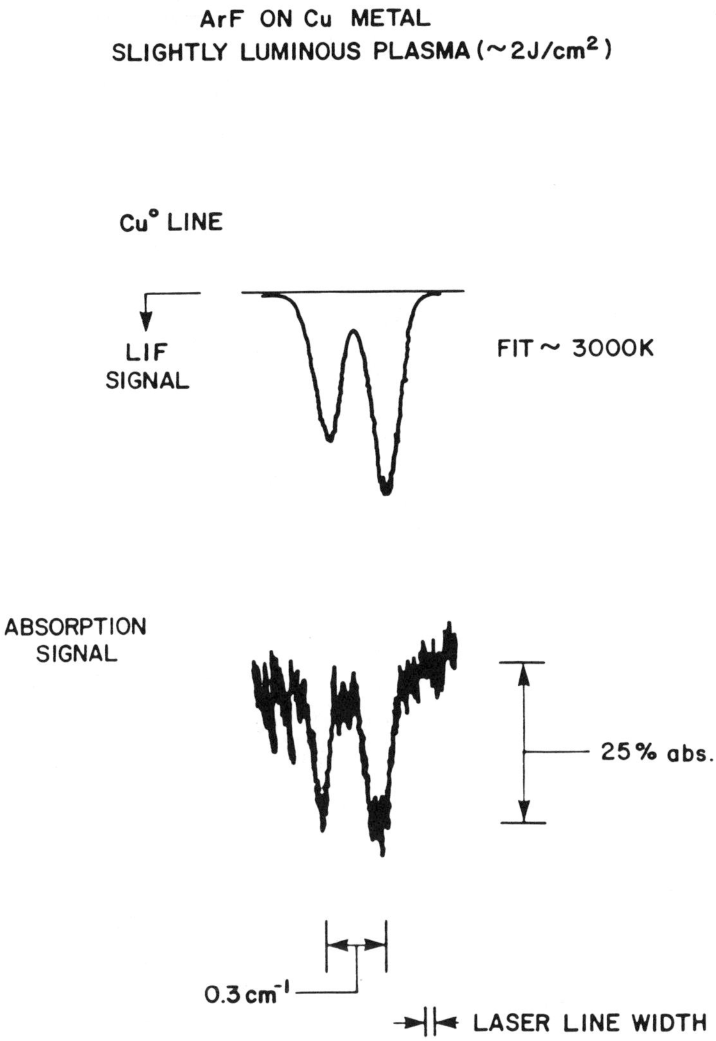

Fig. 3. (**B**) Doppler width of the Cu° hyperfine doublet. The present data is from sufficient fluence to produce a luminous plasma and perpendicular velocities of $>10^6$ cm/s. Nevertheless, the transverse velocities, which give rise to the Doppler width, apparently retain their original low (thermal) value as characterized by an energy equivalent to ~3000 K.

not obvious. One would have expected a $2h\nu_{193nm}$ excitation to have produced the Cu^+; however this finite slope is not evident in the experimental result. One photon excitation followed by thermal ionization is conceivable ($Cu^{\circ} + h\nu_{193nm} \rightarrow Cu^* \xrightarrow{heat} Cu^+ + e^-$).

As the etch fluence and concurrently the Cu^+ ion density increases, the kinetic energy of the Cu^+ also increases significantly. This has been attributed to the fact that charge neutrality requires that the plasma plume maintain a net positive charge (so the electrons do not "diffuse away") (*15*). This plasma potential readily attains values in the 10–100

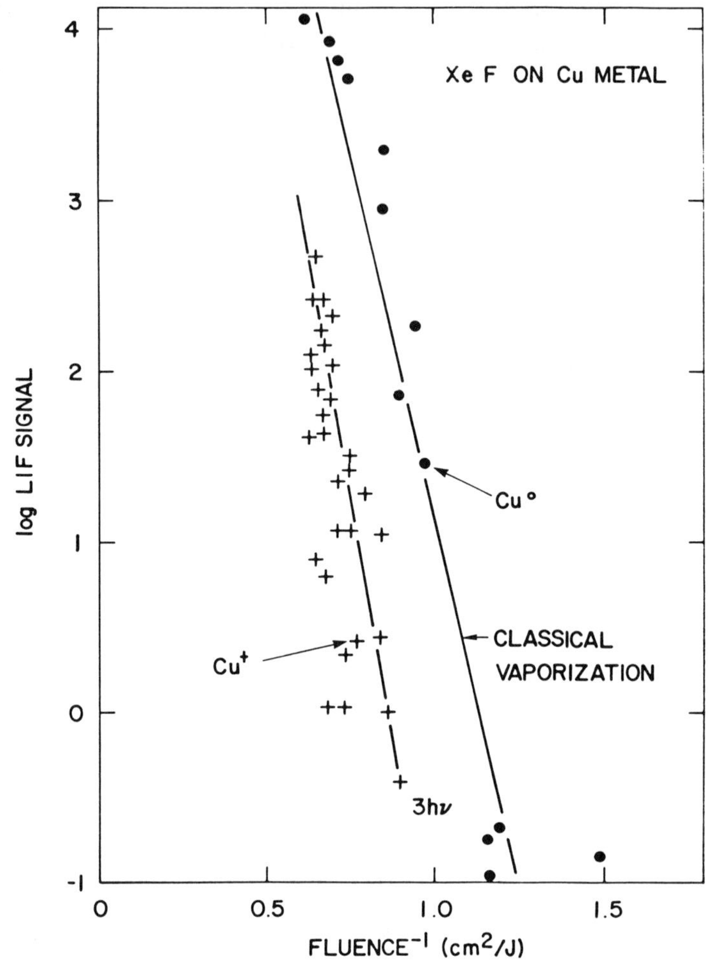

Fig. 4. Magnitude of the Cu° and Cu^+ LIF signals vs the reciprocal of the etch fluence.

volt range and acts as a miniature electrostatic particle (positive ion) accelerator. Not only are the ions accelerated, but the extremely large cross section (> 100 Å^2) for resonant charge exchange means that this ion energy is transferred into kinetic energy of the Cu° atoms also, i.e.,

$$Cu^+(\text{fast}) + Cu^\circ \rightarrow Cu^\circ(\text{fast}) + Cu^+$$

There is also a second method of transferring the (potential) energy of the ions into the neutrals and this is via three body recombination reaction changing ions into neutrals with the excess energy appearing as kinetic energy. This process is estimated to occur at appreciable rates out to $\gtrsim$0.5 cm from the ablated surface when one has electrons in the 1 to 3 eV range (*16*). The above plasma potentials become significant approximately in the 10^9 electrons emitted/cm^2 range. Plasma acceleration effects are thus noticed long before a luminous plasma is formed, which appears to require $> 10^{15}$ electrons/cm^2.

In general, the theories and experiments summarized by Phipps, et al. (*18*) characterize the luminous plasma region in the $> 4 \times 10^8$ W/cm^2 region, or > 4 J/cm^2 for 10 ns pulses. To summarize these results -this theory is applicable when the primary laser absorption occurs in the plasma plume and the energy is transferred to the surface by gasdynamic (shock wave) effects. With our excimer laser etching, the plume absorption is limited to the "sub critical" regime, i.e., the plume electron density is insufficient to obtain plasma "cut off" of UV light. The latter effect requires plume densities of $\sim 10^{21}$ electrons/cm^3 for short UV wavelengths. Nevertheless, the plasma absorption (owing to inverse bremstrahlung (*19*)) still has an absorption of ~30% at 248 nm for the incident plus reflected paths even though the plasma is not opaque. It is worthwhile to note that with the calculated absorption, the etch laser deposits ~l eV/free electron for 2 J/cm^2 and 3×10^{18} electrons/cm^3. Since this is nearly sufficient to cause ionization, and indeed the high energy tail of such a distribution does cause ionization of copper neutrals, it is evident why luminous plasmas commonly set in this $2 - 4 \times 10^8$ W/cm^2 region. One can invoke several other factors to enhance the energy deposited/electron. One can imagine that the ion density in the plume is not uniform, but rather has areas of 3× higher or lower density, i.e., up to $\sim 10^{19}$ and down to $\sim 10^{18}$cm^{-3}. The enhancement comes about because inverse bremstrahlung goes as free electron density (equals on density) *squared*. On the other hand, the free electron damping owing to scattering of neutral copper species has not been included as this has a cross section of probably only 1 to 10 Å^2, but will somewhat aid the inverse bremstrahlung absorption. There also are a number of other laser plasma interactions, but these usually dominate only at high fluence (*20*). Once having attained plasma breakdown (electron avalanche) (*18*) they provide the formalism for calculating surface ablation effects all the way into the 10^{15} W/cm^2 region, i.e., the fusion pellet compression region. As mentioned above, this very high power region is characterized by nearly total UV absorption in the plume and shock wave transfer of this energy to the solid surface. At these high powers, the ablated atoms are rapidly converted to ions by collisions with plasma electrons. The theory (*18*) has wide yet straightforward utility as even over the large range of gigawatts to petawatts, the efficiency of the process is characterized by a single laser parameter, $I\lambda\sqrt{\tau}$; where I is the intensity, λ the wavelength and τ the pulse duration.

In summary, we can now dissect the reasons why the monolayer ablation region is the point of intersection for a number of interactions. At lower fluences and amounts of surface removal, ablation is an isolated event that may depend on local temperatures, morphology or chemistry. As soon as one or more monolayers of surface atoms are removed per pulse, many of the local surface features no longer dominate and now the interaction between the species leaves its imprint on the plume composition (atoms vs ions vs diatomics) and energy. Finally, the dominate interaction resides in the plume at high fluences where it is the UV laser being absorbed by free electrons in the plume that establishes the nature of the laser etch process. Overall, although the plume goes through different phases mentioned above, the utilization of well known cross sections along with incisive experiments, such as LIF, enable one to dissect and to understand a 10 ns surface explosion in terms of specific energy states and vapor composition. The high temperature behavior of both graphite and copper produce interesting, but different behavior. The former because of the resistance of the diatomics to interactions and the latter because the free electron plasma becomes so omnipotent.

REFERENCES

1. Ko, *J. Franklin Inst.* **217,** 173 (1934).
2. Miksch, G. and Weber, H., *Chem. Phys. Lett.* **87,** 544 (1982).
3. *Desorption Induced by Electronic Transition,* M. Knotek and R. Stulen, eds., Springer-Verlag Series on Surface Science, Berlin, Heidelberg, 1987, v. 13, p. 262.
4. Baxter, J., Schick, G., Singh, J., Kobrin, P., Winograd, N., *J. Vac. Sci. Tech.* **A4,** 1218 (1986).
5. Matthias, E. and Dreyfus, R. W., *The Hereaus Seminar on Photoacoustic and Photothermal Phenomena,* Bad-Hannef, FRG, Nov. 22–24, 1988.
6. Srinivasan, R., Braren, B., Seeger, D., and Dreyfus, R. W., *Macromolecules* **19,** 916 (1986).
7. The reaction rate diminishes as t^{-1} owing to the reduced density of the vapor. This t^{-1} is, in turn, correlated to the fact that the expansion must appear to be one dimensional (1D). As soon as the plume dimensions are greater than the laser irradiated area, the expansion becomes 3D and decelerates markedly. Typical dimensions are 1 mm for the average laser spot; hence the primary reactions continue for about 100 ns, but with the t^{-1} term giving a ln (1 mm/0.03 mm) ~3.5 multiplier for the number of collisions; Kelly, Roger unpublished results.
8. Herzberg, G., *Molecular Spectra and Molecular Structure-I: Spectra of Diatomic Molecules,* van Nostrand, New York, 1950.
9. Dreyfus, R. W., Kelly, Roger, and Walkup, R. E., *Nucl. Instru. and Meth.* **B23,** 557 (1987).
10. Becker, K., Tatarczyk, T., and Radic-Perić, J., *Chem. Phys. Lett.* **60,** 502 (1979).
11. Graphite tends not to form luminous plasmas for two reasons: first, it ablates easily for powers $< 10^8 W/cm^2$, and second the C_2 diatomic forms one of the most stable negative ions known.
12. The melting point of graphite is 4530 ± 150 K; *see* Cezairliyan, A. and Miller, A. or J. Shaner, *Bull. Am. Phys. Soc.* **32,** 607–8 (1987). Our SEM micrographs show that there are a small number of misoriented grains in the otherwise planner HOPG graphite surface. These few grains eventually do melt due to a gradually decreasing thermal contact with the bulk graphite.

13. Kelly, Roger and Dreyfus, R. W., *Surface Science* **198,** 263 (1988).
14. Dreyfus, R. W., Walkup, R. E., and Kelly, Roger, *Radiation Effects* **99,** 199 (1986).
15. Bykovski, Y. A. et al. *Sov. Phys. Tech. Phys.* **18,** 1597 (1974).
16. AIP 50th.
17. Dreyfus, R. W. and von Gutfeld, R. J., *Conf. on Lasers and Electrooptics*, Baltimore, MD, April 24–28, 1988.
18. Phipps, C. R., Turner, T. P., Harrison, R. F., York, G. W., Osborne, W. Z., Anderson, G. K., Corlis, X. F., Haynes, L. C., Steele, H. S., Spicochi, K. C., and King, T. R., *J. Appl. Phys.* **64,** 1083 (1988).
19. Johnston, T. and Dawson, J., *Phys. Fluids* **16,** 722 (1973).
20. Kruer, W. L., *The Physics of Laser Plasma Interactions*, Addison-Wesley, Redwood City, CA, 1988.

Author Index

Subject Index